CAMBRIDGE LIBRARY COLLECTION

Books of enduring scholarly value

Physical Sciences

From ancient times, humans have tried to understand the workings of the world around them. The roots of modern physical science go back to the very earliest mechanical devices such as levers and rollers, the mixing of paints and dyes, and the importance of the heavenly bodies in early religious observance and navigation. The physical sciences as we know them today began to emerge as independent academic subjects during the early modern period, in the work of Newton and other 'natural philosophers', and numerous sub-disciplines developed during the centuries that followed. This part of the Cambridge Library Collection is devoted to landmark publications in this area which will be of interest to historians of science concerned with individual scientists, particular discoveries, and advances in scientific method, or with the establishment and development of scientific institutions around the world.

The Opus Majus of Roger Bacon

Roger Bacon, the medieval natural philosopher who broke new ground in promoting scientific method, produced the encyclopedic *Opus Majus* or 'Greater Work' in the mid-thirteenth century. This 1897 publication in two volumes was the first complete edition of the work to appear in print. Written at the request of Pope Clement IV, the *Opus Majus* is the most significant and most influential of Bacon's works, containing his observations of the natural world and theories on knowledge acquisition. Bacon's text appears in the original Latin, and Bridges includes a substantial introduction and brief analysis of each chapter in English, as well as extensive footnotes and an analytical table to aid the reader. Volume 1 contains the first four parts of Bacon's treatise with sections on 'Wisdom and Truth', 'The connection of Philosophy with Theology', 'The Study of Language', and 'Mathematical Science'.

Cambridge University Press has long been a pioneer in the reissuing of out-of-print titles from its own backlist, producing digital reprints of books that are still sought after by scholars and students but could not be reprinted economically using traditional technology. The Cambridge Library Collection extends this activity to a wider range of books which are still of importance to researchers and professionals, either for the source material they contain, or as landmarks in the history of their academic discipline.

Drawing from the world-renowned collections in the Cambridge University Library, and guided by the advice of experts in each subject area, Cambridge University Press is using state-of-the-art scanning machines in its own Printing House to capture the content of each book selected for inclusion. The files are processed to give a consistently clear, crisp image, and the books finished to the high quality standard for which the Press is recognised around the world. The latest print-on-demand technology ensures that the books will remain available indefinitely, and that orders for single or multiple copies can quickly be supplied.

The Cambridge Library Collection will bring back to life books of enduring scholarly value (including out-of-copyright works originally issued by other publishers) across a wide range of disciplines in the humanities and social sciences and in science and technology.

The Opus Majus of Roger Bacon

Volume 1

Edited by John Henry Bridges

Cambridge University Press

CAMBRIDGE UNIVERSITY PRESS

Cambridge, New York, Melbourne, Madrid, Cape Town, Singapore,
São Paolo, Delhi, Dubai, Tokyo

Published in the United States of America by Cambridge University Press, New York

www.cambridge.org
Information on this title: www.cambridge.org/9781108014427

This edition first published 1897
This digitally printed version 2010

ISBN 978-1-108-01442-7 Paperback

THE OPUS MAJUS

OF

ROGER BACON

VOL. I.

HENRY FROWDE, M.A.

PUBLISHER TO THE UNIVERSITY OF OXFORD

LONDON, EDINBURGH AND NEW YORK

THE 'OPUS MAJUS' OF ROGER BACON

EDITED, WITH

INTRODUCTION AND ANALYTICAL TABLE

BY

JOHN HENRY BRIDGES

FELLOW OF THE ROYAL COLLEGE OF PHYSICIANS
SOMETIME FELLOW OF ORIEL COLLEGE

'Induire pour déduire afin de construire'
AUGUSTE COMTE

'Omnes scientiae sunt connexae, et mutuis se fovent auxiliis, sicut partes ejusdem totius, quarum quaelibet opus suum peragit non solum pro se sed pro aliis'
ROGER BACON, *Opus Tertium*

IN TWO VOLUMES: VOL. I

OXFORD
AT THE CLARENDON PRESS
MDCCCXCVII

COLLEGIO REGALI MEDICORUM LONDINENSI

CUJUS OLIM PRAESIDENS

RICARDUS MEAD

LITTERIS PERINDE AC SCIENTIIS ILLUSTRIS

PRIMUM BACONIS EDITOREM CONSILIIS INSTIGAVIT JUVIT

ALTERAM EDITIONEM DEDICO

PREFACE

A SUFFICIENT reason for a new edition of Roger Bacon's principal work would be the extreme rarity of the edition of the *Opus Majus* published by Jebb in 1733, and reprinted seventeen years afterwards in Venice. But a more cogent reason is that this edition is incomplete. The work, as we learn from Bacon's account of it in his *Opus Tertium*, consisted of seven parts; and the seventh part, a discourse on Moral Philosophy, was omitted by the editor.

Why Jebb should have taken this course is not clear. In his preface he speaks of the work as consisting of six parts, 'in sex partes distributum,' and adds, 'tractatum de Morali Philosophia ad calcem adjunxit.' In 1858 a paper was read by Dr. Ingram before the Royal Irish Academy, and was printed in the seventh volume of the Proceedings of this institution, in which the writer showed conclusively the continuity of this seventh part of the *Opus Majus* with all that had gone before. The continuity is marked unmistakably in the very title of the section, *Incipit septima pars hujus persuasionis de Morali Philosophia*, and in its opening words, 'Manifestavi in praecedentibus,' &c. Repeated references to the foregoing parts will be found; and if further proof were wanting, it is supplied in abundance by the two appendages to the *Opus Majus* which were sent by Bacon to Pope Clement IV within a few months of the dispatch of the principal work, published by Professor Brewer in 1859, in the Rolls Series, as *Opera Inedita.* Special mention is made in the *Opus Minus* (Brewer, p. 315) of passages of this seventh section which the author regarded as of special importance. In the *Opus Tertium* (Brewer, pp. 48–52), a brief account is given of each of the six divisions of which it consisted.

Dr. Ingram's paper was carefully studied by Victor Cousin, who had already devoted much time to the study of Bacon's unpublished works. He remarked upon it (*Journal des Savants*, 1859, p. 717), 'Nous croyons qu'il n'y a pas d'exemple dans l'histoire littéraire d'une erreur semblable à celle de Jebb. Elle est vraiment étonnante, mais elle est incontestable.' And in truth the omission is of much greater significance than the mere loss of Bacon's opinions on a subject of importance would imply. Throughout the *Opus Majus* there is an orderly arrangement of the subject-matter formed with a definite purpose, and leading up to a central theme, the consolidation of the Catholic faith as the supreme agency for the civilization and ennoblement of mankind. For this end a complete renovation and reorganization of man's intellectual forces was needed. After a brief exposition of the four principal impediments to wisdom—authority, habit, prejudice, and false conceit of knowledge—Bacon proceeds in his second part to explain the inseparable connexion of philosophy with the highest truths of religion. In primaeval ages both were entrusted to the patriarchs. Subsequently, while the evolution of religious truth was proceeding in Judaea, Greece became the scene of the growth of philosophy. Both were alike ordained in God's providence. In our own times, as in those of antiquity, the study of both should be carried on continuously. But for this purpose it was essential that the wisdom of the ancients should be studied in the language in which it was originally set forth. To limit students to Latin translations is to ensure the multiplication of error. Most of these translations, especially those of the Bible and of Aristotle, are deplorably defective, and have been made by men imperfectly acquainted with the subject treated of. The first condition, therefore, of a renovation of learning is the systematic study of at least three languages besides Latin, namely, Hebrew, Greek, and Arabic.

The second condition was the application of mathematical method to all objects of study, whether in the world or in the Church. Mathematic is the 'gateway and the key to all other sciences'; it raises the understanding to the plane at which knowledge can be distinguished from ignorance. Without it other sciences are unintelligible. It reveals to us the motions of the heavenly bodies, and the laws of the propagation of force in things terrestrial, of which the propagation of light may be taken as a type; without it we are incapable of regulating the festivals of the Church; we remain in ignorance of the influences of climate upon character; of the position of cities and of the boundaries of nations whom it is the function of the Catholic Church to bring within her pale, and to control spiritually. With these subjects the fourth and fifth sections of the *Opus Majus* are occupied; they form the principal bulk of its contents. But mathematical method, though essential, is insufficient. It must be supplemented by the method of experiment. Even a purely geometrical proof is not convincing or conclusive, until the execution of the diagram has enabled us to add ocular, that is to say, experimental, evidence that the demonstration is sound. This method, moreover, will lead us into new regions into which mathematical procedure is not able to penetrate. Experimental science governs all the preceding sciences ('domina est omnium scientiarum praecedentium'), it controls their methods; in prosecuting its own special researches it makes use of their results.

Here then ends the *Opus Majus* as presented in the edition of 1733. A glance at the fourteenth and preceding chapters of the *Opus Tertium*, in which the structure and purpose of the *Opus Majus* are reviewed, will show how disastrously the suppression of the seventh section of the work has mutilated it. 'All these foregoing sciences,' says Bacon, 'are, properly speaking, speculative. There is indeed in every science a practical side, as Avicenna teaches in the first book of his *Art of Medicine*. Nevertheless, of Moral Philosophy alone can it be said that it is in the special and autonomatic sense practical, dealing as it does with human conduct with reference to virtue and vice, beatitude and misery. All other sciences are called speculative: they are not concerned with the deeds of the present or future life affecting man's salvation or damnation. All procedures of art and of nature are directed to these moral actions, and exist on account of them. They are of no account except in that they help forward right action. Thus practical and operative sciences, as experimental alchemy and the rest, are regarded as speculative in reference to the operations with which moral or political science is concerned. This science is the mistress of every department of philosophy. It employs and controls them for the advantage of states and kingdoms. It directs the choice of men who are to study in sciences and arts for the common good. It orders all members of the state or kingdom so that none shall remain without his proper work.'

The seventh part of the *Opus Majus* is for the first time printed in this edition. Unfortunately it is not complete. It consisted, as we learn from the fourteenth chapter of the *Opus Tertium*, of six divisions; and the only two MSS. of it as yet discovered, those of Dublin and Oxford (the first of which, as will be shown afterwards, is copied from the second), stop short before the conclusion of the fourth. We gather, however, that the missing portions are not of primary importance.

Another alteration of considerable importance has been made in the present edition. Professor Émile Charles, in his very important monograph on Roger Bacon (Bordeaux, 1861), pointed out that the treatise *De Multiplicatione Specierum*, which in Jebb's edition of the *Opus Majus* is placed between the fifth and sixth sections of the work, does not in reality belong to it. And indeed the second sentence of the treatise makes this evident. 'Recolendum est,' Bacon observes, 'quod in tertia parte hujus operis tactum est quod essentia, substantia, natura, potestas, potentia, virtus, vis, significant eandem rem.' No such passage is to be found in any part of the *Opus Majus*, least of all in the third part, which deals with Comparative Philology. Here again the *Opus Tertium* comes to our aid. Several references will be found there to a distinct treatise sent to Pope Clement IV simultaneously with the *Opus*

Majus (Brewer, pp. 38, 99, 117, 227). It is a treatise on the propagation of radiant forces, usually spoken of as *Multiplicatio Specierum*, but in one passage entitled *Tractatus de Radiis, quem vobis misi separatim ab Opere Majori.* In the present edition this treatise will be found in its proper place as an appendix to the *Opus Majus.*

It must be added that the text, as edited, is in certain parts of the work far from perfect. This remark applies especially to the third and sixth sections. In the third section several missing pages have been supplied from a Cottonian MS. (Julius D.v.), containing amongst other things a Greek and a Hebrew alphabet. It is the more remarkable that Jebb should have omitted these passages, since it is clear that he had consulted this MS. and had made frequent corrections from it. The sixth section, on Experimental Science, has been so carelessly edited that it seems probable that the editor must have entrusted the work to a less competent assistant. Contractions of the most ordinary kind are misinterpreted (as e.g. 'e converso' is rendered 'ergo,' 'conclusio' is written 'quaestio,' &c.), and in some passages sentences have been omitted. While indicating these shortcomings, it ought at the same time to be added that other parts of the work give proof of great labour and care in the collation of the MSS. consulted. This is especially the case with the fourth and fifth parts, which form a very large proportion of the whole work. Having spent much time over the MSS. used by Jebb, in addition to others, I can bear testimony to the accuracy with which variations have here been noted. All of them which are of more than verbal importance have been noted in the present edition.

A further change has been made in the arrangement of the diagrams, which in Jebb's edition were collected into two tables, and thus rendered inconvenient for comparison with the text, and which, moreover, in many cases are incorrectly drawn. They have been carefully re-copied from a MS. in the British Museum (Royal, 7 F. viii), which Sir E. Maunde Thompson, who was so kind as to examine this and several other Baconian MSS. in the British Museum for me, pronounces to be of the latter part of the thirteenth century, i.e. contemporary, or very nearly so, with Bacon. Each diagram has been placed in its proper context.

It was found that in many cases the reasoning of the author had been entirely nullified by defective punctuation. This has been entirely revised, and the number of distinct sentences and paragraphs has been very largely increased. Students of the *Opus Majus* have found the work difficult to refer to, owing to the multiplicity and diversity of the subjects treated, and the want of such guidance as page-headings or marginal notes can supply. This defect has been remedied, an analysis of the work has been prefixed, and an improved index added.

A few remarks on the MSS. of the *Opus Majus*. Jebb's edition was founded on a MS. in the library of Trinity College, Dublin, which at that time was the only MS. known to contain the complete work. It contains a note in Archbishop Usher's handwriting to the effect that the MS. was presented to the College by Gordian Strowbridge, and that the diagrams were drawn by Sir Christopher Heyden. This MS. is very clearly written in 249 folios of four columns. Sir Christopher Heyden, of Baconsthorpe, in Norfolk, was a well-known astrological writer who died in 1623, which fixes the date of this MS. at the end of the sixteenth or beginning of the seventeenth century. A glance at the MS. shows that the figures were drawn simultaneously with the text, room being left in each page for their insertion. The character of the writing quite corresponds with this date. It is referred to in this edition as D.

In 1825 another MS. of the *Opus Majus* was bought for the Bodleian Library. It belonged at one time to Thomas Allen, the astrologer of Gloucester Hall, who early in the seventeenth century gave twenty MSS. to the Bodleian. This one, however, passed into the possession of Sir Kenelm Digby, whose well-known signature and motto are inscribed on the first page. It is now numbered 235 of the Digby MSS. The greater part of it is of the fifteenth century. But a portion of it (pp. 249–295) is in an older and more beautiful handwriting, considered by Mr. Coxe to be of the fourteenth century. This portion includes a considerable portion of the *Perspectiva*.

These two MSS. have been carefully collated for the present edition. In the course of the collation unmistakable proof was given that the Dublin MS. was a copy of that in the Bodleian. At the close of p. 470, col. 2, of this latter, the sentence breaks off midway, and is continued on the first line of p. 487, col. 1. An error of this kind, analogous to that caused by the transposition of sheets in binding a modern book, is easily explicable. In the Dublin MS. the same rupture of the sentence occurs, but in the middle of a column (fol. 224, col. a, line 12), the sentence being ultimately continued on fol. 229, col. d, line 14. It may be added that these two MSS. exhibit throughout an almost exact correspondence in their errors and in their variants. The Oxford MS. has therefore been taken as the foundation of the present edition, and is denoted by the letter O. It is, however, by no means perfect, especially in the second and third sections of the work: and recourse was had both by Jebb and by the present editor to various MSS. in the British Museum and elsewhere to supply the imperfections. Foremost among these are the two Cottonian MSS., Julius D.v., and Tiberius C.v. (designated here as Jul. and Tib.). The first of these, unfortunately much injured in the fire of 1731, gives the first three sections of the work and a large part of the fourth. The second gives the whole of the fourth.

For the geographical portion of the fourth section, I was allowed to consult the MS. in the possession of Corpus Christi College, Cambridge. It is not, however, believed to be of earlier date than the middle or later part of the fifteenth century. Occasional reference has been made to a MS. of the fifth section of the work (*Perspectiva*) in the possession of Magdalene College, Cambridge, which I was permitted to examine. It is not, however, of earlier date than the Bodleian MS., and its variants are not of great importance.

Of the MSS. of the *Perspectiva* and of the *De Multiplicatione Specierum* which appear to have been transcribed more frequently than other portions of the work, by far the most important is the British Museum MS. (Royal, 7 F. viii), already spoken of as contemporary, or nearly so, with Bacon. This, as I have said, was carefully collated by Jebb, who indeed has in certain places copied on the margin extracts from Combach's printed edition of the *Perspectiva* of 1614. The diagrams of this MS. are of special value. The

Sloane MS. 2156, and the Harleian MS. 80, 60 b, have also been consulted.

The sixth section of the *Opus Majus* (*Scientia Experimentalis*) appears to have been seldom copied. In the third volume of Baconian MSS. presented to University College, Oxford, by John Elmhurst, there is a MS. of this section which is described as copied from Allen's MSS. (see Brewer. p. xliii). It may, therefore, be merely copied from the Oxford MS. of the *Opus Majus*. But it offers some variants, and in one or two passages it has proved serviceable. It is spoken of in this edition as U.

Of the seventh section, here printed for the first time, there is a MS. in the Royal Library (8, F. ii) containing the first two parts and a portion of the third. This has been carefully collated with the corresponding parts of the Dublin and Oxford MSS. The variations will be seen to be of no great importance. The MS. appears to be of the middle of the fifteenth century.

Besides these MSS., others have been consulted which throw light on Bacon's life and work. Chief amongst these is the important MS. of the Mazarin library (formerly numbered 1271, but at present 3576), from which Professor Émile Charles gives copious extracts in his monograph entitled *Roger Bacon, sa vie, ses ouvrages, ses doctrines, d'après des textes inédits* (Bordeaux, 1861). More will be said afterwards of its contents. They offer a considerable instalment of the *Scriptum Principale*, of which the *Opus Majus*, inclusive of its adjuncts, the *Opus Minus* and the *Opus Tertium*, was but the prelude.

Another valuable fragment of this final work is preserved in the British Museum among the Sloane

MSS. (2156). It contains the first book and part of the second of the *Communia Mathematicae.* Attention was called to it both by Brewer and by Charles, and occasional references to it have been made in the present edition.

Yet a third fragment is the elementary work on Greek grammar in the possession of Corpus Christi College, Oxford, which was sent to the Bodleian for my perusal. An imperfect copy, in seventeenth century handwriting, apparently made from the Corpus MS., is contained among the Baconian MSS. of University College.

Bacon's commentary on the *Secretum Secretorum* (Tanner MSS. 116) has also been examined. It throws light on the astrological side of his work.

Among the books consulted for this edition, far the most important are those of Professors Brewer and Émile Charles, already spoken of. It is unfortunate that two such assiduous and able investigators should have worked simultaneously and without communication. Several not unimportant errors might have been avoided, had either of them known of the other's work. Charles had a far more extensive knowledge than Brewer of Bacon's unpublished works; and his extracts from them are so copious as to render it desirable that his monograph, which has already become extremely rare, should be republished. He makes, however, the erroneous statement (p. 62) that the missing portion of the *Opus Majus* (here printed for the first time) had been published in Dublin; and he does not appreciate the distinction, so clearly demonstrated by Professor Brewer, between the *Opus Tertium* (which is at once an introduction and a supplement to the *Opus Majus*), and the far vaster *Scriptum*

Principale projected by Bacon, but only in part executed. On these points, and on the bibliography of Bacon generally, Mr. Little's *Grey Friars in Oxford* is of great value. Other works consulted will be noted as reference is made to them. Attention, however, may be specially called to Govi's recent edition of the Latin translation of Ptolemy's *Optica*, so frequently used by Bacon (Turin, 1885), a publication of the greatest value for the history of science; and to Heiberg's edition of Euclid's *Optica* (vol. vii. of his edition of Euclid, Leipsic, 1895). Wuestenfeld's *Geschichte der Arabischen Aerzte* (Göttingen, 1840), and Cantor's *Geschichte der Mathematik* (Leipsic, 1880–1892), throw light on the mediaeval men of science, Eastern or Western, mentioned by Bacon. Hauréau's *Histoire de la Philosophie Scolastique* (Paris, 1872) defines, with some acerbity, Bacon's position among the schoolmen. Jourdain's *Recherches critiques sur les traductions latines d'Aristote* (nouvelle édit., 1843) contain indispensable information as to the translators of whose shortcomings Bacon so often complains.

For the geographical section of the work, in addition to the classical works of Yule and Bunbury, frequent reference is made to the complete version of the travels of Rubruquis and Carpini, published by the Paris Geographical Society, in the fourth volume of their *Recueil de Voyages et de Mémoires* (1839).

As Seneca occupies so large a place in the seventh section of the *Opus Majus*, it may be mentioned that the edition used for this work is that of Haase (Leipsic, 1887). As to Aristotle, the references are to Didot's edition (Paris, 1848–1873).

It remains for me to express my thanks to the Master and Fellows of Trinity College, Cambridge; to the Master and Fellows of Corpus Christi College, Cambridge; the Master and Fellows of Magdalene College, Cambridge; and the President and Fellows of Corpus Christi College, Oxford, for permission to examine the MSS. of Bacon belonging respectively to these societies.

J. H. B.

March 26, 1897.

FACTS RELATING TO BACON'S LIFE

Contemporary Events.	*Statements resting on later authority.*	*Facts verified by Bacon's statement or by contemporary authority.*
1209. Condemnation of Aristotle's Physic and Metaphysic in Paris. **1215.** Confirmation of this by Papal Legate. (Cp. *Opus Tertium*, cap. 9.) **1222.** Alexander of Hales enters the Franciscan Order, and teaches philosophy in Paris.	**1210–15.** Born near Ilchester in Dorsetshire, or, according to another tradition, in the parish of Bisley in Gloucestershire. (Cf. Brewer, p. lxxxv.)	
1231. Condemnation of Physic and Metaphysic partially removed by Gregory IX. **1238.** Alexander of Hales resigns his post as a teacher of philosophy.		**1230.** Michael Scot introduces his translations of Aristotle. (*Op. Maj.* vol. i. p. 55.) **1233.** Interview of Bacon with Henry III at Oxford, as described by Matthew Paris.
1245–8. First residence of Thomas Aquinas in Paris with Albertus Magnus. **1249.** Death of William of Auvergne (Bishop of Paris).	**1240.** Went from Oxford to Paris about 1240. Probably entered Franciscan Order a few years later.	**1245.** Heard William of Auvergne (Bishop of Paris) lecture on *intellectus agens*. (*Op. Tert.* cap. 23.)
1252. Second residence of Aquinas in Paris of uncertain duration. **1253.** Death of Grosseteste. **1255.** Bonaventura becomes General of Franciscans. **1258.** Bagdad captured from Saracens by Tartars. **1265.** Guy Fulcodi elected Pope Clement IV. **1268.** Death of Clement IV. **1270.** Death of Saint Louis.	**1250–7.** Probably in Oxford. Legend as to Bacon's Tower may perhaps be referred to this period.	**1250.** Saw the leader of the Pastoureaux marching through France in 1250. (*Op. Maj.* vol. i. p. 401.) **1257.** 'Exile' from Oxford to Paris began. (*Op. Tert.* cap. 1.) **1258–67.** His family took the King's side in war with barons. (*Op. Tert.* cap. 3.) **1264–5.** Enters into relations with Guy Fulcodi
1274. Death of Bonaventura; Jerome of Ascoli becomes General of Franciscans. Death of Thomas Aquinas. (?)Birth of Duns Scotus. **1280.** Death of Albertus Magnus. **1288.** Jerome of Ascoli becomes Nicolas IV; Raymundo Galfredi succeeds him as General. **1292.** Death of Nicolas IV.	**1278.** Imprisonment *propter novitates suspectas*, 1278. (See *Summa Historialis* of Antoninus, Archbishop of Florence, a writer of the fifteenth century.) **1292.** Release from prison probably 1292. Died 1292 or 1294. Buried in Franciscan Church in Oxford. Legend as to exposure of his writings to wind and weather told by Wood.	**1266.** Bacon ordered to send his writings to the Pope. **1266–7.** Composition of *Opus Majus*, *Opus Minus*, *Opus Tertium*. **1268.** Death of Clement IV. **1271.** Writes the *Compendium Studii Philosophiae*, denouncing the corruptions of the Church. (See Brewer, p. liv.) **1292.** Writes *Compendium Theologiae*. See MS. of this work (Br M. Royal 7 F vii. fol. 154).

Note.

In the July issue of the *English Historical Review*, 1897, Dr. Gasquet publishes a MS. of Bacon which he has found in the Vatican, and which he inclines to think is a preface to the *Opus Majus*.

There is much to justify this view. The work in question describes Bacon's overflowing gratitude for Pope Clement's message to him; apologizes for the delay in the transmission of his works by pointing out that none of these works were in a complete state; explains the obstacles interposed by the distress of his family, ruined in the civil wars, and by the restrictions of his Order; introduces his disciple John, who had been for seven years under his tuition; and finally concludes with a brief summary of the contents of the *Opus Majus*. This he describes, not as his principal work, but as a *Persuasio*. It has seven parts. After briefly noting the contents of the first two, Bacon passes to the seventh (published for the first time in this edition) and then comments successively on the sixth, fifth, fourth, and third.

It will be observed by readers of this short treatise that it contains little that is not set forth with much greater fullness in the *Opus Tertium*, which is to be regarded as the real Introduction to the collection of writings sent by Bacon in 1267 to Pope Clement IV. The first chapter of Dr. Gasquet's MS. is almost exactly identical with pp. 7–12 in Brewer's edition of *Opus Tertium*, the latter, however, having certain sentences not contained in the former. The fifth chapter is a repetition of *Opus Majus*, pt. i. cap. 16. One or two sentences, however, of this newly published work deserve attention. We learn from it that Bacon's life in Paris between 1257 and 1267 was a time of comparative inaction: *a decem annis propter languores multos et infirmitates varias occupationibus exterioribus studii non vacavi.* He had written, he says, much before entering the Franciscan Order, with a view to the instruction of youth (*multa in alio statu conscripseram propter juvenum rudimenta*); and of late years he had sent fragments of his works to friends (*aliqua capitula nunc de una scientia nunc de alia ad instantiam amicorum aliquando more transitorio compilavi.* No treatise, however, on any department of philosophy had been issued in a complete form.

On the whole I am inclined to think that the short work edited by Dr. Gasquet is a first draft of what was afterwards expanded into the *Opus Tertium.* Bacon tells us that he was in the habit of writing his discourses several times over until they were brought into satisfactory shape. *Sentiens meam imbecillitatem nihil scribo difficile quod non transeat usque ad quartum vel quintum exemplum antequam habeam quod intendo.*

J. H. B.

July 21, 1897.

INTRODUCTION

I. BACON'S LIFE.

IN considering the little that is known of the life of Bacon, it is well to give precedence to the few facts that are fixed with perfect precision by his own statement. We know with entire accuracy the date of the composition of the *Opus Majus*, and of the two subsidiary works, the *Opus Minus* and the *Opus Tertium*. Pope Clement[1] IV's instructions to him to transmit the results of his labours were issued June 22, 1266 from Viterbo. Within the year that followed, the *Opus Majus*, with its supplement, the *Opus Minus*, and its introduction, the *Opus Tertium*, had been completed and sent to the Pope. At this time he speaks of himself as an old man, and he says that he had been studying language, science, and philosophy for nearly forty years (*Opus Tertium*, cap. 20). From this it may be supposed that he was born between 1210 and 1215. But the place of his birth cannot be said to be fixed with certainty.

One, and only one, notice of his name occurs in a con-

[1] Guy Fulcodi (or Foulques), who succeeded to the Papacy in 1265 as Clement IV, was born at Saint Gilles in Languedoc. He began his career by studying law, in which he achieved great distinction. He was married and had several children. He seems to have acted for some time as a private secretary to Louis IX. After his wife's death he entered the Church, was made archbishop of Narbonne in 1259, and cardinal bishop of S. Sabina in 1261. (See Fleury, *Hist. Ecclésiastique*, liv. 85, whose spelling of the name Guy Fulcodi is here adopted.) Brewer conjectures (pp. xi et seq.) that he entered into relations with Bacon on the occasion of his mission to England as Papal legate in 1263 or 1264. But Bacon was then in Paris, and had been there for several years. Guy Fulcodi had far better opportunities of hearing about Bacon in Paris than could have occurred during the time of his stormy and ineffectual legation to England.

temporary writer. Matthew Paris relates, under the year 1233, that Henry III convoked the counts and barons of the kingdom to a council at Oxford. Their animosity against Pierre des Roches, Bishop of Winchester, the king's chief adviser, who had surrounded his person with a body-guard of Poitevins and filled England with these foreigners, led them to refuse the summons. While the king was debating what measures to take against the recalcitrant barons, a Dominican preacher, Robert Bacon by name, told him frankly that there would be no hope of permanent peace in the kingdom so long as the Bishop of Winchester and his son, or kinsman, Peter of Rievaulx, retained power. Robert Bacon's opinion was echoed by others, and the king was induced to listen to it patiently. 'Then a certain clerk who was present at the Court, Roger Bacon by name, a man of mirthful speech, said with pleasant yet pointed wit, "My lord king, what is that which is most hurtful and fearful to those that sail across the sea?" "Those know it," the king replied, "who have much experience of the waters." "My lord," said the clerk, "I will tell you; stones and rocks"; meaning thereby Pierre des Roches.' It has been thought that the date of the dialogue was too early to refer to the Roger Bacon with whom we are here concerned. But since he might well be more than twenty years old at the time, the doubt seems hardly founded.

What is certain from Bacon's own statement is that his family was one of some wealth, since he himself had been able to spend much money on experimental research. It appears also that this family had taken the royal side throughout the disputes between Henry and his barons, and had suffered pecuniary loss and exile for their loyalty. He tells Pope Clement that, being in sore distress for the money necessary for the transcription and conveyance of his MSS., 'I wrote to my brother, a rich man in my country. But he, belonging as he did to the king's party, was in exile with my mother, brothers, and the whole family. Ruined and reduced to utter poverty, he was unable to help me, and up to the present day he has sent me no reply.' (*Op. Tert.* cap. 3)

The forty years of study, of which he speaks in 1267, may

be divided into two periods, apparently of nearly equal length; the periods before and after his admission into the Franciscan Order. In the seventeenth chapter of the *Opus Tertium* he speaks of having devoted more than twenty years to the study of languages and of science. 'I sought,' he says, 'the friendship of all wise men among the Latins; and I caused young men to be trained in languages, in geometrical figures, in numbers, in the construction of tables, in the use of instruments, and in many other necessary things. . . During this time I spent more than two thousand pounds in those things and in the purchase of books and instruments.' We may presume that the pounds were French, which at that time would correspond to between 600 and 700 pounds sterling. The sum was a large one. And whether large or small, it would be quite incompatible with the profession of an Order specially devoted to poverty. It may be inferred, therefore, that since he had studied independently for some twenty years, it was not till some time between 1245 and 1250 that Bacon became a Franciscan.

Among the men distinguished for their learning whose friendship he cultivated at this part of his career may be counted, in all probability, Adam de Marisco; Edmund Rich, afterwards Archbishop of Canterbury and ultimately canonized; Thomas Bungay, whose name was one day to be associated with his own as a worker of magic; Thomas, Bishop of St. David; John of Basingstoke, scholar and traveller; John Peckham, afterwards Archbishop of Canterbury; Hermann, one of the principal translators of Aristotle; Shirwood, the treasurer of Lincoln; and last and greatest, the illustrious Bishop of Lincoln, Robert Grosseteste. In Bacon's earlier years of study, Grosseteste had not plunged into the arduous and absorbing work of his episcopate. *Novit scientias*, Bacon says of him. He was *rector scholarum*, and also Chancellor of Oxford, and in 1224 was the rector of the Franciscans recently established there. The terms in which Bacon bears testimony to his encouragement of philology, to his attempts to apply mathematical method to the study of physical phenomena, to his disregard of the philosophy of the schools as founded on bad translations of Aristotle (Brewer, *Compend. Studii*, cap. 8), would be conclusive as to his personal contact with this great man, even though it were not confirmed by reference to Grosseteste's scientific writings, in which Bacon's debt to him is unmistakable. His treatise *De Physicis Lineis, Angulis, et Figuris* contains passages as to the spherical radiation of force, and as to the change in its direction by reflection and refraction, which bear a close resemblance to the language used many years afterwards by Bacon.

It would appear that, at the beginning of the thirteenth century, there was a stronger impulse towards scientific study in Oxford than in Paris. In the eleventh chapter of the *Opus Tertium*, when speaking of the science of Optics, Bacon observes, 'On this science no lectures have as yet been given in Paris, nor anywhere among the Latins, except twice at Oxford.' It is not stated that the lecturer was Grosseteste; but we may well believe it. It may be supposed that the influence of Adelard of Bath, the first translator of Euclid, had left its traces. Twenty years before the close of the twelfth century we hear of two Englishmen, Alexander Neckham and Alfred Sershall, lecturing in Paris on the Physics of Aristotle, then recently introduced from the school of translators from Arabic directed by Archbishop Raymond of Toledo.

But the University of Paris, placed nearer the centre of the spiritual forces that swayed mediaeval society, had grown up under the dialectical influences of theological controversy; and when Bacon went there, perhaps about 1240, he found what is called, vaguely and inaccurately enough, the scholastic philosophy in the fullness of its growth, with the enlarged scope given to it by the recent permission to study the Physics, Metaphysics, and Psychology of Aristotle. Its two most prominent representatives were at this time Alexander of Hales and William of Auvergne. Of the methods and the controversies then current Bacon made himself a master, and received the title of doctor. To be able to speak the language of the schools with authority was the first condition of obtaining a hearing. But he was not slow to perceive that the men who taught this philosophy were, for the most part, wholly destitute of positive knowledge. They knew no language but Latin. Beyond the shreds of arithmetic, mensuration, and astronomy taught in the manuals of the Quadrivium, they were ignorant of mathematics. Of the possibility of applying mathematical knowledge to the facts of nature they had formed no conception whatever. Their philosophy was a tangle of barren controversies reducible, for the most part, to verbal disputes. It bore no relation to the facts of real life. It held out no hope of raising the Catholic Church to the position of intellectual domination needed for establishing her authority over the Asiatic world, from which dangers were looming of appalling magnitude.

It was in Paris that Bacon came into contact with a remarkable man of whom very little would be known to us but for Bacon's eulogies, Peter of Maricourt[1], a native of Picardy. From the description given of him in the thirteenth chapter of the *Opus Tertium*, he would seem to have been an unambitious man, anxious only to pursue his researches in private, regardless of the metaphysical turmoil around him. Speaking of experimental research, Bacon says: 'One man I know, and one only, who can be praised for his achievements in this science. Of discourses and battles of words he takes no heed: he follows the works of wisdom, and in these finds rest. What others strive to see dimly and blindly, like bats in twilight, he gazes at in the full light of day, because he is a master of experiment. Through experiment he gains knowledge of natural things, medical, chemical, indeed of everything in the heavens or earth. He is ashamed that things should be known to laymen, old women, soldiers, ploughmen, of which he is ignorant. Therefore he has looked closely into the doings of those who work in metals and minerals of all

[1] There is some doubt as to the orthography of the name, though none can now be left as to the identity of the person indicated. Émile Charles (pp. 16-17) mentions a MS. in the Paris library (Bibliothèque Nationale, Manuscrits Latins, 7378) in which the only known work of Peter Peregrinus is spoken of as 'Epistola Petri Peregrini de Maricourt ad Sygerium de Fontancourt de Magnete.' Charles adds that there is a village called Mehariscourt in Picardy near the abbey of Corbie. The Latin form of the word in one MS. of the *Opus Tertium* is written Maharncuria, but in others Mahariscuria. Cf. vol. ii. p. 203 of the present work; see also Bertelli's *Declinazione Magnetica* (Rome, 1892).

kinds; he knows everything relating to the art of war, the making of weapons, and the chase; he has looked closely into agriculture, mensuration, and farming work; he has even taken note of the remedies, lot-casting, and charms used by old women and by wizards and magicians, and of the deceptions and devices of conjurors, so that nothing which deserves inquiry should escape him, and that he may be able to expose the falsehoods of magicians. If philosophy is to be carried to its perfection and is to be handled with utility and certainty, his aid is indispensable. As for reward, he neither receives nor seeks it. If he frequented kings and princes, he would easily find those who would bestow on him honours and wealth. Or, if in Paris he would display the results of his researches, the whole world would follow him. But since either of these courses would hinder him from pursuing the great experiments in which he delights, he puts honour and wealth aside, knowing well that his wisdom would secure him wealth whenever he chose. For the last three years he has been working at the production of a mirror that shall produce combustion at a fixed distance; a problem which the Latins have neither solved nor attempted, though books have been written upon the subject.'

Of this remarkable man little is known but what Bacon tells us in the foregoing and other passages of the *Opus Tertium*, and the *Opus Majus*. But what we know is not inconsistent with Bacon's eulogy. Libri, in a note contained in the second volume of his *History of Mathematics*, transcribes a letter written by Peter Peregrinus of Maricourt to a certain Sigermus of Fontancourt, which is a treatise on the properties of the magic stone on the relations of its poles to those of the heavens and earth, on the way to find these poles; on the repulsion in two magnets of poles of the same name, and the attraction of those of different names; and on the construction of a globe which should revolve with the revolution of the heavens, and thus supply the place of the ordinary observation by the astrolabe. This is, no doubt, the invention of which Bacon speaks in the sixth part of the *Opus Majus*. Gilbert, in his great work on Magnetism, makes frequent mention of this treatise of Peter Peregrinus; and a careful comparison of the two works, separated as they are by an interval of more than three centuries, shows undoubted and weighty obligations of Gilbert to his predecessor. In the construction of globular magnets (the 'terrella,' or model of the earth), in the mode of finding their poles, the procedure, and indeed the very language of Peter, is closely followed by the later inquirer.

To a mind so original as Bacon's, trained in scientific method by Grosseteste and other members of the English mathematical school, the influence of an experimental thinker like Peter of Maricourt must have been stimulating in the extreme. Bacon was thirsting for reality in a barren land infested with metaphysical mirage. From the horse-load of verbal controversies contained in the *Summa* of Alexander of Hales, from the interminable series of tedious commentaries on Aristotle, of which so great a master as Albert was setting the first fatal example, he took refuge in the visions of the harvest of new truth that was to be reaped by patient observation of Nature, by submission of her processes to experimental questioning, by following the lowly paths used by plain men in their daily avocations. 'The wiser men are,' he said, 'the more humbly will they submit to learn from others; they do not disdain the simplicity of those who teach them; they are willing to lower themselves to the level of husbandmen, of poor women, of children. Many things are known to the simple and unlearned which escape the notice of the wise. I have learned more important truth beyond comparison from men of humble station, who are not named in the schools, than from all the famous doctors. Let no man therefore boast of his wisdom, or look down upon the lowly, who have knowledge of many secret things which God has not shown to those renowned for wisdom' (*Opus Majus*, vol. i. p. 10).

Assuming that Bacon entered the Franciscan Order about 1247, he would be at that time still in Paris. The degree of doctor, rarely conferred before the age of thirty-five, was probably received about the same time. He tells us (*Opus Tertium*, cap. 23) that he heard William of Auvergne lecturing to the University on the 'active intellect.' This must have been before 1248, the date of William's death. We know that he must have been still in France in 1250, for in that year the revolt of the Pastoureaux broke out; and Bacon tells us that he 'saw their leader walking barefoot in a troup of armed men carrying something in his hands with the care with which a man carries a sacred relic' (*Opus Majus*, vol. i. p. 401).

For some time between this date and 1257 he was probably in Oxford. Whether he lectured there publicly we do not know. But that he incurred the suspicion of his superiors in the Franciscan Order is certain; whether by audacity in speculation, by experiments looked upon as magical, or by frank exposure of the ignorance of professorial magnates, cannot be said with certainty. His old friends and teachers, Edmund Rich and Adam de Marisco, had passed from the scene. Grosseteste, his revered master, was dead, or died (1253) shortly after his return, in despair at the corruption of the Papacy, and half doubting whether Rome had not become the seat of Antichrist. No one was left to promote the study of Greek, which for aught we know died out in Oxford till Erasmus witnessed its revival. In 1255, John of Fidanza, better known as Bonaventura, became General of the Franciscan Order, a man of exalted and aspiring mysticism, eager to revive the spirit of St. Francis, and not likely to care much for new learning that might lead he knew not whither. Perhaps it was by his direction that Roger Bacon, about 1257, was removed from Oxford, and placed under close supervision in the Paris house.

What degree of restriction was placed upon his liberty is not very easy to define with precision. He was not forbidden to write, although he implies that he had not availed himself of the power to do so to any considerable extent. To multiply books by copyists was impracticable; first, because copyists outside the Order could not be trusted to make an honest use of the copies at their disposal; and secondly, because a strict prohibition was laid down and enforced against communicating any manuscripts to those who were not members of the Order. When Pope Clement's message reached him requiring him to transmit his works with the least possible delay, these works for the most part were still unwritten. Nevertheless there were exceptions. He had compiled, he tells us, from time to time, certain chapters on various subjects at the instance of friends (*Opus Tertium*, cap. 2). Among these chapters is probably to be reckoned the treatise *De Multiplicatione Specierum*, which was sent to the Pope by the same messenger who conveyed the *Opus Majus*, though it does not, strictly speaking, form a part of that work. Careful examination shows it to be a portion of the more complete philosophical treatise to the completion of which Bacon always aspired, till the time came, ten years afterwards, when his philosophical career was fatally arrested. Its style is different from that of the other three treatises, *Majus*, *Minus*, and *Tertium*. It is not like these a *Persuasio*, that is a more or less popular discourse addressed to a reader like Clement IV; a reader of keen understanding doubtless, but at the same time the busiest man in Christendom. The *Multiplicatio Specierum* is a fragment of a systematic work written with full observance of philosophic language and of the dialectic of the schools.

Whatever the discipline imposed during this period of his life, one important sphere of activity undoubtedly remained open to him. For many years he had been striving to form a school of young men, who should carry on the work which he had begun. We have seen in the treatise which throws so much light on the details of his life (*Opus Tertium*, cap. 17), that he had been engaged for a long time in instructing young men in languages, in geometry, in arithmetic, in the construction of tables, and in the use of scientific instruments. From this part of his work he was evidently not cut off during his life in Paris from 1257 to 1267. The messenger whom he selected to convey his manuscripts to Pope Clement was a poor lad whom he had been training in this way for five or six years. On the whole it seems probable that the restrictions placed on his liberty at this period of his life were not of extreme severity.

Of the reception given to Bacon's manuscripts in Rome we know absolutely nothing. A few months after their arrival Clement IV died; and the papal see remained vacant for three years. The Pope elected in 1271 (Gregory X) was a Franciscan. Owing his elevation to St. Bonaventura, he was not likely to show favour to a suspected member of his Order. Yet it was in this year or shortly afterwards that Bacon wrote the work known as *Compendium Studii Philosophiae*[1], an introductory discourse, perhaps, for the encyclopaedic *Srriptum Principale*, at the completion of which he was always aiming. In this treatise Bacon plunged into stronger invective against the intellectual and moral vices of his time than he had ever used before. In no previous writing had the moral corruption of the Church, from the court of Rome downwards, been so fiercely stigmatized; 'the whole clergy is given up to pride, luxury, and avarice. Wherever clergymen are gathered together, as at Paris and Oxford, their quarrels, their contentions, and their vices are a scandal to laymen.' Unbridled violence among kings and nobles, fraud and falsehood among tradesmen and artificers were the inevitable result. Progress in wisdom was hopeless when the moral condition of those who should promote it was so far below that of the teachers of the pagan world Unless sweeping remedies were applied by a reforming Pope, there was no prospect but the advent of Antichrist in the near future (Brewer, pp. 399–404).

Perhaps even these denunciations roused less antagonism than the sweeping attacks on the scholastic pedantry of his contemporaries, their false conceit of wisdom, and their preference of metaphysical subtleties and verbal strifes to the pursuit of real knowledge. Of these charges his previous writings had been full, but they were now renewed and emphasized. Aristotelian study, which at the beginning of the century had been the great stimulant of thought, was already becoming the great obstruction, and was preparing for the next century a reign of darkness. Based on false and ignorant translations, it were better, Bacon said, to do away with it altogether than that it should be carried on by men ignorant of the language in which Aristotle wrote, and destitute of the scientific training which alone could qualify them for explaining him (Brewer, pp. 469–473).

[1] Contained in Brewer's work, pp. 393–519.

The storm of indignation had long been gathering: and in 1277 it broke. In that year Jerome d'Ascoli, who four years before had succeeded Bonaventura as General of the Franciscan Order, held a chapter in Paris. Bacon was summoned on account of 'certain suspected novelties.' He was condemned, and thrown into prison. What were the 'novelties' that constituted his crime we do not know. His works abounded in them. It was not perhaps difficult to show that he had gone too far in connecting changes in religious faith with conjunctions of Jupiter and Mercury; and in hinting that underneath the jugglery of the magicians, valuable truths might sometimes lie concealed. The real motives for stifling his voice lay far deeper.

That he should have held the history of Greek philosophy to have been under the keeping and guidance of Providence no less than the history of Judaea; that he should have regarded the teaching of the Stoics on personal morality as superior to that of any Christian teacher; that he should have dwelt with such frequent emphasis on the ethical value of Mohammedan writers like Alfarabius, Avicenna, and Algazel—these were things likely to startle even the most tolerant and thoughtful of his contemporaries, much more the common average of his Order, who had suspected him of unsound views for twenty years. Not indeed that his career would have been impeded by the fact that the founder of the Franciscans had shown disregard, if not dislike of worldly knowledge. Alexander of Hales had joined the brotherhood before the death of St. Francis, and had dominated the schools of Paris long before the voice of Albert had been heard there, and while Aquinas was a child. To a man of ordinary temper, addicted to bold speculation, the protection of so powerful a corporation as the Franciscans had become when Bacon joined them would have been invaluable. But Bacon threw his chances away. He attacked the celebrities of his own Order as severely as those of its rival. His fiery and impatient spirit was to be bound by no shackles of prudence. He had come to Paris fresh from the teaching of men like Grosseteste, eager for the promotion and diffusion of science, no less than for the reform of the Church. He found the great university immersed in dialectical controversy. Many of the controverted questions were of momentous importance, and Bacon was prepared to take his part in them. But they were prosecuted by men devoid of scientific training, unprepared therefore to distinguish truth from error, verbal subtleties from fundamental realities; unwilling even to take the trouble to study Aristotle and the Bible in their original language. He saw that philosophy without science could not fail to degenerate (as history, ancient and modern, shows that it always has degenerated) into academic pedantry, and would confirm that one of the aberrations of intellect which he looked on as the worst and the most fatal, the false conceit of knowledge. Against ignorance under the cloak of wisdom he urged, like Socrates, a lifelong war; and, like Galileo, he met with a worse fate than that of Socrates, the martyrdom of enforced silence.

No crusade has been conducted by blameless crusaders. It cannot be denied that Bacon's indiscriminating zeal included, with pedants and obscurantists who were his lawful prey, two men who were his equals, one of them, perhaps, his superior. Albert was a student of nature as well as a philosopher. Aquinas, as a student of man and of society, and as the constructive thinker who gave coherency to the vast fabric of Catholic discipline, achieved results which, judged at the distance of six centuries, Bacon neither equalled nor approached. Jealousy of the rival Dominican Order, of which these men were the chief ornaments, cannot account for Bacon's failure to recognize their value; for the Irrefragable Doctor, Alexander of Hales, was a Franciscan, and was criticized more harshly than either. In their failure to appreciate duly the importance of scientific culture as a basis of Catholic action on a doubting and unbelieving world, the doctors of the Paris schools were all alike involved in his unmeasured strictures. We may understand, though we cannot justify, his impatience. He has bitterly expiated it by many centuries of neglect.

It can hardly be doubted that the seclusion consequent

on his condemnation in 1277 was effective and rigorous. Appeals to the Pope had been anticipated by Jerome, who took care to impress on the court of Rome the expediency of confirming his decision. All hopes of completing the *Scriptum Principale* were shattered. He remained a prisoner, so it is thought for fourteen years. Jerome meantime had become Pope Nicholas IV. After his death in 1292, a chapter of the Franciscans was held in Paris, at which Raymond Gaufredi, then General of the Order, set free some of those who had been condemned in 1277. It may be looked upon as nearly certain that Bacon was of the number. Certain at least it is from his own words that in that year he was again at work, on his last treatise, the *Compendium Theologiae*, in which the date 1292 is expressly mentioned. Whether he died in this year, or two years afterwards, is uncertain. He was buried in the Franciscan church in Oxford.

The legend that his works were nailed to the walls of the library and allowed to perish ignominiously may be dismissed. But that his life-long efforts to establish a Catholic school of progressive learning utterly failed, there can be no doubt whatever. Such men as Rich, Grosseteste, and Bacon, were not seen at Oxford in the fourteenth century. Greek, mathematics, and experimental science were overwhelmed in the paralyzing mists of Scotian dialectic. Nevertheless it would be an error to suppose that his life-work was a failure. Here and there throughout Europe the tradition of the Doctor Mirabilis survived as a stimulating force, and kept the embers of scientific study alive till the time of the Renascence.

In proof of this, three instances may be given :—

1. Peter d'Ailly, in his *Imago Mundi*, written early in the fifteenth century, discussing the relations of the extreme east and west of the habitable globe, has a long passage treating of the probable proximity of Spain and India. For all that appears in the work this passage is his own. But in fact it is a verbal quotation from the fourth part of the *Opus Majus*, vol. i. p. 290. And it has a history worth recording. For it is cited in 1498 in a letter from Columbus to Ferdinand and

Isabella, as one of the authorities that had put it into his mind to venture on his great voyage.

2. John Dee, in a memorial addressed to Queen Elizabeth in 1582 on the reformation of the Calendar, speaking of those who had advocated this change, says[1]: 'None hath done it more earnestly, neither with better reason and skill, than hath a subject of this British Sceptre Royal done, named as some think David Dee of Radik, but otherwise and most commonly (upon his name altered at the alteration of state into friarly profession) called Roger Bacon: who at large wrote thereof divers treatises and discourses to Pope Clement the fifth (*sic*) about the year of our Lord 1267. To whom he wrote and sent also great volumes exquisitely compiled of all sciences and singularities, philosophical and mathematical, as they might be available to the state of Christ his Catholic Church.' Dee proceeds to give extracts from Bacon's works in proof of these assertions; and remarks that Paul of Middleburg, who was much occupied with the question of the Calendar, and had treated of it in his work *Paulina de recta Paschae celebratione*, had made great use of Bacon. 'His great volume is more than half thereof written (though not acknowledged), by such order and method generally and particularly as our Roger Bacon laid out for the handling of the matter.' When we remember that it was Paul of Middleburg by whom Copernicus was urged with a view to this very problem to construct more accurate astronomical tables, we shall gladly acknowledge that here, too, Bacon's labour was not lost.

3. No part of Bacon's work was more frequently transcribed than his *Perspectiva*. Based as it was upon the great work of Alhazen, which was itself a development of the Optics of Euclid and Ptolemy, and claiming indeed to be but an abridgement or condensation of the truths laid down by his predecessor with wearisome copiousness, it was in fact much more than this. It selected from a mass of propositions,

[1] Dee's memorial is contained among the Bryan Twyne MSS. in Corpus Christi College, Oxford. The supposition that Roger Bacon changed his name on entrance into the Franciscan Order appears to rest on no authority but that of John Dee's very erratic imagination.

many of them mere displays of geometrical ingenuity, precisely those which aimed at the interpretation of nature, and at the adaptation of the laws of luminous radiation to human purposes. He was aware of what was unknown to Ptolemy and Alhazen, the concentration of parallel rays from reflecting surfaces formed by revolutions of a conic section; though how far he was indebted for this knowledge to Peter Peregrinus or to Vitello cannot be stated with certainty. Of the magnifying powers of convex lenses Bacon had a clear comprehension. He imagined, and was within measurable distance of effecting, the combination of lenses which was to bring far things near, but which was not to be realized till the time of Galileo.

In 1614, four years after the invention of the telescope, Combach, professor of philosophy in the University of Marpurg, published this great work of Bacon, 'viri eminentissimi.' It would be interesting to know whether the allusion in the *Novum Organum* (lib. i. 80) to the work of an obscure monk ('monachi alicujus in cellula') has reference to this work. The *Cogitata et Visa* was written before Combach's edition was published; but examples of the *Perspectiva* were numerous, and it can hardly have been unknown to Francis Bacon. In any case it must have been known to Descartes, to whose epoch-making researches on *Dioptrique* it assuredly contributed a stimulating influence. This at least they have in common, that light is looked upon as correlated with other modes of propagation of force through the Ether.

With the scientific Renascence of the sixteenth century, Roger Bacon's name slowly emerged from the darkness which had enwrapped it for three centuries. Astrologers like Dee, Heyden, and Allen hailed him as a champion of their outworn creed. Men of greater mark and sounder judgement, like Selden and Mead, were struck by his emancipation from the pedantry of the schools, and by his forecasts, made at so remote a time, of an age of industrial and scientific discovery. His central aim, the enlistment of progressive intellect in the cause of moral and religious renovation, was appreciated by none. But since the publication of his principal work in the eighteenth century, his name has gradually ascended towards its permanent position, on the lofty summits which were the earliest to 'take the morning' of European thought.

II. Bacon's Position in the Metaphysical Controversies of the Thirteenth Century.

It is too often forgotten that Bacon was a schoolman; trained in scholastic methods, and ready to take part in the philosophic discussions which interested his contemporaries. It is not perhaps surprising that this side of his work should have been ignored; for in the *Opus Majus*, though visible enough to an attentive reader, it is thrown into the shade by the prominence given to positive science, and by the practical application of science to political and religious purposes. Certain chapters of the *Opus Tertium*, which supplement too hasty or imperfect treatment in the larger work (chapters 38–52), afford better illustrations of Bacon's aptitude for metaphysical discussion. Nevertheless, the position of Bacon in the scholastic controversies of the thirteenth century remained an unknown quantity till the appearance of Professor Charles's monograph. His comprehensive survey of Bacon's unpublished works includes a careful study of, and copious extracts from the important fragment of the *Scriptum Principale*, entitled 'Communia Naturalium,' of which copies exist in the Mazarine library in Paris and in the British Museum.

Hauréau's comprehensive work on *Scholastic Philosophy* has made it easy to refute the illusion, still, however, not entirely dissipated, that scholasticism implies a special set of philosophical tenets or an uniform method of treatment. Philosophical writers in the thirteenth century differed from one another no less than philosophical writers in the nineteenth; though in either case a certain similarity in the subjects considered, and in the mode of handling them, was impressed by the circumstances of the time. Scholastic philosophy means simply philosophy taught in mediaeval schools. And between the schools of the twelfth, of the thirteenth, and of the fourteenth centuries, there were great and essential differences.

To pass from the reading of the *Polycraticus* of John of Salisbury, who knew nothing of Aristotle but his logic, and that imperfectly, to a treatise of Albert or of Aquinas seems, and is, a transition quite as abrupt as to exchange a volume of Addison or Swift for one of Schopenhauer or Carlyle. In the one case as in the other, a tide of revolution had swept between the centuries. For it was nothing less than a revolution for the western mind to receive very suddenly from the Mohammedan world the results of three centuries of Arabian learning, including as it did all the more serious part of Aristotle's work, enriched with keen-witted and audacious comment, and accompanied by the scientific results of the schools of Alexandria: the *Syntaxis* of Ptolemy and the biology of Galen.

Isolated thinkers like Adelard of Bath, the first translator of Euclid into Latin, had already entered this field of study, when Raymond, archbishop of Toledo, established in the middle of the twelfth century a systematic school of translators from the Arabic, of whom the Jew, John Avendeath (otherwise known as Johannes Hispalensis), Dominic Gundisalvi, archdeacon of Segovia, the translator of Algazel, and Gerard of Cremona, best known by his translations of the Almagest and of Alhazen, were the most prominent representatives[1]. Their translations of Aristotle, including his *Physics*, *Metaphysics* and *Psychology*, were not long in finding their way across the Pyrenees. Alexander Neckham, afterwards abbot of Cirencester, lectured upon them in Paris in 1180. His junior contemporary and countryman, Alfred of Sershall, pursued a similar course. Neither of these men roused suspicion. But the case was far otherwise with David of Dinant and Amaury of Bennes. Though we know little

[1] See Jourdain, *Recherches sur l'âge et l'origine des traductions Latines d'Aristote*, pp. 107-124 (ed. 1843). The history of mediaeval translations from Greek into Arabic, sometimes through intermediate Syriac versions, and from Arabic into Latin, deserves more elaborate treatment than it has yet received; provided always that the writer of such a history combined the two conditions so constantly insisted on by Bacon: knowledge of the languages concerned and knowledge of the subjects treated. Meantime much useful preliminary work has been done in this direction by such writers as Wuestenfeld and Jourdain.

of either, except through the criticism of their opponents, notably through that of Albert and Aquinas, yet such criticism is too detailed and definite to admit of doubt that their deductions from Aristotle and from his Arabian commentators led them to the assertion of the unity of substance; in other words, to the ultimate identity of matter, mind, and God. As quoted by Albert, the language of David was: 'It is manifest that there is one sole substance, not only of all bodies, but also of all souls, and that this is nothing but God himself. God, matter, and mind, are one and the same sole substance' (Albert. *Summa Theolog.* part II. tract. xii. quaest. 72, memb. 4, art. 2). David kept himself within the limits of philosophic theory. He is said to have been personally intimate with Innocent III; and at least during his lifetime his heresies escaped notice. It was otherwise with his contemporary Amaury of Bennes, who, maintaining the same opinions, was condemned by the Pope and forced publicly to disavow them. But they survived in his disciples, who used them in ways directly hostile to Catholic faith and discipline. A Council was held in Paris in 1210. Amaury's body was disinterred and buried in unconsecrated ground; several of his followers were burnt. It was at this council, the decrees of which were confirmed and enforced five years afterwards by Robert de Courçon, the papal legate, that the study of the Physic and Metaphysic of Aristotle was prohibited, on the mistaken supposition that the ultimate source of these heresies was to be found there; a mistake due probably to the comments of Averroes, with which the first translations of these works into Latin were accompanied [1].

How to deal with the problem of matter so as to give no countenance to pantheistic error, was therefore an urgent and momentous question, to which the schoolmen of the thirteenth century, and Albert especially, devoted their full powers.

[1] See Jean de Launoy's work *De varia Aristotelis in Academia Parisiensi fortuna liber* (Paris, 1653), in which seven stages are noted, from the condemnation of Aristotle in 1209, to the condemnation of his opponents by the Parlement of Paris in 1624. Cf. Hauréau, *Hist. de la Philos. Scolast.*, Part II. vol. ii. pp. 73–119.

Terrestrial substance, said Aristotle, was made up of matter and form. Apart from form, what then was matter? A pure essence, having the capacity, *potentia*, to become the subject of form, was the reply. How, then, distinguish matter from this *potentia*? Yet, if this be so, if matter is potentially the subject of all possible forms, we have in matter something that underlies all substance. Suppose all forms destroyed, matter holding in itself all the conditions of existence still remains. How, then, distinguish matter from God?

Albert's attempted solution of the problem is involved and obscure in the extreme, and it must not occupy us here. We are concerned with Bacon's. Bacon attacked the problem in his own way, and with a full sense of its importance. His conclusions are expressed in the seventh chapter of the fourth part of the *Opus Majus*, and in the thirty-eighth chapter of the *Opus Tertium*; and a still further exposition of them is found in the unpublished work of Bacon already mentioned, entitled 'Communia Naturalium.' This treatise on Physical Philosophy consists of four parts, of which the discussion of Matter occupies the second.

Substance, Bacon maintains, can be predicated neither of matter nor of form; but only of the compound which results from their union. 'Compositum habet rationem per se existendi in ordine entium: non sic materia et forma.' Matter and form are not substances: substance results from their union. Proceeding from above downwards through the hierarchy of being in the order of increasing speciality, we have, as the genus generalissimum, 'Substantia composita universalis.' This may be corporal or spiritual. Corporal substance may be terrestrial or celestial. Terrestrial substance may be a mixture of elements, or a single element. Mixed substance may be animate or inanimate. Animate substance may be sensitive (i.e. animal) or vegetal. Animal substance may be rational or irrational.

To each of these grades in the hierarchy of substance belong corresponding grades, not merely in the hierarchy of form, but also in the hierarchy of matter. 'Matter,' says Bacon, 'is not what most teachers of philosophy maintain it to be, "una numero." In the descending scale from general to special, each grade of matter, like each grade of form, is distinct from the preceding. One kind of matter is separated from another by specific differences, just as form is separated from form. The difference between an ass and a horse is not a difference of form only; it is a difference of matter' ('Commun. Natur.' Part II. Dist. ii. ch. 6).

Bacon has condensed these views in the diagrammatic form shown in the subjoined schedules, which I have copied from the Mazarine MS. pp. 23, 24. (They have been collated with those of the Br. Mus. MS. Royal, 7 F. vii. fol. 91 and 92. The variants in this MS. for the schedules of *substantia composita* and *forma* are unimportant. Those of *materia* are omitted; this MS. being in other respects less perfect than that of the Mazarine library.)

How are we to estimate these speculations? It is obvious in the first place that they stand in marked opposition to, or at least in distinction from, theories current among Bacon's contemporaries. To judge rightly of them we must bear in mind that throughout the greater part of the thirteenth century questions were being agitated of even greater importance than the controversy between realism and nominalism. The pantheistic tendencies discernible in Averroes and other Arabian thinkers had been diffused, as we have seen, by men like Amaury and David of Dinant They were responsible, as some thought, for the disastrous anarchy which early in the century had devastated southern France. Bacon was quick to perceive the danger of maintaining the unity of matter. It had been defended, as he points out (*Op. Maj.* vol. i. p. 144), by passages from Aristotle which he wishes to believe had been badly translated. In any case, he says, 'the error is enormous, as great as any that can possibly be found in speculative questions. If it be granted, it is impossible to comprehend the generation of things, and the whole course of nature will be misunderstood. And what is more, if this error be looked at closely, it will be found to tend towards heresy, or rather to be the profanest of heresies, since the inevitable result of it is to endow matter with the

DIVISIO SUBSTANTIAE COMPOSITAE UNIVERSALIS.

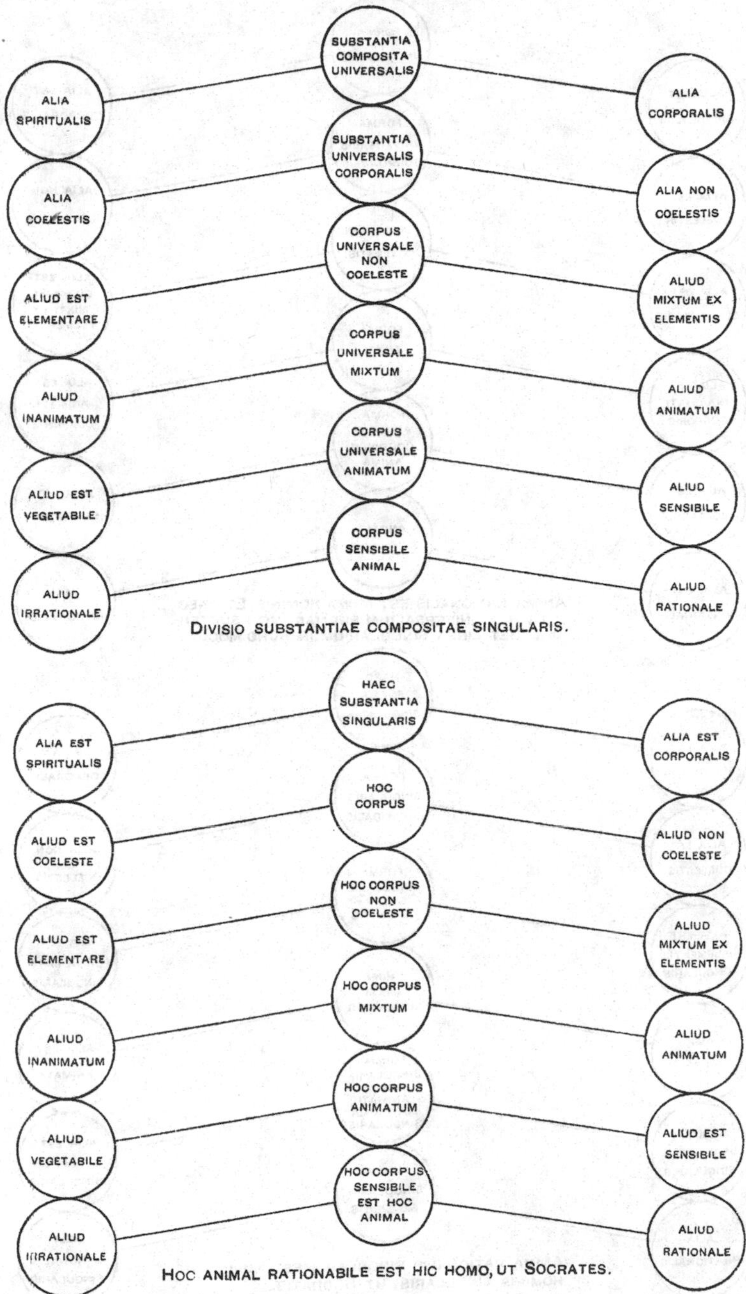

HOC ANIMAL RATIONABILE EST HIC HOMO, UT SOCRATES.

[*Vol. I. To face p.* xl.]

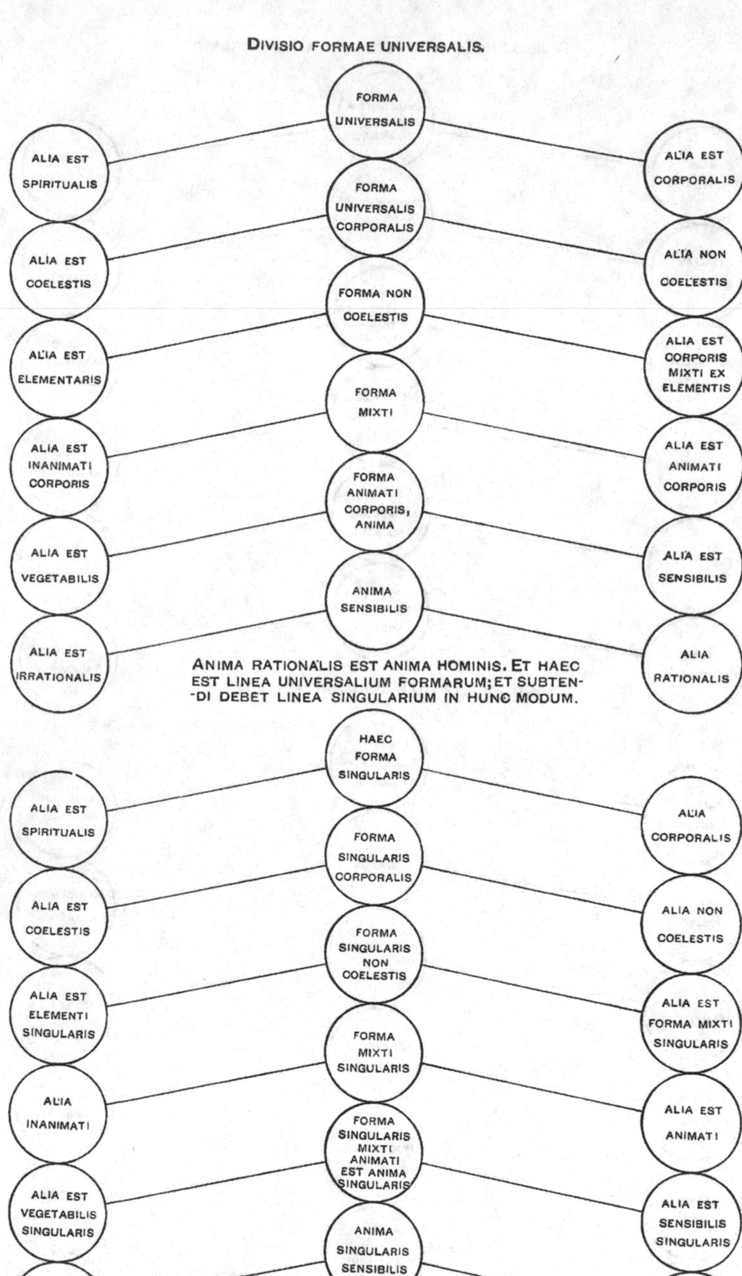
DIVISIO FORMAE UNIVERSALIS.
FORMA UNIVERSALIS
ALIA EST SPIRITUALIS
ALIA EST CORPORALIS
FORMA UNIVERSALIS CORPORALIS
ALIA EST COELESTIS
ALIA NON COELESTIS
FORMA NON COELESTIS
ALIA EST ELEMENTARIS
ALIA EST CORPORIS MIXTI EX ELEMENTIS
FORMA MIXTI
ALIA EST INANIMATI CORPORIS
ALIA EST ANIMATI CORPORIS
FORMA ANIMATI CORPORIS, ANIMA
ALIA EST VEGETABILIS
ALIA EST SENSIBILIS
ANIMA SENSIBILIS
ALIA EST IRRATIONALIS
ALIA RATIONALIS
ANIMA RATIONALIS EST ANIMA HOMINIS. ET HAEC EST LINEA UNIVERSALIUM FORMARUM; ET SUBTEN-DI DEBET LINEA SINGULARIUM IN HUNC MODUM.
HAEC FORMA SINGULARIS
ALIA EST SPIRITUALIS
ALIA CORPORALIS
FORMA SINGULARIS CORPORALIS
ALIA EST COELESTIS
ALIA NON COELESTIS
FORMA SINGULARIS NON COELESTIS
ALIA EST ELEMENTI SINGULARIS
ALIA EST FORMA MIXTI SINGULARIS
FORMA MIXTI SINGULARIS
ALIA INANIMATI
ALIA EST ANIMATI
FORMA SINGULARIS MIXTI ANIMATI EST ANIMA SINGULARIS
ALIA EST VEGETABILIS SINGULARIS
ALIA EST SENSIBILIS SINGULARIS
ANIMA SINGULARIS SENSIBILIS
ALIA IRRATIONALIS
ALIA RATIONALIS SINGULARIS
ANIMA RATIONALIS SINGULARIS EST HUJUS HOMINIS SINGULARIS, UT SOCRATIS.

DIVISIO MATERIAE PRIMAE UNIVERSALIS.

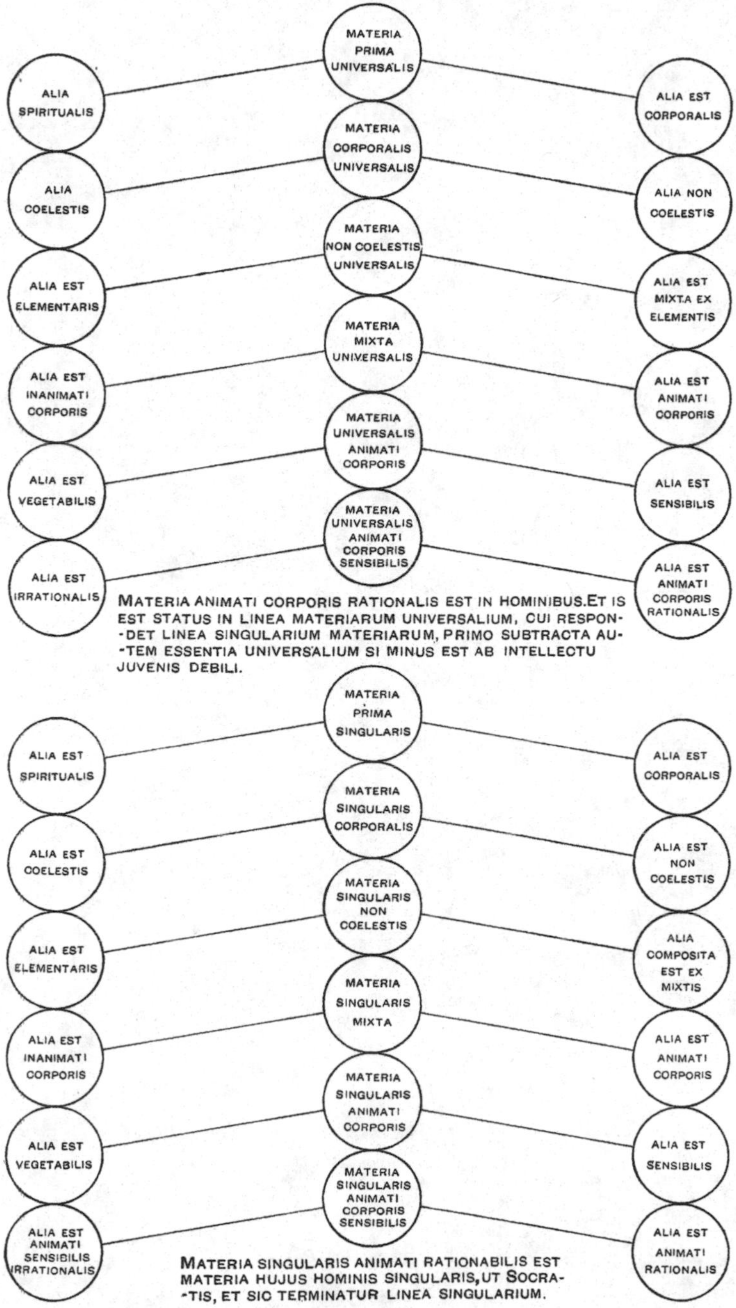

creative power of God.' Whatever dangers were involved in the unity of matter, Bacon met by a bold denial of such unity. 'Divide et impera,' he said in effect; matter, thus split up into sections, is no longer to be feared.

Looking at Bacon's theory by the light of subsequent centuries, it is not difficult to see that its value lay in its solvent and destructive power. His aim from beginning to end of his career was to draw men away from verbal subtleties and concentrate them on the realities of life, as plain men understand them. 'You ask me,' he would say to the young students around him, 'what is this matter which remains apart from all form, with capacity for receiving all? But who told you that it was one and indivisible? There are as many kinds and degrees of matter as there are of things. Look at the things, try them, see how they act on you, how you can act on them. As to the matter and form that may underlie them, leave that to God.'

Bacon's part in the great controversy between realism and nominalism will lead us to a similar conclusion. It was a less burning controversy in the thirteenth century than in the days of Roscelin and Abelard, or than it became afterwards in the days of Duns Scotus and William of Ockham; and it was debated by Albert and by Aquinas with the far larger and deeper understanding of its complications, that might be expected from men who were not merely trained like their predecessors in the study of Aristotle's Logic, but had become conversant with the problems raised in his Physic and Metaphysic. Both these thinkers rejected the independent existence of universals *in re* as clearly as Aristotle had done. They were clear that universals had no existence except in the mind. 'Non est universale nisi dum intelligitur' (Albert. *Met.* lib. v. tract. vi. cap. 7). 'Una et eadem natura quae singularis erat et individuatur per materiam in singularibus hominibus efficitur postea universalis per actionem intellectus depurantis illam a conditionibus quae sunt hic et nunc' (Aquinas, *Tractatus primus de universalibus*). Nevertheless, both of them left a place for the universal *ante rem*, not indeed in the fantastic world of Ideas which Plato had portrayed, but as radiations centred in the primal form, the mind of God.

Turning to Bacon, who discusses the question of universals at considerable length and with extreme independence, we find the same tendency to emancipate himself from bondage to words, entities, and verbal discussions, and to dig down to a foundation of solid fact. One individual, he says[1], is of more account than all the universals in the world. A universal is nothing but the similarity of several individuals; 'convenientia plurium individuorum.' 'Two things,' he goes on to say, 'are needful for the individual. The first is absolute: it is that which constitutes his existence, as when we say, "This man is made of soul and body." The second is that in which he resembles another man, and not an ass or a pig. This is his universal. But the absolute nature of an individual is of far more importance than his related nature. It is fixed and absolute by itself. Thus the singular is of more account (*nobilius*) than its universal. Experience leads us to this conclusion, and so also does theology. God has not created the world for the sake of the universal man, but for the sake of individual persons.' 'Individuum est natura absoluta et fixa habens esse per se; et universale non est nisi convenientia individui respectu alterius.'

In some passages Bacon appears to go much further in the direction of nominalism than Albert and Aquinas. 'The prevalent view,' he remarks, 'is that universals exist only in the mind. Yet two stones would be like one another, even though there should be no mind to perceive them. But it is precisely this likeness of the two stones that constitutes their universal' ('Commun. Natur.' Part II. Dist. ii. ch. 10).

Closely allied with the controversy as to universals was the question of Individuation. Are things individualized by form or by matter? Albert and Aquinas took the latter view, Bonaventura the former. 'Individuorum multitudo,' says Albert (*De Coelo*, tract. iii. c. 8), 'fit omnis per divisionem materiae. Formae quae sunt receptibiles in materia indivi-

[1] On the question of Universals, and also on that of Individuation, cf. the extracts from the *Communia Naturalium* given by Émile Charles, pp. 383–386.

duantur per materiam.' (Cf. Aquinas, *Summa Theol.* i. quaest. iii. cert. 2.) Aquinas was obliged, however, to add that this materia must be 'signata': must be quantified. 'Signatio ejus est esse sub certis signationibus quae faciunt esse hic et nunc.' This addition went far to neutralize the Thomist view of Individuation; for as his opponents at once rejoined, 'What determines quantity if not form?'

In opposition to Aquinas, Bonaventura maintained that 'species est totum esse individui.' Substance consisting of the union of matter and form, matter was uniform in all: the form was that which distinguished, individualized.

Bacon ('Commun. Natur.' Pt. II. Dist. ii. ch. 9), in opposition to either view, maintained that the question was meaningless and foolish. All substances, whether universal or singular, have their own constitutive principles. Soul and body make man. This soul and this body make this man. In the intention and procedure of nature, 'this man' is prior to 'man'; 'man' comes in as something subsidiary, 'extra essentiam ejus, similis accidenti,' as the means of comparison with other individuals. There is no more reason for inquiring what causes individuation than for inquiring what causes universality. There is no answer to such a question, except that the Creator makes everything as its nature requires. Individual matter and form is made in one way: specific or generic matter and form is made in another. 'Stultitia magna est in hujusmodi quaestione quam faciunt de individuatione.'

III. Bacon's 'Scriptum Principale.'

The foregoing remarks, which it would be easy, but not, in this place, justifiable to prolong, will illustrate Bacon's position as a schoolman, thoroughly versed in the *technique* of scholastic controversy. But he was a schoolman whom a long and laborious study of the realities of life, whether in nature or in man, had taught to distinguish things from words: solid facts from subtle figments. He was not alone in this. Albert and Aquinas were solid thinkers like himself. Less versed in natural science than Bacon, they had more than he to do with the science of man; they had to face the difficult and urgent problems connected with the spiritual government of mankind. Their philosophy, like his, dealt with real things. And if theirs was less positive, less free from metaphysical figments, it is only that the complications of human nature were less adapted for positive treatment than the physical phenomena to which Bacon devoted so large a share of his attention.

But in contrast with these three great schoolmen stand the weavers of word-systems, like Alexander of Hales, Henry of Ghent, and Duns Scotus, wasting their own and other men's time and energy in defining, dividing, and refining with infinite ingenuity, and with such result as when children build sand-castles on the shore. With such men it may have been needful to fight, yet fighting was but beating the air. Of what avail to discuss Individuation with dialecticians who explained it by '*haecceitas*[1]'?

In all Bacon's discussion of scholastic problems, the solution he reached was of a kind to favour the falling off of the metaphysical husk, and to bring to light the real and positive problem which lay beneath it. His scholastic theories are therefore for us, and in all probability were for him, of far greater negative than constructive value. But his central aim lay in another direction above and beyond scholasticism. We shall best learn how to appreciate it by looking at the programme of the encyclopaedic work, the 'scriptum principale,' often spoken of in the *Opus Majus* and the *Opus Tertium*, but

[1] This word is believed to be due rather to the disciples of Duns Scotus than to the master himself. Happily for Bacon's peace of mind, he did not live to witness the triumphal career of the Doctor Subtilis. Of Hauréau's careful appreciation of his work the final words may be quoted: 'Cette philosophie n'explique pas la nature, elle l'invente; substituant l'ordre rationnel à l'ordre réel, elle dispense, il est vrai, de l'étude des choses; mais, quand après avoir admiré l'économie d'un système si complet, si habilement ordonné, on abaisse ses regards vers ces choses dont on a jusqu'alors dédaigné de s'enquérir, on soupçonne dès l'abord qu'on vient d'achever un rêve, et bientôt, devant le spectacle qu'offre la réalité, s'effacent, s'évanouissent l'une après l'autre toutes les abstractions décevantes, toutes les chimères dont la création appartient au système, à lui seul.' [*Hist. de la Philos. Scholastique*, Part II. vol. ii. pp. 171-259, ed. 1872-80.]

of which the persecutions and imprisonment of his later life never allowed him to execute more than a few fragments. And of these fragments many are lost.

This 'scriptum principale,' as he tells us in the first chapter of the unpublished work entitled 'Communia Naturalium,' consisted, or was intended to consist, of four volumes. The first volume dealt with Grammar and Logic, the second with Mathematic, the third with Physic, the fourth with Metaphysic and Morals.

The second chapter of the 'Communia Naturalium' is entitled, 'De universo ordine scientiarum naturalium.' He distinguishes eight natural sciences. The first treats of the principles common to Natural Philosophy. The others are:—(1) Perspective or Optic, (2) Astronomy, (3) Barology, (4) Alchemy, (5) Agriculture, (6) Medicine, (7) Experimental Science. The general principles of Physics form the subject of the first treatise here spoken of. Of the seven special sciences, the first three form part of what would in the present day be called Physics. Under Astronomy is included not merely the study of planetary motions, but the scientific determination of terrestrial positions, in other words, Geography, and also the influence of the stars and the sun on the earth and man: that is to say, the study of climate and of astrological forces. As to the third of the special sciences, 'scientia de elementis,' or as he also calls it, 'scientia de Ponderibus,' what in the present day would be called Barology, it is not without interest to find it thus set apart as a distinct department of speculation. The fourth, Alkimia, corresponds, so far as the description of its purpose goes, very nearly to the modern science of Chemistry. It deals, says Bacon, with the 'mistiones elementorum,' with the generation of liquids, gases, and solids ('humores et spiritus et corpora'), with all inanimate substances, including organic products ('usque ad partes animalium et plantarum inclusive').

The title of the fifth science, *Agricultura*, would be misleading, if Bacon had not given us a clear explanation of its purpose[1]. It is the science of living bodies, vegetal and

[1] Cf. the long extract given by Charles, pp. 370–374. The unfortunate rarity

animal; reserving, however, the subject of man's physical nature for subsequent treatment under the head of medicine. Before man be properly investigated, we must know the nature and surroundings of other animate things. First we must distinguish the soil fit for different kinds of plants, arable land, forest land, pasture land, garden land. We then examine the whole subject of plants which has been left incomplete in the treatise attributed to Aristotle, *De Vegetabilibus.* But as lands cannot be tilled without domestic animals, and as forests pastures, and deserts depend for their value on the wild animals they contain, the science we are now speaking of embraces the full consideration of animal life on which, as Bacon believed, Aristotle wrote far more volumes than have come down to us In the sixth science we proceed to the study of the animal possessing reason, the study of Man.

of Charles' work is my excuse for citing the portion of this extract relating to the study of living bodies, based, as Bacon explains, on the preliminary study of *Alkimia speculativa.* 'Deinceps de plantarum natura et animalium specialis scientia et maxima constituitur, scilicet de omnibus animatis praeterquam de homine, de quo propter nobilitatem suam et dignitatem constituitur scientia propria quae dicitur medicina. Sed in ordine disciplinae prima est scientia animatorum praecedentium hominem et ejus usui necessariorum, quae primo descendit ad omne genus agri et terrae, distinguens quatuor species agrorum, propter vegetabilia e terra nascentia in eis. Est enim ager in quo serunt segetes et legumina; est ager consitus arboribus, ut nemus; est ager pascivus, ut prata et deserta; est ager qui hortus dicitur, in quo domesticae arbores et caules et herbae et radices tam nutritivae quam medicinales parantur. Haec igitur scientia extendit se ad perfectam considerationem omnium vegetabilium quorum notitia nimis imperfecta traditur in libro *De Vegetabilibus* Aristotelis; et ideo necessaria est scientia sufficiens de plantis et animalibus supplens defectus librorum communium Aristotelis vulgatorum apud Latinos, qui vocantur *De Plantis et Animalibus.* Sed cum agrorum cultura non potest fieri sine copia animalium domesticorum, neque utilitas agrorum, praecipue consitorum arboribus, et pascuorum et desertorum, posset haberi nisi nutrirentur animalia sylvestria, ideo extendit se haec scientia ad plenam considerationem animalium omnium; et ad horum cognitionem misit Aristoteles plura millia hominum per regiones mundi, et fecit illa praeclara quinquaginta volumina prius memorata. Haec autem scientia traditur in libris Plinii, in libro Palladii *De Agricultura*, et in libro Georgicorum Virgilii non ignobili, cum expositione egregii commentatoris ejus.

Scientia septima est de animali rationali, scilicet de homine, et praecipue de sanitate et infirmitate ejus; et ideo de ejus compositione et generatione illius, sine quibus sanitas et infirmitas ejus non possunt intelligi nec doceri. Constat vero quod homo est res naturalis, et ideo scientia ejus naturalibus constituta erit inter naturales comprehensa.

Our aim being to understand the conditions of his health or disease, we have first to examine his structure and development, without which health and disease cannot be understood or spoken of.

Finally, to complete the whole, comes Experimental science. It is, he says, a final judge of the assertions and reasonings put forth in all the foregoing sciences. More than this: it gives directions to those engaged in other sciences as to the construction of instruments by which their conclusions are to be tested, in the same way in which a navigator instructs a shipwright as to the building of a ship. Thus, for instance, it instructs the geometer to make a mirror such that the rays reflected from it shall converge in a single point. It scrutinizes every natural, every artificial force. It sifts the artifices of magic, as logic sifts the reasonings of the sophist, so as to dissipate falsehood and error, and leave nothing but truth remaining.

How Bacon would have treated this part of the subject we have no means of judging, other than the sixth section of the *Opus Majus*. But even the summary exposition there given is enough to show how large was his conception of experimental method, and at the same time how carefully he steered clear of the danger of undervaluing the mathematical or deductive process of discovery. So far as was possible the two should be pursued simultaneously and in close alliance. Euclid's demonstration of his first proposition would, he says, fail to carry complete conviction unless visual evidence of it were forthcoming in the construction of the figure. And on the other hand, we see that his inductive investigations of the rainbow were controlled at every step by deductions from astronomy.

With these general remarks, we may now pass to each of the principal divisions of the 'Scriptum Principale,' which in the main correspond to the order followed in the *Opus Majus*. First comes Language, as the channel through which the thoughts of other men are handed down to us; then follows Mathematic, embracing the four branches of the Quadrivium, geometry, arithmetic, astronomy, and music. Thence we pass to Physic, which included the study of the propagation of force, specially illustrated by the radiation of light and heat. Next comes Alkimia Speculativa, not the mere metallurgy of the gold-seekers, but the study of the transformation of matter from its simplest to its most complicated state. The study of living matter followed, ending with Medicine, the science dealing with the physical structure of man. Finally, the edifice of the sciences is crowned by Ethic and Metaphysic.

Of this comprehensive scheme let us see what fragments are forthcoming.

IV. BACON'S PHILOLOGY.

In urging that the comparative study of language should form part of the University curriculum, Bacon stood nearly alone. He does indeed full justice to those among his contemporaries who had promoted the translation of Greek books into Latin; and, first among those, to the illustrious bishop of Lincoln, his forerunner and counsellor. But though Grosseteste had caused many books to be translated for the sake of their contents, it does not appear that he or any one else had proposed to carry the study of language, as such, beyond the routine of grammar presented in the Trivium; the Latin accidence and syntax of Priscian or Donatus.

What Bacon proposed was the systematic and comparative study of Hebrew, Arabic, and Greek, with the dialects belonging to each. With Hebrew went Chaldaean, and, in more distant relationship, Arabic: with Greek its various dialects, which were, he tells us, comparable to the Picard, Norman, or Burgundian dialects of French, or to the northern, southern, eastern, and western dialects of English. 'I do not mean,' says Bacon, 'that every one should learn these languages as he learns his mother tongue, so as to be able to speak them as we speak English, French, and Latin; nor again that we should content ourselves with being able to translate into our own language the Latin versions. There is an intermediate degree of attainment quite easy to those who have teachers. We should know enough to be able to understand how these languages should be rendered in Latin. The point is that a man should be able to read these languages, and understand their grammatical structure ('accidentia partium orationis,' *Compendium Studii*, Brewer, p. 433).

What Bacon's linguistic attainments were cannot be precisely decided. No woık of his, published or unpublished, that I am aware of, affords evidence of knowledge of Arabic. His own words in the twenty-fifth chapter of *Opus Tertium* are scarcely decisive on the point. 'De Arabica tango locis suis; sed nihil scribo Arabice, sicut Hebraee, Graece, et Latine, quia evidentius et facilius ostenditur propositum meum in his. Nam pro studio theologiae parum valet, licet pro philosophia multum, et pro conversione infidelium.' Some pages printed for the first time in this edition show acquaintance at least with the Hebrew alphabet. An elementary Greek grammar, in the possession of Corpus Christi College, Oxford, testifies to his knowledge of Greek, which indeed is sufficiently apparent in the present work, and still more in the ninth and following chapters of the *Compendium Studii* (Brewer, pp. 495–519). This grammar is incomplete, dealing chiefly with the alphabet, with the Greek system of accentuation, aspiration, and quantity, and with the numeral system. It concludes with the paradigm of the verb τύπτω. Its opening sentence seems to indicate that it formed a part of Bacon's encyclopaedic work. 'Here begins the first book of the volume on the grammar of languages other than Latin. This book deals with Greek grammar.' 'I have already,' he continues, 'spoken of the advantage to the Latin world of knowing the four languages, Greek, Hebrew, Arabic, and Chaldaean; and in the preface to this volume devoted to grammar I have explained the division of subjects and their order. I now proceed to consider Greek grammar, beginning with such rudiments as boys are taught in Latin in order that they may read, write, and construe simple passages, and may pass thence to points of greater difficulty.'

A point of interest presents itself as to Bacon's pronunciation of Greek. Much attention is given to the transliteration of the Greek alphabet into its Latin equivalents. The Lord's Prayer, the Salutation to the Virgin, and the Apostles' Creed are written out in Latin, underlined first with the Greek words in Roman character, and secondly with the same words in Greek. The second of these is here given as an example:

Ave	Maria	gratiosa	Dominus	cum	te	benedicta
Chere	Maria	kecharitomeni	ho Kyrios	meta	su	eulogimeni
Χαῖρε	Μαρία	κεχαριτωμένη	ὁ Κύριος	μετὰ	σοῦ	εὐλογημένη

tu	in	mulieribus	et	benedictus	fructus	ventris	tui.
sy	en	gynexi	ke	eulogimenos	ho karpos	tis kilias	su.
σὺ	ἐν	γυναιξὶ	καὶ	εὐλογημένος	ὁ καρπὸς	τῆς κοιλίας	σοῦ.

Amen.
Amen.
Ἀμήν.

It is evident from the transliteration of vowels and diphthongs here adopted, with which may be compared pp. 75–76 of the *Opus Majus*, printed for the first time in this edition that these were pronounced as in modern Greek [1]. It appears also in the subsequent discussion on accents, that accents were considered, no less than quantity, in pronunciation. Bacon may not improbably have learnt the language from one of the Greeks who had been invited into England by Grosseteste. Some of these, he tells us, had become permanent residents. (In *Compend. Studii*, Brewer, pp. 495–514. the same subject is treated.)

In urging so strongly the study of language, Bacon had two main purposes in view: an improved text of the Bible, and an

[1] In the Corpus Coll. Grammar, a systematic scheme of transliteration and of pronunciation is also given. We learn from it that the second letter of the alphabet was pronounced like the modern English *v*; and that there was no single letter rendering the sound of our *b*. 'Item π post μ vel ν, sive in eadem dictione, sive in diversis, dummodo sine intervallo proferantur, sonum nostri *b* habet, quem aliter non habent, ut λαμπάς, ἄμπελον. Similiter τ post μ vel ν sonat nostrum *d*, quod aliter non habent, ut ἀντίχριστος.' All this is in accordance with modern Greek pronunciation. The transliteration of the diphthongs *αυ* and *ευ* was a matter of some difficulty owing to the confusion between *u* and *v*. Bacon usually renders them as *af* and *ef*. But in modern Greek it is only before θ, κ, ξ, π, σ, τ, φ, χ, ψ, that they are thus pronounced: before other letters they would have the sound of *av* or *ev*. With regard to accents, Bacon's language (both in the Corpus MS. and in the *Compendium Studii*) puts it beyond all doubt that they governed his pronunciation of the language.

intelligible translation of Aristotle. Under both these heads the minor works, edited by Brewer, contain much for which in the *Opus Majus* Bacon had not found room. With regard to the first, the valuable memoir published in 1888 by Abbé Martin may be consulted [1]. It appears that, towards the end of the twelfth or the beginning of the thirteenth century, a text of the Bible had become generally current in Paris under the title of 'Textus Parisiensis.' Bacon, writing in 1267, speaks of it as having been hastily compiled, about forty years before, by second-rate theologians and carelessly copied by uncritical booksellers (*Opus Minus*, Brewer, p. 333). It abounded in errors and in interpolations inserted from patristic quotations, from liturgies, and from the works of Josephus. Many of these errors attracted notice, and attempts were made, principally by members of the mendicant orders, to correct them. But these attempts, in Bacon's judgement, only resulted in making the matter worse. Each critic worked independently and without adequate critical apparatus. Not merely did Franciscan differ from Dominican, but the members of each Order differed amongst themselves. Successive corrected versions were put forward, each worse than the preceding. By the middle of the century the Paris text had fallen into hopeless confusion; and it had become, in Bacon's judgement, far the lesser evil to use the uncorrected text than any of those which had been so uncritically amended. Of these strong remarks he gives many pointed illustrations.

So devoid were these successive editors, not merely of linguistic knowledge, but of the critical spirit, that they seem to have been entirely unaware of the origin and history of the Vulgate. Bacon's history of the various Biblical versions, ending with that of Jerome, as given in the *Opus Tertium*, pp. 334–349, is not one of the least interesting portions of his work. His principal result was to show that, before Jerome's translation from the Hebrew, the version regarded as authentic by the Church was the Septuagint; although theologians had felt themselves at liberty to correct that version from that of Aquila, Symmachus, and above all of Theodotion.

[1] See note on vol. i. p. 77.

After the time of Jerome, the translation from the Septuagint continued to be used in the Psalter; but, with that exception, Jerome's translation from the Hebrew constituted the Vulgate, and was received as authentic by the Church. Bacon is careful to add that Jerome's version is by no means free from error, due partly to over-haste, partly to his unwillingness to offend his contemporaries by making too many changes in the text hitherto accepted.

With Aristotle the case was even worse than with the Bible. The brilliant hopes with which the century had opened, of re-entering the temple of Greek wisdom, and listening to the voice of the greatest of ancient thinkers, had been falsified by the failure of Aristotle's translators to comply with the two elementary conditions of translation; knowledge of the language in which, and comprehension of the subject about which, the book was written. Something has already been said of the Toledo school of translators instituted by Archbishop Raymond in the twelfth century. A new and vigorous impulse was given forty or fifty years afterwards by the Emperor Frederic II, whose preference of Mahommedanism to Christianity, had he occupied a humbler station, would assuredly have subjected him to a worse fate than that of Bacon. Leaving out of account translations of Aristotle's *Organum*, parts of which were familiar to the western world from the times of Augustine and Boethius, the translators of Aristotle's philosophic and scientific work in the twelfth and thirteenth century, to whom Bacon calls attention, were five: Gerard of Cremona, Alured or Alfred of England, Michael Scot, Hermann the German, William of Moerbeke, otherwise called the Fleming. Of these, Gerard, Scot, and Hermann translated from Arabic versions. Gerard spent many years in Spain, attained a thorough knowledge of Arabic, and translated Ptolemy's *Almagest*, and Aristotle's *Meteorologics*, also the astronomy of Alfraganus, several works of Alkindi, and, almost certainly, the *Optics* of Alhazen. He died in 1187. Michael Scot flourished in the first half of the thirteenth century. He was a friend of the Emperor Frederic II, under whose patronage he visited Spain, and translated from the Arabic many of Aristotle's works, with the comments of Averroes. Albertus Magnus says of him that he was ignorant of natural things, and that he did not thoroughly understand Aristotle's books. Bacon, who speaks of the impression produced in the schools when he appeared in 1230, with translations of Aristotle's metaphysical and scientific treatises, says that he was ignorant of words and of things, and that the greater part of his work was due to Andrew the Jew. Scot's translation of Avicenna's treatise *De Animalibus*, as I have remarked in a note (vol. ii. p. 85), certainly seems to bear out this severe judgement. Hermann, the German, was personally known to Bacon; he worked in Spain, and with the help of Arab interpreters, produced translations of the Rhetoric, Poetic, and Ethic of Aristotle. He mentions incidentally (*Jourdain*, p. 140) that Grosseteste had produced a more complete rendering of the Ethic directly from the Greek. Hermann, in answer to some questions put to him by Bacon as to Aristotle's logical works, frankly confessed his ignorance of logic. 'Nor was he well acquainted,' Bacon continues, 'with Arabic, being rather an encourager of translations than a translator himself; the principal part of his work was done by Saracens in his employment' (*Compend. Studii*, cap. 8).

William the Fleming (of Moerbeke) had the advantage over these men that he translated directly from the Greek. His work is believed to have been done at the special request of Thomas Aquinas, who made use of it in his Commentaries on Aristotle. 'But it was notorious in Paris,' says Bacon, 'that William of Moerbeke was totally ignorant of science, and his translations are consequently full of errors.' On the whole, he concludes that it would have been better that Aristotle should never have been translated, rather than that such a mass of error should be propagated under the shelter of his name. Had I the power of disposing of these works, I would have them all burnt: it is a waste of time to study them, a source of error and of diffusion of ignorance greater than can be described.' 'Aristotle's works,' he continues, 'are the foundation of all wisdom, but they must be studied in the original to be of any profit' (*Compend. Studii*, Brewer, p. 469).

Every one who considers Bacon's efforts in promoting the study of language must agree with Professor Brewer (p. lxii) that 'his labours in this respect have attracted less attention than they deserve. . . . It is as creditable to his discernment as to his courage that he should have seen, better than Lord Bacon did, the paramount importance of philology, and urged it repeatedly on his contemporaries. It is amazing to hear a scholar of the thirteenth century insisting on the necessity of constant references to original authorities as the only sure foundation of sacred criticism.'

It may be that Bacon's exhortations, reiterated as we feel sure they would be, not in writing merely, but in conversation with the young men whom he gathered round him, were not entirely without effect on the following generation. In the council convoked in 1312 by Clement V at Vienne, one of the provisions, says Fleury (*Hist. Eccl.* book 91), was 'the establishment in the Roman Curia, and in the Universities of Paris, Oxford, Bologna, and Salamanca, of teachers for the three languages, Hebrew, Arabic, and Chaldaean, two for each. They were to be maintained in Rome by the Pope, in Paris by the King of France, and in the other cities by the prelates, bishops, and chapters of the country.' This subject has been carefully studied by Mr. Rashdall in his important work on the Universities of Europe in the middle ages. He gives strong authority for the belief that Greek was included; and if so, the avowed purpose of the ordinance, which was the conversion of the Mahommedans and Jews, may not have been the only purpose; some faint echo of Bacon's exhortation to study Aristotle and the Bible in the original, with the view of understanding them better, may have been still audible. Few and short-lived were the attempts made to carry the decree of this Council into effect. In 1320 we hear of a rate levied upon benefices in the province of Canterbury for the support of a converted Jew alleged to be teaching Greek at Oxford. But Oxford was already passing under the spell of the enchanter. The fine webs of Duns Scotus, which the sword of Ockham might cleave but could not dissipate, were paralyzing her energies. Five generations were to pass before she could again begin to promote the study of 'languages other than Latin[1]'; and even then not in the comprehensive spirit which Bacon had advocated. It is tempting, though painful and perhaps useless, to imagine how far European culture might have advanced had schools of Oriental languages, concurrently with those of Greek and Latin, been instituted and continuously maintained from the thirteenth century.

V. Bacon's Mathematics.

In the *Opus Majus*, though much is said of the importance and necessity of mathematical method, there is very little display of mathematical knowledge. Frequent references are made to Euclid, whose Elements had been introduced to the western world early in the previous century, by Adelard of Bath, and more completely in the thirteenth century by Campanus of Novara. Archimedes and Apollonius are rarely mentioned. But in his Optics Bacon shows that he was acquainted with the properties of parabolic concave mirrors. and of their power of causing parallel rays to converge after reflection to a focus. In this respect he was in advance of his principal teachers in Optic, Euclid, Ptolemy, and Alhazen.

Of the Calculus, arithmetical or algebraical, Bacon has but slight occasion to speak in the *Opus Majus*. It has always to be remembered that this work, with its appendices, the *Opus Minus* and *Opus Tertium*, was not, properly speaking, a philosophical treatise, but an exhortation addressed to a statesman, absorbed in ecclesiastical and political struggles, to exert his authority for the revival of learning. Hence it is uniformly spoken of as a *Persuasio*. It contains just as much learning and science as was thought needful to convince the Pope that learning and science were capable of strengthening the Church. It is but the preamble to the 'Scriptum

[1] The organized teaching of Greek in Oxford is due to Richard Foxe. the founder of Corpus Christi College (1515-16 . But when Erasmus was in Oxford about twenty years earlier, such men as Thomas Linacre and William Grocyn had already become Greek scholars, under the teaching, perhaps, of Cornelio Vitelli. (Cf. Hallam, *Lit. Hist. of Europe*, part i. ch. 3.)

Principale,' which there is reason for thinking that Bacon had already begun, but which he regretfully expresses his inability to send at such short notice. Hence though it deals, often very cursorily, with every department of knowledge then recognized, we must not infer Bacon's ignorance of a subject from the fact that this provisional treatise makes no mention of it.

Among the fragments of the 'Scriptum Principale' which have come down to us, is a portion of the first book on Mathematics, preserved among the Sloane MSS. (2156). This first book contained three parts. We have the first part of this book, and a considerable part of the second. A few fragments more are to be found in the Bodleian (Digby MSS. 76). As far as I am aware nothing more is extant.

The first part deals with preliminary principles ('quaedam preambula ad interiora mathematicae').

It has five divisions or distinctions. The subjects dealt with are the relation of mathematic to metaphysic: its distinction from magic; the hindrances to its culture offered by the four causes of error, viz. false conceit of wisdom, authority, custom, and popular prejudice; the utility of mathematics, its importance to the preliminary studies of logic and grammar. The final chapter of this section is curious. The final purpose, says Bacon, of logic is conviction. But conviction is not reached by argumentative process alone, but by the arts of rhetoric and poetry, which are therefore in a true sense departments of logic. But these arts are governed by the laws of music, which is a branch of mathematical science.

The second division deals with the definition of the parts of quantity. Certain general terms, such as simultaneity in space and time, limit, continuity, infinity, dimension, are explained. The distinction is drawn between continuous and discrete quantity. Continuous quantity in one, two, and three dimensions is defined. Discrete quantity is distinguished into what is permanent, as number; what is not permanent, as sound.

The third division expounds the distinction between the speculative and the practical departments of geometry and of arithmetic. The section on practical or applied geometry is of much interest as illustrating Bacon's enlarged views of scientific training. He indicates eight departments of this branch of science. (1) Agriculture, in a far wider sense than is usually given to the word, comprising mensuration, architecture, civil, mechanical, and military engineering. (2) The fabrication of astronomical instruments. (3) Of musical instruments. (4) Of optical instruments. (5) Of barological instruments. (6) Of instruments of experimental science. (7) Of medical and surgical appliances. (8) Of chemical apparatus.

In connexion with the practical branch of arithmetic, after speaking of the use of the Abacus, he mentions 'vias algorithmi, scilicet quomodo conjugantur numeri et dividuntur, secundum omnem speciem algorithmi, tam in particularibus fractionibus quam in integris.' In this connexion he speaks of 'Algebra quae est negotiatio, et almochabala quae est census.' How far Bacon had assimilated the work of Mohammed ben Musa[1], whose surname, Al Chwarismi, is incorporated in the word Algorithm, we cannot tell. But with the work of one of the two great mathematicians of the thirteenth century, Jordanus Nemorarius, he was certainly familiar, as may be seen by reference to vol. i. pp. 158, 169 of the *Opus Majus*. Among other branches of practical arithmetic he includes the construction of astronomical tables, mensuration, alloys and coinage, partnership, and other operations of commerce. These things are treated of at great length in the *Liber Abaci* of Leonard of Pisa, the other great mathematician of the time, whose work, dedicated to Michael Scot, Bacon

[1] Muhammed ibn Mûsâ Alchwarismî was born in the first quarter of the ninth century. He constructed astronomical tables for the Caliph Al Mamoun, which were translated into Latin by Adelard of Bath. Of more importance, however, are his Arithmetic and his Algebra. The first of these remained for a long time unknown. But it was discovered in Cambridge in 1857, and is included among the *Trattati d'Aritmetica* published by Boncompagni. A full account of this work and of the Algebra, translated and edited by Rosen (London, 1831), will be found in Cantor, vol. i. pp. 611-629. Cantor (p. 612) explains clearly the passage of the word Alchwarismi into Algorithm. Bacon's interpretation of the words Aldschebr walmukâbala, which Alchwarismi uses, is incorrect. Dschebr means Restoration, mukâbala means opposition.

had possibly seen and studied; though he makes no mention of it, in any work known to us[1].

Astrology and astronomy come next. The first is the speculative branch, dealing with planetary motions, with the figure of the earth and of its various regions. Astronomy, the practical branch, has to do with the construction of tables and with the forecast of future events. Bacon admits that this use of the words has not been universally adopted, but maintains its correctness. 'Astrologia componitur ex hoc nomine astron quod est stella, et hoc nomine logos quod est verbum, vel ratio, vel sermo, quia est sermo de stellis. Astronomia vero dicitur lex stellarum et nomos est lex. Unde quia lex universaliter sonat in practicum, ut in morali philosophia lex est ipsa practica, ita similiter Astronomia est practica astronomiae.'

In the fourth division music is considered. This includes not merely sound but gesture. Audible music is considered under the two heads of vocal and instrumental. In the vocal division every branch of elocution is included. Finally, the effect of music on the temper and health both of men and of animals should be systematically studied.

Abstraction is the subject of the fifth division. First we have the abstraction common to all science, since science deals with universals, not with particulars. There is then the abstraction of a first cause from secondary causes and of spirit from body, which the metaphysician deals with. Mathematical abstraction has to do with the study of quantity apart from the substance to which it belongs: apart from all natural changes such as growth, diminution or change of place.

This first part closes with an explanation of the difference between axioms, postulates, and definitions.

The second part begins with the study of whole numbers and fractions: passing from this to the subject of arithmetical, geometrical, and harmonic ratio, and to the question of proportion generally. Continuous and discontinuous proportion are considered; and Euclid's definition of proportion is carefully considered.

[1] To each of these two mathematicians Cantor devotes a full chapter. Cf. *Gesch. der Math.* vol. ii. pp. 3-79.

Here the portion of the work contained in the Sloane MSS. ends. We find it continued, however, in a somewhat fragmentary way in the Bodleian Digby MSS. No. 76. The author proceeds to the consideration of geometrical truths, professing his intention to select those which were of paramount importance, since it was obvious that the number of possible problems in geometry was infinite.

'Illae (veritates),' he says, 'sunt eligendae quae possunt vocari radices et elementa respectu ramorum et foliorum, quorum fructus vadit in infinitum.' Proof is given in this part of the work that Bacon was acquainted with the geometry of Apollonius as well as with that of Euclid. After defining the cone ('pyramis rotunda') he mentions its three sections, presenting curves of a different form from the circle, one of which was of use in the construction of mirrors capable of rendering rays convergent to a point. He promises to deal with these curves later in the work.

Of the whole, so far as the fragment of his mathematical work preserved to us enables us to judge, it would seem that Bacon had made himself acquainted with the highest mathematics of his time; though no evidence is forthcoming to show that he contributed personally to the advance of the science, otherwise than by strongly insisting on its culture, and by pointing out new fields for its practical application, in the better government of the Church, and in the development of industry. His interest, like that of Galileo, lay in applied rather than in abstract mathematics. Whether the study of equations as carried on by the Italian algebraists of the fourteenth and fifteenth centuries would have interested him is doubtful. But he would have eagerly welcomed the invention of logarithms, as facilitating the construction of astronomical tables.

VI. Bacon's Astrology.

The transition from Mathematics to Physics supplies the best opportunity for a few remarks on the subject of Bacon's Astrology, on which something is also said in a note to vol. i. p. 269. Bacon dwelt frequently and emphatically on the unity and the correlation of the sciences. In passing from Mathematics to the direct study of Nature, he found a connecting link in the imaginary science of Astrology, which he studied zealously. That the fixed stars and the planets exercised a powerful influence on all earthly things and not least on man; that the careful observation of their position at the moment of birth would do much to reveal the hidden springs of character, and make it possible to form a forecast of the ensuing life, that the influences radiating from them acted with greater or less potency according as the course of the rays was perpendicular or oblique, and that in this way an explanation could be given of climate, temperament, and of the thousand complex chances and changes of mortal life, was a belief firmly held by Bacon, and it operated powerfully over his whole view of man's position in the world. He has been much reproached for holding it; and it has been supposed to be an explanation, if not an excuse, for the disastrous repression exercised over him by his superiors, and for the popular discredit attaching to his name.

But this would be an entire misconception of the beliefs current in Bacon's time. The influence of the stars over human life was a belief almost universally held by all instructed men from the thirteenth to the sixteenth century; and abundant traces of it are visible throughout the seventeenth, not to speak of still later times. The *Divina Commedia* is full of it. Beatrice, admonishing Dante at her first meeting with him in the Earthly Paradise, speaks of the rich endowment with which he came into the world,

> 'Per ovra delle ruote magne,
> Che drizzan ciascun seme ad alcun fine,
> Secondo che le stelle son compagne [1].'

[1] *Purgat.* xxx. 109-11. Even more significant is the passage, *Parad.* viii. 127-32:—

> 'La circular natura, ch' è suggello
> Alla cera mortal, fa ben sua arte
> Ma non distingue l'un dall' altro ostello.
> Quinci addivien ch' Esaù si diparte
> Per seme da Jacób, e vien Quirino
> Da sì vil padre che si rende a Marte.'

By Dante's master in theology, Thomas Aquinas, the reality of astrological influences is laid down with perfect clearness in *S. T. Pars Prima* (quaest. 115, art. 3 and 4). In the first instance the question is put whether heavenly bodies are a cause of things that take place in terrestrial bodies. This after the usual statement of reasons, for and against, is answered emphatically in the affirmative. 'Celestial bodies causally affect all the varied motions of terrestrial bodies[1].' The second question is, Are heavenly bodies a cause of human actions? The authoritative conclusion here is, Directly speaking, they are not, but indirectly they are[2]. 'Indirectly, and by accident, impressions of celestial bodies may reach intellect and will, since both intellect and will receive somewhat from inferior faculties which are bound up with bodily organs. A distinction is, however, to be made between will and intellect. The intellect is of necessity affected by the lower apprehensive faculties of imagination, thought, or memory, and when these are stirred, the intellect is stirred likewise. But the will does not of necessity follow the promptings of the lower appetite. For although the passions of anger and desire have a certain power of moving the will; yet it remains in the power of will to follow passion or to repudiate it. Thus the influence of celestial bodies, so far as it produces change in the lower faculties, has to do rather with intellect than with will; and will is the proximate cause of human action.'

Finally Aquinas remarks that 'most men follow passions, which are motions of the sensitive appetite; and with these heavenly bodies may have to do. For few are the wise who withstand such passions. And thus it is that astrologers may often foretell truly; but for the most part rather in general than in special, since nothing hinders any one man

[1] The words are: 'Corpora caelestia cum tantum mobilia sint secundum lationis motum, causa sunt omnium eorum quae in his corporibus inferioribus variis motibus aguntur.'

[2] 'Cum intellectus et voluntas, quae humanorum actuum principia sunt, corporeis organis vires alligatae minime sint; non possunt corpora ipsa caelestia humanorum actuum causae directe esse, sed indirecte, agendo per se in corpora quae ad utriusque potentiae opera conducunt.'

from withstanding passions by free will. Hence, astrologers themselves say that the wise man governs the stars, in so far, namely, as he governs his own passions.' Further on, in *Prima Secundae*, quaest. ix. art. 5, this subject is again discussed. The question asked is, 'Can will be influenced by a heavenly body?' The conclusion is, 'Since will is a faculty absolutely immaterial and incorporeal, it can only be influenced by heavenly bodies indirectly.' And in his comment Aquinas observes, 'So far as will is influenced by any outward object, it can evidently be influenced by heavenly bodies: since all external bodies, which, when presented to the senses, move will, and even the very organs of sensitive faculties, are influenced by the motions of the heavens. But there is no direct action of heavenly bodies upon the will. For the will, as Aristotle says (*De Anima*, lib. iii) resides in reason; and reason is a power of the soul not bound to a bodily organ. . . . On the other hand, he adds, 'Sensitive appetite is the function (actus) of a bodily organ. Wherefore nothing hinders impressions of celestial bodies from rendering some men apt to anger, or to lust, or to some passion of this kind; and thus from natural complexion many men follow passions, and wise men alone withstand them. And so, in a general way, are verified those things that are foretold of the actions of men in accordance with the consideration of heavenly bodies.'

Now the view taken by Bacon coincides precisely with that of Aquinas. Confusion, he says (*Opus Majus*, p. 150), had arisen in the matter in consequence of the equivocal meaning of the word, Mathematics, sometimes held to be derived from *μαντική*, sometimes from *μάθησις*. The characteristic, he says, of false mathematic, was to assert that through the powers of the constellations all things took place of necessity. No place was left for contingent matter, for judgement, for free will. Such a view of nature was condemned not only by theologians but by philosophers. Aristotle and Plato, Cicero and Pliny. Avicenna and Albumazar were unanimous in holding that free will remained uncoerced by the motions of heavenly bodies. 'True mathematicians and astrologers lay down no necessity, no infallibility, in their predictions of contingent events. . . . What they do is to consider the way in which the body may be affected by celestial things, and the way in which the body may act upon the mind in private affairs or public, always without prejudice to the freedom of the will. For although the reasonable soul is not coerced to any future actions, yet it may be strongly stirred and induced, so as freely to will those things towards which celestial force may incline it; as we see men in community taking counsel, or through fear or love, and feelings of this kind, freely choosing what before they would not, though not forced to do so; like the sailor who to save himself from drowning throws precious merchandise into the sea. We see, indeed, that impressions from things on earth may so act upon sense as to stir men to will what before they had no care for, so that they take no account of death or disgrace or fear, if only they accomplish their desire, as with those who see and hear the onset of their enemies, and are borne onwards at all hazards to avenge themselves. Far more potent than the impressions of earthly things are those of the heavenly upon bodily organs, which being strongly moved men are led on to actions of which they had not thought before, yet always with full reservation of the freedom of the will.'

There are perhaps few fictitious creeds for the origin of which it is so easy to account as for the belief that the position of the planets with regard to one another and to the constellations of the zodiac were of significance to man and his environment. With populations whose religion was astrolatric rather than polytheistic, taking shape in worship of the heavens rather in that of invisible but manlike gods, astrology would be an easy and almost inevitable deduction from their creed. The immense majority of the Asiatic population, whether Semitic or Mongol, were, unlike the Indians and the Greeks, not polytheists but astrolaters. When the Arabs received and enlarged their inheritance of Ptolemaic astronomy, their astrologic beliefs, far from being dissipated, were strongly confirmed. Of the seven wanderers of the sky, the influence on earthly things of two, the Sun and the Moon, was too obvious to be disputed. The one swayed the tides, the other brought summer and winter. Why should the rest be supposed inert? Was it not probable that the successive and infinitely varying connexion of each of them, singly or combined, with the fixed groups of the starry vault, indicated changes or tendencies to change here below which careful and prolonged study might at last interpret?

So it was that with the growth of knowledge, and with increasing strength of the conviction that all nature was under the dominion of fixed laws, astrology came to be regarded as the key to the understanding of all that was specially contingent and variable in man's environment; the phenomena of temperament and of disease; the revolutions of states, and even of religions. The boundary of its lawful application was drawn differently by different thinkers. Apart from charlatans and miracle-mongers, few stretched it farther than Bacon. But by him, as strictly as by Aquinas, the saving clause, 'salva arbitrii libertate,' was always added. Outside influences might suggest motive and kindle passion; they could never trench upon the sacred domain of the freedom of the will [1].

What is strange is not that the belief in the convergence of

[1] Comte has pointed out (*Philosophie Positive,* vol. iii. pp. 273–280, ed. Littré) that in order to appreciate astrology with any approach to justice, it is needful to keep steadily in view the very real connexion between the sciences of astronomy and biology. On the relations of mass and of distance between the sun and earth, involving as they do the familiar facts of weight, equilibrium of fluids, temperature, life on our planet is obviously dependent. If we consider the period and velocity of the earth's rotation, the degree of ellipticity of her orbit, the angle at which the axis of rotation is inclined to the plane of the orbit, the same truth is impressed upon us even more strongly. 'In the early stages of the human mind these connecting links between astronomy and biology were studied from a very different point of view; but at least they were studied and not left out of sight, as is the common tendency in our own time under the restricting influence of a nascent and incomplete positivism. Beneath the chimerical beliefs of the old philosophy in the physiological influence of the stars, there lay a strong though confused recognition of the truth that the facts of life were in some way dependent on the solar system. Like all primitive inspirations of man's intelligence this feeling needed rectification by positive science, but not destruction; though unhappily in science, as in politics, it is often hard to reorganize without some brief period of overthrow.' This was written in 1836. Much has been done since by Mr. Spencer and others to familiarize the European mind with the dependence of life on its astronomical conditions. But the injustice in our historical judgement of mediaeval astrology still remains.

stellar influences towards the central point of a closed universe should have arisen, but that it should so long and so persistently have survived the discovery that the universe was not closed but boundless. That Francis Bacon, who rejected or doubted the Copernican theory, should have retained his belief in astrology is not surprising. But we should have expected that with men like Kepler and Campanella, it would have vanished like the morning mist. Yet it was not so.

VII. The Propagation of Force.

Bacon's views of stellar influences must be taken in connexion with his speculations as to the transmissions of force through space. These are set forth briefly in the second and third Distinctions of the fourth part of the *Opus Majus*; and more in detail in the special treatise *De Multiplicatione Specierum*, which in this edition is given as an appendix.

'Species' is the word chosen by Bacon to express the emanation of force which he conceives to be continually proceeding from every bodily object in all directions. Body of every kind is endowed with force which indeed is identical with its substance or essence. The first result of this force, resembling it in character, is its species, otherwise called likeness, or image, or intention or impression In other words, body is a centre of activity or force radiating in every direction. Species is the first result of this force, the ray proceeding from the body. Tracing back this doctrine to its origin, we find it expounded in the fourth book of *Lucretius*, in Diogenes Laertius' account of the system of Epicurus, and in the traces that remain to us of older philosophers, notably of Democritus. Aristotle, in his short treatise on Divination by dreams, alludes to the theory of Democritus that εἴδωλα and ἀπόρροιαι were continually emitted from objects which in the stillness of the night were capable of affecting the sleeper. By Epicurus, in his letter to Herodotus quoted in his biography by Diogenes Laertius, the theory is more fully detailed. 'There are moulds,' he says, 'corresponding to all solid bodies preserving the same shape and arrangement as these bodies which emanate from them, and are conveyed through space with incredible velocity. These may be called images. Their flow from bodies is continuous so that they are not separately perceived.' The description of them by Lucretius is more definite and better known. 'Pictures of things and thin shapes are emitted from things off their surface; these are like films or may each be named a rind, because each image bears an appearance and form like to the thing, whatever it is, from whose body it is shed and wanders forth' (Lucretius iv. 40, Monro's translation). And again, 'Many idols are begotten in a short time, so that the birth of such things is with good reason named a rapid one. And as the sun must send forth many rays of light in a short time in order that all things may be continually filled with it, so also for a like reason there must be carried away from things in a moment of time idols of things, many in number, in many ways, in all directions round. . . . As soon as ever the brightness of water is set down in the open air, if the heaven is starry, in a moment the clear radiant constellations of aether imaged in the water correspond to those in the heaven. Now do you see in what a moment of time an image drops down from the borders of heaven to the borders of earth' (Lucretius iv. 159 and 211). He goes on to explain that not the sense of sight only, but all the senses, are affected by these emanations.

But it would be an entire misapprehension of Bacon's views as to the propagation of force to identify them with the crude physics of Epicurus.

In the first place, Bacon wholly rejects the notion that the species is something emitted from the agent, or acting body (*De Mult. Spec.* pp. 432–438). If it were so, the agent would be weakened and ultimately destroyed by the emission, which is not the case. Nor again does the agent create the species out of nothing. Nor does it collect the species from surrounding space and send it on into the body on which action takes place—the patient. Nor, as some have supposed, does the agent impress the patient as with a seal.

What happens is that the agent stimulates the potential activity of the matter of the patient. The species is generated out of the matter acted on. 'Fit species de potentia activa materiae patientis.' The agent acts on the first part of the body of the patient, and stimulates its latent energy to the generation of the species. That part thus transmuted acts on the part next succeeding; and so the action proceeds (*De Mult. Spec.* p. 457).

While the agent acts on the patient, the patient re-acts on the agent. 'Omne agens physice patitur et transmutatur insimul dum agit, et omne patiens physice agit' (*De Mult. Spec.* p. 439). Heavenly bodies as they act on one another, so do they receive emanations of force from terrestrial bodies. Not that they are so affected by them as to be destroyed, being incorruptible. Nevertheless there is in this way an interchange of force between all parts of the universe (p. 448).

The ray, or species, is of corporeal nature; but this corporeal nature is not distinct from that of the medium; it is generated from the substance of the medium, and is continually re-formed out of successive portions of the medium occurring in the line along which the force is propagated (p. 505). If wind is driving the air transversely to the line of force, this in no way affects this line. The species is formed and reformed from particles of the medium presented in the line of propagation, and from no others.

Finally the propagation of rays occupies time (vol. ii. pp. 67–72 and 525–9), though its velocity is such that the time occupied in passing through so vast a space as the diameter of the universe is imperceptible to sense.

It will be seen from the foregoing how wide is the divergence between Democritean and Baconian physics. Though Bacon retains the word 'species' in his theory, the word has almost entirely lost the significance attached to it by Lucretius. We are no longer dealing with the notion that bodies emit from their surface films or moulds which are transmitted through space. Like the word 'ray,' which is retained by the modern physicist who accepts the undulatory theory, 'species' for Bacon has become a mere word to denote the propagation of force in certain definite directions. Indeed the multiplication of species as defined by him has much in common with the undulatory theory. He formally rejects the contrasted theory of emission. The species, like the wave, is a motion or change in successive portions of the aerial or ethereal medium; occupying time in its transit: propagated so long as the medium be homogeneous, in direct lines; liable to deflection when the medium alters its character.

In Bacon's theory of the radiation of forces two very important points are to be noted. The first is his clear grasp of the principle that time was occupied in their transmission. He discusses, in the passages already cited, the view of Aristotle and others that the propagation of light differed from that of sound and odour by being instantaneous. We might admit, Aristotle had said, that light could pass through short spaces without our being able to detect any interval of time during the passage. But when light passes from east to west through the universe, the space is so vast that if time were occupied we could not fail to detect it. Bacon's conception of the subject is far more scientific. Our inability to perceive minute intervals of time is no evidence, he said, for their non-existence. Imperceptible time, he remarks, has many degrees. There is, first, the interval of time occupied by a single propagation of force (or, as we should say, undulation) followed by the interval of rest before the next propagation begins. Take such a multiple of that interval as would suffice for the whole distance between the extremities of a diameter of the universe, and that multiple may still remain below the limits of our power of perception. It is interesting to compare with this passage the speculations of the second Bacon on the same subject (*Nov. Org.* ii. 46). Francis Bacon had formed the conjecture that the transit of light from the stars occupied time. But he did not grasp this conjecture with the same firmness as Roger Bacon, and he follows it up with ingenious arguments which explain it away.

Radiant force, in Bacon's view, proceeded independently of man's power to perceive it. Opaque bodies, he observes, offered resistance to the passage of a luminous ray (*De Mult. Spec.* p. 478; see also vol. i. p. 114). But 'no substance is so dense as altogether to prevent rays from passing. Matter is common to all things, and thus there is no substance on which the action involved in the passage of a ray may not produce a change. Thus it is that rays of heat and sound penetrate through the walls of a vessel of gold or brass. It is said by Boetius that a lynx's eye will pierce through thick walls. In this case the wall would be permeable to visual rays. In any case there are many dense bodies which altogether interfere with the visual and other senses of man, so that rays cannot pass with such energy as to produce an effect on human sense, and yet nevertheless rays do really pass, though without our being aware of it.' Recent discoveries have given significance to this remarkable passage; which, not merely to his contemporaries but to succeeding generations, must have seemed in the highest degree fantastical.

VIII. Bacon's Optics.

The most striking illustration of laws governing the transit of force through space was obviously to be looked for in the science of Optics (*Perspectiva*). The fifth section of the *Opus Majus*, amounting to about one-fifth of the whole, is devoted to this science; and much supplemental matter is added in the treatise *De Multiplicatione Specierum*. Optics had been studied by the Greeks to much purpose. The works of Euclid, Theon, and Ptolemy were translated into Arabic, and were carefully studied by Arabian men of science, notably by Abû 'Alî al Hasan ibn al Hasan ibn Alhaitam, better known to Occidentals under the name of Alhazen. Their principal results are embodied in Bacon's work.

Euclid, or the author passing under his name[1], was aware that light proceeded in straight lines, and that visual rays were reflected from plane mirrors in such a way that the angles

[1] Heiberg, in his recent edition of Euclid's *Optica* (forming the seventh volume of the complete edition of Euclid, edited by Heiberg and Menge, remarks (p. xxviii of *Prolegomena*), 'Optica qualia hic e codice Vindobonensi maxime primo loco repetivimus, Euclidis esse, non est cur dubitemus. Sed cum recentiores tantum extent codices, mirum non est locos nonnullos tam corruptos esse ut verba Euclidis restitui nequeant.'

made with the surface on each side were equal. He conceived the assemblage of rays as a cone having its apex in the eye, and its base in the boundary of the object seen: that the apparent magnitude of the object depended on the magnitude of the angle of the cone. Thence followed the ordinary principles of perspective, as that of equal magnitudes at unequal distances; those nearer to the eye appeared larger, and so on. In the *Catoptrica* (attributed to Euclid, but probably due to Theon), from the equality of the angles of reflexion and incidence in plane mirrors was deduced the convergence of rays falling on a concave speculum.

Ptolemy[1] carried the science much further than Euclid. To the study of reflected light he added that of refraction. The chief interest of his work lies in the application to the subject of the experimental method, an instance of it unique, if we except the Pythagorean experiments in acoustics, in the history of Greek science. Using an extremely simple but ingenious apparatus, he discovered, not merely that the luminous ray in passing from one medium to another was deflected, but, within certain limits, he ascertained the amount of deflexion and its dependence on two distinct factors, the angle of incidence, and the nature of the two media concerned. Ptolemy distinctly describes and explains the error introduced by refraction into astronomical observations. The fact that in his great astronomical treatise there is no mention of refraction had led to the conclusion that the *Almagest* and the *Optics* must be attributed to distinct authors. The *Optics*, however, may be a later work. We know it only from a translation from the Arabic into Latin, made in the twelfth century; it has been recently edited by Gilberto Govi, of Turin. The researches of Euclid, Ptolemy, and others on Optics, engaged the attention of the Arabian schools from an early period.

[1] On Ptolemy's *Optics* there is a very interesting chapter in Delambre's *Astronomie Ancienne*, vol. ii. pp. 411 430, ed. 1817. (See also a note on p. li of Preface to vol. i which modifies some of his conclusions.) All our knowledge of Ptolemy's optical work comes from an imperfect Latin translation from the Arabic made in the twelfth century by Admiral Eugenius of Sicily. There are late MSS. of this work in Paris and in the Bodleian Library. But Govi's recent edition is from a much older MS. in the Ambrosian Library of Milan.

Their knowledge of the subject is summed up in the work of Alhazen, whose remarkable work, *Thesaurus Opticae*, written perhaps in the eleventh century, was translated in the twelfth into Latin; as Jourdain thinks, by Gerard of Cremona, the translator of Ptolemy's *Almagest*. Alhazen was the writer on whom Roger Bacon principally relied; though he makes frequent use of the optical treatises of Euclid, Ptolemy, Tideus, and Alkindi.

Alhazen's work is copious in the extreme; in some parts extremely tedious. Its value as a document in the history of science is, however, very great. It consists of seven books. The first begins with a brief exposition of the nature of light and colour, and proceeds to explain the anatomy of the organs of vision. The second deals with the function of vision and with the physiology of perception. The third, with imperfections and illusions incident to vision. The fourth, fifth and sixth are devoted to the subject of reflexion. Seven kinds of mirrors are discussed, plane, spherical, cylindrical and conical; the convex and concave forms of the three last being separately considered. The multiplication and position of the images formed is treated with inordinate length, but with such geometrical skill as to secure for him an abiding place in the history of pure, no less than of applied mathematics. 'His investigation,' says Cantor (*Geschichte der Mathematik*, vol. i. p. 677), 'of the problem: In a spherical concave mirror, to find the point from which an object of given position will be reflected to an eye of given position, is one which, analytically handled, leads to an equation of the fourth degree.' Alhazen solved it, as Govi remarks (Ptol. *Opt.* p. xix), by the use of an hyperbola.

The seventh book of the *Thesaurus Opticae* deals with refraction. A very elaborate description is given of the instrument for measuring it, part of which Bacon quotes. Moreover, an attempt is made to explain the cause of refraction which is substantially identical with Bacon's, as may be seen by comparison of Alhazen vii. 8 with *De Mult. Specierum*, Part II. cap. 3. The apparatus for measuring the angle of refraction, which was more accurately designed than that of Ptolemy, enabled a series of observations to be made of the angle of refraction, in different media, on which the true law of the variations of refraction at different angles and in different media might ultimately be based. Vitello[1], Bacon's contemporary, drew up a table of refractions, as Ptolemy had done before him, for the three media of air, water, and glass. It was soon seen that the angle of refraction did not vary in accordance with the angle of incidence. But more than three centuries were to pass before the discovery of the law of sines, that is to say, the law that the ratio of the sines of the angles of incidence and refraction is constant for refraction in the same medium, was effected by Snell and Descartes.

It might seem, at first sight, that the optical work of Bacon was little more than an abridgement of that of Alhazen[2]. But this view would render Bacon but scanty justice. Problems of great importance were indicated by him which Alhazen had entirely neglected. In considering the point on the axis of a spherical concave mirror to which rays were reflected, Bacon remarks that this point would be different for rays reflected from each concentric circle traced round the centre of the mirror. Such a mirror failed therefore to produce complete

[1] Of Vitello, or Witelo, very little is known. He describes himself in his dedication to William of Morbeta (identified by Cantor as William of Moerbeke) as filius Thuringorum et Polonorum. In lib. x. 74 of his work he speaks of Poland as his country, and other passages (x. 42 and 67) show that he travelled in Italy. His work on Optics was edited with great care, and with many emendations, by Risner, and published at Bale in 1572 in the same volume that contained Risner's edition of Alhazen. Indeed, it may be described with little exaggeration as a revised edition of Alhazen's work: with many additions certainly from other authors, but with none of those acknowledgements of his principal teacher of which Bacon's *Perspectiva* is full. Vitello's tables of refraction have excited much admiration. They prove, however, on careful examination to be an almost exact repetition of those of Ptolemy. Whether Bacon and Vitello ever came into contact there is no evidence to show. Bacon was always ready to mention the sources of his knowledge. Not so Vitello. If he borrowed from Bacon, he would not have said so.

[2] It must be owned that where Bacon differed from Alhazen, the advantage was not always on his side. Alhazen contended vigorously against the view of the older oculists that vision took place by visual force issuing from the eye, maintaining that the ray proceeded to the eye from the object. Bacon (vol. ii. pp. 49 53) makes a fruitless attempt to conciliate these opposite views.

convergence of rays. For such convergence the curvature must be other than spherical, it must be that produced by the rotation of a conic section.

Bacon, moreover, is distinguished from the Arabian optical writers, and from other investigators of his own time, by his sedulous endeavours to turn the discovery of the laws of reflexion and refraction to practical account. Neither in Alhazen nor in Vitello is there any attempt to construct instruments for the purpose of increasing the power of vision. With Bacon this object was always held steadily in view. Of the simple microscope he had a perfectly clear conception. His scientific imagination played freely with the possibilities of bringing distant objects near, and of indefinitely magnifying minute objects, by giving suitable directions to refracted rays, and by the use of appropriate media. It would be, however, an entire exaggeration of his achievements to speak of him as the inventor of the telescope. No evidence is forthcoming for his having effected the simple combination of two convex lenses, or of a convex with a concave lens, on which the power of telescopic vision depends. All that can be claimed for him is that he was the first definitely and explicitly to bring the problem forward, leaving it for after generations to solve. In truth, his conception of an optical image, as constructed by the assemblage of foci of rays proceeding from each point of the object magnified, though in the main correct, was not always clearly grasped. Of the distinction between virtual and real images, his notion was entirely in default.

Nor, again, had Bacon a clear conception of the conditions of distinct vision. He examined to much better purpose than Alhazen had done the structures of the eye; and he was aware of the refraction produced by the curved surface of the cornea, and by the doubly convex crystalline lens. But what he failed to grasp [1] was the necessity of a clear image of the object defined on the retina; that image being produced by the focussing on the retina of rays proceeding from each point of the object. The phenomena of accommodation, produced by the action of the ciliary muscle, which, by altering the curva-

[1] See vol. ii. p. 159.

ture of the lens, enables rays from near objects to be accurately focussed, were unknown to him. But this is only to say that he had not anticipated the physiological knowledge of the nineteenth century.

It must always be borne in mind that, in Bacon's view, the radiation of light through space did not stand alone. It was a type of other radiant activities, such as colour (then supposed to be distinct from, though dependent on, light), heat, sound, and odour. (With regard to sound, however, certain reserves were made.) It is interesting to note Bacon's handling of an important problem, not to be solved but by a more potent calculus than any in his possession, how these various actions, crossing one another's paths in their passage through space, retained their distinctness [1].

IX. Bacon's Alchemy.

It will be remembered that among the various branches of knowledge regarded by Bacon as falling under the head of Physics, was Barology (*Scientia ponderum*). The treatise of Jordanus Nemorarius, *De Ponderibus*, to which reference is made, vol. i. p. 169, had perhaps suggested the treatment of the phenomena of gravity as a distinct branch of science. No treatise by Bacon upon this subject, so far as I am aware, is extant; and the few remarks in the fourth section of the *Opus Majus* (pp. 167–174) contain all that we know of his speculations on the theory of gravitation.

Nor is anything known to us of the way in which Bacon treated, if indeed he ever attempted, the science which he called 'Agricultura,' which, as we have seen, was intended to include the study of living bodies, vegetable and animal. But the case is otherwise with the science regarded by him as preparatory to the study of life, 'Alkimia Speculativa.' On the subject of Alchemy, very little is said in the *Opus Majus*; and the omission was supplied in the provisional way, which alone was possible under the hurry of compilation to satisfy Pope Clement's orders, by the *Opus Minus*, the first of the

[1] See note on vol. ii. p. 46.

two appendages to that work. Unfortunately the only text of the *Opus Minus* which we possess has come down to us, not merely incomplete, but in so corrupt a state as to render it often very difficult to decipher Bacon's meaning. Enough remains, however, to show the large and comprehensive spirit in which Bacon regarded the subject.

The contempt expressed in much modern writing for mediaeval alchemy might be well retorted on its authors. Admit that some prosecutors of the occult art were deceivers as well as deceived, and that others were impelled by wild hopes of gain, has the pursuit of physical science in modern times been wholly free from similar taints? Electricity applied to medicine has been a fertile field for impostors. And will any one maintain that the pursuit of chemistry has not been stimulated by hopes of industrial profit? Yet such things are not allowed to cast a shade on the names of a Lavoisier, a Dalton, or a Faraday. Alchemy was chemistry in its prescientific period. Under the guidance of hypotheses which were not nearly so wild or crude as they at first appear, it attacked, like the true science which gradually grew from it, the important problem of the transmutation of matter by artificial agencies. It took for granted that metals were compound bodies, the elements of which might be separated and recomposed. This was no unreasonable supposition. Indeed, until modern spectrology had shown that the vapour of many metals existed undecomposed in the intense heat of the sun's atmosphere, there was no adequate reason for abandoning the attempt to decompose them. It would be hard to find in alchemy any conjecture more baseless than that of Phlogiston, the subtle spirit of flame, the loss of which by combustion made the oxide heavier than the metal. Yet Priestley accepted this hypothesis, and a Lavoisier was needed to destroy it.

Alkimia, as conceived by Bacon, fell into two great divisions—speculative and operative. Under the latter was included the metallurgy of the gold-seekers, and generally all the practical and industrial processes pursued, with more or less wisdom, by men who had a definite purpose in view—

the transmutation of metals, the discovery of the philosopher's egg, or the elixir vitae. But Bacon was one of the few who saw that the empirical proceedings of the honest mystics or scheming charlatans, who were toiling at their royal road to wealth or longevity, covered speculations of a far deeper kind; the study of the transition of matter from the four Aristotelian elements, through increasing degrees of complexity, up to the highly compound forms exhibited by organized bodies. The 'Alkimia Speculativa' of Bacon was, indeed, not alchemy at all as commonly understood: it was nothing less than chemistry. 'Alkimia Speculativa,' he says, in the twelfth chapter of the *Opus Tertium*, 'treats of the generation of things from their elements, and of all inanimate things—as of the elements and liquids (*humores*) simple and compound; common stones, gems, and marbles; gold, and other metals; sulphur, salts, pigments, lapis lazuli, minium, and other colours; oils, bitumen, and very many other things—of which we find nothing in the books of Aristotle; nor are the natural philosophers or any of the Latins acquainted with these things. And being ignorant of them, they can know nothing of what follows in physics, that is, of the generation of animate things—as vegetables, animals, and man—because knowing not what is prior, they must remain ignorant of what is posterior. For the generation of men, and of brutes, and of plants, is from elemental and liquid substances, and is of like manner with the generation of inanimate things. Wherefore, through ignorance of this science, neither can natural philosophy, commonly so-called, be known, nor the theory, and therefore neither the practice, of medicine; not merely because natural philosophy and theoretical medicine are necessary for the practice, but because all simple medicines are derived from inanimate things by this science.'

Of such fundamental truths of chemical science—as the composition of air and water, the theory of combustion, and the chemistry of carbon—he, like his contemporaries, was ignorant; but the ignorance was shared by the second Bacon with the first, and was not to be dissipated for five centuries. All that could be done in the meanwhile was to collect empirical information as to a few metals and their oxides, some of the principal alkalis, acids, and salts. On all these things the Arab investigators, from Geber downwards, had accumulated a considerable mass of material. It is not easy to define the results of each inquirer, owing to the prevalent habit of describing their procedure and results in mystical language. Self-defence against charges of magic and imposture was probably their motive. And that the danger was real, the history of Bacon's life suffices to show. His efforts to refute the charge of magic were incessant. In his treatise, *De Secretis Operibus Artis et Naturae et de nullitate Magiae*, he describes in detail the various procedures of the magician, sleight of hand, ventriloquism, pretended movements of inanimate things in dim light, the aid of accomplices, utterance of mysterious formulae, invocation of spirits. We gather from the description that the lapse of six centuries has done little to change the character of charlatanism. But Bacon was aware that the charlatan was often in possession of valuable secrets. 'Many books are held to be magical,' he says, 'which are not really so, but which contain important truths; which are of this kind, and which are not, it is for the experience of the wise man to decide. If he find in them any result of natural or artificial forces (opus naturae vel artis), let him accept this: otherwise let him reject them as worthless.' Bacon carefully guards himself against denial of the mystical force of words uttered under solemn conditions, as in the daily miracle of the Eucharist, or in the solemn invocations that protected the innocent when exposed to judicial ordeal. Such powers might be exerted for good as for evil; and the unlawful use of them was strictly and severely to be condemned.

But to whatever extent Bacon may have shared the illusions of his time with regard to the practical operations of alchemy, it is a striking proof of his scientific discernment that under the head of Speculative Alchemy he should have formed a clear, though distant survey, of chemical science as the intermediate link between Aristotelian Physics and the science of living bodies. As Physics followed on Mathematics, so did Chemistry, in Bacon's arrangement of the sciences, succeed

Physics. After chemistry came the study of living bodies, on which Bacon, while assigning to this science its natural place in the series, has said little or nothing. But on the study of plants and animals was based, under the name of Medicina, the study of the physical structure and functions of man. Here Bacon had for his guides, not Galen indeed, to whom his references are few, but Avicenna, Haly, and a host of Arabian professors of medical art, to whom Galen had supplied a very substantial foundation of anatomical and physiological knowledge. Bacon's short treatise, to which reference is occasionally made in the *Opus Majus*, *De retardandis senectutis accidentibus*, will sufficiently illustrate his views on this branch of science.

X. Experimental Science.

Last among the series of the natural sciences comes that which Bacon denotes as 'Scientia Experimentalis.' The sample of it, for it can hardly be regarded as more than a sample, given in the sixth section of the *Opus Majus* indicates that it was connected in Bacon's mind with no special department of research, but was a general method used for the double purpose of controlling results already reached by mathematical procedure, and of stimulating new researches in fields not as yet opened to inquiry.

In some respects this is the most original part of his work. Not that experiment was a new thing. Experiments without number had been made by man from the time of his first appearance on the planet. The Greeks towards the end of their marvellous scientific career had begun to use experiment in their investigations of natural truth. Galen had applied it in his researches into the nervous system; Ptolemy had arrived by its means at his remarkable discovery of the refraction of light. The Arab astronomers, far more skilful mechanicians than the Greeks, had constructed extremely elaborate apparatus for the same purpose, and also to verify the equality of the angles of incidence and reflection. But no one before Bacon had abstracted the method of experiment from the concrete problem, and had seen its bearing and importance as a universal method of research. Implicitly men of science had begun to recognize the value of experiment. What Bacon did was to make the recognition explicit. Experiment took its place as a distinct department of philosophy.

What makes this result peculiarly remarkable is that it was reached by a thinker who was so profoundly penetrated by the mathematical spirit. In this matter Roger Bacon compares favourably with his illustrious namesake of the seventeenth century, who wholly failed to appreciate the import of mathematical method. He rises to the level of one greater than either—the author of the *Discours sur la Méthode.* For Descartes as for Roger Bacon, mathematics was *clavis scientiarum*, the key to the temple of science. But it was held by both alike that experiment was needed to carry out the researches which mathematical deduction had suggested; and that, as each science grew, the share taken by experiment in its progress was to become more and more predominant.

XI. Moral Philosophy.

Last in order, both in the *Opus Majus* and in the *Scriptum Principale*, comes the science the study of which is the keystone and crown of the whole work—the science of life and conduct. All the other sciences lead up to this. Their conclusions form its point of departure.

The analysis which has been given of this, as of other parts of the work, renders it unnecessary to cover the ground a second time. But a few remarks may be made on its salient features. In the first book, which treats of man's relation to God, Bacon follows the procedure common to Aquinas, indeed to most of the schoolmen, of pushing metaphysical reasoning as far as it can be made to go in support of the articles of the Catholic faith. Theology, says Aquinas (*S. T.* Pars I. Quaest. i. art. 5), uses other sciences as her handmaids and assistants. Man is more easily led on to things above reason, if he begins with things which reason can demonstrate. It is true that unassisted reason is incompetent to discover and demonstrate the doctrine of the Trinity. 'Impossibile est,' he says (Quaest. xxxii. art. 1), 'per rationem naturalem ad cognitionem Trinitatis divinarum personarum pervenire.' But he goes on to explain that there are two modes of employing reason. One is to discover and prove a principle: as in physics we prove the uniformity of the motion of the heavens. The second mode is, when the principle is admitted, to show that certain observed effects are consistent with and follow from it. So, for instance, assuming the reality of our hypotheses as to eccentrics and epicycles, we can show that the movements of the planets take place in accordance with these hypotheses. It is this latter form of reasoning that we use in reference to the Trinity. 'Trinitate posita, congruunt hujusmodi rationes.' We find analogies with this doctrine when we consider what passes in our own minds. 'Ipse conceptus cordis de ratione sua habet quod ab alio procedat, scilicet a notitia concipientis' (Quaest. xxxiv. art. 1). 'Quanto aliquid magis intelligitur, tanto conceptio intellectualis est magis intima intelligenti et magis unum . . . unde cum divinum intelligere sit in fine perfectionis . . . necesse est quod Verbum divinum sit perfecte unum cum eo a quo procedit, absque omni diversitate' (Quaest. xxvii. art. 1). Similarly (art. 3), the procession of the Third Person is likened to the operation of the will which we call in human beings love. 'Processio Verbi attenditur secundum actionem intelligibilem. Secundum autem operationem voluntatis invenitur in nobis quaedam alia processio, scilicet processio amoris, secundum quam amatum est in amante, sicut per conceptionem verbi res dicta vel intellecta est in intelligente. Unde et, praeter processionem Verbi, ponitur alia processio in divinis, quae est processio amoris [1].'

Bacon, as we might expect, was not less eager to find the mysteries of revelation foreshadowed by human reason. Holding, as he has fully explained in the second part of the

[1] It is perhaps hardly necessary to refer in this connexion to Hampden's lectures on *The Scholastic Philosophy considered in its relation to Christian Theology*. Compare p. 81 (second edition), 'The object of the Scholastic Theology was to detect and draw forth from the Scripture, by aid of the subtle analysis of the philosophy of Aristotle, the mystical truths of God on which the Scripture Revelation was conceived to be founded.'

Opus Majus, that the rise and progress of Greek philosophy was no less a part of divine providence than the succession of the priests and prophets of Judaea, he found without surprise that Aristotle, Plato, Porphyry and others had apprehended, more or less dimly, some of the fundamental truths of Christian theology; among them being the Trinity, the Incarnation, the existence of angels and the resurrection of the body. Moral philosophy, as Bacon conceived it, was in every respect concurrent with theology. 'De iisdem negotiatur quibus theologia, licet alio modo.' It is perhaps more surprising that he should have gathered these truths not merely from Greek and pre-Christian writers, but from the great Mahommedan teachers, such as Albumazar, Avicenna, and Algazel. Some of the most remarkable passages in the first part of his moral philosophy are quotations from Avicenna. More than once he refers to the passage in which Avicenna, speaking of future life in the unseen world, observes: Our present relation to that life is like that of the deaf man who never listened to the delights of harmony, though he never doubted that such delights existed. Or again: We are like the palsied man to whom delicious food is offered which yet we cannot taste till the palsy be healed. Avicenna tells us how the soul's vision is clogged by bodily impulses, and limited by the obtruding influences of the visible world; and he insists on the need of purging the soul from sin, of concentration of its forces on invisible things, and of acceptance of revealed truth. We may well believe that the attempt to level up Mahommedan philosophers to the level of Christian teachers was among the *novitates* for which Jerome d'Ascoli cut short Bacon's philosophical career.

The second part of the Moral Philosophy, dealing with the laws of civil and social life, is summarily disposed of in two short chapters. Possibly a reason for this cursory treatment may be found in Bacon's aversion to the introduction of Roman law, which finds vehement expression in the twenty-fourth chapter of the *Opus Tertium*, and again in the *Compendium Studii* (Brewer, pp. 84–87, and 418)[1].

[1] Something additional on this subject was probably said in the missing sixth

We are here brought face to face with the failure, and the cause of the failure, of Bacon's social and political ideal. He was aiming at an enlarged and renovated Catholicism which should bind together and incorporate all that was best and noblest in Hebrew, Greek, and Arabic tradition in the fabric of the Christian Church, for the spiritual government of the world. The keystone of the fabric was supplied by the mistress-science, theology, resting on Mosaic and Christian revelation, consolidated by Aristotelian philosophy, and penetrated by the vital and progressive spirit of natural science. A progressive papacy, carrying on in continuous and harmonious development the work which Mosaic law and Greek intellect had begun—such was Bacon's vision: and the marvellous upheaval of thought in Paris and elsewhere during the thirteenth century seemed to bring that vision within reach of fulfilment.

But while Paris was building up its systems of philosophic theology, south of the Alps, in the rival university of Bologna, work of another kind was going on. The study of the civil law of Rome, which had never wholly ceased in the cities of North Italy, had been stimulated early in the twelfth century by the teaching of Irnerius and others; and from that teaching the university of Bologna gradually arose, as the university of Paris had arisen from the teaching of Abelard. It was a momentous event in the history of Europe. Civil law was a study as secular as the Roman empire itself. Clerical and lay students sat at the lectures side by side. 'Very early in the twelfth century men of mature age, men of good birth and good position, beneficed and dignified ecclesiastics, or sons of nobles, flocked from the remotest parts of Europe to the lecture-rooms of Bologna' (Rashdall, *History of Universities*, i. 124). The civil law embraced the entire system of man's social relations, and dealt with them on principles with which theology had no concern.

The Church felt the danger, and coped with it in the only

part of the *Moralis Philosophia*. But his language on the subject does not warrant the belief that the subject was fully dealt with. Cf. *Op. Tert.* cap. xiv. Brewer, p. 52.

way that was possible, by borrowing weapons from her lay rival, and arranging her own system of law in a form not less comprehensive and systematic. Irnerius had hardly finished his lectures when a fellow-citizen, the monk Gratian, in 1143 published his great text-book of canon law known as the *Decretum*, to which, in 1234, Gregory IX added five books of Decretals.

Nominally the situation was saved, but at the cost of secularizing the Church. For the canon law was in reality based on the civil law. 'Everything in the canon law was Roman which was not of directly Christian or Jewish origin.' 'After the age of Gratian the studies even of ecclesiastics took a predominantly legal turn. Speculative theology was abandoned in favour of the canon and even of the civil law; while the estrangement of the canon law from theology kept pace with the increasing closeness of its union with the faculty of civil law' (Rashdall, i. p. 138). In 1219 Honorius III formally prohibited the study of civil law in Paris on the ground that it threatened to extinguish the study of theology in the one great theological school of Europe. But prohibitions that were powerless to exclude Aristotle were equally impotent against the invasion of Ulpian and Justinian.

Bacon's pages reflect very vividly the conflict of clerical with secular influences. 'More praise,' he says, 'is gained in the Church of God by a civil jurist, though he may know nothing but civil law and be utterly ignorant of canon law and theology, than by any master in theology, and he is more quickly promoted to high ecclesiastical positions.' 'Oh that the canon law might be purged from the superfluities of civil law, and be ordered by theology,' he exclaims, 'then would the government of the Church be carried on honourably and suitably to its high position' (*Opus Tertium*, ch. 24).

He recurs to the same subject in a later work. 'For the last forty years the abuse of the civil law of Italy has been undermining not merely the study of philosophy, but the Church of God, and all the kingdoms of Christendom.' 'They monopolize,' he proceeds to say, 'every office of emolument, so that students of theology and philosophy are deprived of the means of following their studies. And besides this, the study of civil law is obliterating the distinction between clerical and lay professions. The doctors of law of Bologna call themselves clerks and masters, though they have not the tonsure, though they take to themselves wives, have families, and in every respect adopt the ways and practices of laymen . . . If clergymen and laymen are to be subject to the same law, at least let it be the law of England for Englishmen, and of France for Frenchmen, and not the law of Lombardy' (*Compendium Studii*, cap. 4)[1].

When Bacon appealed to the Pope to arrest the diffusion of civil law, he was like one who should attempt to stop the tide or the courses of the stars. He was fighting against the laws of historical evolution. It was written that the constitution of society should be settled on a human and secular, not on a theological basis; and the study of civil law, radiating in the twelfth and thirteenth centuries from Bologna into every part of Christendom, was one of the most significant among many signs that the function of the Catholic Church, as the organizer of political society, was gone.

Widely different was the future of that Church in all that related to personal morality. Yet here too there was much to be desired. In the third section of the work this subject is discussed with great fullness. 'On virtue and vice,' says Bacon, 'the ancient philosophers have spoken so wonderfully that a Christian man may well be astounded at men who were unbelievers thus attaining the summits of morality.' 'On the Christian virtues of faith, hope, and charity,' he adds, 'we can speak things of which they knew nothing. But in the virtues needed for integrity of life, and for human fellowship, we are not their equals either in word or deed. Blameworthy and shameful in us that it should be so.' Acting on this view, Bacon has composed this third part almost entirely of selections from

[1] This was written (as Brewer shows, p. lv) in 1271, three years after the death of Clement IV. Guy Fulcodi, before his ecclesiastical career began, had been a distinguished lawyer, and would hardly have tolerated such strong language.

Aristotle, Plato, Cicero, and above all, from Seneca, adding the fewest words of his own that were needed to mould them into systematic shape.

He begins by adopting Aristotle's general scheme of the moral virtues as means between opposing vices. From these he passes to the consideration of special vices in the order of the seven mortal sins of Catholic theology. Six of these deal with man's conduct under prosperous circumstances; the seventh, anger, with his conduct in adversity. Dealing briefly with the first class, Bacon devotes much more attention to the subject of anger. His reason for doing so lay in the disturbing influences of this passion on the whole of man's life public and private, and also that, in seeking remedies for its ravages, we are led up to the state of inward peace and resignation under outward trials which forms the highest plane to which the soul can aspire.

Nearly the whole of Seneca's three dialogues on Anger are quoted, but with complete rearrangement, in pursuance of the aim in view. Beginning with a picture of this passion and its disastrous effects on the highest qualities of the soul, such as clemency, pity, and joy, he enlarges on examples of self-restraint, and thence proceeds to consider remedial action; patient inquiry, time allowed for the mood to pass by, and constant remembrance of human fellowship with the offender.

This leads him to the wider subject of fortitude under calamity, of forgiveness of injury and insult, of recognition of the truth that whom God loves He chastens. He concludes the section with long selections from the dialogues on consolation under bereavement, exile, and poverty, on the shortness of life and the state of inward bliss and spiritual peace. It appears that though other parts of Seneca were well known, especially the series of letters to Lucilius, these dialogues had escaped notice till Bacon called attention to them. The apocryphal correspondence between Seneca and St. Paul shows that an affinity between Seneca and Christian teaching had been widely recognized in the Church. Nowhere is this affinity so strongly marked as in the dialogues *De Providentia*,

De Vita Beata, and *De Tranquillitate Animi*, from which Bacon has quoted so largely.

The fourth section of the Moral Philosophy contains the first attempt ever made at the comparative study of the religions of the world. These Bacon ranges in six classes: Pagans, Idolaters, Tartars, Saracens, Jews, Christians. What specially called attention to this subject in Bacon's time were the events proceeding in Central Asia, and already seriously affecting European politics. Mongol hordes had swept over Russia and South Eastern Europe, and were threatening the Western kingdoms. Franciscan and Dominican missionaries had been sent by Pope Innocent IV and by Louis IX to investigate the danger at its source. The reports brought back by these missionaries, especially those of Carpini and Rubruquis, brought the religious problem before the view of the leaders of the Church in all its magnitude. It was seen that beyond the Christian world, beyond the Mahommedan world which bounded it, there lay regions of unsuspected magnitude in the extreme East, where other creeds prevailed. One of these was Buddhism, recently imported into Central Asia from Tibet, with its elaborate monastic system, its image-worship, and its complicated liturgy. This creed was always spoken of by Rubruquis and Carpini, as by Marco Polo in the succeeding generation, as Idolatry. Christianity of the Nestorian type was widely disseminated; though not, it would seem, in its most highly militant form. Side by side with these were tribes whose religion was of a lower grade, not rising above the rudest fetichism; these were spoken of as Pagans. Between these various modes of faith the Tartar chiefs held a doubtful and almost neutral attitude. If these could be brought within the pale of the Catholic Church, Mahommedanism, crushed between the forces of the West and the extreme East, would cease to be a danger. The issue remained undecided in Bacon's time. But we can imagine with what interest he would confer, as he tells us that he did, with Rubruquis on his return to Paris, and listen to his story of the Parliament of Religions, Saracen, Christian, and Buddhist, held at Kara Korum at the suggestion, and under the presidence, of Mangu Kaan. (Cf. vol. ii. p. 389.)

In Bacon's demonstration of the superiority of Christianity to other religions, use is made of this singular experiment. The majority of those who took part in it accepted the unity of God. The Pagans were few in number. The Buddhists (spoken of as Idolaters) raised the question of the origin of evil as an objection to a single ruler of the Universe; but they allowed the question to be evaded. The Tartars, though somewhat indifferent on religious matters, were disposed to side with the Mahommedans and Christians in maintaining the unity of God. On the whole, the conclusion to which this conference tended was a fair sample, in Bacon's judgement, of the preponderating voice of mankind.

Appeal is then made to Aristotelian reasoning as to the necessity of a First Cause. The attributes of wisdom and goodness are shown to follow from omnipotence. Man's duty being to do God's will, how is man to know it? Evidently by revelation. And which revelation is true? There can be but one: for if there were more the human race could not be united. 'The unity of the Church follows from the unity of God. If there were more Gods than one, more worlds than one, and more mankinds than one, then there might be more revelations than one, but not otherwise.' Which, then, is the true revelation? On a comparison of the six religions before us, three, the Pagan, the Buddhist, and the Tartar, are at once ruled out. Of the three that remain, the Jewish, the Saracen, and the Christian, philosophic reasoning, external and miraculous evidence, and ethical purity combine in giving preference to the last. The book, as we have it, closes with some ardent and rapturous words on the Sacrament of the Altar, as the means whereby Christ always remains present with His Church.

Of the missing books we are not left in entire ignorance. We know from the fourteenth chapter of the *Opus Tertium* that the purpose of the fifth book was, to insist upon such modes of setting forth moral truth as were likely to impress, not merely the intellect, but the emotions and character of the hearer. The art of preaching, Bacon thought, was one demanding the most serious and systematic study. Rhetoric was no mere field for the gratification of vanity by ornamental display. It was a part of logic, and the most important part, since by its means truth was so conveyed to the listener that 'he is seized unawares and lifted above himself and filled with thoughts beyond his power to control, so that if evil he is absorbed by the love of good, if imperfect he receives the spirit of perfection, not through violence, but through the strong and gentle power of speech.' Rhetoric thus conceived implied the study of music in its widest sense, the study of rhythm and metre, the management of voice and of gesture (*Opus Tertium*, ch. 75).

The sixth and final section of Bacon's moral philosophy treated, he says, of lawsuits and of justice. He implies, however, that he dealt with this subject cursorily.

XII. General Characteristics of the *Opus Majus*.

The question presents itself, How far can the *Opus Majus*, with its two appendices, the *Opus Minus* and the *Opus Tertium*, be accepted as the final exposition of Bacon's philosophy and polity? It is spoken of by the author throughout as a *persuasio praeambula*. It is a hortatory discourse addressed to a busy statesman (for Clement IV, like most other popes of the thirteenth century, may be so called), urging him to initiate a reform of Christian education, with the direct object of establishing the ascendency of the Catholic Church over all nations and religions of the world.

A fundamental principle with Bacon was that truth of whatever kind was homogeneous. 'All the sciences,' he said, 'are connected; they lend each other material aid as parts of one great whole, each doing its own work, not for itself alone, but for the other parts: as the eye guides the whole body, and the foot sustains it and leads it from place to place. As with an eye torn out, or a foot cut off, so is it with the different departments of wisdom; none can attain its proper result separately, since all are parts of one and the same complete wisdom' (*Op. Tert.* ch. 4). Much light is thrown by passages like these, and there are many such, on the varied and at first sight heterogeneous character of the *Opus Majus*. A glance at the Index of this edition will give some notion of the multiplicity of the topics treated. History of philosophy, comparative philology, mathematics, astronomy, geography, optics, the physiology of sensation, all find a place; and all are subordinated to the service of the Catholic Church as the guardian of the highest interests of man. All these topics are handled so far and in such a way as to convince the Pope, or others in authority, of the width of the field to be cultivated, and of the importance of the object in view. Bacon's procedure is like that of a traveller in a new world, who brings back specimens of its produce, with the view of persuading the authorities of his country to undertake a more systematic exploration. To that further and more complete inquiry he proposed to devote the remainder of his life. He speaks of it in several passages of the present work under the title of *Scriptum Principale*. But, as we have reason to believe, of the twenty-five years of life that remained, more than half were sterilized by his imprisonment. When released, though he persevered, like Galileo, indomitably to the end, he was too old to think with his former vigour, and was capable only of such inferior work as the *Compendium Theologiae*, or the Commentary on the *Secretum Secretorum*. There remain the years between 1268 and 1278. They produced the *Compendium Studii* (published by Brewer), the *Communia Naturalium*, the *Communia Mathematicae*, and other fragments of the *Scriptum Principale*. But, making large allowance for what may have been lost through neglect or through malignant hostility, or for what may yet remain to be discovered, the balance of probabilities indicates clearly enough that the *Scriptum Principale* was never brought to completion. The *Opus Majus* remains the one work in which the central thought of Bacon is dominant from first to last; the unity of science, and its subordination to the highest ethical purpose conceivable by man.

Another characteristic of Bacon's philosophy, to which it

seems to me that sufficient attention has not yet been called, is the sense of historical continuity by which it is pervaded. Not indeed that Bacon stood alone in this respect. Comte, in a remarkable passage of his appreciation of the mediaeval Church, called attention, perhaps for the first time, to the awakening of the historic sense which the very constitution of that Church involved; rising as it did from the threefold root of Roman law, Greek thought, and Hebrew theocracy (*Philosophie Positive*, vol. v. p. 247, ed. Littré). As an example of this influence, he proceeds to quote the example of Bossuet, one of the first of European thinkers to form, in however imperfect a way, a broad and definite conception of the unity of history. But the example of Roger Bacon, writing four centuries earlier, is even stronger and more startling. Two centuries before the Renascence, he states explicitly what others may have implicitly thought, but would have shrunk from avowing even to themselves, that the whole course of intellectual development of mankind from the beginning of the world was not multiple but one, not discrete but continuous. He takes pains to synchronize the demi-gods, the heroes and the thinkers of Greece with the kings and prophets of Judaea. In his conception, philosophy, science, and religious truth had a common origin with the patriarchs: though separated in later centuries, they pursued a parallel course in Judaea and in Greece. The growth of science, no less than the growth of religion, was a process of continuous evolution, taking place under divine guidance. It may be said that traces of such a doctrine as this may be found here and there in the early fathers, and especially in the writings of St. Augustine. But a comparison of the ninth and tenth books of *De Civitate Dei* with the second and seventh sections of the *Opus Majus*, will reveal a profound difference in the mode of treatment, even more than in the conclusions reached. What the earlier writer looks at as concessions wrung from an opponent, the later hails as the testimony of a friend. Augustine dwells on the points that separate the Christian from Porphyry and Seneca; Bacon on the points of union.

There are students of history even yet surviving to whom

the centuries following the fall of the Western Empire seem a chasm hard to pass; so that they prefer, with Vico, to conceive of an ancient civilization which has run its course, and a new cycle as beginning. For Roger Bacon the apparent breach of continuity was in great part filled up by the long series of thinkers and students, who kept the torch of science alive in the Mahommedan schools of Mesopotamia and Spain. A glance at the Index to this edition will show the use which Bacon made of such men as Thabit ben Corra, Alfarabius, Alfraganus, Alkindi, Alhazen, Albumazar, Avicenna, Hali, and Averroes. They are spoken of, and most truly, not merely as the principal channels through which Greek philosophy and science were introduced to the Western world, but as having increased the treasure entrusted to them; a treasure which the Westerns of the thirteenth century, 'unless they are dolts and asses,' will regard it as their duty to transmit with due interest to their posterity.

At the close of these introductory remarks, some attempt may be made to assign Bacon's position in the history of human thought. It appears on the surface that he belongs to the order of thinkers, typified by Pythagoras rather than by Aristotle, who engage in speculation, not for its own sake alone, but for social or ethical results, that are to follow. His protests against the intellectual prejudices of his time, his forecasts of an age of industry and invention, the prominence given to experiment, alike as the test of received opinion and the guide to new fields of discovery, render comparison with his great namesake of the sixteenth century unavoidable. Yet the resemblance is perhaps less striking than the contrast. Between the fiery Franciscan, doubly pledged by science and by religion to a life of poverty, impatient of prejudice, intolerant of dullness, reckless of personal fame or advancement, and the wise man of the world richly endowed with every literary gift, hampered in his philosophical achievements by a throng of dubious ambitions, there is but little in common. In wealth of words, in brilliancy of imagination, Francis Bacon was immeasurably superior. But Roger Bacon had the sounder estimate and the firmer grasp of that combination of deductive with inductive method which marks the scientific discoverer. Finally, Francis Bacon was of his time; with Roger Bacon it was far otherwise.

M. Hauréau, the historian of Scholastic philosophy, and also M. Renan, have suggested a parallel (or, it may be, have adopted it from Littré) between Roger Bacon and Auguste Comte. Some anticipation of the *Philosophie Positive* there assuredly is in Bacon's subordination of metaphysic to science, in his serial arrangement of the sciences, and in his avowal of a constructive purpose as the goal of speculative inquiry. But it is well not to push such comparisons too far. We shall best understand Bacon's life and work by regarding him as a progressive schoolman. Like the other great schoolmen of the thirteenth century, he set before himself the purpose of strengthening the Church in her work of moral regeneration, by surrounding her with every intellectual resource. But the forces that he brought to bear were not limited, like theirs, to the stationary dialectic of Aristotle; they were also, in great part, drawn from the progressive culture of natural and historical science. As compared with his successors of the Renascence, his purpose was loftier; for, in urging the continuous advancement of knowledge, he had higher things than knowledge in view. His aim, pursued in no spirit of utilitarian narrowness, yet steadily concentrated on the moral progress of mankind, was, *Induire pour déduire afin de construire.*

ANALYSIS OF THE 'OPUS MAJUS.'

PART I.

THE FOUR GENERAL CAUSES OF HUMAN IGNORANCE.

CHAPTER I.

VOL. I. PAGES

True wisdom implies—1. Sound methods of gaining knowledge; 2. The application of knowledge to important purposes; as the government of the Church, the conversion of the heathen, and the repression of evil-doers. Such wisdom is hard to obtain, and there are four principal causes of failure. These are—1. Subjection to unworthy authority. 2. The influence of habit. 3. Popular prejudice. 4. False conceit of our own wisdom. We will begin by dealing with the three first: the fourth demands separate treatment 1-4

CHAPTER II.

Opinions of philosophers on these defects.—Citations from Aristotle, Seneca, Cicero, Averrhoes, Abelard, Jerome, Chrysostom . . 4-6

CHAPTER III.

Leaving authority, let us look to experience. Authority, habit, and prejudice may sometimes lead to truth, but the probabilities are much against it. We may take an illustration from arithmetic. Of perfect numbers, there is but one among the first ten, one between 10 and 100, one between 100 and 1,000, one between 1,000 and 10,000. Among men the number of the truly wise is even smaller. Even of philosophers few are to be counted among them. Aristotle, the wisest of all, was not perfect 6-8

CHAPTER IV.

Of the three first causes of error, popular prejudice is the most potent. The truth that pearls should not be cast before swine is confirmed by Aristotle and other philosophers 9-11

CHAPTER V.

CHAPTER VI.

CHAPTER VII.

CHAPTER VIII.

CHAPTER IX.

CHAPTER X.

CHAPTER XI.

VOL. I. PAGES

other ways, should be credited with no authority whatever in things of which they know nothing 23-24

CHAPTER XII.

Many things known to the ancients we neglect, as the study of mathematics and languages, from sheer ignorance of their value. In other cases the fact that subjects are not studied by the fathers is held a good reason for passing them by. We forget that the saints and fathers were justified by the circumstances of their time, and moreover that they were not infallible. Augustine found much fault with Jerome, and so in other cases 24-26

CHAPTER XIII.

We must remember that the best Greek work was not known to the Latin fathers. Plato indeed was translated and carefully studied; but Aristotle, from the very fact of being Plato's opponent, was neglected. Yet Augustine had translated the Categories, a work highly valued by Alcuin. Boethius also translated some of the logical works. But if the greater works of Aristotle had been known to them, they would have gladly received these, and not have troubled themselves about the ashes of his philosophy 26-28

CHAPTER XIV.

The early Church made no use of Greek science except for the purpose of regulating its calendar and its music. The explanation of this neglect of ancient learning is fivefold. 1. Philosophy was the foundation of law and government to all the nations of antiquity except the Hebrews. 2. Therefore it was that philosophy resisted Christianity. 3. Moreover these nations not merely studied philosophy, but practised augury and oracular magic. 4. They persecuted Christians. 5. The Church, finding her enemies occupied on the one hand with the study of philosophy, and on the other with the study of magic, associated these two things, and thus came to despise and dislike philosophy. The truth is however that philosophy, so far from being hostile to the Church, is capable of yielding it indispensable support 28-30

CHAPTER XV.

Later ecclesiastical authorities have followed a similar course, though without the excuse which justifies the early fathers. Though Greek philosophy is no longer untranslated, they study only its most trivial productions, neglecting the great works of science and ethic 30-31

CHAPTER XVI.

VOL. I. PAGES

Let me not be misunderstood. I am not proposing to your Holiness any violent change in the ordinary course of studies. I am merely suggesting free access to a land of plenty for those who care to avail themselves of it. If once the leaders of thought become more enlightened, the rest will soon follow 31-32

PART II.

THE CONNEXION OF PHILOSOPHY WITH THEOLOGY.

CHAPTER I.

Theology is the mistress-science. All truth is contained in the Scriptures; but to elicit truth we need the help of the canon law and of philosophy. Wisdom comes from one God, is given to one world, for one purpose. Itself therefore is one. It cannot be inconsistent with itself 33-34

CHAPTER II.

The canon law has its root in Scripture. From that root spring branch and fruit; the teaching of the fathers and doctors of the Church and the rules finally laid down by them 34-35

CHAPTER III.

So with philosophy. Augustine dwells at length on the importance of taking from it all that is precious, as the Hebrews of old borrowed jewels from the Egyptians. He included in the word ethics, history, knowledge of the arts, logic, and grammar 35-37

CHAPTER IV.

Jerome and Bede used similar language. Bede remarks that, as Solomon when building his temple called foreign workmen to his aid, so has Christ availed himself of heathen philosophers in building his Church. Paul, too, quotes heathen poets 37-38

CHAPTER V.

Wherever truth is found it belongs to Christ. We must distinguish between *intellectus agens* and *intellectus possibilis*. The first is no part of our nature. Our minds are capable of receiving it and of being stirred to action by it, but it comes from without. This can be

CHAPTER XI.

CHAPTER XII.

CHAPTER XIII.

CHAPTER XIV.

VOL. I. PAGES

it was afterwards hidden from men; that Thales and his successors down to Aristotle, aided by some trace of primitive tradition, revived its culture. Philosophy, therefore, has its place in the divine government of the world: its conclusions must be demonstrated, diffused, and enlarged. It is a component part of that perfect wisdom which is contained in the Scriptures 64-65

PART III.

THE STUDY OF LANGUAGE.

Having seen the essential unity of wisdom as contained in Scripture, the canon law, and philosophy, we pass to the divisions of the subject. These are five. The first is the study of grammar, not so much of Latin, as of the languages from which Latin received its culture. 1. The quality of one language can never be perfectly reproduced in another. This is true even of the dialects of a language: far more true as between different languages. A literal translation of Homer into Latin, and thence into a modern language, results in utter absurdity 66-67

2. Secondly, Latin is altogether wanting in many of the necessary words for the things described by foreign authors 67

3. Thirdly, the translator must not only be perfectly acquainted with his subject, but also with the two languages with which he deals. Boethius and Robert Grosseteste fulfilled these conditions. But most translators have failed, and especially the translators of Aristotle. Jerome and Augustine have pointed out the errors of the Septuagint version and of the other versions. Nevertheless Jerome has left very many uncorrected 67-70

4. Fourthly, in the Latin texts now in our possession there are vast omissions, as well as many confused and corrupt passages. E. g. the third and fourth books of Maccabees are wanting: also the books of Samuel, Nathan, and Gad. We need also good translations of Josephus's book of Antiquities, of many of the Greek fathers, of the second and third books of Avicenna, and of many essential treatises of Aristotle, on metaphysics, on physical and mathematical science, and on that part of his logic which deals with practical reason, the part that is of the greatest importance in the guidance of human life 70-73

5. Fifthly, owing to our neglect of foreign languages, we fail to understand the allusions to them contained in many writings of

PART IV.

MATHEMATICAL SCIENCE.

First Distinction.

CHAPTER I.

CHAPTER II.

VOL. I. PAGES

There is high authority for this estimate of mathematics. In the study of divine things and also of man's social life Boethius and Ptolemy show that it is of great service. The various modes of proportion have their analogue in civil polity. It is needed both in grammar and logic, as Alpharabius and Cassiodorus have shown. For prosody depends entirely on arithmetical relations. Logic, on its practical side, has the same purpose as poetic and rhetoric, which depend on harmony. Further, the subject-matter of logic is intimately connected with mathematic. This is obvious in the Categories of quantity, of time, and of place. In the Category of quality much belongs to the mathematical domain, as e. g. geometrical form. The same may be said of the Category of relation. Spiritual substances can only be known through the medium of corporal: and the first step to the knowledge of body is the study of the heavenly bodies. The dependence of astronomy on mathematics is obvious 98–103

CHAPTER III.

We reach the same conclusion by reasoning. (*a*) In all other sciences we use mathematical examples, because they illustrate the point without confusing complications; e.g. in explaining the difference between augmentation and change, we add a gnomon to a rectangle, augmenting its magnitude without changing its shape. (*b*) Mathematical knowledge is innate in us and only needs drawing out, as Cicero explains, *Quaest. Tusc.* book i. (*c*) Mathematical truths are discovered prior to others. (*d*) They are simpler of comprehension than others. (*e*) Students attain them who are incapable of proceeding further. (*f*) We see the same thing with children, who easily appreciate the simple arithmetical relations on which music depends. (*g*) We acquire our knowledge of things known to ourselves more easily than of things known to nature (i.e. intrinsically simpler; as, for instance, the truths of theology). But mathematics have the double character of being both relatively and absolutely simpler. (*h*) In mathematics demonstration is more complete: its cogency has the force of necessity, which is not the case either in physics, in metaphysic, in ethic. (*i*) In other sciences the uncertainty of the premisses involves uncertainty in the conclusion. These principles require verification by some science more perfect than themselves: i. e. by mathematics. (*k*) Finally, the subject-matter of mathematics is more directly cognizable by our senses. It deals with quantity, which lies at the root of all knowledge. The simplest process of intellect implies continuous quantity,

VOL. I. PAGES

i. e. time. (*l*) Finally, we are confirmed by the experience of all who have most distinguished themselves in science. They owe their results to the mathematical foundation of their studies . . 103–108

Second Distinction.

CHAPTER I.

Leaving method, and passing to the objects of study, we find it impossible to make progress without mathematics. The number and motions of the heavens, planetary motions, eclipses obviously require this science. And things terrestrial no less, since they are governed by things celestial. The sun, acting on the surrounding medium, diffuses light. This may be taken as a type of the propagation, or multiplication, of species, or forces, by other agents. Let us consider how it takes place 109–111

CHAPTER II.

Rays passing from a rarer medium to a denser, if they impinge upon the latter perpendicularly, pursue a rectilinear course. Otherwise they are refracted, i. e. diverted towards the perpendicular drawn to the surface of contact. In passing from denser to rarer they are diverted away from this perpendicular. Hence if the denser body be spherical, the solar rays passing through it converge at a given point beyond it. The convergence of many rays produces heat. If the second medium is so dense that the 'species' cannot pass through, or at least that its transit cannot be appreciated by human vision, it is reflected. If the ray falls perpendicular to this medium, reflexion is in the reverse direction of incidence. If otherwise, the angle of reflexion is equal to the angle of incidence. By concave reflectors solar rays may be concentrated. If the reflector is spherical, those rays will be focussed which impinge on points in it corresponding to a circle placed at right angles with its axis. But reflectors can be devised of such form that all the rays shall fall on the reflecting surface at equal angles, and thus be reflected to the same focus. We have further to consider the diffusion of light impinging on objects not directly from the sun but indirectly. Constant exposure to direct rays would be destructive. Lastly, we have to consider the effect of light on the nerves of vision, when there is no question of its following any rectilinear course, its path being modified by the vital principle . . . 111–117

CHAPTER III.

Rays issuing in infinite number from a point in every direction find their termination on hollow surface of a sphere. Each point of

VOL. I. PAGES

the surface acted on is the vertex of a cone of rays, of which the base is the whole surface of the agent 117–119

Third Distinction.

CHAPTER I.

Light and other forces not merely propagate themselves by multiplication of species, but work ulterior effects; light produces heat, heat putrefaction, putrefaction death, and so on. To these effects the same law applies. Rectilinear action is more effective than curvilinear; perpendicular than oblique. In refraction the effect is greater than in reflexion, because in the latter the reflected and incident rays neutralize each other. In refraction action is stronger where the second medium is denser than the first, because the ray is deflected towards, not away from, the perpendicular. In reflexion more is done by oblique rays than by perpendicular. With the latter there is neutralization of incident and reflected rays. Further, there can be but one perpendicular ray, but infinite numbers of oblique; and these may all be made to converge . . . 119–123

CHAPTER II.

The rays of which a natural action consist form, as we have seen, a cone. In the shorter cones the strength of the action is promoted first by greater proximity of the agent, secondly by the greater proximity of the conterminal rays after intersection. On the other hand in the longer cones, the rays before intersection are the nearer to each other, and in this respect the action will be more potent. But the first of these conditions will outweigh the second . . 123–124

CHAPTER III.

When two equal spheres interact, the half of each which is averted from the other is unaffected; the extreme rays from each can only embrace the half of the other sphere. But with unequal spheres, the less receives rays from less than the hemisphere of the greater, which touch more than its own hemisphere. From each point of a sphere rays issue into space outside of the tangent plane. Of these rays only one is perpendicular to the surface. This is the potent ray. Rays vertical to a sphere are divergent. But when the object of vision is very remote, as in the case of stellar bodies, they appear to us to be parallel: just as the walls of a house seem parallel, although their lines of direction converge to the earth's centre. 124–127

Fourth Distinction.

CHAPTER I.

VOL. I. PAGES

By the foregoing principles and others akin to them, which for want of space are here omitted, all natural actions are to be explained. A few illustrations may be given. The planets receive their light from the sun. Hence when the earth is between the sun and moon, the moon is eclipsed by the earth's shadow. The cone of this shadow not reaching however to the other planets, these are not eclipsed. Again, that the eye and the stars can mutually transmit their emanations through the media of the planetary orbits, proves these media to be of great rarity, to be invisible, and non-luminous. The sphere of fire also is neither luminous nor visible. A planet differs from the sphere in which it moves by greater condensation of celestial substance; hence its luminosity. This, though ultimately derived from the sun, is not due to the sun's reflected rays. The diffusion of moonlight proves the moon to have independent luminosity. Owing to the magnitude of the sun being 170 times greater than that of the moon, much more than half of the earth is illuminated by the sun; and the same is the case with the moon and planets. Other affections of the moon's light depend on her varying conjunctions with planets and with constellations . . . 127–130

CHAPTER II.

The same principles may be applied to disprove the alleged simplicity of cosmic structure. A star on the meridian is seen to be further from the pole than at its rising. In the latter position refraction displaces it. This shows the world to consist of distinct substances of varying density; for a ray would not be refracted while passing through the same substance, even though its parts should be of different density 130–132

CHAPTER III.

Hence it is that the temperatures of the various zones are explained. Beneath the poles the cones of rays are prolonged, and therefore feeble: capable only of raising vapours from the earth and sea, so that the air of those climates is heavy and cold, and unfit for living things. Nevertheless, owing to the length of the days and of the twilight in summer, the sun being never far distant, there may be places in those regions, favoured by the position and inclination of certain mountains, where the rays are so reflected that the climate is temperate 132–135

CHAPTER IV.

VOL. I. PAGES

Passing to the torrid zone, it would seem that the region under the equinoctial circle must be the hottest, since twice in the year the sun's rays there are vertical. But this is over-weighed by the fact that under the tropical signs the sun is nearly stationary. The matter is further complicated by the eccentricity of the solar orbit. . 135-137

CHAPTER V.

The emanations from the stars affect not merely climate, but character; implanting on the new-born child dispositions to good or evil, to quick or to dull apprehension: though free will, God's grace, temptations of the devil, or education may modify these innate tendencies 137-139

CHAPTER VI.

Our theory may be applied to the tides. These evidently depend on the moon. When the rays fall obliquely on the surface, their effect is only to raise vapours from the surface and create ebullition and a consequent flow of water till the time comes when the rays fall vertically, and with force enough to extract the vapour; and then the reflux begins. This however leaves it unexplained why the same thing happens in the hemisphere averted from the moon. We must suppose the ninth or starry heaven to be solid and impenetrable, and that the vertical rays of the moon are reflected from it, these producing the same effect as the incident rays 139-142

CHAPTER VII.

The application of these principles to the preservation of life and health is obvious. Protection must be sought against the vertical rays of injurious emanations, as of the moon at night, of Saturn and Mars, of persons infected with contagious disease, of the evil eye; and we must adapt our bodies to the reception of emanations known to be salubrious 142-143

CHAPTER VIII.

We may give mathematical disproof of what is one of the greatest errors in philosophy: namely, that matter is of one kind only, the differences of substance resulting only from difference in form. The consequence of this error would be to elevate matter to equality with

VOL. I. PAGES

God. Nor is it enough to say that matter is infinite potentially, but not in essence. Nor that it is potentially infinite in the sense in which this is said of continuous quantity. For to attribute to matter existence in indefinite numbers of substances is to attribute to it infinity, not merely potentially, but in act. The contradiction in which this lands us may be set forth geometrically. Nothing infinite can have finite power, and conversely nothing finite can have infinite power 143–148

CHAPTER IX.

When two spheres are brought together, and the straight lines from their centres to the point of contact are continuous, the question arises whether these lines become one, or whether we are to regard them as two. Averroes maintained a distinction between mathematical quantity and natural quantity. But this distinction is untenable. The lines in question are two, although they have the effect of one, and for convenience of speech may be spoken of as one. Against the separability of different masses of matter it is argued that if two circular planes are brought into contact and then separated, air will penetrate into the outer portion before the inner, hence for a moment there will be a vacuum in the central part. But the answer is that the separation is not simultaneous throughout the whole surface of the plane, so that the air penetrates gradually. From the divisibility of matter, it is not to be argued that the world is composed of an infinite number of material particles, as Leucippus and Democritus maintained. Were this so, it might be inferred that the diameter of the square was commensurable with its side; which Euclid in the seventh proposition of his tenth book has shown to be impossible 148–152

CHAPTER X.

On geometrical grounds the shape of the universe can be inferred to be spherical. No other form would preclude the possibility of a vacuum in the course of its revolution. Cylindrical or lenticular form would suffice if revolution took place round a certain axis. With the spherical form revolution round whatsoever axis would avoid vacuum. Looked at from within, it must be concave and spherical; otherwise lines drawn from the centre of the earth to the extremities of the universe would not be equal. Further, the sphere is that form which under a given surface has the greatest content. It is the simplest and noblest of forms. The water, the air, and the fire surrounding the earth concentrically, are of similar form . . 152–157

CHAPTER XI.

VOL. I. PAGES

Suppose two vessels similar in shape and equal in size; one placed at a higher level than the other. More water can be placed in the lower, for its surface will be a portion of a smaller sphere: the diameter of the rim of the vessel being equal in both cases . 157–159

CHAPTER XII.

The Platonic school maintained that heaven and the four elements corresponded to the five regular solids. For there can be no more than five. Since in the dodekahedron the other four can be inscribed, this was regarded as representing heaven: fire was identified with the tetrahedron, air with the octahedron, earth with the cube, water with the icosihedron. But the difficulty in this theory is that, though solid masses can be built up of tetrahedra and of cubes without leaving vacua, this is not the case with the other three . 159–164

CHAPTER XIII.

There can be only one universe. For, on the supposition that there were two, both being spherical would touch in one point only, so that a vacuum would be left, which is impossible. Further, the universe cannot be infinite, otherwise two infinite lines, from one of which a given portion was cut off, would be equal; i.e. the part would be equal to the whole, which is impossible . . . 164–165

CHAPTER XIV.

Unity of time does not imply unity of matter. Nor is it needful to suppose plurality of ages (aeva). The subject of time is not matter, but motion. The subject of motion is not matter, but body composed of matter and form. Motion is of linear dimension. Prior excludes posterior, past excludes future. But as to the present being a point having no dimension, there is no such exclusion; one point does not exclude another: many points occupy the position of one. One present moment suffices for all present moments. Hence time is one. And so to the conception of aevum the same applies. It is single and not multiple 165–167

CHAPTER XV.

In a body falling to the earth's centre, a strain is involved: since, though the central point of the body tends directly towards the centre, the extreme points are prevented from doing so. From this strain heat results. This is shown by experiment to be the fact. But on geometrical grounds the reason of this fact appears; and our knowledge of it becomes thus more complete 167–169

CHAPTER XVI.

VOL. I. PAGES

All these, apart from their literal importance, have a distinct spiritual signification. The river symbolizes the world; the Dead Sea, hell; Jericho, the flesh; the Mount of Olives, spiritual life; the valley of Jehoshaphat, humility; Jerusalem, the soul in the enjoyment of peace, or again the Church militant and triumphant. Minute research will reveal numberless intermediate meanings . . 183-187

The third head relates to sacred chronology. Scripture presents to us a succession of times, with regard to which precise knowledge can only be given by mathematical astronomy. The starting-point is the creation of the world. Was this in the autumnal or the vernal equinox? From what is said in the Old Testament as to the Feast of Ingathering [Exodus xxiii. 16] we should infer the former. Yet Jewish and Christian commentators adopt the latter view. It will be for astronomy to decide this difficult point. Further, the question of the longevity of the ancient patriarchs has to be considered. One mode of accounting for it may be the more favourable position of the sun and the planets in primitive times. Again there is the problem of the Deluge. The right interpretation of Josephus points to November as the month in which it began. Lastly, did night come before day, or the converse? The former would seem to be the true view 187–195

The fourth head deals not with chronology in general, but with the definition of periods. How is the beginning of a lunation to be fixed? by astronomical calculation, or by the moment when the new moon is visible? The actual lunation is variable. The average lunation must be used. The Jews use the Metonic cycle of nineteen years, or 235 lunations. This gives twenty-nine days, twelve hours, and $\frac{793}{1080}$ of an hour for the mean lunation. They take a period of thirteen lunar cycles or 247 years, within which all their festivals recur at the same moment. The lunation is considered to begin with the sunset immediately following the computed time. These considerations may be applied to the date of the Creation, of Noah's issue from the ark, of the Passover, and finally of the Passion. The current belief in the Latin Church is that Christ was born in the second year of a lunar cycle, and died on March 25 (a. d. viii Kal. Aprilis), the moon being at the fifteenth day (the Greeks holding that it was the fourteenth day). Against this much may be urged. It implies that it was in the thirteenth year of a cycle. From the computation of S. Dionysius this would involve the Passion taking place on a Sunday, which is impossible. A table is appended showing one solution of the difficulty. This would show the date of the Passion to be April 3 (a. d. iii Non. Aprilis), on the fifteenth year of the lunar cycle, Christ being then thirty-two years old. This view is offered to the Pope for consideration 195–210

VOL. I. PAGES

Fifth head. This relates to geometrical forms; and again it has to be premised that the spiritual meaning of Scripture is not to be grasped, unless the literal meaning be first understood. The precise form of the ark, the tabernacle, the temple should be presented to us with mathematical accuracy; it will then be possible to interpret their mystical signification. A remarkable case in point is the rainbow, of which we are told that it is a symbol of God's promise as to the Deluge. The meaning of this is misunderstood for want of understanding the geometry of the rainbow, which is produced by the solar rays striking on the raindrops and being refracted or reflected thence. Only by geometry can such a text as that of the threefold burning of mountains by the sun be understood. The mountains receive direct rays. The rays reflected by them are focussed in the air and contribute to warm them. Lastly, there are the doubly refracted rays passing from the sun into the clouds and from the clouds into the air. Again, the rays may fall either vertically, or obliquely, or horizontally, producing different effects in each case. The laws as to the passage of light through space apply equally to the passage of all other forces. Of all these things there is a spiritual interpretation. Direct vertical rays may be compared to the action of grace on the righteous; they neither reflect nor refract it. In the wicked we see the light driven away, reflected: in the imperfect it is bent aside, refracted. It is to be noted here that the geometrical form of the triangle is specially adapted to symbolize the Trinity. Each angle is distinct yet each embraces the whole space. Again, our geometrical principles as to the action of forces may be applied to the estimation of the forces of temptation, which act in proportion to their proximity. Hence the need of keeping the tempted as far removed as possible from the objects which tempt 210–219

Sixth head. This relates to number. (*a*) As with geometrical forms, so with number there is a spiritual meaning behind the literal meaning. (*b*) For the understanding of chronicles it is necessary to know the different systems of numeration, and to be able to convert one into another. (*c*) There are various arithmetical operations in the Jewish law requiring a knowledge of the subject. (*d*) There are many corruptions in the text which only an arithmetician can unravel. (*e*) The perfections inherent in the number three can only be understood by arithmeticians. (*f*) Arithmetic is necessary in astronomy, which we have seen to be needful to the theologian. Some instances of its value in astronomy are here given. We have to define the length of a line on the earth's surface corresponding to a degree. We must take a mile as 4,000 cubits: a cubit as 1½ feet. How many miles must we walk northward from a given point to find the pole-star a degree higher in the sky? We shall find the result to be fifty-six miles, 2,984$\frac{8}{63}$ cubits. From the diameter of the earth we obtain its circumference and its surface. The distance of the heavenly bodies may be measured in semidiameters of the earth = 3,250 miles. Alfraganus estimates the distance of the starry sphere to be 20,110 semidiameters; which gives the diameter of this sphere as 130,715,000 miles. From this we can calculate the circumference and surface. The longest distance of Saturn is the semidiameter of this sphere, or 65,357,500 miles, of Jupiter 46,816,250 miles, of Mars 28,847,000 miles, of the Sun 3,965,000 miles, of Venus 3,640,000 miles, of Mercury 542,750 miles, of the Moon 208,541$\frac{2}{3}$ miles; the shorter distance of the preceding planet being always equal to the longer distance of the succeeding. As to the ninth and the tenth heaven we have no scientific knowledge. As to the height of the atmosphere there is great uncertainty. For measuring the apparent diameters of the Sun and the Moon recourse has been to water-clocks; the number of drops issuing from the beginning to the end of the Sun's rising have been compared with the number issuing during a revolution of the heavens. But there are more accurate methods by astrolabes or quadrants. From these, and from observations of eclipses, the moon's diameter is estimated at $\frac{5}{17}$ of the Earth's diameter. The Earth is therefore about 31$\frac{1}{4}$ as large as the moon. Similarly the Sun will be found to be 170 times as large as the Earth. Mercury is $\frac{1}{22000}$ of the Earth. In a similar way the relative magnitude of the other planets as compared with the Earth may be determined. The fixed stars are 1,022 in number, and are divided into six groups according to this magnitude. Those of the first magnitude are 107 times as large as the Earth; of the second, ninety times the Earth's magnitude: of the third, seventy-two times, of the fourth, fifty-four times, of the fifth, thirty-six times, of the sixth, eighteen times. But besides these there are infinite numbers of other stars whose magnitude cannot be determined 219–236

Seventh head. Music. The theologian should be acquainted with the theory, if not with the practice, of vocal and instrumental music. Music covers the whole ground of recitation, punctuation, accent, things necessary for prose as well as poetry. Of instruments again many are spoken of in Scripture, and each has a spiritual as well as literal meaning. Nor must dancing, which may be called visible music, be forgotten 236–238

We have now to consider certain objections that have been raised with regard to mathematical science. Mathematic has been confounded with magic, and put on the same level as fortune-telling, witch-craft, and the preparation of charms and incantations. It has been credited with the doctrine that human actions are absolutely

VOL. I. PAGES

determined by stellar influences. But this doctrine is explicitly condemned by all philosophers, both of Greece, Rome, and Arabia. They maintain free-will against any such supposed physical necessity; and they reprobate utterly the use of charms and incantations. The prejudice against mathematics was strengthened by the strife of heathenism against Christianity, in which magic was used by the former, and in which Christian miracles were accounted for by magic. The reproach of denying free-will remained, when the suspicion of magic had passed away. But examination of Ptolemy's works shows clearly that he never fell into this error. He, and his principal Arabian commentators, held that the forecasts of the future revealed by astronomy indicated general tendencies only, and did not impose any specific compulsion on individual actions. But this distinction has not been always perceived in the attacks made by Christian writers on mathematical science. They have confounded true science with false 238–249

What is true is that the influences of the stars implant certain tendencies to good or evil action, always at the same time leaving free scope to human will. It is evident that physical temperament is one of the factors in human action. Temperament is itself a result of the influence of the stars. That climate affects character is obvious to every one. And this influence is to be traced *in minimis*. From every part of the sky to every point of the earth's surface a cone of forces proceeds, affecting more or less every community and every individual, though not in such a way as to override free-will. These things would be studied by physicians if they were better acquainted with astronomy. In the case of a commonwealth, or of the statesmen who direct commonwealths, it is more easy to discern the nature and degree of such influences, and to forecast their results. And great advantage will accrue to the community from doing so 249–253

We now arrive at the practical applications of mathematical science to Church government. The first of these is the proof offered by astrology of the superiority of Christianity to other religions, and the insight given into the nature of Antichrist. It must be understood that the heaven is to be divided into twelve Houses; that is, taking the quadrants formed by the intersection of the meridian and horizon, each may be divided into three. To five of the seven planets two Houses are allotted; to Mercury and the Moon one each. The association of Jupiter with each of the other planets has a special meaning. If with Saturn, the reference is to Judaism. With Mars, to Chaldaeism. With the Sun, to Egyptian religion. With Venus, to the Saracen worship. With Mercury, to Christianity 253–258

VOL. I PAGES

But there is another and more essential meaning of the word House. If we divide the ecliptic into twelve parts, and through the divisions conduct circles intersecting at the poles of the ecliptic, the regions between any two of these circles is in the true sense a House; bearing the name of its zodiacal sign. The principal House of each planet is that in which it was created. Leo is the House of the Sun, Capricorn of Saturn, Sagittarius of Jupiter, Virgo of Mercury. There are certain signs for each planet which are called its Exaltation; for the Sun Aries, for Saturn Libra, for Jupiter Cancer, for Mars Capricorn, for Venus Pisces, for Mercury Virgo. The signs are divided into four groups: three, Aries, Leo, Sagittarius, are hot and dry: three, Taurus, Virgo, and Capricorn, are cold and dry: three, Gemini, Libra, Aquarius, are hot and moist: three, Cancer, Scorpio, and Pisces, are cold and moist. A planet being in any of the groups to which its House belongs is said to be in its Triplicity. We have further to consider Boundaries and Aspects. With regard to Boundaries, they vary for each planet and for each sign. With regard to Aspects, each sign is divided into three equal portions of ten degrees, and a portion is assigned to each planet in sequence 258–261

It will appear in the result that Mercury has manifold and strong connexions with the sign of Virgo. As Mercury is connected with Christian faith, so is the Moon, with its irregular motions, connected with the corruption of that faith. The conjunctions of the planets, and especially those of Jupiter and Saturn, throw light on important epochs of history. These are of three kinds, occurring respectively in periods of 20, of 240, and of 960 years. Judging from what took place at previous periods, it may be inferred that the Mahometan faith will not be of much longer duration. There yet remains the period of Antichrist, on the date of which study of these astronomical periods may throw light 261–269

A second application of mathematics to the service of the Church is the correction of the Calendar. The Julian Calendar fixed the length of the year at $365\frac{1}{4}$ days. Hence the arrangement of an additional day every fourth year. But this estimate is known to be too great by the $\frac{1}{130}$ part of a day. Therefore in every 130 years there is an accumulated error of one day 269–271

Again, it has been assumed that the equinoxes and solstices occurred on fixed days. These days in the beginning of the Church were fixed thus: the winter solstice on December 25, the vernal equinox on March 25, the summer solstice on June 24, the autumnal equinox on September 24. Subsequently the vernal equinox was transposed to March 21. Hence the earliest Easter, being the first Sunday after the full moon succeeding to the vernal equinox, was

March 22. But the winter solstice and the vernal equinox are no longer on these days. At the present time the winter solstice is on December 13, the vernal equinox on March 13, the summer solstice on June 15, the autumnal equinox on September 16. In every 125 years the error of another day accumulates. The consequence of this error is that Easter is observed at the wrong time, in the third and in the fourteenth years of the lunar cycle. The error will be far more serious in future centuries, so that ultimately the period when Lent should be observed will be at a time when every one is eating meat 271-274

Further, there is a serious error in the computation of the lunar cycle. Successive periods of nineteen years differ in length, some containing four, others five leap-years. The only satisfactory period would be thirty Arab years of twelve lunations, making 10,631 days. It is true that the Council of Nice adopted the lunar cycle; and 120 years afterwards it was confirmed by Pope Leo. The error at this latter period was not more than a day. Since that time astronomy has been in disrepute, for reasons already mentioned, and no one has been found who could clearly indicate the error, with sufficient authority; although in the time of Pope Hilary some consideration was given to the subject. At the present time the error is so great as to attract the ridicule of Jewish and Arab astronomers, and deserves the serious attention of the ruler of the Church . 275-285

Geography.

We now pass to the influence of the heavens upon things terrestrial. In all things that are brought forth on earth, whether for good or evil, the sun and the heavens are the moving cause. We have therefore to consider the different ways in which different parts of the earth's surface are affected by these agencies. Imagining that surface divided by the equator and the equinoctial colure into four equal portions, we have specially to consider the portion contained between the equator and the poles, bounded east and west by the colure. What proportion does land bear to water? Ptolemy thought one-sixth, but other authorities think the proportion much greater. Seneca and Pliny look on the space of ocean dividing the west of Spain from the east of India as inconsiderable. Under the word Spain we must include a vast tract extending westwards across the Straits of Gibraltar, in the direction of Atlas. On the whole, it seems probable that the land known to us from east to west extends over more than half the earth's circumference. Further, it seems probable

VOL. I. PAGES

that the disposition of land and water may be similar on the other side of the northern hemisphere to that of this side; and the same may be argued of the two divisions of the southern hemisphere, especially as the sun in its annual course comes nearer to it . 286–294

Speaking of the parts known to us, Ptolemy and others have distinguished seven *climata*—marking them by the increasing length of the longest day. The position of each place referred to is defined by the intersection of the line of latitude and longitude belonging to it. The zones, or *climata*, of Ptolemy are marked according to the increase of a quarter of an hour in the length of the longest day up to the sixty-first degree of latitude; thence to the sixty-fourth degree, by half an hour's increase; thence to the sixty-sixth by one hour's increase. Beyond this point we come to the region where in the summer season the sun is always above the horizon, in winter always beneath it. Here the divisions must be marked according as the longest day is one month, or two or three, up to six. As to longitude, it should be measured not from any arbitrary point, but from the true east and west on the equator. The neglect of this precaution has led to much confusion in the tables of Toledo; and generally a far more accurate determination of the latitude and longitude of towns and states is needed, such as can only be instituted by apostolical, imperial, or at least regal authority 294–301

A right understanding of locality and climate concerns alike the interpretation of Scripture, the propagation of the faith, and the welfare of commonwealths. Missionaries in particular should know the distribution of the various religions of the world; where the lost tribes of Judea are to be looked for, where the incursions of Antichrist are most to be dreaded. Our knowledge of these matters, derived from Pliny, Ptolemy and other writers of antiquity, has been recently enlarged by the travels of William Rubruquis in Central Asia 301–305

At the tropic of Cancer we begin to find regions where the sun at the summer solstice casts no shadow. Southwards to the tropic of Capricorn the same phenomenon occurs for each place twice in the year. And we have knowledge of regions south of the tropic of Capricorn where the noonday shadow is always to the south. And though in these last the sun approaches so nearly at our winter time, and recedes so far at our summer time, as to involve great extremities of heat and cold, yet these evils may be lessened by the configuration of mountain and plain, so that the region is habitable, as we know in the case of the island of Taprobane 305–309

We may begin our description of the habitable world with India. Its southern coasts are washed by a branch of the Atlantic Ocean, so vast that from the mouth of the Red Sea to the south of India is

VOL. I. PAGES

Babylon. In Babylonia dwelt Noah and his sons after the deluge. Both the Tigris and the Euphrates rise in Armenia. The Tigris flows for some space underground before reaching Nineveh and joining the Euphrates. The Euphrates swells in summer as does the Nile. To the east of the Euphrates lie North Arabia and Syria, the southern part of which is the Holy Land. Here a more detailed description becomes necessary 332-335

Beginning with the coast, we find Gaza on the confines of Egypt and Palestine, then proceeding northward Ascalon, Joppa, Azotus, Caesarea, Acon, Tyre, Sarepta, Sidon, Barut, Gibeleth, Tortosa, Laodicea. From this last to Antioch is two days' journey. From Antioch to Tarsus in Cilicia three days' journey 335 336

Passing to the interior, we find Beer-sheba at the southern boundary. Twenty miles to the north is Hebron, the place of sepulture of the patriarchs. Near Hebron is Carmel, and a little to the east, the town and mountain of Ziph; fourteen miles to the north is Bethlehem, which is six miles to the south of Jerusalem. This city is twelve leagues from Joppa, and nine leagues to the east of it is Jericho. Tekoa, the country of Amos, is twelve miles to the south-east, and here we come to Pentapolis, the region of the Dead Sea, where nothing lives, where bodies that usually sink in water float, lumps of bitumen are found on the surface; here too are found the apples of Sodom 336-339

The Jordan flows into the Dead Sea. Its origin is by two branches from Lebanon, and it passes through the lakes of Gennesareth and Tiberias. North of Jericho is Scythopolis or Bethsan. Westward and to the north of Jerusalem is Anathoth the birthplace of Jeremiah. Thence northward, twelve leagues from Jerusalem, is Samaria, now called Sebaste. North-eastward is the plain of Megiddo; north of this and east of Acon, at seven leagues' distance, is Nazareth. Two leagues further east is Mount Tabor, and the city of Tiberias, and the lake of that name, containing the purest waters of Judaea; it is near but distinct from the lake of Gennesareth. To the north, separated by a desert region, are Bethsaida and Capernaum. East of Acon, and two leagues north of Nazareth, is Cana of Galilee. Still further north is Corazaim. We are thus brought into the region of Lebanon, whence fertilizing waters descend to the neighbourhood of Tyre and Sidon . . 339-343

Beyond Jordan, north of the Dead Sea, is the citadel of Macheron. Here was the tribe of Reuben. North is Pella, at the boundary of Palestine on that side of Jordan. Eastward were the kingdoms of Og and of Sihon. On their boundaries is Ramoth-Gilead, not far from the range of Anti-Lebanon. In this neighbourhood is the city of Damascus, situated four days' journey from Jerusalem . 343-344

VOL. I. PAGES

Summing up these details, we may speak of Syria as including the space from the Euphrates to the Mediterranean, and from Cilicia to Egypt. It is divided into the provinces of Syria Comagena, Syria Caele, Phoenician Syria, and the three divisions of Palestine, Galilaea, Samaria and Judaea. Comage is the capital city of the first, Antioch of the second. Phoenician Syria extends from the south of Lebanon to the north of Palestine. It contains the cities of Damascus, Tripolis, Tyre and Sidon; and beyond the Jordan, Pella and Mount Hermon and Mount Gilead 344-347

The three regions of Galilaea, Samaria, and Judaea belonged to the Hebrews, who held the region beyond the Jordan from the Dead Sea to Mount Hermon. In the north part of Galilee is Decapolis. South of this Ithuraea or the tetrarchy of Trachonitis. The northern part of Galilee contained a mixed race of Jews and heathen. Lower Galilee begins with the lake of Tiberias. South of Galilee is the district of Samaria, and south of Samaria is Judaea 347-348

We see then that the amount of territory possessed by the Jews was but small. From Dan to Beer-sheba is but 170 miles, from Joppa to the Jordan, not seventy. To this, their possessions east of the Jordan are to be added 348-350

Passing to other regions of the world, we find the range of Mount Taurus extending from the Indian Ocean, separating Parthia, Mesopotamia, and Syria, from the Scythian regions, and from Armenia and Cappadocia, and finally reaching Cilicia. In its course this range receives various names, as Caucasian, Caspian, Hyrcanian and many others. Media, Persia, Parthia, lie between the Indus and the Tigris east and west, having the Caspian Sea and Caucasian mountains to the north, and the Persian gulf to the south. The Parthian empire coincides with what was once the Persian 350-352

India lies east of the Indus, bounded on the north by the Seric sea and by the mountains which extend east from the Caucasian range. Vast as the Indus is, the Ganges flowing from the Caucasus into the Eastern Ocean is yet greater. On the Ganges live the Brahmins, living a life of extreme temperance and chastity, and attaining great longevity. North of India is the Scythian ocean and the Caucasian range already spoken of 352-353

Westward from India, at the confines of India and Parthia, we find the Caspian gates on the south shores of the Caspian or Hyrcanian Sea. This sea is entirely inland, formed by great rivers flowing from the north. West of Parthia is Hyrcania, and then the greater Armenia divided by the Euphrates from Cappadocia. The lesser Armenia is identified with Cilicia. This region from south to

VOL. I. PAGES

north is of about four days' journey, bounded by Lycaonia, now called Turkia, and including several ancient provinces as Lydia and Phrygia. The whole of this country is now called by the Greeks Anatolia, otherwise Asia Minor 353–356

Dividing this from Europe is the Arm of St. George, with Constantinople on its left shore, leading to the Pontic Sea, which extends 1,400 miles from east to west. From north to south its narrowest part is between Sinopolis and the province of Cassaria (Crimea), on the east of which is the shallow sea of Maeotis formed by the mouth of the Tanais (Sea of Asof). This river flows from the Riphaean mountains in the extreme north. From the Tanais to the Danube is a vast desert plain, which it takes the Tartar horsemen two months to traverse. It is bounded by Poland and Hungary. To the north of it is Great Russia—bounded to the west by the Baltic Sea, beyond which lie Denmark, Sweden, and Norway. Then across a great expanse of sea Scotland and England lie to the west, and beyond them Ireland 356–358

At the north of the Baltic Sea is Esthonia, eastward is Livonia, and south of it Courlandia, then Prussia and Pomerania. On the confines of Dacia and Saxonia is the port of Lubec. In the Baltic lies the island of Gothland. South of Prussia is Poland; south of this Bohemia, then Austria; west of Austria is Alemannia, France and Spain; eastward is Hungary bounded to the north-east by Albania, which extends as far as the river Don, having Bulgaria, and Constantinople, and the province of Cassaria to the south, and Russia to the north. North again of Russia are the Hyperboreans, a peaceful long-lived race, with a climate far more equable than might be expected from its position. Among these northern nations is a great variety of religions. The Livonians, Curlandians, Prussians, Esthonians are pagans. The Ruscenes are Greek Christians, though in common with the Poles, Bohemians and many others, their language is Sclavonian. The Tartars, who have subjugated almost all nations from the Danube to the extreme east, lead a wandering pastoral life. Some of the tribes subject to them are pagans, others follow the Mahometan law 358–361

On the north-eastern border of the Pontic Sea is the land of the Georgians, and Corasiminians, where of old Amazons used to live. South of these are Cappadocia and Armenia. From the mountains of Armenia flow the Euphrates and Tigris. Here Noah's ark rested. SS. Bartholomew, Judas, and Thaddeus suffered martyrdom here; and at one time there were 800 Christian churches; though Rubruquis found but two, and these small. The city of Naxuana, the principal town of Armenia, now destroyed by the Tartars, was visited by him, as he passed up the river Araxes. In the mountains to the east he

VOL. I. PAGES

found the Alani who are tolerant and orthodox Christians, and east of these the Lelgi who are Saracens 361-364

This brings us to the southern shore of the Caspian Sea, where Alexander erected the Caspian gates, as a protection against the incursion of barbaric tribes from the north. Rubruquis passed through these gates. This being the region of Gog and Magog, who after ravaging the world will become at last the foes of Antichrist, it is of the highest importance to the Church that its geography should be known. The Caspian Sea, extending from west to east, is as large as the Pontic Sea, and to compass it is a four months' journey. Rubruquis traversed its northern shore in going to the Tartar emperor, and its western shore in returning. It has no connexion with the ocean, being formed by the Ethilia (Volga) and other rivers. From the Don to the Ethilia was the land of old called Albania, till lately occupied by the Cumani. East of the Volga also the Cumanians dwelt till dispossessed by the Tartars. Northward is Asiatic Bulgaria, whose inhabitants are of the same race and language as those on the Danube. East of this region is the land of the Huns 364-367

These regions extend as far as Caracathaia or Black Cathay. Here it is that the emperor of Tartary dwells, changing his abode according to the season. It has sometimes been said that Prester John lived here, but his kingdom, formed of a Nestorian tribe, was far more westerly. His power was established on the death of Coir Cham. Prester John was succeeded by his brother Unc, who took the title of Cham, and who reigned for a time in Cara Corum, the principal Tartar city. But Unc Cham was destroyed by the Tartar leader Chinghis. His grandson, the son of Unc's daughter, is Mangu Cham to whom Rubruquis was sent by the King of France. The name given to these Tartars by themselves is Moal. They have by this time conquered nearly all Asia; they hold Russia, Bulgaria, and the neighbourhood of Constantinople 367-371

Eastward of Black Cathay are several tribes with peculiar habits; and beyond them, to the extreme east, is Great Cathay, the nation of the Seres, whence comes silken clothing. The inhabitants of this country are skilful workmen, and acquainted with the medical properties of plants, and carefully study the changes of the pulse and other symptoms of disease. They use paper-money. They are for the most part idolaters (Buddhists); but Saracens are mixed with them, and also Nestorians, who preach and practise a corrupt form of Christianity. Their patriarch professes willingness to acknowledge the authority of the Roman See. Of these various peoples some write from above downwards and from right to left. Others write as we do, and with our letters. The inhabitants of Eastern Cathay paint

VOL. I. PAGES

rather than write, forming groups of letters, each group representing a sentence. The geography of Western Europe does not require any detailed description 371–376

ASTROLOGY.

This geographical description should be followed by an account which I have not been able to complete, (*a*) of the stars, (*b*) of the planets, (*c*) of the way in which the climatic disposition of each place is affected by them, (*d*) thus making true astrological judgements possible, (*e*) with the result of promoting the safety of the state. All that I can do is to offer a few remarks on these points. Each of the fixed stars has properties of its own. Those of the twelve signs are specially important to us. A special treatise would be needed to describe these characteristics adequately and to point out the way in which we are affected by them. Each planet has also its special character, modified by the zodiacal sign in which they are situate, by their aspects, that is by any two or more of them being in the same, in opposite, or in intermediate directions, by the position reached in their epicycle or eccentric, and by the House of the heavens which they may be occupying. This again requires a special treatise . 376–379

From this we pass to the effects produced by the heavens on places and things upon the earth. The sun may be said to beget no less than the parent: and the sun's influence endures through life. Each spot on the earth's surface is the centre of a distinct horizon, and is thus affected differently from every other. We have to consider its distance from the pole and the equator, what stars are in the zenith, and what signs are dominant there. On all these points the Hebrew astronomers are specially instructive. Special parts of the body are affected by different constellations, and by the position of the moon in them. Each hour of the day is under the control of a particular planet. On medical art especially but also on all other departments of human activity, the bearing of all this is obvious. The quarters of the moon must be of course studied, and also the daily transit of the moon through a portion of the Zodiac 379–385

Besides the revolutions of the moon and sun, attention must be given to those of the planets, especially to those of longer period. The comet for instance of 1264 was due to Mars, and was related to the wars in England, Spain, and Italy of that time. All planetary conjunctions are important; but of special importance to man's welfare are the conjunctions of the Moon with stars of various nature. Position in the orbit must always be examined. In the apsis planetary force is greatest: weakest in the opposite point. We

VOL. I. PAGES

must observe too in what house each planet may be; for on this the whole complexion of the year may largely depend. On the foregoing data we form our judgements of events past, present, and future. As a first step to this we must form accurate astronomical tables exhibiting the position of the heavenly bodies at any given date. Comparing these positions with the political events of such a date, we obtain methods for forecasting future events. Our best authorities on this subject are Albumazar, Ptolemy, Haly, and the Hebrew astrologers. A work of Aristotle on celestial impressions should also be translated 385-390

Next comes the question, How far can this knowledge guide our action? We cannot do away with the force resulting from any special position of the stars: but we can modify our own action so as not to be adversely affected by it. If we know that great cold is coming, we can make provision against cold. Similarly the trained astronomer will provide against pestilence, or any other calamity which he sees impending. Of such modifying influences Moses and Solomon have given examples recorded by Josephus. Another instance is the advice given by Aristotle to Alexander as to the treatment of certain perverse tribes: Change their atmosphere and thus change their morals. 390-394

What deters students from the examination of this subject is fear of the imputation of magic. Doubtless astrology may be turned to a bad account; but so may the most useful of tools. Men do not dispense with weapons because malefactors use them, nor cease to go to law because some lawyers are dishonest. So, again, there is a strange force in certain words, uttered with full consciousness of their meaning, which may co-operate with stellar forces. This too has been abused for purposes of magic and witchcraft, and discredit has been thrown on such agencies, which yet in wise hands are useful. Like the stars, the human soul, which is nobler than they, emits forces from itself, which acting in harmony with stellar forces may produce great results. The objection often made that, when the stars act on any object, they must act equally on other objects in the same horizon, is of no account: for no two objects are in the same horizon 394-398

Fascination, again, is a word with a bad sound. Yet important truth is contained in it. Influences radiate from certain persons, which may be good or bad, and which may be concentrated by a strong will and directed to a special object. Words used with this concentrated purpose may have great power. The exercise of miraculous power by the saints has been accompanied by words. We need knowledge of all these forces to resist the coming power of Antichrist. It is certain that the Tartar conquests have been

VOL. I. PAGES

largely assisted by confidence in their astrologers. Should they make common cause with the Saracens, the results to Christendom may be disastrous. The children's crusade in the last century, and the fascination exercised by the Shepherd in our own time, are proofs of what hostile influences may do if not arrested promptly. What is needed is that all these influences, instead of being contemptuously ignored, should be carefully studied, and used under the direction of the Head of the Church as a defence against the inroads of Antichrist 398-403

ANALYSIS OF THE 'OPUS MAJUS.'

PART V.

OPTIC. PART I.

VOL. II. PAGES

In this section there are three parts. The first explains the general principles of vision. The second deals with direct vision. The third with reflected and refracted vision 1-166

In Part I there are twelve Distinctions.

FIRST DISTINCTION, DEALING WITH THE GENERAL CHARACTER OF OPTICAL SCIENCE, AND WITH THE ORGANS OF VISION, PSYCHICAL AND CEREBRAL 1-12

CHAPTER I.

The superiority of vision to other senses has been pointed out by Aristotle and other authors. Vision reveals the difference of things. It supplies a knowledge of the heavenly bodies. It is the channel of experimental knowledge. Hearing can give faith, but not proof. Of the other senses we need not speak. Therefore the sense of vision becomes the object of a special science 1-3

CHAPTER II.

The first thing is to determine the organs of the sensitive soul, which are to be found in the brain. Others have told us that the visual nerves issue from the brain; but the account given is defective. The brain has two membranes—the Dura Mater, lining the cranium; the Pia Mater, enfolding the brain. The brain has distinct chambers, or cells, each of which has its own function. In the anterior part of the first cell is sensus communis. This takes cognizance of, and distinguishes, the impressions brought by each special sense. But it is unable to retain these impressions, being loose and slippery. In the back part of the same cell there is therefore the organ of Imagination, which, being neither too moist or too dry, can retain and store up the material received by sensus communis. The combined operation of these two organs is called Phantasia 4-5

CHAPTER III.

VOL. II. PAGES

The attributes, or properties, of which sense takes cognizance may be reduced to twenty-nine heads. Nine of these are apprehended by some special sense; as colour by sight, heat by touch, sound by hearing. The other twenty may be called sensibilia communia, because apprehended by all, or by more than one, of the special senses. Distance, position, figure, number, motion, rest, are among these . 5-7

CHAPTER IV.

But animals, though they have not intellect, have other faculties than those implied in the apprehension of these twenty-nine properties. First there is the instinct prompting flight from an animal of a dangerous species, or approach to one of its own, independently of experience. Each object in nature has its own constitution or complexion, and radiates corresponding impressions (species) which concur with those of like complexion, or disagree with those of unlike. They operate on special and on common sense, but are taken cognizance of by a higher faculty of the sensitive soul, which we may call the estimative sense, and may locate in the posterior cell of the brain. But as in the case of sensus communis, it does not retain its impressions, and needs another faculty acting as its storehouse, the memorative faculty. Midway between these two divisions or cells of the brain is the cell devoted to the cogitative faculty, the mistress faculty in brutes which have not true reason. By this faculty the spider weaves its web, the bird its nest. In man the rational soul, coming from without, uses this cogitative faculty as its chief instrument 7-9

CHAPTER V.

That we do not find this view of the threefold division of the brain and of animal faculty in Aristotle is due to imperfect translation. It can be got from Avicenna, who has been better translated. And Avicenna is the greatest philosopher since Aristotle. In any case it will not do to confound the faculty which stores sensations with the faculty which stores judgements, under the common name of memory. It must further be stated that while placing these faculties in the brain, the substance of the brain is not itself sensitive, as Avicenna points out, respectfully correcting Aristotle on this point. We must remember however that the sensitive soul has a twofold organ—the brain and the heart. The latter is the true seat of all life, as Aristotle maintains: the former is that which first receives impressions and in which the various operations of the senses are separately manifested 9-12

VOL. II. PAGES

SECOND DISTINCTION 12-18

CHAPTER I.

On the origin of the optic nerves. Clearly the function of vision cannot be understood without a knowledge of the structure of the eye. Most writers on Perspective have ignored this part of the subject, or contented themselves with a bare reference to work on medicine. I hope however to make the matter clear by reference to three authorities—Alhazen, Constantine, and Avicenna. There are two cavities in the fore part of the brain called ventricles, one on the right, the other on the left. From these the optic nerves issue. They meet and cross; the right nerve passing to the left eye, and conversely. In the hollow of the eye the nerve spreads itself out in spherical form. Each nerve consists of three coats, the innermost derived from the pia mater; outside this is one coming from the dura mater; these two are wrapt in a third, coming from the lining of the skull 12-15

CHAPTER II.

Thus the eye itself consists of three coats; containing three humours, and a structure like a spider's web (tela araneae). The innermost is called by Avicenna and others the retina, and is supplied with veins, arteries, and slender nervous fibres; the second, coming from the dura mater, is called the uvea; it has an opening in front through which light passes. Behind, it is highly vascular: in front it is transparent and is called cornea. The third and outermost coat is the sclerotic, which is firm and solid. It is sometimes called consolidativa: it does not extend so far forward as to cover the cornea . . 15-16

CHAPTER III.

The uvea contains three humours, and also a web-like structure, arising from its anterior part. A crystalline or glacial substance fills the greater part of the cavity of the eye; it consists of two parts. The hinder part, in contact with the terminal expansion of the nerve is like melted glass, thence called vitreous; the anterior part, which has received various names, is like ice, hail, or crystal; it is somewhat whiter than the vitreous. Anterior to the web, occupying the space between it and the cornea, is the third humour, somewhat like the white of egg. On the mode in which these parts receive nourishment there is some difference of opinion. Vision is dependent on the crystalline humour, though not so essentially as on the optic nerve 17-18

VOL. II. PAGES

CHAPTER I.

The eye approaches the most perfect of geometrical forms, the sphere. But between the various humours there are important differences of form. The crystalline is a portion of a sphere of different curvature from the vitreous. The centre of curvature of the vitreous is situated further forward on the axis of the eye; that of the crystalline further backward. The cornea has the same centre as the globe of the eye. The same may be said of the liquor albugineus. The centre of the uvea is anterior to that of the cornea . . . 18–20

CHAPTER II.

It must be remembered that, in speaking of these bodies as spherical, we have only to deal with sections of spheres, relating to those parts of the eye concerned with the passage of light. In other parts these structures are not necessarily of spherical form 20–21

CHAPTER III.

The centre of the consolidativa is generally placed further back than that of the other parts. The truth is that it is not of strictly spherical form, neither is its inner surface concentric with the outer. It is of irregular spheroid form, somewhat prolate on the forward surface. But the central points of all these ocular structures lie on one straight line, as the figure shows 21–25

CHAPTER I.

The cornea closes the opening in the uvea, and prevents the escape of the aqueous. Being transparent it allows the passage of luminous impressions. It is strong enough to resist inroads of air and of dusty particles from without, being made up of four layers. The aqueous is also transparent: its moisture prevents the crystalline and iris from getting too dry. The uvea is usually black, so that feeble impressions of light and colour may produce their full effect. It is however, from various causes, sometimes grey. The colour of the eye depends on this structure . . . 26–27

CHAPTER II.

The anterior glacialis [lens] is connected very specially with the function of vision. It is moist, for luminous impressions would not

VOL. II. PAGES

affect dry surfaces so readily; it is of delicate structure, suited to subtle influences of light. It is transparent, yet not too transparent, so that the passage of light may leave sufficient traces to allow of the exercise of judgement. Its surface forms part of a larger sphere than that of the vitreous, as already explained 27–28

CHAPTER III.

The vitreous is denser than the crystalline, so that refraction of the entering ray takes place towards the normal. Through it visual impressions are continued to the expanded fibres of the optic nerve. This nerve is the direct channel of communication with the brain. The consolidativa (sclerotic) is moist so as to adapt itself to the inward structures; firm, so as to keep them in their places; and white, so as to contribute to the beauty of the face . . . 28–29

CHAPTER IV.

The eyelids protect the eyes during sleep, and from invasion of irritating particles; thus the eyes have rest from the toil of receiving impressions. The lashes moderate the intensity of impressions. Creative goodness has supplied two eyes, in case of accident to one; and to add beauty to the face. The eyes are spherical, so as to admit of easy and swift motion in all directions; and also, because had their surface been plane, rays from the points of a large object would not have been normal to the surface. As it is, perpendicular rays impinge upon their surface from nearly a fourth part of the horizon at once 29–30

FIFTH DISTINCTION 30–34

CHAPTER I.

We have now explained the instruments of vision. We pass to the function itself; considering in the first place rectilinear vision. First, we find that impressions must emanate from the visible object, as Aristotle has shown to be the case in every kind of sensation. Vision is a virtus passiva; but the passive must resemble the active, which in this case is the visible object. Thus the species, or impression, is a likeness of the object. These species are thrown out as from a centre along every possible radius. Vision then takes place by the radiation of species, especially by those of light and colour. Colour has much to do with vision. A strong colour impression remains in the eye after the object has ceased to act on it. The species of light are even more essential than those of colour 30–32

CHAPTER II.

VOL. II. PAGES

Vision is not completed in the eye, but in the commissure of the optic nerves within the brain. Here the impressions coming from each eye coincide, so that the object is seen as single. If one of the eyes be forcibly misplaced, this coincidence does not take place, and the impression of duality is produced 32–33

CHAPTER III.

The terminal point of the visual process would seem to be the optic commissure: in the same way as the olfactory bulbs terminate the olfactory process. This in no way conflicts with what has been said as to the location of *sensus communis.* We are here speaking of special sensations. The function of vision then is partially carried on within the eye; and is completed in the point of junction of the optic nerves 33–34

SIXTH DISTINCTION, CONTAINING EXPLANATIONS OF DIFFICULTIES IN THE FUNCTION OF VISION 35–46

CHAPTER I.

A difficulty is suggested by the smallness of the pupil. Although cones of rays come from every part of the object to every part of the ocular surface, yet we need only consider for practical purposes one cone, made up of those rays which are normal to this surface. As to the oblique rays we shall see afterwards. Now it can easily be shown by geometry that from the base of a triangle, however large, any number of lines can be drawn to the vertex, which will all pass through any line subtending the base however short. Matter is infinitely divisible: as many divisions can be made in a grain of millet as in the diameter of the world 35–37

CHAPTER II.

But independently of the cone whose rays fall perpendicularly on the eye, there is an infinity of others whose rays fall obliquely. It would seem that confusion must result. The explanation is that the perpendicular rays come with much greater force, and obscure the effect of the others, as the sun's light conceals the light of the stars. It has further to be observed that of these oblique rays many are made to converge by refraction with those which fall perpendicularly on the central part of the eye. There are thus various degrees in vision. The central rays and those which converge with them are seen well: some rays result in imperfect vision; some in none at all 37–39

CHAPTER III.

VOL. II. PAGES

A third difficulty presents itself. The species from colours must mix throughout the medium; black and white becoming grey, and so on; the mixture of species having the result as the mixture of their originals. The cone reaching the eye would thus present no distinctions of colour. The reply given by Aristotle, Averroes, and others is that these species have only a spiritual existence, and do not observe the laws of material forms. Any number of them therefore may meet in the same point without being mixed. But this view is profoundly erroneous. We shall find the true solution by recurring to what has been said before; the preponderating force of the rays that fall vertically over all the rest. Suppose the object to be black in the centre, white and red at the extremities. The black in their course will come into contact with white rays and red rays: but these latter will be oblique, and will therefore not affect the result. So with the rays from the white and the red parts of the object 39–42

CHAPTER IV.

Those who have spoken of species as spiritual, do not use that word as it is used of God or the Soul: they mean imperceptible. But such a use of the word is confusing. The species resembles that of which it is the image; it is therefore material. It passes through a material medium; is therefore material. It produces material results; being related to that result as the incomplete to the complete, as the embryo to the fully formed animal. Species are indeed invisible, except it be by accident; when an opaque body intercepts the ray of light or colour, we become sensible of it. Or again, weakness of sight may make us sensible of the passage of the ray, though we should not be so otherwise. The same principle applies to objects perceived by the other senses. It may be asked, How is it that three candles placed near a small opening appear as three on the other side? There is true mixture of the species in this case, for one illumination results. Still if the eye be applied to the opening, the distinction of the principal rays proceeding from each will make itself felt 43–46

CHAPTER I.

Importance of the difference in density between crystalline [lens] and vitreous. If the vitreous were of the same density as the lens, the rays passing from the latter to the former would follow their

CHAPTER II.

VOL. II. PAGES

The third condition is the Position of the object in respect of the eye. They must be connected by straight lines either direct, refracted, or reflected. Why is this the case with vision, and not so in the case of hearing, or of the sensation of heat? Aristotle here fails us. The difficulty is obvious. A man can hear his own voice: why cannot he see his own face? It would seem that sound generates something more than species of sound: it must generate sound itself. The trembling of the air following on that trembling of the object struck which constitutes sound, is propagated in all directions, and is therefore perceived as sound everywhere. In the same way an odorous body not merely sends out species, but subtle material particles, which being diffused through the air themselves emanate in every direction. It has been already explained that there are certain qualities capable of completing their species, by generating fresh sources of radiation. Fire is one of those 56-58

CHAPTER III.

The fourth condition of vision is that the object be of suitable Magnitude. Each point of the object must be represented on the surface of the lens. And although that surface, like all others, is infinitely divisible, yet, for purposes of sensation, the division must not go so far that the points become undistinguishable. Again, the object must not be too large. What are the limits? If the extreme angle of vision be a right angle, an eye placed at the earth's centre would see one fourth part of the heavens; at the earth's surface something less, as geometry shows. The view that the limit of the angle of vision must be a right angle rests on no sound basis, either of theory or experiment. Examination of the structure of the eye furnishes geometrical proof that it is less than a right angle. And experiment makes this certain 58-61

NINTH DISTINCTION. OTHER CONDITIONS OF VISION . . 62-74

CHAPTER I.

The fifth condition of vision is that the Density of the object shall exceed that of the air and of the heaven. This is why we see water, glass, and other transparent bodies. Air we can only see when accumulated in great masses, which produces the same result as density. So the most transparent water becomes opaque if the depth be very great, shadows being cast by each particle of water on that which succeeds to it. Further, we have to note that the

VOL. II. PAGES

power of the eye to send out radiations, which we have seen to be an essential factor in vision, is limited. And as such limitation of penetrating power implies visibility, we have in extreme distance a reason why extremely rare bodies become visible. The blue colour of the sky is to be explained as in the case of deep water. Blue is the colour which most nearly approaches blackness; it results here from the shadows thrown by each particle of air on succeeding particles. If the heavenly spheres were opaque, it might be supposed that they would be visible objects. But theological reasoning shows that they are not opaque. Beyond the eight stellar spheres is the sphere of water, and beyond that again the tenth. It may be asked how is it that we see a ray of light passing through a window, though in this case the air being warmed by the sun is rarer than the air surrounding it? The reply is that what we see is not the luminous ray but the boundaries of the denser air which it penetrates. As to the heavenly spheres, it has been already shown that they are themselves transparent, the star placed in each of them being opaque 62–66

CHAPTER II.

The sixth condition of vision is Rarity of the Medium. It is objected that flame is rarer than air, and yet that a flame placed between the object and the eye impedes vision. But it is a mistake to suppose that flame is rarer; and Alhazen, who has been quoted as saying this, is misinterpreted. Whether a lynx sees through a wall or not, the human eye which we are here considering certainly does not. On the other hand, we are not to suppose a vacuum between heaven and earth. Radiation of species would be impossible in that case. A vacuum is a mere mathematical abstraction 66–67

CHAPTER III.

The seventh condition is Time. Time, as Aristotle shows, is needed for an act of memory: much more for an act of sensation. It has been maintained that radiations from the object and the eye take place in an indivisible instant of time. For if not, then the particles of time, however small, would form a perceptible aggregate while the radiation passes from the east of the heavens to the west. This view is held by all but Alhazen, and by him is contested on insufficient grounds. Some however of his arguments against it are sound, resting as they do on Aristotle's doctrine that finite energy requires time for its operation, the time being inversely as the energy. Suppose the energy infinite, then, and then only, does the time become zero. The notion that the ray is spiritual not

VOL. II. PAGES

material, and therefore not amenable to physical laws, has been already disposed of. It is material, and as such cannot be in several places at the same time 68–71

CHAPTER IV.

It may be further observed that the time occupied by a luminous radiation may be so small as to be imperceptible to our senses even when the distance traversed is very great. Aristotle's supposed denial of it was a denial of the view of Empedocles that a corporeal body was transmitted across space. But the luminous radiation is not a body, but a form continually renewing itself out of the particles of the medium, as it travels. It is true, as Aristotle says, that there is a difference between the transmission of light and that of sound and other sensory impressions. With sound there are three distinct displacements of material particles, viz. motion of the body struck, tremor of the air, and rarefaction of the air. Nothing of this kind takes place with light, for though light makes heat and rarefies the air, yet it is transmitted through celestial spaces where rarefaction and heat-production is impossible. In the transmission of odours, again, there are special differences. But all need time: though not the same time. We see the stroke of a distant hammer before we hear it. Lightning reaches us before thunder.

The eighth condition is a healthy state of the eye. A final condition, relating to the axis of vision, will be spoken of afterwards (p. 97) 71–74

TENTH DISTINCTION 74–82

CHAPTER I.

We may now inquire with greater precision of what matters vision takes cognizance, what degree of certainty it brings, and to what errors (limiting ourselves to direct vision) it is exposed. Knowledge is either given to us directly and immediately by Sense, or indirectly and *per accidens*. Supposing the eight conditions described to be present in just degree, the eye will perceive the twenty-two qualities previously noted, with or without the help of the other special senses, and of *Sensus Communis* and *Imaginatio* (pp. 4–5). By qualities indirectly sensible I mean those which come under our cognizance through sensation, but which are appreciated by other faculties, as judgement; as when a lamb seeing an object of the form and colour of a wolf, knows it to be an enemy. When I see a man, I see an animated object and a substance; thus in an indirect way substance may be spoken of as sensible. Again, the sensations peculiar to one

VOL. II. PAGES

of our senses are indirectly appreciated by another, as when we see iron to be hot, or ground to be wet. For cold, heat, moisture and dryness cannot be directly appreciated by the eye . . . 74–76

CHAPTER II.

The question arises, Do all sensible things of which the eye takes cognizance propagate rays to the eye, or light and colour only? The latter is the true view. These other things are magnitudes or properties of magnitudes, and belong to matter, which is passive, not active. The air which is the medium of sound is itself soundless, and the medium of colour, colourless. They are called sensible, not because they emit species to the sense, but because the sense appreciates them. All that is really necessary for vision are the rays of light and colour, combined with the ray directed to the object from the eye 76-79

CHAPTER III.

Knowledge comes to us in three ways: (1) by sensation pure and simple, as when a colour strikes the eye without remembrance of any previous colour; (2) by similitude of the present sensation with a past sensation remembered; (3) by a ratiocinative process, as when we judge an object to be transparent because we are able to see an opaque body behind it. But, although the process by which our perception takes place is one of reasoning, it takes place so instantaneously that we are not aware that we reason. We are logicians instinctively, without having names for the various steps which the mind takes. These three modes of knowledge have been inappropriately named, Sensation, Science, and Syllogism . . 79-82

OPTIC, PART II.

SPECIAL CONSIDERATION OF DIRECT VISION 83–129

FIRST DISTINCTION. ON VISION AS DETERMINED BY THE STRUCTURE OF THE EYE 83-91

CHAPTER I.

Those whose eyes are deep-set see further than those whose eyes are prominent. First, because the eye is nearer to the brain. Secondly, because it is better preserved from external injury. Thirdly, because the visual force being compressed within a narrower channel issues with greater energy and directness. We often prolong this channel by holding the hollow of the hand

from the eye are strong enough to penetrate it when near; not so when distant. We must conceive the pencil of rays from the object and that from the eye as having a common axis drawn through the various ocular centres; this central line being specially potent in vision. The point at which it strikes the object is seen with clearness; adjacent points also, in proportion to their proximity. The axis of the eye is directed successively to each point in the object. The eyes act in concert, their axis being directed to the same point 92–94

CHAPTER II.

When one of the eyes is pushed by the finger from its proper position, the angle of vision ceases to be the same for both of them, and the object appears double. So too when the eyes are fixed on a given point, an object placed between that point and the eye, or beyond that point, will appear double. It is easy to demonstrate this by experiment 94-95

CHAPTER III.

Further experiments on double vision may be made. If the finger be held between the eye and a candle, and the eyes be fixed on the candle, the finger will appear double. If the right eye be closed, the left image will disappear. If however the light be very distant, the result is different. The right image vanishes on closing the right eye 96–97

CHAPTER IV.

Again, the axes of each eye may make such an angle with the common axis of vision as to produce a double image. Or the direction of any point in the object of vision may make such an angle with the axes of the eyes that the point is seen double 97–99

THIRD DISTINCTION 99–129

CHAPTER I.

Let us consider further the three modes of perception, through sense, through recollection, and through argument, in reference to the eight conditions of vision. Light and colour are appreciated through sense only: and this without error, so long as the right conditions of vision are fulfilled without excess or defect. Starlight is not seen by day because of excess of solar light. Diminish the latter, as by descending to the bottom of a well, and the stars are seen. The Milky Way, consisting of a multitude of small stars clustered, produces the erroneous impression of a continuous light; this is due only to distance. This luminous impression is caused by the passage of rays from these small stars through the sphere of fire; the medium in the stellar spaces would be too rare, that of the sphere of air too dense. We have also to take into account the refraction of rays in the sphere of fire; but of this afterwards. The light of dawn raises a question. Why do we not see it earlier? Why, since outside the earth's shadow the whole sky is illumined, does the sky appear dark to us? The reply is, first, that the spheres of heaven are very distant; also that the heaven, apart from the dense bodies contained in it, is too rare to have fixed light; light passes through it without affecting the eye. When the sun's rays touch that portion of air which is comparatively near to us, this illuminated air becomes perceptible. It may happen that a luminous body of small magnitude appears far larger than it is on account of rapid motion. Sparks from a fire produce this effect. Shooting stars are probably bodies of small magnitude. Colour, like light, is apprehended by pure sense; and correct apprehension of it depends on the conditions before mentioned. Extreme transparency of the object, great intensity of the colour, too great remoteness of it, rapid succession of different colours, will interfere with accurate perception 99–104

CHAPTER II.

Of Perception through knowledge (recollection). Under this head are to be classed distinctions of characters, whether general or special, in visible things. The moon's light outside the earth's shadow is clear and white, in the upper part of the shadow red, in the lower part is invisible. These differences, though perceived by sight, become known to us only after repeated observation. The explanation is not easy. The moon's light, like that of other stars, comes from the sun. When the moon is in conjunction, the hindrance to illumination of the part of the moon turned to us is the moon itself; the solar rays diverge widely on either side, and do not reach us, even secondarily, by dispersion. But when the moon is in the earth's shadow, the earth being distant from the moon, the moon is near the vertex of the cone, and accidental solar rays entering within it produce the reddish light seen in a lunar eclipse. The degree of this light will vary with the distance of the moon from the earth at the time of eclipse, which may be greater or less. Another problem, raising a similar difficulty, occurs with colour. If we look through a very thin piece of particoloured cloth at an object of given colour, we may see that colour if the apertures in the web are large enough; if otherwise, we see a mixed colour. The quality of the medium will modify the result.

The rays, in this case, coming from the threads of the web are so nearly coincident with the rays coming from the object, as to be confused with them 104-106

CHAPTER III.

Of Perception through reasoning. Many examples can be given of this. The most striking is our perception of distance. An object may be so distant, as to subtend so small an angle in the eye that vision ceases. Short of that limit, the degree of distance is determined by a continuous series of objects between the object and the eye. In a flat country we have no means of judging the height of the clouds, which we can do when we see them on the sides or summits of mountains. [Clouds would appear to be of no great height, though, as we know from the fact of twilight, exhalations other than clouds may rise fifty-one miles. Such exhalations are not aqueous: being dry they retain the sun's heat better, and thus rise higher.] For judging, then, of distance, we must have an intermediate series of objects, each of which shall be appreciable by the eye with sufficient accuracy. These limits are soon exceeded. A line of trees appears continuous, though there may be a great interval between each of them. So planets seem to be in the same surface as fixed stars, though the difference of remoteness is immense. So an equilateral figure of many sides becomes at a distance undistinguishable from a circle. A circle may be taken for a straight line, a sphere for a plane figure. When a circle is held sideways before the eye, the part nearer to the eye will be recognized as nearer if the distance is moderate: if it is very far off, the difference of distances in the points bears so small a proportion to the whole as not to be recognized. Thus it is that, when the moon is in her first or third quarter, the circular line defining the light part from the dark appears as a straight line. So too the sun and moon seem to us flat, though they are spherical 106–108

CHAPTER IV.

All this is exemplified in the study of the Moon's Phases. The base of the cone of solar light occupying the moon's surface appears to us twice in the month as a straight line; otherwise as curved: a fact unexplained in the Latin translations of Aristotle and Averroes. Here we have to leave the region of sense, and penetrate to the real facts, which, but for the remoteness of the sun and moon, we should be able to see. The boundary of rays proceeding from one eye to the moon is a great circle of that body; that of rays proceeding from the sun to the moon is also a great circle, or nearly so. At conjunction and at full moon these circles coincide, the lunar surface

VOL. II. PAGES

presented to the earth being wholly concealed or wholly illuminated. At other times the illuminated surface is represented by the space left between the intersections of two great circles, the points of intersection being at the extremity of a lunar diameter. Regarding these points as the lunar poles, the inner boundary of the illuminated part may be regarded as a meridian circle. A circle drawn through the lunar poles so as to divide the lunar surface visible to us into two equal parts may be called the lunar colure. The inner boundary of the luminous portion becomes identical with the colure on the seventh and twenty-first day, and appears to us as a straight line. Before the fourteenth day this circle presents its convex side to the sun, after that day its concave side. In the latter half of the month the phases are seen reversed till complete occultation returns.

So too by a reasoning process we become convinced that though the heaven appears to us as a plane surface, it must in reality be spherical, or at least polyhedral, so as to be undistinguishable from a sphere. Were it plane, the stars would all disappear below the horizon or rise above it at the same moment 108–114

CHAPTER V.

As distance is appreciated only through a reasoning process, so it is with magnitude. The mere magnitude of the angle subtended by the object does not suffice, for objects of the same magnitude at different distances appear of different magnitude, and conversely. As in the case of distance, there must be intermediate objects familiar to the eye with which mental comparison can be made. Extreme distance often renders such a comparison impossible . . 114–116

CHAPTER VI.

The eye looking at a sphere sees somewhat less than the half of it, as the third book of the *Elements* shows. Heavenly bodies look larger at rising and setting than when at the meridian. Why is this? When looking either to east or west, the sky seems a plane, or nearly plane, surface stretched overhead. What is over the head seems nearer to us, and subtending the same angle, is judged smaller. Further, owing to the interposition of terrestrial objects between ourselves and the rising or setting star, we are better able to appreciate its distance, and thus are led to imagine it larger. Our perception of motion and rest depends also on a reasoning process. Motion is apprehended by a change in the position of a body relatively to another, the change involving a lapse of time. Rest is the absence of such a change. Hence illusions arise. When clouds are driven across the moon, the moon seems to be moving through them. The clouds must be numerous and almost continuous for this effect to be

VOL. II. PAGES

produced. Again, while we are walking at night in any direction, a star to the right of us will seem to move with us, since it holds very nearly the same position relatively to us at the end of our passage as at the beginning: the star being far off, the difference made by our change of place is imperceptible. When the sun is on the meridian, a long line of men extending from east to west will all see him directly in front and the shadows of their bodies will appear parallel; the divergence of the shadows is too small to be perceptible. The motion of the planets is imperceptible by direct inspection. Rapid motion in a circle, followed by rest, gives the sensation of movement in surrounding objects. In this case when the man stands still, the humours of the eye still continue in motion. In a moving ship, trees and houses on the bank seem to move, especially if distant: if the eyes converge on a near object it will seem stationary 116–119

CHAPTER VII.

One of the most difficult problems is that of the scintillation of the fixed stars. Aristotle remarks their contrast in this respect with the planets, and attributes it to their greater distance; the eye, being more strained, is tremulous. Scintillation is different from the tremor sometimes seen in the sun and other planets at rising and setting. If distance were the sole explanation, we should expect Saturn to scintillate, which it does not. Further, it is only the larger of the fixed stars that do so. Therefore strength of light must be a condition. And yet, since the sun at noon does not scintillate, the light must not be too strong. One of the causes would seem to be the internal strain of the eye at very distant objects. The planetary bodies are easily perceived to be near, and with them there is no strain. Moreover, the fact of extreme distance of itself weakens the visual rays. It may be objected that the strain is greater in the case of small stars than of large: but here the condition of sufficiently strong light is wanting. Again, it is objected that if ocular strain be one of the factors, this depends upon each observer's choice and will. But this is not so. It is one of those actions which have become involuntary through habit. But how is this difference of distance between the planets and the stars to be known? Its quantity doubtless is not known. But the fact that there is a difference is a matter of sensation. Has the density, or has the motion of the medium, anything to do with scintillation? Perhaps both contribute, as Averroes has suggested; and perhaps also at these immense distances the visual power is exhausted, and acting only intermittently, produces the tremulous impression. By motion of the medium we are to understand here, not violent motions of the

VOL. II. PAGES

lower air like wind, but subtle motions caused by the revolutions of the heavens and of the exhalations which pervade them. A combination of the foregoing causes may sufficiently explain this complicated problem 120–126

CHAPTER VIII.

We may now sum up the results of our study of direct vision. We have spoken of vision as operating by sensation, by recollection, and by a syllogistic process. If these two last are to be interpreted as in the schools, they would imply the intervention of Reason. But it is evident that brutes have the power of recognizing and distinguishing both universals and particulars. A dog recognizes a particular man, and also recognizes men in general from dogs, trees or other objects. Therefore this faculty must be a function of sensitive life, and cannot imply reason in the strict sense of that word. Further, animals pass through a train of mental processes analogous to syllogistic reasoning, though they cannot put it into a logical figure. They have a storehouse of mental impressions. They can generalize, and they can draw conclusions; though they are not conscious of doing so, and cannot give an account of what is passing within them . . 126–129

OPTIC. PART III.

FIRST DISTINCTION: ON REFLECTED VISION . . . 130–146

CHAPTER I.

It is unnecessary to repeat what has been already said of the structure of the eye and the functions of sensation. Bodies that impede the passage of visual radiations may be either wholly opaque, as a stone wall; or partially opaque, as water, glass, crystal. Opaque bodies do not destroy the radiation: they merely alter its path. An opaque body may be rough or smooth. When the surface is rough, the parts being unsymmetrical scatter the radiations irregularly, and there is no image. With smooth surfaces the parts all act alike, and the radiation comes back to the eye uninjured, though feebler than in direct vision. Proof is subjoined of the equality of the angles of incidence and reflexion, and this proof will hold, whether the surface be plane or spherical: for the ray may be regarded as falling on the tangent plane 130–132

CHAPTER II.

The mirror contains nothing. What is seen is the real object, only the radiation from it has followed an angular course. The eye

VOL. II. PAGES

must be situated at the extremity of that course, whether it has been straight or angular, in order to receive the impression. It follows that the radiation produces no durable effect on the reflecting surface. From this again it follows that the moon and the stars shine with their own light, not with light reflected from the sun: otherwise we should see the sun's image reflected in them. Neither, again, is a comet the reflexion of solar light from the surface of a star. The case of the rainbow will be considered afterwards. The position in which we judge the reflected object to be is, in plane mirrors, at the intersection behind the mirror of the prolongation of the reflected ray with that of the perpendicular let fall on the mirror from the object. But in mirrors other than plane the apparent position varies very greatly 132–134

CHAPTER III.

There are seven kinds of mirrors, spherical, conical, cylindrical, or plane. Of the first three classes each may be either convex or concave. Geometrical proof is given that in plane mirrors the image and the object appear equidistant from the plane of the mirror on opposite sides. Nothing really takes place either in the mirror or behind it. The effect produced is a mental impression caused by the peculiar path taken by the radiation from the object. In spherical convex mirrors, the intersection of the visual ray with a line drawn from the object to the centre of the sphere fixes the apparent position of the object. This intersection may be beyond the mirror, in the mirror, or on the same side as the object. Usually the image appears less than the reality, because the rays come from a smaller area than in the case of the plane mirror: in the spherical most of them are dispersed. The image is erect but distorted: the outer rays of the cone touching points more distant from the eye than the median rays. Only when the image of a right line crosses the centre of the sphere is that image rectilinear. Moreover, in convex mirrors the image is nearer to the mirror than the object is, because the intersection of the visual ray with the perpendicular occurs sooner in spherical than in plane mirrors. In cylindrical mirrors the errors are even greater than in spherical, except in the one case where a line in the object is equidistant from the axis of the cylinder. Only if this line has any breadth, its transverse magnitude will be distorted. In convex conical mirrors the same errors occur as in convex spherical 134–137

CHAPTER IV.

The greatest illusions are those that occur in concave mirrors, both as to the size of the image, and the number of its repetitions; some

VOL. II. PAGES

which is denser than the air, towards the normal; with the exception of such rays as fall vertically on the cornea. The same point may emit a vertical ray and also an infinite number of oblique rays, these latter co-operating in the act of vision 146–148

CHAPTER II.

The intersection of the perpendicular directed from a point in the object with the visual ray determines the apparent position of that point. If the eye be in the rarer medium and the object be in the denser the point appears nearer than it really is; and conversely 148–149

CHAPTER III.

The case is complicated if the surface separating the two media be not plane but spherical. Here we have eight distinct cases. I. When the curved refracting surface is concave to the eye. 1. The eye being in the rarer medium, (*a*) the eye is between the centre of curvature and the object, (*b*) the centre of curvature is between the eye and the object. 2. The eye being in the denser medium, cases (*a*) and (*b*) as before. II. When the curved refracting surface is convex to the eye. 1. The eye being in the rarer medium, (*a*) the object may be between the centre of curvature and the eye, (*b*) the centre between the eye and the object. 2. The eye being in the denser medium, cases (*a*) and (*b*) as before 150–153

CHAPTER IV.

Examples of the foregoing. The submerged part of an oar illustrates case 1 of ch. 2. If the eye were under water the upper part would illustrate case 2. If an object be placed in a basin, and the observer step backwards till it becomes invisible, the object will again become visible by pouring water into the basin. This again illustrates case 1. The sun and moon when near the horizon, where vapours abound, seem larger than usual. The vapours form a lens with the concavity towards the eye: the eye being between the centre of curvature and the object. The object therefore seems larger and nearer. Why then do the sun and moon not appear larger when they are high up and the sky is cloudy? Because in this case there is no refraction, or but little. So there is no apparent increase of size when, the horizon being free from vapour, there is no refraction. The angle under which the celestial object is seen is of more importance in forming our judgement than the apparent distance, since the absence of intervening objects makes judgement of distance difficult. A crystal lens if consisting of a small portion of a sphere, with the convexity turned to the eye, will magnify small objects placed beneath it, in

PART VI.

VOL. II. PAGES

EXPERIMENTAL SCIENCE 167–222

CHAPTER I.

Having laid down the general principles of wisdom so far as they are found in language, in mathematics, and in optics, I pass to the subject of experimental science. There are two modes of acquiring knowledge—reasoning and experience. Reasoning guides us to a sound conclusion, but does not remove doubt from the mind until confirmed by experience. A man who has never seen fire may read the proofs that fire burns, but will not be satisfied of it till he has been burnt. Even in geometry the demonstration of the first proposition of Euclid fails to carry conviction till the figure has been inspected. When Aristotle speaks of knowledge of the cause as a higher kind of knowledge than that gained by experience, he is speaking of mere empiric knowledge of a fact; I am speaking of experimental knowledge of its cause. There are numerous beliefs commonly held in the absence of experiment, and wholly false, such as that adamant can be broken by goats' blood, that the beaver when chased throws away his testicles, that a vessel of hot water freezes more rapidly than one of cold, and so on. Experience is of two kinds: (1) that in which we use our bodily senses aided by instruments, and by evidence of trustworthy witnesses; and (2) internal experience of things spiritual, which comes of grace, and which often leads to knowledge of earthly things. The mind stained with vice is like a rusty or uneven mirror, in which things seem other than they are. Without virtue a man may repeat words like a parrot, and imitate other men's wisdom like an ape, and all to no purpose. The intellectual effect of a stainless life is well illustrated in the young man who is the bearer of this treatise. The degrees of spiritual experience are seven. (1) Spiritual illumination; (2) virtue; (3) the gift of the Holy Spirit described by Isaiah; (4) the Beatitudes; (5) spiritual sensibility; (6) Fruits, such as the peace of God which passes understanding; (7) states of Rapture 167–172

CHAPTER II.

It is solely by the aid of this science that we shall be able to disabuse men of the fraudulent tricks by which magicians have imposed on them. As compared with other sciences, this science has three characteristics ('praerogativas'). Of these the first is, that it constitutes a test to which all the conclusions of other sciences are to be subjected. In other sciences the principles are discovered by experiment, but the conclusion by reasoning. An instance of this is afforded by the rainbow, and by other phenomena of a similar kind, as haloes, &c. The natural philosopher forms a judgement on these things: the experimenter proceeds to test the judgement. He seeks for visible objects in which the colours of the rainbow appear in the same order. He finds this the case with Irish hexagonal crystals when held in the sun's rays. This property, he discovers, is not peculiar to these crystals, but is common to all transparent substances of similar shape, similarly placed. He finds these colours again on the surface of crystals when slightly roughened. He finds them in the drops that fall from the rower's oar, when the sun's rays strike them, or from a water-wheel, or in the morning-dew on the grass. They may be seen again in sunshine when the eye is half opened, and in many other cases 172–174

CHAPTER III.

The shape in which the colours are disposed will vary. Sometimes it is rectangular, sometimes circular 174

CHAPTER IV.

Armed with these terrestrial facts, the experimenter proceeds to examine the celestial phenomenon. He finds, on examining the sun's altitude and that of the summit of the bow, that the two vary inversely. The bow is always opposite the sun. A line may be drawn from the centre of the sun through the eye of the observer and the centre of the circle of which the bow is an arc to the sun's nadir. As one extremity of this line is depressed, the other is elevated. It becomes thus possible to compute the altitude of the sun beyond which no rainbow is possible, and also the maximum altitude of the bow. It will be found both by calculation and experience that this altitude in the latitude of Paris is forty-two degrees 175–178

CHAPTER V.

Still further investigating the shape of the iris, and the portion of it that can be seen, the experimenter conceives a cone of which the apex is the eye, the base is the circle of the iris, the axis being the line already described drawn from the sun's centre through the eye to the sun's nadir. In cases where this cone is very short, the whole of the base may be above the horizon, as may often be seen in the spray of a waterfall. In the sky however the cone is too elongated to admit of this: the base is bisected in various proportions by the plane of the

VOL. II. PAGES

horizon. The arcs visible are not portions of the same circle. When the sun is high, and a small arc is visible, it belongs to a larger circle than the arc seen when the sun is rising or setting. A bow can be seen when the sun is just below the horizon; but owing to terrestrial vapours, only the crown of the arch is usually seen . . . 178–181

CHAPTER VI.

In some latitudes there can be no rainbow at noon even in the winter solstice. When the latitude (i. e. the distance from the zenith to the equator) is 24° 25′, the sun's altitude at noon in the winter solstice will be 42°, therefore there can be no bow. Passing north from this latitude, there can always be a noon rainbow till we come to latitude 66° 25′, when at the winter solstice there is no sun. Similar calculations can be made for other latitudes 181–185

CHAPTER VII.

We have now to inquire whether the iris comes from incident, reflected, or refracted rays. Is the bow an image of the sun? Are the colours on the clouds real? Why is the iris of circular form? Here we call experiment to our aid. We find on trial that if we move in a direction parallel to the rainbow it follows us with a velocity exactly equal to our own. If we approach it, it recedes: if we recede, it follows. The same phenomenon occurs with respect to the sun. We have seen that the sun is always opposite the rainbow; the line between the centre of the bow and the centre of the sun passing through the eye of the observer. If the sun were apparently stationary, this would involve the bow moving much faster than the observer, the latter moving through the same angle, but at less distance from the apex. But this is not so. Therefore there is an apparent motion of the sun concurrently with that of the bow. The case is analogous to what happens when a hundred men are ranged in line facing the sun. Each sees the sun in front of him. Their shadows seem parallel, though we know that in reality they must diverge, yet owing to the vast distance of the sun this divergence is imperceptible. We are thus brought to the conclusion that, supposing a rainbow to occur, each of the hundred men, facing backwards, would see a different rainbow, to the centre of which his own shadow would point. The rays causing the iris are therefore not incident rays, otherwise the colour would appear fixed in the cloud. And for the same reason they are not refracted rays, for in refraction the image does not follow the change of place of the observer, as is the case here. One condition of the phenomenon is that the atmosphere shall be more illuminated at the standpoint of the observer, and less

VOL. II. PAGES

These colours appear to have some relation to the various structures of the eye. In addition to the problem of the rainbow, there is the problem of haloes and coronae. On this I give the best explanation that as yet occurs to me. I do not however pretend that it is satisfactory. Far more careful experiments, made with properly constructed instruments, are needed before an adequate explanation can be given 196–201

THE SECOND PREROGATIVE OF EXPERIMENTAL SCIENCE . 202–215

In all sciences Experiment is able to reveal truths quite unconnected with the discussion of principles, and with regard to which it is useless in the first instance to assign a reason. The initial state of mind should be readiness to believe; this should be followed by experiment: reasoning should come last. I subjoin examples of my meaning.

1. The astronomer constructs his spherical astrolabe, by which he can observe the precise longitude and latitude of heavenly bodies at different times. But it is not inconceivable that experiment may devise means of bringing this instrument into such relation with the revolution of the heavens that it should follow their course. The motion of the tides, the periodic changes in certain diseases, the diurnal opening and closing of flowers, are facts tending to belief that such a discovery is possible. If effected it would supersede all other astronomical instruments 202–203

2. My next example relates to the act of prolonging human life. As yet we have nothing to rely on but ordinary rules of health. These are observed but by few, and usually not till the close of life, when it is too late. If a suitable regimen were observed by all, no doubt life would be much prolonged. But there are special remedies unknown as yet to medicine, but to be found by experiment, which may extend the period of life much further. Observation of the habits of certain animals may guide us to truths on this matter which are as yet hidden. Other indications are given in the works of Aristotle, Pliny, Artephius, and others. A combination of gold, pearl, flower of sea-dew, spermaceti, aloes, bone of stag's heart, flesh of Tyrian snake and of Aethiopian dragon, properly prepared in due proportions, might promote longevity to an extent hitherto unimagined 204–208

3. A third example may be found in Alchemy. The problem here is not merely to transmute the baser into the more precious metals, but to promote gold to its highest degree of perfection. In this perfected gold we should probably have a further aid to the prolongation of life 214–215

In this we leave altogether the domain of the sciences now recognized, and open out entirely new departments of research. At present the influences exerted on us by the stars can only be known through difficult astronomical calculations. Experimental science may enable us to estimate them directly. It may be possible for us to act on the character of the inhabitants of any region by altering their environment. Inventions of the greatest utility may be discovered, as perpetual fire, or explosive substances, or modes of counteracting dangerous poisons, and innumerable other properties of matter as yet unknown for want of experiment. The Magnet, of which use is already made, is but a type of other mutual attractions exerted by bodies at a distance. For instance, if a young sapling be longitudinally divided and the two divisions be brought near together, held each by the middle, the extremities will bend towards each other. In conclusion, I may point out the influence which the possessors of this science may exercise in the promotion of Christianity among the heathen, whether in subduing their pride, in disabusing them of false beliefs in magic, or in overcoming their material force . 215–222

PART VII.

We have now considered Philology, Mathematics, and Experimental Science, and have seen their intrinsic importance, and their value to the Church. I come now to a fourth science, which stands on a higher level; that which relates to the practical conduct of life. Other sciences relate to action of various kinds; but this to those actions by which we become good or bad. It is the science which instructs Man as to his relations to God, and to his neighbour and to himself. It deals therefore with the final purpose of all human wisdom. It is closely related to Theology, to which it supplies important aid, sharing therefore its dignity. Proceeding to the divisions of the subject, the first point to note is that the conclusions of the previous sciences form the principles of ethical science. The others have prepared the way for this science as their mistress. We find ethical principles everywhere diffused through them, which are now by this highest of the sciences to be gathered together and find

VOL. II. PAGES

their proper place. It is called by Aristotle Moral Science; by others Civil Science, as laying down the obligations of citizens and states; the city standing in old times in the place of the modern state or empire. It falls under two divisions: first, the establishment of laws of conduct; secondly, exhortations towards their fulfilment 223–225

The first division consists of three parts: (1) duty to God; (2) duty to our neighbour; (3) duty to ourselves. These three divisions are indicated both in the Old and in the New Testament. We may begin by laying down certain principles either held by this science in common with Metaphysics, or reached through metaphysical methods which here would be out of place. These are, (1) that God exists; (2) that He is naturally known to man; (3) that He is of infinite power and wisdom; (4) that He is One; (5) that He is also Trine; (6) that He is the author of Nature; (7) of angelic substances; (8) of human souls; (9) these are immortal; (10) the highest good is in the future life; (11) man's capacity for this good; (12) the moral government of the world; (13) future rewards and punishments; (14) the worship of God; (15) man's duty to his neighbour and to himself; (16) the need of revelation; (17) of mediation between God and man 225–228

There are other principles of which Metaphysics can take no cognizance—relating to the nature of God and the angels and to future life. Of these the first is the Triune nature of God. Something of this has been revealed to the ancient philosophers, having been received by them from the patriarchs, as explained in the second part of this work. Plato and Porphyrius are instances of this; examples may be found also in Aristotle and Avicenna 228–232

The mutual relations of the three Persons of the Trinity constitute the first foundations of Moral Science. Next to this is the Incarnation. Here too many traces of knowledge of the truth are discoverable in the ancient philosophers, both Greek and Arabian. There are also facts in natural history pointing in the same direction . 232–234

The coming of Antichrist is also an article of faith, of which some knowledge is to be observed in the writings of Greek and Arabian philosophers, and which some have thought is to be fulfilled in the actual Tartar invasions, though this is uncertain. It will be followed by the final punishment of the Evil one 234–235

The creation of the world and of the human race is also a principle laid down in the writings of Aristotle, Albumazar, and others. We find also the doctrine of the existence of angels, good and bad. The former direct the motions of the heavenly bodies; and one of these is specially attached to the destiny of each human being as his guide through life. Bad angels may tempt him to evil . . 235–238

VOL. II. PAGES

Of the immortality of the soul much is to be found in Aristotle, Plato, Cicero, Hermes Mercurius, and Avicenna; and they were not ignorant that the body must be regarded as inseparable from the soul. They had knowledge also of a future state of rewards and punishments; though, as Avicenna insists, the cares, enjoyments, and occupations of this world hide this knowledge from us. In shaking off these earthly trammels and receiving spiritual enlightenment true wisdom consists, which Aristotle in the sixth and tenth book of his Ethics holds up as identical with true happiness. Theophrastus his successor, and Cicero, confirm what he has said. Thus that highest good of which Aristotle speaks consists in participation of the life of God. Nor were the ancient philosophers unaware of the future misery that awaited the bad, as the writings of Socrates, Cicero, and Hermes show 238–246

We now pass to the obligation to worship God, in thankfulness for our creation, in reverence for His infinite power, in consideration of our future bliss. On these points Avicenna, Porphyry, Plato, and Cicero have spoken explicitly. St. Augustine has accepted Cicero's explanation and definition of religion. Avicenna has compared our approach to the presence of God to our entrance into the presence-chamber of a great king. And Hermes has enlarged eloquently on this subject. As to the ceremonies of heathen nations, they were for the most part superstitious and useless, and there is no need to dwell upon them. They were practised by these ancient writers not for their intrinsic value, but in order to conform to popular prejudice and custom. 246–249

SECOND PART: CIVIC MORALITY 250–253

CHAPTER I.

The first subject under this head is the regulation of the propagation of the race by the law of marriage. Next comes the subordination of ranks, both in the state and in the family. Thirdly, the appointment of educators and judges. A definite position and function should be allotted to each citizen. Provision should be made for the treatment of criminals, and for the replenishment of the treasury. Laws must be made for testamentary dispositions and for contracts of all kinds. Occupations injurious to the state must be prohibited. The defence of the country by an organized force must be secured . 250–252

CHAPTER II.

It is further necessary that orderly succession in the government should be arranged. The ruler should choose his successor with the

VOL. II. PAGES

consent of his nobles and of the people. Any subsequent pretender should be outlawed. These are the principal divisions of the subject; which however includes the whole of Civil Law as received among the Latins, to whom the Greeks transmitted it 252-253

THIRD PART: PERSONAL MORALITY 254-365

CHAPTER I.

This third division is subordinate to the two preceding. Our duty to God occupies of course the first place. Secondly comes public good; which takes precedence of private good, as Aristotle has maintained. Love, Peace, Justice fall in this second division. Man is a social animal. A hermit living by himself is neither good nor bad. We are born to make our lives useful to others, as Cicero, Seneca, and the Stoics have so often said. On the subject of personal conduct and character admirable truths have been laid down by heathen writers, which may put Christians to shame. We will begin with those relating to virtue and vice in general; passing afterwards to special branches 254-255

CHAPTER II.

Aristotle has defined two kinds of virtue. The first consists in the subordination of feeling to reason; to this the name of moral virtue is appropriated. In his *Ethics* he has spoken of twelve moral virtues, each regarded as the mean between two opposite vices. These are fortitude, chastity, liberality, munificence, magnanimity, public spirit, gentleness, friendliness, sincerity, gaiety, modesty, justice. The second kind of which Aristotle speaks is virtue of the Intellect. Its branches belong to the region of speculation, except so far as they are directed to the knowledge of divine things, or of matters useful to the state. Aristotle has distinguished innate from acquired virtue. All the philosophers of antiquity speak of virtue as the only real good, notably Seneca, Apuleius and Cicero, and have also insisted on the inconsistency of those who theorize about virtue without practising it. They have spoken of it as the beauty and the health of the soul, and have pointed out that it is only to be acquired by long and arduous practice. By this means even inveterate vices may at last be eradicated 255-262

CHAPTER III.

The ancients have spoken of vice in the same spirit as of virtue. Algazel has said that vice acts on the soul like rust on a polished

VOL. II. PAGES

mirror, hindering the perception of the highest truth. Sin blinds the soul; and hence the view of Socrates that evil actions result from ignorance. There is a natural shame at sin for its vileness; a consciousness that it lowers to the level of the beasts, each race of whom typifies some one among human vices. We take our tone from those around us for good or evil: hence it is of the utmost moment with whom we associate. This truth also is illustrated by the lower animals, as Boetius has pointed out. Much importance is attached by Seneca and others to self-examination at the close of each day, as to the control or the encouragement that has been given to vicious impulse. Such control, as Aristotle and Seneca insist, should be exercised from childhood 262–266

CHAPTER IV.

From virtue and vice in general we pass to special virtues and vices. We find that the teaching of ancient philosophers relates to Avarice, Pride, Lust, Gluttony, Anger, Envy, and Sloth—in other words, to the seven mortal sins. Of these, all but one, Anger, relate to pleasure and prosperity; Anger only has to do with adverse circumstances. Let us begin with the former class. Disregard of wealth was preached and practised by Aristotle and a long series of philosophers. Seneca and Ptolemy have pointed out that this earth is but an infinitesimal part of the universe, and that the disputes of contending nations for its possession are like the quarrels of ants for an anthill. Apuleius, in his study of Socrates, has insisted strongly on the distinction between the inner man and the casual accidents of health, strength, or wealth that surround him. Sallust and Seneca have spoken of the general corruption that has followed the pursuit of wealth 266–270

CHAPTER V.

As to sexual desire, Archytas and Cicero have said much of its disturbing influence on reason. Aristotle in his exhortations to Alexander dwelt on this, and on the degradation of man through lust to the level of the brutes. Seneca and Cicero have dwelt on the same subject. On Gluttony and Drunkenness, Seneca has enlarged in his discourse to Helvia, and in his other letters. He speaks of it as throwing the door open to lust, cruelty, and other vices. Cicero quotes a letter of the philosopher Anacharsis upholding the advantage of simplicity of life. Plato, Avicenna, and others have said much of the necessity of setting the mind free from the trammels of the body 270–275

CHAPTER VI.

VOL. II. PAGES

Having spoken of the vices connected with pleasure and prosperity, we pass to the discomfiture of the soul in adversity by Anger. In resisting anger, the first step is to see it as it really is, with all its disastrous consequences to spiritual and temporal welfare. By nature man is mild and gentle, inclined to help his fellow-man. Giving way to anger he exhibits the physical symptoms of wild beasts, distortion of the countenance, agitation of the limbs, swelling of the veins, foaming at the mouth, gnashing of the teeth, spasm of the breath. The effect on the rational faculties is no less disturbing. It has often utterly destroyed the mental balance, and the temporary insanity caused by it has become permanent. No less pernicious is its effect on practical conduct. All the virtues are inter-connected, so that the loss of one involves injury to the rest. Anger is destructive to the noblest virtues, such as clemency, magnanimity, pity, natural affection, tranquillity and joy. Clemency is specially characteristic of man, and to the rulers of men it is peculiarly appropriate. It may be noted that among bees the king alone is without a sting. The noblest of the Roman emperors have been distinguished for this virtue. Closely connected with it is the virtue of magnanimity, which leads its possessor to forgive injuries, and to be unconscious of their existence. But this state of mind is wholly incompatible with anger. As that part of the heavens which is nearest the stars is free from clouds, so is the magnanimous spirit free from the disturbing influences of anger. The swelling and exaltation of spirit which anger produces is no sign of vital energy, but rather of morbid flatulence and weakness. Besides magnanimity, other virtues are impaired by anger, as mercy, patience, and joy. Mercy is that which lifts man nearest to the level of God. Nothing is nobler than forgiveness and forbearance, and readiness to find extenuating circumstances in every offence, remembering how prone we are to give offence ourselves. To avenge ourselves on brute animals is what no one thinks of: and should we be less indulgent to men than to brutes? Readiness to forgive a personal injury is stimulated by cultivating indulgent feelings towards human nature. But such a state of mind as this is wholly incompatible with anger. The destructive effects of Anger on such virtues as Piety and Peace are too obvious to dwell upon . . . 275–288

CHAPTER VII.

In addition to these things, anger makes a man reckless of his own life and that of his friends, as there are many signal examples in history to show. Under its influence a man becomes careless of wealth or reputation, and is led to blasphemous rebellion against

CHAPTER VIII.

CHAPTER IX.

CHAPTER X.

VOL. II. PAGES

that God sends these sufferings to strengthen their character. In the training of children the father has more regard for vigour of character than the mother, and acts accordingly. So it is with God. Those whom He loves He chastens, lest through ease and comfort their forces should decay. There can be no proof of heroism except wrestling with calamity, therefore those are to be pitied who have no such trial of their force. A brave man is proud that his general sends him to the post of danger. The sailor, the ploughman, are hardened by their calling: and it is so with character. Brave men face their trials willingly, and are purified by them as gold in the furnace. Thus the true blessings of life are given; the false appearance of happiness is seen for what it is, a superficial veneer veiling the vice and worthlessness within 298–303

CHAPTER XI.

Seneca proves further that the wise man cannot sustain injury or insult. That he will be attacked is certain, but he will not give way to attack; it will not injure him; he will be as asbestos in the flame or rocks in the sea-foam. We may distinguish injury from insult; the former being the graver, though the latter is often more dreaded. The injurer seeks to inflict hurt. But for the wise man there is no hurt but sin. What is external, what is inflicted by fortune, cannot touch him. He can lose nothing: for what is truly his own he carries with him. It may be urged that such a man is impossible. Rare he is doubtless: but Cato proves the possibility. Injury can only be inflicted by the strong on the weak; and how can the bad man be stronger than the good? Obviously the bad man may intend a wrong: but it does not follow that the wise man will suffer it. He is beyond the reach alike of injury or benefit; he is raised to the level of God, except that he is mortal. He bears all assaults gladly, looking on them as trials of his strength. Insult is a smaller matter, to be complained of rather than avenged. It implies some mistrust in the recipient of his own worth, from which a wise man would be free. We are not to suppose him insensible. Bodily pain, the loss of friends, political disaster, he will feel like other men; but of mere insult he will be simply unconscious, or will treat it as a passing dream. He will not take offence at it any more than a mother at the petulance of her child. For those who would insult him are but grown-up children. If he restrains and punishes them, it is not because himself has been hurt, but for their own welfare. The physician is not angry with his delirious patient. If rich men pass him by unnoticed, he reflects that they are but beggars or worse, since the beggar wants little, they much. It would be doing them too much honour to be offended by them. The current of opinion is nothing to him: he

VOL. II. PAGES

honour in bearing it. It is hard to be poor. All depends on who is the poor man. The birds and beasts do not feel their poverty. The rich man is but a money-chest. The crowds that follow him are but as flies looking for honey, or wolves round a carcass . 318–319

CHAPTER XIV.

Writing to a lady who had lost her son, Seneca continues, The mastery of life is only to be seen when trial tests it. Your son stood this test nobly. Natural affection has its natural limits. The grief of animals for the loss of those dear to them lasts but a short time. Men alone grieve long; and even with them it is the weaker sex, or the more barbarous tribes, who are crushed by it. Was this sorrow unexpected by you? Remember that none are exempt from it. Of no earthly possession have we a secure hold. We have them as actors have their stage properties; the scene ended, they pass back to the manager. What we have we should be ready to give up at a moment's call. In loving it think of it as even now passing away. Why weep one part of life when the whole of it is sad? Man is but a frail helpless thing that the most trivial accidents may destroy. His longest life is but a few years, and may be cut short at any moment. There are noble instances of constancy under such calamities as yours, as of the priest engaged in sacrifice who, hearing of his son's death, went on calmly with the service. Cornelia, mother of the Gracchi, Livia and Octavia the sister and the wife of Augustus, bore their bereavement nobly. Reflect further that life is so uncertain that the sudden close of it may mean avoidance of calamity which would otherwise have followed. Nothing is sure but what has been already done. The sooner we are disentangled from earthly things the speedier will be our flight to the realms above . . 319–322

CHAPTER XV.

My reasons for treating this third part of Moral Philosophy at such length are two. First, that although we are far in advance of heathen nations in knowledge of spiritual life, we are far behind them both in word and deed, as to public and private life in this world. Secondly, the works of Seneca from which I quote have been hitherto unknown to me, in spite of a long search for them, and probably to others 322–323

CHAPTER XVI.

I proceed therefore with selections from his Essay on the Brevity of Life. Men complain that life is short. It would be long enough if

VOL. II. PAGES

they spent it wisely. But they waste it like spendthrifts, in luxury, vice and sloth, or at the best in restless activity. Men cherish a dream of retirement and quiet reflection at the close of life. But they put off realizing it till too late. We know how Augustus longed for such a time of peace. The art of life is no easy matter to be learnt in a moment. Life itself is not too long for it. The wise man knowing this guards his time against the encroachments both of pleasure and business. Thus whatever his span of life he truly lives. For length of years is no proof of life. The shipwrecked sailor, tost hither and thither, has not made a voyage. Men crave for life when they see the end of it at hand. But while it lasted they lavished it recklessly. Much too is wasted in schemes for the morrow, the immediate work before us being left undone. Time must be seized as it flies. Of the three parts of time, past, present and future, the first alone is our inalienable possession. Yet how few care to look at it. The ambitious, cruel or intemperate man hates his memories. The one part of life which is really his own he flees from ; and yet men cling to life when it threatens to escape them. Few understand what true leisure is. Most men's pursuits even when harmless are but laborious futility. Learned men occupy themselves with the collection of trivial and useless facts : [though be it said in passing that sometimes these facts have philological interest ;] wise men have the whole expanse of history open to them. They can escape from the littleness of the present to companionship with the great thinkers and teachers of past ages. All time belongs to such men. They recall the past : they act in the present : they arrange the future. They see things to which other men are blind, and they escape other men's miserable anxieties as to what may befall them. It is well therefore to spare a few years for quiet meditation before our faculties are enfeebled 323–333

CHAPTER XVII.

It is well for us not to be exempt from the common lot. Great wealth is great slavery. Life at the best is but a stormy voyage, and the only haven is death. Life has been lent us ; let us return the loan ungrudgingly. And for the death of those dear to us, though we must feel like men, let us also bear like men 333–335

CHAPTER XVIII.

I now pass to Seneca's discourses on the Blessed Life, and on Peace of Mind ; for they are well adapted to strengthen the spirit and render it careless of outward prosperity. Men wish for happiness, but know not where to find it. It is worse with them than with travellers who have lost their way. These can be set right by natives of the place. But in life it is the broadest and most trodden road that leads farthest astray. We must leave the crowd if we would go right. We must look within; the mind must be a law to itself. Nature must be our guide. If our mind is attuned to nature it will not fail us. We shall thus substitute for the distractions of pleasure and fortune the joy and tranquillity of inward peace. He is to be called truly blessed to whom the only good or evil is good or evil of the soul, the only pleasure, to be free from slavery to pleasure. Virtue is undying, pleasure perishes in the act of fruition. Pleasure may be the lot of good men and of bad: but for the good it is a servant, not a master; they are undisturbed by its absence; like the universe, like the Deity, they retain self-mastery. They live and act without strife, without vacillation, in inward harmony and peace. Yet virtue is not to be sought for the joy that it brings, any more than fields are tilled for the flowers that may spring up amidst the corn. It is an end in itself; there is nothing beyond it. With vicious pleasure virtue is incompatible. Lawful pleasures it accepts, as using but not abusing them. Unless raised above them, the wise man could not withstand the trials of life. Pleasure is but an addition to his life, not the governing principle. To confound these things, to regard pleasure and virtue as inseparable, is a fatal error, leading inevitably to the triumph of the baser element. There must be no compromise. The highest good ceases to be the highest if mixed with something alien to itself. The very foundation of virtue is undermined when it is made to depend on outward circumstance. It will no longer inspire deeds of heroism and self-sacrifice. When such are called for, they will be done unwillingly, and not with the prompt obedience of the faithful soldier. The reward of our fidelity is true freedom. Nothing can be done to us against our will, for our will is at one with that of God. But who, it may be asked, comes up to this standard? whose life is not utterly inconsistent with it? Admit this, and yet we need not be deterred from aiming at it. It is best to have high aims. If we resolve to be fearless in the presence of death, not to be the slaves of fortune, to acknowledge our bond to our fellow-men, to act when alone as though the world beheld us, to set restraint on appetite, to bow in all things before God's will, we are at least on the upward path, though we may not attain. Those who scoff at us will profit little. A man may be wealthy and prosperous, yet good. He may use his wealth to relieve the needy or to support the state. He will not throw it away; yet, if it be taken from him he will be content. The difference between him and others is that his riches belong to him, they belong to their riches. If innocently gained they cause him neither pride nor shame. He will hold them lightly, and will use

VOL. II. PAGES

them worthily. We must distinguish between those who are struggling towards true wisdom, and those who have attained it. We can at least resolve not to bow down before wealth as a blessing, not to be the slaves of it, always to hold ourselves in readiness to abandon it. Strange indeed is the tendency in bad men to malign men better than themselves, instead of attacking one another; to pick out the specks in these, heedless of their own hideous eruptions. Yet they only bring into greater prominence the virtue which they assault 335-347

CHAPTER XIX.

We must pursue our steady course, not diverted from it by the varying blasts of opinion. We shall incur the reproach that a philosophic life involves abandonment of the duties of a citizen. But in retirement we remain active, still working at what will be of use to those who come after us. The turbulent politics around us are not our only field of action 348-349

CHAPTER XX.

I conclude with extracts from Seneca's discourse on Peace of Mind, without which there can be no happiness. We are as ready to flatter ourselves as others. We are too timid to tell ourselves the truth. Yet without uprooting our secret vices there can be no inward peace. Sometimes we yield to the desire for incessant change; sometimes we sink into inertness. Or we have not attained full mastery over our passions, and are kept in perpetual struggle; or again we are still tormented by ungratified ambition. These disappointed hopes prey on the mind, and we become envious of the success of others. We shift and turn in hope that change may bring relief, we lose all power of forbearance, and end with disgust for life and the world. The remedy for this is not, as Athenodorus advises, to re-enter the arena of public life, but to choose such occupations in retirement as shall make our lives useful to others. We may serve the state by advising and directing the young, as well as by holding public offices. We are not to flee from men and live only for ourselves. We remain citizens; but of the world, not merely of a single state. We act on those around us by the influence of example. Socrates under the Thirty Tyrants was a tower of strength to his fellow citizens. It is well carefully to measure our own powers, which we are apt to overrate. Defects of temper, health, fortune, often unfit us for public life, and point to retirement as the better course. We should not begin what we cannot carry through. Great care should be taken in the choice of friends; not insisting on perfection, but above all avoiding the

VOL. II. PAGES

querulous. Consider next the question of wealth, perhaps the most fruitful source of our miseries. The less wealth we start with the less we can lose: and the rich bear losses quite as badly as the poor. Of this Diogenes was well aware, when he took to a life of voluntary poverty. There can be no more doubt of his happiness than of the happiness of God. When his one slave left him he merely remarked that if the slave could do without him surely he could do without the slave. Our wisdom is to reduce the number of our wants, to be content with a moderate fortune, and to live well within it. Even in intellectual pursuits there should be the same moderation. Vast collections of books are of small avail, monuments of luxury rather than aids to study. In all the perplexities of life courage first, then custom, will teach us to bear the yoke. Custom is nature's sovereign remedy for all cares and sorrows. Life, whatever its station or degree, is a servitude. Custom adapts us to it and we cease to complain. Let us leave the unattainable, and attempt things within our reach. The lofty are liable to the deepest fall. These counsels are for ordinary men, not for the man of perfect wisdom. For he can walk more boldly: to him not fortune only, but his own life, is but a temporary loan, which he is ready to surrender at call and gratefully. The test of a good life is readiness to die. Knowing that he was born into a world of trouble, no evil can take him unawares. Let us avoid the crowds of men who rush hither and thither like insects on the trees in restless and purposeless activity. Idle conversation is apt to be worse than idle. Let us refrain from too many occupations; our disappointment will be the less. Let us take the work that comes to us, neither too eager nor too fickle. Let us possess our souls in patience, all else being external to us and indifferent. The behaviour of Theodorus the philosopher, and of Canus Julius when sent to his death by Caligula, are noble examples. Further, we must not indulge in too much grief at the errors of mankind. Laughter is more to the purpose than indignation; though calm observation is a better attitude than either. Nor need the painful death of brave men afflict us, since they themselves endured it gladly 349–362

CHAPTER XXI.

Since the body acts upon the mind, time for relaxation and recreation must be given. Saintly men have always admitted this and practised it, following the example set by St. Benedict. Seneca, to whom much divine truth has been revealed, has insisted on this point strongly. We must not always be alone, he says, nor always in a crowd. There must be a change of occupation; occasional amusement and gaiety. Hours of work must be limited; exercise in the open air, occasional change of scene, are good; even indulgence in wine is not to be wholly forbidden. The mind is stirred by such influences and raised above its usual level. To these may be added the inspiring influence of music, of which the prophet Elijah availed himself 362–365

MORAL PHILOSOPHY : FOURTH PART 366–404

The beauty of the subject, and the rarity of the books treating of it explain the length of the preceding part. I now pass to the grounds for accepting the Christian religion, which points the way to happiness in a future life, and thus gives a meaning and a purpose to moral philosophy. God has never left men without the means of salvation : hence we find ancient philosophers, and especially Aristotle, considering what are the principles which preserve or destroy states 366–367

The religions now existing in the world are those of Saracens, Tartars, Pagans, Idolaters (Buddhists), Jews, Christians. They are distinguished not merely by opinion, but also by difference of moral aim. The Saracens permit excessive indulgence in sexual pleasures. The Tartars err from lust of power; as we learn from the travels of William Rubruquis. Their mode of life is coarse and rude. The Pagans live by custom rather than reason, and suppose that their present enjoyments will be continued in a future life. The Idolaters resemble them in this, except that their priests practise chastity and abstinence. The Jews combined spiritual with temporal blessings; the latter as well as the former belonging to the future life. The Christians, while accepting temporal wealth in the present life, are wholly independent of it in the future . . 367–370

Of these sects we may place the Pagans lowest; they are guided by no priesthood, and each follows his own way. The Idolaters have a priesthood and a ritual; but they have a multiplicity of gods. The Tartars come third; they worship one God, though with many superstitious and cruel observances. Next in order come the Jews, of whom the more spiritual attained to the knowledge of the true Christ. Fifth come the Christians, who practise the Jewish law spiritually. Finally, there is the law of Antichrist, which for a time overwhelms all others. And each of these has its own moral principle—pleasure, wealth, ambition, fame, or blessedness in a future life. In the mathematical section of this work it has been shown that there is a connexion between these sects and the various planetary influences, which incline men's characters in certain directions, though without depriving them of free-will. As these influences change, each of the sects may be modified by the others. So the Saracens, though mainly under the influence of Venus, are modified by Jewish and

VOL. II. PAGES

Christian law. The Tartars, though governed by Mars, in like manner are modified by Mercury, which implies Christian influence. The Pagans, again, and the Idolaters may differ according as the influence of Mars or of the Sun may prevail. With the Pagans, terrestial things as well as celestial may be objects of worship . 370–372

We have now to consider the means of showing the truth of the Christian religion. We may appeal to miracles; or we may take the ground of reason and philosophy common to us and the other sects. Though the Christian should not place his chief reliance on reason, yet he should be able to render an account of the faith that is in him. And as the heathen do not accept this faith, we must challenge them on philosophical ground. Philosophy is given to men for the very purpose of leading them to truth 372–373

In what follows I am appealing to the wise among them rather than the simple. There are three kinds of knowledge: that coming from our own study or experience; that which is learnt from others, and that which is natural, in the sense of being shared by the whole species. It is no less natural to act on such truth than to know it. Practical and speculative reason have the same source, and are essentially the same, as Aristotle has taught . . . 373–375

The existence of God is one of these natural truths, as Cicero has said. The necessity of demonstrating it comes from the weakening effect of sin on our faculties. The divine unity is not known by nature; and even those who accept it are in error as to God's attributes. These therefore have to be explained, beginning, as the mathematician begins, with elementary principles. That God is an eternal first Cause of infinite wisdom, power, and goodness is accepted by the Tartars, Saracens, Jews, and Christians. Pagans cannot deny this truth when presented to them; as the recent conference in Tartary between Christians, Saracens, and Buddhists has shown. Their resistance elsewhere to Christianity is due to attempts to impose on them a foreign yoke 375–377

The Christian advocate may further plead that in tracing causes we cannot go back endlessly. There must, as Aristotle has laid down, be a first Cause, that has always existed, and must exist for ever, unchangeably. Boethius has shown that imperfection implies perfection, to which it is an approximation. Perfect power, as Aristotle shows, is boundless power: and if the power of God be infinite, *a fortiori* is His essence infinite. It follows that His goodness is infinite; and from infinite power and goodness follows the attribute of infinite wisdom. Such a cause is capable therefore of creating this world, and is disposed to govern it in the best way. If it be said that the world is eternal, this is to raise it to equality with God. If more than one Cause is asserted, then none can be

VOL. II. PAGES

infinite. In one world at least there can be only one. And if there should be more worlds than one, the same infinite God is capable of governing them. But plurality of worlds is an hypothesis which Aristotle has dissipated ; for they would all tend to the same central point, and lose their separate existence 377-381

God being one and infinite, man is bound to yield Him boundless reverence, in thankfulness for his creation and for his hopes of future happiness. Even the Pagans believe in a future life ; so do the Saracens, although their belief is tainted with thoughts of animal enjoyments. The Jews have the same belief. Thus the Buddhists have universal opinion against them. This present life is full of misery, from which we must believe that the future life will be free. In our glorified body and soul we shall participate in the divine nature. So too will the punishment of the evil be infinite. This then is the twofold motive for doing the will of God. But man cannot of himself know that will ; as the difference of his religions show, differences which exist even within the pale of Christianity. Of the true nature even of material things we are profoundly ignorant ; much more of things immaterial 381-383

Therefore Revelation is necessary. Aristotle has said that in these things the human mind is as the eye of the owl or bat to the light of the sun, or, as Avicenna says, as the deaf man to musical harmony, or as the untaught child to the highest truth. Man, says Seneca, is too mortal for immortal things. And apart from intrinsic inability to discern infinite truth he is blinded by sin. Man must therefore be taught by divine authority 383-385

All sects claim such authority. We have to consider to which of them has true revelation been made. Evidently to one only, since God is one ; there is only one world and one human race. Avicenna and Alpharabius fully recognize this. We have therefore to choose between the six religions of which I have spoken. The Pagans and the Idolaters, who take creatures for God, or who assert plurality of gods, may be at once eliminated. When confronted by the Tartar emperor with Christians and Mahometans, they at once succumbed. The Tartars, while accepting the unity of God, worship fire, and their teachers are mere magicians. It is clear therefore that they are not in possession of the authoritative truth which we seek. This they themselves have confessed. We must therefore examine the remaining three religions—Jewish, Mahommedan, and Christian . 386-388

The superiority of the Christian law is proved first by the authority of philosophy, which as I have before shown has given its sanction to the principal doctrines of Christianity. No such testimony has been given to Judaism or Mahommedanism. Seneca has attacked the Jewish creed ; and Mahommedanism has been asserted by

VOL. II. PAGES

Avicenna and Albumazar to be imperfect and transitory. We may add to this the testimony of the Sibyls. Coming more into detail, we find the Jewish prophets bearing testimony for Christ, and Josephus supplying proof that Judaism would give way to Christianity. Other evidence is contained in the books of Esdras and of the twelve Patriarchs. The authorship of these books is unknown, but they are of weighty authority. Further, the Jewish religion taken by itself, with its incessant slaughtering of cattle, is extremely repulsive; and is so spoken of in the Psalms and prophetical books. Similarly the Koran uses very strong language in praise of Christ. Mahommedan writers say much in dispraise of their own sect, and admit that the life of its founder was stained by many vices. Accepting then the testimony of each sect as standing on the same level, we find none of them, except the Christian, testifying in favour of their own founder, to whom both Jewish and Mahommedan authorities, as well as Christian, bear witness 388–393

The credibility of Christian writers rests on six grounds: personal sanctity; wisdom; miraculous powers; firmness under persecution; uniformity of faith; their victory in spite of humble origin and estate. Besides these grounds for accepting Christianity, there are the miracles wrought by Jesus, especially the forgiveness of sins, the surest proof of His Divine nature. For neither Moses nor for Mahomet was divinity claimed. There is the yet further proof of holiness of life, which both in Mahommedanism and Judaism is found wanting. In neither is there any recommendation of poverty, chastity, and obedience. Nor is the sanction of a future life clearly set forth in them 393–396

Having thus proved that the Christian faith is to be accepted as true, we need not examine each one of its articles. There is however one of them as to which much difficulty has been experienced, the Sacrament of the Altar. I wish therefore to give reasons why it should be willingly and ardently accepted, containing as it does the essence of the whole. First, it is accepted by the whole Christian Church. Secondly, it is manifestly set forth in Scripture, as in the Gospel of St. John, and in the Book of Wisdom, and in the Epistle of St. Peter. Thirdly, it is confirmed by the unanimous authority of the saints. Fourthly, we may add the testimony of innumerable miracles, of which I here record two of undoubted authenticity and recent date. Finally, as the Creator is omnipresent throughout the world, so by this sacrament is the Redeemer omnipresent to those who are in a state of grace. Without the Creator's presence the creature would cease to exist. So without the Re-creator's presence would the re-created fall from the state of grace. It consists with the infinite power and goodness of Christ that this infinite good should be

ANALYSIS OF

'THE MULTIPLICATION OF SPECIES.'

PART I.

CHAPTER I.

VOL. II. PAGES

(*a*) As to the name and the essential meaning of species.

By species is here meant the first effect of any natural agent. Thus light passing through space is the species of the light in the sun. If light passes through coloured glass, the colour shed by the light is called the species of the originating colour. Species is here used in the sense of likeness or image. In a dream the hallucination is a species. Intention, form, virtue, impression, passion, are other words used to express it 407-410

This first effect of the agent resembles the agent. The agent transmutes the patient into its likeness; thus fire transmutes what it acts on into fire, heat into heat, light into light, and so on. The subsequent effects have not the same likeness to the agent. This first effect is of the same specific nature as the agent, otherwise we should be at a loss to assign its category; for it cannot belong to the category of accidents; and there is nothing intermediate between substance and accident. But, although of the same nature, its essence is incomplete; just as we should say of the embryo, only that the incompleteness is far greater. Some agents produce more complete species than others, as light, colour, heat. Generally speaking, the higher the rank of agents, the more incomplete are their species 410-413

(*b*) There can be only one such resembling effect for each agent. This we call univocal; the multiple secondary effects may be called equivocal. While the agent is acting on the patient, the effect or virtue is called species; afterwards, when the effect is completed, it takes the name of the agent. Fire, when it has taken hold of the wood, is called fire. This, however, is only true when the patient is corruptible; when the patient is incorruptible, the case is otherwise. The sun does not communicate its nature to the moon and planets by shining on them. The species of light is completed in them, but not the species of the sun. And generally it would be truer to say that the patient was assimilated through the species to the nature of the agent, than to speak of the patient losing its identity in that of the agent 413–416

(*c*) Natural agents produce the same species whatever be the nature of the patient; in other words, their action is uniform. This does not apply to agents possessing free-will, except in so far as they may resemble natural agents. The difference in the patient may bring about a different result, but the species is the same . . 417–418

CHAPTER II.

(*a*) We have now to inquire what are the things which generate species. First, all qualities that act on sense, producing some change in the sense-organ. There may be a doubt as to sound. Here there is a tremor in the part of the object struck: this tremor is communicated to the adjacent part, and so again to the next. The first tremor makes a sound; so does the second and the third; but the second sound does not come from the first sound; it comes from its own tremor 418–419

(*b*) Do substances generate species? Assuredly, since substances are nobler than accidents. An accident is not generated until its subject be first generated. Heat, the accident, is posterior to fire, the substance. When heat appears in water, fire must pre-exist, and the substantial nature of fire, and this we call the species of the substance fire. Such species are not cognizable by the five senses, or by *sensus communis*. But they are cognizable by a mental process, akin to that which causes the sheep instinctively to flee from the wolf which it has never before seen, feeling something injurious in its proximity. It is sometimes said that substance can have no contrary: but there is a sense in which substances, or substantial forms act in opposition 419–423

(*c*) The species of substance is of the whole compound; i.e. of matter and form, not of form alone 423–424

(*d*) Every corporeal substance can produce species, therefore also the organs of sense. The species generated by the organs of sense are concerned in sensation 424–425

(*e*) With regard to matter, which is passive, it may be affirmed that it does not generate species, except in so far as compounded with form 425–427

(*f*) Light and colour generate species, but other properties of body,

CHAPTER V.

CHAPTER VI.

VOL. II. PAGES

heavenly bodies moved by spiritual intelligences are unchangeable.) We shall find that the four elemental substances can carry their action to completion, and especially the element of fire . . 449-451

But many compound substances are higher in rank than the elemental. Why cannot they do the same? There resides in them an aptitude for doing this, but they are practically debarred from doing it. Were it otherwise, the higher substances would transmute the lower into their own substance, and the lower would disappear, and thus the order of the universe would be subverted. Consequently spiritual and celestial substances, while generating their species, cannot bring them to actual completion. Such compound substances as the heavens or as man, if their species were made complete, would do away with all lower substances. Consequently, the carrying of species to their complete effect is limited to very few substances, such as the elements, and especially the element of fire . . . 451-455

The same thing is to be said of attributes or accidents. Some of these, as hot, cold, moist, dry, can complete their species, especially the accident of heat. Light also can do this in bodies adapted to it. Of colour, odour, savour, and sound the same cannot be said . 455-456

PART II.

CHAPTER I.

We now have to consider the mode in which the radiation of species takes place. The first part of the patient, when acted on, becomes itself a force acting upon the second part, and so onward. These successive parts of the patient are of equal dimension. From the starting-point in the agent, or, which is the same thing, from the point of the patient first affected, emanate rays in all directions in straight lines. Those emanations, of whatever kind they be, are called rays, by analogy with what we see emanating from a star or any luminous point. Their rectilinear direction continues until modified by an opaque surface which reflects them, or a change of medium which refracts them. These lines are not merely of one dimension: they have breadth, and depth, as well as length . 457-460

CHAPTER II.

There are five kinds of rays to be considered. (*a*) Rectilinear rays, where the medium is uniform. (*b*) Refracted rays. The ray passing from a rarer to a denser medium is deflected towards the perpendicular drawn to the surface at that point; and conversely, when passing from denser to rarer. (*c*) Reflected rays. When the ray meets an opaque surface, it is turned back in a direction making with the perpendicular to the surface an angle equal to the angle of incidence. (*d*) In the case of animate bodies, the ray follows a tortuous path along a nerve towards the centre of sensation. (*e*) Secondary, or accidental, rays may be emitted, not from the luminous point, but from the principal ray: as in the case of rays coming through an opening into a dark room, from which secondary rays are dispersed through the room; although the eye placed in one of those rays cannot see the luminous point. Rays from the sun of this kind are of greater utility to living beings than direct rays 460–465

CHAPTER III.

Whether the ray falls on a line or on a surface, whether the surface be plane, convex or concave, the law is the same. The mode of measuring the incident and the reflected angle is explained elsewhere. In refraction from the rarer to the denser medium the ray is deflected towards the normal. From the denser to the rarer away from the normal. The accompanying diagrams will illustrate this. Perpendicular rays, coinciding with the normal, are not deflected 465–468

What is the cause of reflexion and refraction? All media, however transparent, offer a certain amount of resistance to the passage of the ray. This is more completely overcome by the perpendicular than by the oblique ray. The oblique ray, passing into the denser medium, follows the course of the perpendicular ray as far as possible, i.e. is deflected towards it, as being the path of least resistance. On the contrary, when passing into the rarer medium, the energy which it has been hitherto exerting in the denser, now finds a vent in carrying it away from the normal, i.e. along a path of greater resistance 468–470

That these are the respective paths followed, is shown by the double refraction taking place when rays impinge on a globular body filled with the denser medium. The diagram shows that the rays issuing from the globe converge to a point, and that at this point combustible substances are ignited 470–472

CHAPTER IV.

To what extent does refraction take place in the heavens? The celestial orbs are of the same degree of translucency; in these therefore there is no refraction of rays. Rays passing from the sphere of ether to that of air (which latter includes that of fire) are refracted; since ether is the more subtle medium. The proof of this is that the

VOL. II. PAGES

apparent polar distance of a star is less when rising than when passing the meridian. At the zenith, the ray, impinging on the sphere of air perpendicularly, is unrefracted. At the horizon it is refracted. With the circumpolar stars the polar distance appears less, when they pass the meridian nearer the horizon, than when they pass it nearer the zenith; for the same reason. The precise quantity of this refraction can be measured by astronomical instruments. Ptolemy first indicated this source of error, and Alhazen has verified it. It has to be taken account of in lunar tables 472–476

There is no subsequent refraction between the sphere of fire and that of air, as the one passes into the other by imperceptible degrees 476–477

Rays falling from the planets on regions of the earth beyond the tropics must all fall obliquely and therefore be refracted . . 477–478

CHAPTER V.

(*a*) No bodies are so dense that rays may not pass through them, though these may not always be perceptible by human organs. Some bodies of intermediate density may partly refract, and partly reflect the rays, as in the case of water 478–479

(*b*) Hardness and solidity are to be distinguished from density. Glass and crystal, though hard, must yet be rare, i. e. their particles must lie somewhat far apart, since rays pass freely through them. Hardness depends not on the number of the particles but on stability and freedom from moisture. (*c*) Reflexion depends on the smoothness of the surface that reflects. A rough surface disperses the rays in different directions 479

(*d*) We must not suppose that the ray produces any effect on the surface that reflects it. No secondary or accidental ray is generated at the point of reflexion. Were it so, such an effect would be visible, whatever our position might be with regard to the mirror. All that the reflecting surface does is to change the direction of the ray. The reflected ray is strictly continuous with the incident ray . 479–481

CHAPTER VI.

In reflexion, the angles made by the incident and the reflected ray with the perpendicular to the surface are equal. This may be shown experimentally by an instrument constructed for the purpose, and the diagrams annexed give geometrical proof of it. The rule holds, whether the reflecting surface be plane, concave, or convex . 481–486

CHAPTER VII.

Rays reflected from a plain surface are divergent. Rays reflected from a spherical concave surface converge at various points on the diameter which forms the axis of the mirror; those from the circumference of any circle drawn on the mirror in a plane perpendicular to the axis will meet in a point. The smaller the circle the greater will be the distance of the point of convergence from the mirror; and conversely. (Such circles are, of course, merely imaginary; they merely indicate a series of points similarly placed with regard to the axis) 486–490

But another form of concave mirror has been devised in which the curvature is such that all the rays shall converge to a single point. This cannot be explained here, but it may be mentioned that the curvature is that of a conic section 490–491

CHAPTER VIII.

Rays from the object proceed equally in all directions, as radii from the centre of a sphere. It is objected that rays passing through an angular orifice illuminate an angular not a circular space. But this depends on the distance of the illuminated surface from the orifice. If the distance be sufficient, the illuminated surface is circular. The smaller the orifice, the less the distance required to obtain this result. Much depends also on the strength of the rays; those of the sun at noonday, being stronger, produce the result more readily than the feebler rays of morning light. For the same reason, at noon a smaller portion of the solar surface will effect the result than in the morning; in the latter case the illumination not being circular, much will be dispersed and lost. That the shape of a flame is not spherical but conical, is due to the ascending force of the particles. In the sphere of fire, where there is no such force, the fire assumes its natural (spherical) shape. The centre of the illuminated surface is most strongly illuminated; the surrounding parts being illuminated by intersecting rays, and therefore more feebly. Since all regular figures can be inscribed in a circle, the spherical form of emission of rays includes all the others, and especially the conical . . 491–494

CHAPTER IX.

Species proceed from every point in the agent, not from one only. We see this in the case of shadows. If an illuminated body be less than the illuminating, the shadow formed is conical and limited; if the two are equal, it is cylindrical and unlimited. If the illuminated body be greater, the shadow is divergent and unlimited. We are supposing that the agent is homogeneous. If it be heterogeneous, we must consider each homogeneous part separately. In dealing with the cone of rays proceeding from any point, we have to remember that it is the axis of the cone, directed vertically to the surface acted

thing, but a continuous succession, each generating that which follows. It has been said that light is body, or an emanation from body. But Aristotle denies this, remarking that in that case we should perceive its translation from east to west at sunrise. The remarks on the subject in the Topics are not to be taken as conveying Aristotle's real opinion 504–507

CHAPTER II.

The second question is whether the species of a corporeal agent should be regarded as corporeal or spiritual. Clearly the former: since it is not a soul, or an intelligence, or a first cause; and these only can be considered as spiritual. The species must be of like nature with the agent, and of like nature with the completed effect. It cannot be of nobler nature than that which generates it. Some hold that, though in its essence corporeal, yet its mode of existence is spiritual. But there cannot be this disparateness between the essence and the mode of existence. The essence of the species has dimension, derived as we have seen from the medium; it is therefore corporeal. It is said that it has not material essence (material being here taken for corporeal). But even spiritual substances have material essence: they are compounded of form and of matter: much more, therefore, are species material. It has been already shown that the species of corporeal substance resembles the total compound. In the medium not merely is the formal essence generated but the material, although the essence be incomplete. I therefore conclude that the species of a corporeal object is corporeal 507–509

The contrary view has arisen from imperfect translations of Averroes, Avicenna, and Aristotle. In these the word insensible has been mistaken for immaterial. And because spiritual things are insensible, therefore what is insensible has been taken to mean spiritual and immaterial. It is objected that species are not always insensible, as when a ray passes through a window or through coloured glass. But this is only accidental. Some expressions of Avicenna as to the reception of impressions in the brain have been wrongly interpreted. They must be taken in connexion with what Aristotle has said on the same subject 509–511

CHAPTER III.

I pass to the question of the mingling of species in the medium. The question is one of much difficulty, but Ptolemy and Alhazen throw great light upon it. Species of the same kind melt into one, and if opposite, then the stronger overwhelms the weaker. Those that are contingent, i.e. have no relation to each other, co-exist

PART IV.

CHAPTER I.

VOL. II. PAGES

When the direction of radiation is from terrestrial to celestial, it might be thought that the increasing rarity of the medium would compensate the weakening of the species from dispersion; but the second cause operates at a faster rate than the first. For radiation in the reverse direction, from heaven to earth, there can, of course, be no such compensation 518–521

CHAPTER II.

Supposing space were unlimited, could multiplication of species go on indefinitely? No; for, as we have seen, the species grows continually weaker the further it is from its source. If the supply of matter on which it could operate were infinite, then no doubt the action would be infinite; but this is not the case. It might be said that, since matter is infinitely divisible, there would always remain something on which the species, however enfeebled by distance, could act; that therefore the duration of the action would be infinite. But this seems a forced and unsatisfactory conclusion. Again, it might be urged that the medium may grow continually more subtle so as to correspond to the continual weakening of the species. But this does not in fact take place. As a fact, the resistance, in our world, which is finite, is always sufficient to prevent the species from passing beyond a certain distance 521–524

Supposing the existence of a vacuum between earth and heaven, no species could radiate; for no natural action can take place in a vacuum. It follows also that beyond the ninth and tenth heavens no species from the stars can radiate. For beyond these there is nothing, not even empty space 524–525

CHAPTER III.

The radiation of force takes place in time. For it involves change in the medium; and change involves time. When the path of propagation is refracted, more time is taken than when the path is direct. Finite force cannot act instantaneously, for then it would be equal to infinite force. As a point is to an instant, so is a line to time. The passage of a point is instantaneous, in the passage of a line time is occupied. Translation implies before and after . . 525–526

It is said that the transit of light is instantaneous. But all that is meant is that the time required is insensible. Aristotle, in saying this, is arguing against Empedocles, who regarded light as a substance flowing from the luminous body. In this case no doubt we should be able to see its passage from east to west in the heavens. But the view here taken of light is not that it is a body translated, but a movement continuously propagated. Aristotle no doubt dis-

VOL. II. PAGES

tinguishes the propagation of light from that of sounds or odours. But the distinction is of degree; light being more subtle, its motion is more rapid 526-528

Alkindi urges that, though through short distances the transit of light may be imperceptible to our senses, yet that the summation of short periods would amount, in the transit from east to west, to something that would be perceptible. But of imperceptibility there are many degrees; and the propagation of light is so rapid that the summation of a vast number of imperceptible periods may still not reach the point at which perception of time by human sense is possible 528

It is further maintained that in radiation of species, as here described, more time would be occupied than in translation of a body through space: because between the formation of any one species and the formation of that which follows it, there is an interval of rest. The fact is, however, that the resistance of the medium to a body carried through it would be far greater than it would be to the formation of species, which involves an exceedingly slight disturbance of the medium; so that the motion in this latter case will be far more rapid, and the interposition of intervals of rest will pass unperceived . 528-529

PART V.

CHAPTER I.

In propagation of species we have to consider, first, that which we call *univocal*, as when light makes light; secondly, the *equivocal*, as when light makes heat, or any other effect of a different kind. As to the first, we remark that the strongest are those which proceed by direct lines, as contrasted with those which are refracted or reflected. We may omit for the present those which proceed on tortuous paths, as in the case of nervous action; since here a different agency, that of life, comes into play. The species called accidental, i. e. which proceed, not directly from the object, but secondarily from the species, are the feeblest of all 530-534

CHAPTER II.

The path of some species falls perpendicularly on the object; these are the strongest; and this whether the object be a plane surface or curved. In either case the angles made are equal. That which impinges on the convex surface of a sphere, is stronger than that which falls on the concave surface, as being necessarily shorter. The

VOL. II. PAGES

refracted ray is stronger than the reflected, its path lying nearer to the direct perpendicular ray. The ray refracted by passage from a denser to a rarer medium is less potent, for the same reason, than the ray refracted from rarer to denser. In the case of reflected rays, those that make an angle with the vertical line produce a greater effect than those that fall vertically; not merely because they diverge less from the direct ray, but because they can be accumulated by suitably shaped mirrors. Some of these are described, though I have not myself seen them, which bring rays to a focus both in front and behind them 534-539

CHAPTER III.

Cones of rays come from the agent to each part of the surface acted on. The shorter the cone, the stronger is the action. The shortest cone is thus defined. Let a circle be drawn in the agent sphere dividing its operative from its inoperative part. From any points in the circumference of this circle let diameters of the sphere be drawn. Tangents to the sphere at the extremities of these diameters will form the shortest cone. Other points connected with this subject are illustrated by the accompanying diagrams . 539-543

PART VI.

CHAPTER I.

As species are generable, so are they corruptible. And this in two ways: (1) from exhaustion of their own strength; (2) from tendency in the substance acted on to resort to its own natural condition 544-545

CHAPTER II.

When the generating agent is removed, does the species at once disappear? Not instantaneously, but gradually, from the causes stated in the previous chapter. The apparent permanence in some cases of the species, after the agent has disappeared, is to be explained. The semen retains its force in the absence of the male parent; but here the female supplies the place. A heated stone retains its heat. Here the species, i. e. incomplete effect, has passed into a more complete effect 545-547

CHAPTER III.

Are the reiterated actions of the agent on the patient to be regarded as numerically identical, or as distinct? They must be regarded as

CHAPTER IV.

FRATRIS ROGERI BACON

ORDINIS MINORUM, OPUS MAJUS.

PARS PRIMA

HUJUS PERSUASIONIS,

In qua excluduntur quatuor universales causae totius ignorantiae humanae, habens quatuor distinctiones. In prima sunt quatuor capitula. In primo data totius persuasionis intentione reprobantur illae quatuor causae universali sermone.

CAPITULUM I.

Sapientiae perfecta consideratio consistit in duobus, videlicet, ut videatur quid ad eam requiritur, quatenus optime sciatur; deinde quomodo ad omnia comparetur, ut per eam modis congruis dirigantur. Nam per lumen sapientiae ordinatur Ecclesia Dei. Respublica fidelium disponitur, infidelium conversio procuratur; et illi, qui in malitia obstinati sunt, valent per virtutem sapientiae reprimi, ut melius a finibus Ecclesiae longius pellantur quam per effusionem sanguinis Christiani. Omnia vero quae indigent regimine sapientiae ad haec quatuor reducuntur; nec pluribus potest comparari. De hac igitur sapientia tam relate quam absolute scienda nunc, secundum tenorem epistolae[1], quicquid possim circa

Definition of wisdom.

[1] Epistolae praecedentis, Cottonian MSS., Jul. D. v. (hereafter referred to as Jul. . This was the letter of Pope Clement IV. It ran thus:

'Dilecto filio Fratri Rogerio dicto Baccon ordinis Fratrum Minorum.

'Tuae devotionis literas gratanter recepimus, sed et verba notavimus diligenter persuasionem ad praesens Vestrae Beatitudini praesentare conabor. Quoniam autem illa, de quibus agitur, sunt grandia et insolita, gratiam et favorem humanae fragilitatis requirunt. Nam secundum Philosophum septimo Metaphysicae, ea quae sunt maximae cognitionis secundum se, sunt minimae apprehensionis quoad nos. Involuta enim veritas in alto latet et in profundo posita est, ut dicit Seneca septimo de beneficiis, et quarto Naturalium. Et Marcus Tullius in Hortensio dicit, quod omnis noster intellectus multis obstruitur difficultatibus, quoniam ipse se habet ad manifestissimum in sua natura, sicut oculus noctuae et vespertilionis ad lucem solis, ut Philosophus dicit secundo Metaphysicae, et velut surdus a nativitate ad delectationem harmonicam, sicut undecimo Metaphysicae dicit Avicenna. Quapropter sufficit nobis in inquisitione veritatis proprii intellectus imbecillitas, ut quantum possumus causas et occasiones erroris extraneas longius a debilitate sensus nostri relegemus.

Four causes of error: authority, custom, popular prejudice, presumption of knowledge.

Quatuor vero sunt maxima comprehendendae veritatis offendicula, quae omnem quemcumque sapientem impediunt, et vix aliquem permittunt ad verum titulum sapientiae pervenire, videlicet fragilis et indignae auctoritatis exemplum, consuetudinis diuturnitas, vulgi sensus imperiti, et propriae ignorantiae occultatio cum ostentatione sapientiae apparentis. His omnis homo involvitur, omnis status occupatur. Nam quilibet in singulis artibus vitae et studii et omnis negotii tribus pessimis ad eandem conclusionem utitur argumentis, scilicet, hoc exemplificatum est per majores, hoc consuetum est, hoc vulgatum est; ergo tenendum. Sed oppositum conclusionis longe melius sequitur ex praemissis, sicut per auctoritatem et experientiam et rationem multi-

quae ad explanationem earum dilectus filius G. dictus Bonecor miles viva voce nobis proposuit tam fideliter quam prudenter. Sane ut melius nobis liqueat quid intendas volumus et tibi per Apostolica scripta praecipiendo mandamus quatenus, non obstante praecepto praelati cujuscunque contrario vel tui ordinis constitutione quacunque, opus illud quod te dilecto filio Raymundo de Lauduno communicare rogavimus in minore officio constituti scriptum de bona litera nobis mittere quam citius poteris non omittas, et per tuas nobis declares literas quae tibi videnter adhibenda esse remedia circa illa quae nuper esse tanti discriminis intimasti, et hoc quanto secretius poteris facias indilate. Datum Viterbii x Kalend Julii anno ii.' (Wadding, *Annales Minorum*, vol. iv, p. 265.)

pliciter probabo. Si vero haec tria refellantur aliquando magnifica rationis potentia, quartum semper in promptu est et in ore cujuslibet, ut quilibet suam ignorantiam excuset; et licet nihil dignum sciat, illud tamen magnificet imprudenter, ut sic saltem suae stultitiae infelici solatio veritatem opprimat et elidat. Ex his autem pestibus mortiferis accidunt omnia mala humano generi; nam ignorantur utilissima et maxima et pulcherrima sapientiae documenta, et omnium scientiarum et artium secreta; sed pejus est, quod homines horum quatuor caligine excaecati non percipiunt suam ignorantiam, sed cum omni cautela palliant et defendunt, quatenus remedium non inveniant; et quod pessimum est, cum sint in tenebris errorum densissimis, aestimant se esse in plena luce veritatis; propter quod verissima reputant esse in fine falsitatis, optima nullius valoris, maxima nec pondus nec pretium obtinere et e contrario falsissima celebrant, pessima laudant, extollunt vilissima, caecutientes, aliud esse omnem sapientiae fulgorem, fastidientes[1] quae magna facilitate possunt adipisci. Et propter stultitiae magnitudinem ponunt summos labores, consumunt tempora multa, magnas expensas effundunt in iis, quae nullius utilitatis vel parvae sunt, nec dignitatis alicujus secundum judicium sapientis. Et ideo necesse est ut violentia et malitia harum quatuor causarum omnis mali cognoscantur in principio, et reprobentur, et longe a consideratione sapientiae relegentur. Nam ubi haec tria dominantur, nulla ratio movet, nullum jus judicat, nulla lex ligat, fas locum non habet, naturae dictamen perit, facies rerum mutatur, ordo confunditur, praevalet vitium, virtus extinguitur, falsitas regnat, veritas exsufflatur. Et ideo nihil magis necessarium est considerationi, quam certa damnatio istorum quatuor per sententias sapientum electas, quibus non poterit contradici.

Quoniam vero sapientes tria prima simul collidunt et reprobant, et quartum propter singularem stultitiam propriam limaturam desiderat, ideo primo trium malitiam aperire conabor. Sed quamvis auctoritas sit unum de istis, nulla ratione loquor de solida et vera auctoritate, quae vel Dei judicio collata est

[1] Caecutientes . . . I restore the omission from B. et D., 'omnem sapientiae fulgorem, fastidientes.'

Ecclesiae, vel quae ex merito et dignitate proprie nascitur in sanctis philosophis et perfectis prophetis, qui juxta humanam possibilitatem in studio sapientiae experti sunt; sed de illa auctoritate loquor, quam sine Dei auxilio violenter usurpaverunt multi in hoc mundo, nec ex merito sapientiae, sed ex propria praesumptione et desiderio famae, et quam vulgus imperitum multis concessit in pernicionem propriam judicio Dei justo. Nam secundum scripturas propter peccata populi multotiens regnat hypocrita; de sophisticis enim auctoritatibus multitudinis insensatae loquor, quae aequivocae sunt auctoritatis, sicut oculus lapideus aut depictus nomen habet oculi, non virtutem.

CAPITULUM II.

The first three recognized by Cicero, Seneca, Adelard, Averrhoes and others.

Sed tria reprobat sacra scriptura, sacri doctores condemnant, jus Canonicum vetat, philosophia reprehendit; sed propter rationes prius tactas de philosophicis allegandis, et quia minus vulgatae sunt sententiae Philosophorum circa haec tria, eas principaliter introducam. Similiter vero omnes has tres pestes Seneca libro secundo epistolarum prope finem uno sermone condemnat. Dicit igitur, 'inter causas malorum nostrorum est, quod vivimus ad exemplar, nec ratione componimur, sed consuetudine abducimur; quod si pauci facerent nollemus imitari, cum plures facere ceperint, quia frequentius quam honestius, illud facimus, et recti locum tenet apud nos error, ubi publicus factus est[1].' Philosophus vero per totam philosophiam suam persequens auctoritatem indignam secundo Metaphysicae causas humani erroris asserit praecipuas consuetudinem et populi testimonium[2].' Et iterum Seneca libro de vita beata dicit, 'Nemo sibi solum errat, sed alieni erroris causa et auctor est, versatque nos et praecipitat error, et alienis perimus exemplis[3].' Et in secundo libro de Ira,

[1] The passage is in the 123rd letter (lib. xx. Ep. 8, ed. Haase). Part of the quotation is slightly condensed from the original, which runs, 'Cum plures facere coeperunt, quasi honestius sit quia frequentius, sequimur.' In the last sentence Jebb substitutes certi for recti.

[2] *Met.* lib. i. (minor), cap. 3 *αἱ δ' ἀκροάσεις κατὰ τὰ ἔθη συμβαίνουσιν· ὡς γὰρ εἰώθαμεν οὕτως ἀξιοῦμεν λέγεσθαι*, et seq.

[3] Seneca, *Dialog.* vii. cap. 1.

'propter malitiam consuetudinis,' ait, 'difficulter reciduntur vitia, quae nobiscum creverunt[1].' Et in libro de vitae beatitudine contra vulgi sententias refert, 'Nulla res majoribus malis nos implicat, quam quod ad rumorem componimur, optima rati ea quae magno assensu recepta sunt, nec ad rationem sed ad similitudinem vivimus. Inde est ista tanta coacervatio aliorum supra alios ruentium. Quod enim in strage homini magna evenit, cum se populus ipse premit, nemo ita cadit ut non alium in se attrahat, primique sunt exitio sequentibus.' Et iterum dicit in eodem, 'Stat contra rationem defensor mali sui populus'; et infra, 'Non tam bene cum rebus humanis agitur, ut meliora pluribus placeant,' et sequitur, 'argumentum pessimi turba est[2].' Et Marcus Tullius tertio de quaestionibus Tusculanis ait, 'Cum magistris traditi sumus, ita variis imbuimur erroribus, ut vanitati veritas, et opinioni confirmatae ipsa natura cedat[3].' Atque in Lucullo dicit, 'quidam obsecuti amico cuidam aut una cum aliquo quem audierunt oratione capti, de rebus incognitis judicant, et ad quamcumque sunt disciplinam quasi tempestate delati, tanquam ad saxum inhaerescunt; plerique errare malunt, eamque sententiam quam amaverunt defendere, quam sine pertinacia perquirere quid constantissime dicant[4].' Et propter pravitatem consuetudinis queritur in primo de divina natura, 'non pudet speculatorem naturae ab animis consuetudine imbutis petere testimonium veritatis[5].' Et contra vulgi sensum dicit proemio secundi quaestionum, philosophia 'est paucis contenta judicibus, multitudinem ipsam consulto fugiens, eique suspecta et invisa[6],' et in eodem libro ait, 'laudabiliora mihi videntur omnia quae sine populo teste fiunt[7].' Sed alii divisim haec tria persequuntur. Nam in libro Quaestionum Naturalium Alardi[8] quaeritur de auctoritate fragili: 'Quid est

[1] Seneca, *Dial.* iv. 18.

[2] These passages are from Seneca, *Dial.* vii. 1 and 2.

[3] Cic. *Tusc. Disput.* lib. iii. cap. 1.

[4] *Academ. Priorum*, lib. ii. 3. This is a selection of two passages from the chapter. The quotation is not entirely accurate.

[5] *De Deorum Natura*, i. 30.

[6] *Tusc. Disp.* lib. ii. cap. 1, § 4.

[7] Lib. ii. cap. 26, § 64.

[8] Jebb has Mardi, a mistake which, if he had had the Oxford MSS. before

aliud auctoritas hujusmodi quam capistrum? Ut bruta quippe animalia capistro ducuntur, nec cui nec quo ducantur discernunt, sic nos paucos bestiali crudelitate captos ligatosque auctoritas ipsa in periculum ducit.' Et in libro de Dei aeternitate, 'qui elegerit alteram partem quaestionis propter amorem consuetudinis non potest recte discernere veram opinionem[1].' Et Averroes in fine secundi Physicorum dicit, 'Consuetudo est maxima causa impediens a pluribus rebus manifestis. Quemadmodum enim consueto ad aliquas opiniones licet nocivas, erunt ei faciles, et ideo credit eas esse utiles; similiter cum fuerit assuetus credere sermones falsos a pueritia, erit illa consuetudo causa ad negandum veritatem, sicut quidam tam assueti fuerint comedere venenum, quod factum est eis cibus.' Et idem Averroes vult secundo Metaphysicae 'quod opposita principiorum cum fuerint famosa sunt magis receptibilia a multitudine consequentibus testimonia multorum quam ipsa principia.' Atque Hieronymus in prologo quinti libri super Jeremiam asserit, 'veritatem paucis contentam esse, et hostium multitudine non terreri.' Johannes quoque Chrysostomus super Matthaeum dicit, quod a veritate nudos se esse professi sunt, qui multitudine se armaverunt.

CAPITULUM III.

They are shown by practical experience.

Quod per auctoritates probatum est experientia cujuslibet certius dijudicatur. Nam experimur in nobis et in aliis quod haec tria, ut in pluribus amplectuntur mala, frequentius falsis

him, he would have avoided. The writer referred to is Adelard of Bath, who lived in the early part of the twelfth century. He fills an important place in the history of mediaeval science. He was the first translator of Euclid into Latin; not, however, from Greek but from Arabic. A more complete translation was made in the following century by Campano. [See Weissenborn, *Abhandlungen zur Geschichte der Mathematik*, Drittes Heft, Leipsic, 1880, pp. 141-166.] Adelard studied in the Schools of Tours and Laon; and subsequently travelled in Greece and Asia Minor. The passage here cited is from the sixth chapter of his *Quaestiones Naturales*, a work full of audacious and original speculation. In Bacon's mathematical treatise, as yet unpublished, he is frequently mentioned, always under the name Alardus. [See Sloane MSS. 2156, ff. 72–95.]

[1] I do not know what work is here referred to.

adhaerent. Quod si aliquando bonis et veris implicentur, fere semper sunt imperfecta, et debilem gradum continent sapientiae. Matris quidem opera ut in pluribus sequitur filia, patris natus, domini servus, regis subditus, praelati bajulus, magistri discipulus. Quia familiare est filiis Adae sibi auctoritatem vindicare et sua exempla spargere in lucem. Nam omnes homines secundum Aristotelem quarto Ethicorum amant sua opera, ut parentes natos, et poetae metra, et sic de aliis. Et ideo multi nimia licentia scribendi usi sunt, adeo ut pravis et bestialibus viris insinuare non dubitaverunt, cur chartas non impletis, cur a tergo non scribitis. Et hi sunt sicut pastor claudus et caecutiens cum ovibus multis, quas errantes per devia falsitatis nec possunt nec sciunt ad saniora sapientiae pascua revocare, et sunt similes avibus, qui optant sine alis volare, prius de magisterio praesumentes, quam boni discipuli gradum adepti sunt, quum necessario tot incidunt in errores, quod otiosi comparatione ipsorum reputant se felices; sicut quando multi currunt in stadio, ille quem desperatio currere non permittet, quantumcunque sibi videatur pernitiosum, se tamen felicem reputat comparatione illius, qui currens cadit in foveam invisam. Et ideo videmus fide oculata quod pro uno exemplo veritatis tam in scientia quam in vita sunt plus quam mille falsitates. Mundus enim plenus est hujusmodi exemplis, atque unum exemplum verae perfectionis decem millia imperfecta invenit de facili. Natura enim in numeris formavit nobis perfectionis et imperfectionis documentum. Nam tum numerus perfectus[1] dicitur, cujus partes aliquotae ipsum praecise constituunt, et non est nisi unus infra denarium, viz. senarius; et unus inter 10 et 100, ut 28; et unus inter 100 et 1000, ut 496; et unus inter 1000 et 10000 ut 8128; et sic ultra; et utinam sic esset inter homines, et hoc humano generi sufficeret. Sed nunquam fuit hoc nec in vita nec in scientia, neque erit usquam in finalem peccati destructionem, quoniam non solum est raritas eorum, qui in omni veritate et scientia perfecti sunt, sed eorum qui ad perfectionem unius veritatis vel scientiae devenerunt. Primi sunt et erunt et

[1] When 2^{n-1} is a prime number, then $2^{n-1}(2^n-1)$ is a perfect number, as 6, 28, 496, 8128, &c.

fuerunt semper paucissimi. Nam sunt vere perfecti, sed de 10000 hominum unus non invenitur sic perfectus nec in statu vitae neque in professione sapientiae; utinam de secundo genere perfectorum infra denarium unus esset, et sic ultra, ut perfectio numerorum in hominibus conservaretur. Sed non est ita, imo longe aliter invenitur. Similiter de consuetudine probamus per experientiam in actibus nostris, quod nunc datum est in exemplis. Revolvat quilibet vitam suam ab infantia sua, et inveniet quod in pluribus ejus operibus facilius mala et falsa duxit in consuetudinem. Nam in bonis et veris identitas humanae fragilitati mater est satietatis, et delectatur miser homo in varietate utilium secundum sententiam auctoritatum quas in principio allegavi; e contrario quidem in malis et falsis ac nocivis sibi et aliis. Nam ut in pluribus actibus, nisi gratia specialis et privilegium divinum in aliquibus perfectis obtinet, humana corruptio diligenter continet ea, quae veritati et saluti sunt contraria; nec taedio afficitur in peccati continuatione, nec fastidium de facili invenit in rebus variis. Quod si aliquis a juventute applicetur ad vitae et scientiae veritatem, hic ut in pluribus actibus continet imperfectionem, et in ea jocundatur, perfectio enim contristat eum frequentius; nam paucissimos delectat, et maxime in virtutum et scientiarum plenitudine, et ideo accidit quos aetas juvenilis vix cavet ab errore, et senectus cum summa difficultate ad perfectionem in aliquo transcendit. De vulgo enim idem judicium est. Nam multitudo generis humani semper erravit in Dei veritate, et paucitas Christianorum recepit; atque scimus quod plebs Christianorum imperfecta est. Nam paucitas sanctorum hoc ostendit. Similiter de philosophica doctrina, vulgus enim semper caruit sapientia philosophiae. Brevis enim philosophorum numerus hoc declarat. Et vulgus quidem philosophantium imperfectum semper remansit. Nam de famosis philosophus solus Aristoteles cum sua familia vocatus est judicio omnium sapientum, quoniam ipse omnes partes philosophiae digessit secundum possibilitatem sui temporis, sed tamen ad finem sapientiae non pervenit, sicut satis manifestabitur inferius.

CAPITULUM IV.

Of these three popular prejudice is the most potent.

Sed tamen quantumcunque fragilis sit auctoritas, nomen habet honoris, et consuetudo violentior est ad peccatum quam ipsa; utraque autem earum impetuosior est sensus vulgi. Nam auctoritas solum allicit, consuetudo ligat, opinio vulgi obstinatos parit et confirmat. Consuetudo autem est altera natura, ut dicit philosophus libro de memoria et reminiscentia[1], et in libro Problematum, et ideo majorem violentiam quam auctoritas inducit. Unde philosophus 10mo Ethicorum sententiae Jeremiae de pelle Aethiopum adhaeret, dicens, 'quod impossibile est vel difficile eum, qui per consuetudinem malam induratus est, mutari ad bonum.' Et Sallustius in Jugurthino sensum Solomonis profert, cum ait, 'ubi adolescentiam habuere, ibi senectutem agunt.' Sententia quidem vulgi caeteris improba magis est. Nam, ut ait Seneca tertio Declamationum libro, 'multitudo semel mota modum servare non potest'; propter quod Johannes Chrysostomus super Matthaeum dicit 'convenerunt ut multitudine vincerent quem ratione superare non poterant.' Sic considerandum est diligenter quod vulgus imperitum non solum violentius est ad persuasionem mali quam reliqua duo, sed stultum est et magis elongatum a fine sapientiae. Nam exemplum perfectionis trahitur ab aliquo in consuetudinem, sed vulgo sufficit ut non erret. In nullo enim statu Ecclesiae requiritur ut ejus perfectionem teneat multitudo. Nam et apud religiosos paucitas figitur in centro perfectionis suae et multitudo vagatur in circumferentia. Sic est in statu multorum secularium, ut cernimus ad oculum evidenter. Turba enim sicut cum Moyse non ascendit in montem, sic nec cum Christo in transfiguratione Christi multitudo discipulorum assumpta est, sed tres electi specialiter. Et cum magistrum perfectionis Christum turba secuta est per duos annos praedicantem, postea demisit eum, et in fine clamavit, 'Crucifige.' Nam vulgus nihil perfectum continuare potest, utinam enim vita, nec in studio errasset. Et sicut nos hoc videmus in veritate fidei, sic in professoribus philosophiae. Nam semper sapientes contra

[1] *De Memoria*, cap. 2 ὥσπερ γὰρ φύσις ἤδη τὸ ἔθος.

vulgus divisi sunt, et arcana sapientiae non toti mundo sed plebi philosophantium revelaverunt. Quoniam sapientes Graeci nocturnis vigiliis congregati vacabant sine vulgo collationibus sapientiae, de quibus A. Gellius scribit in libro Noctium Atticarum, i. e. collationum sapientiae nocturnalium, quas sapientes Attici, i. e. Athenienses celebrabant ut multitudinem vitarent. In quo etiam dicit 'stultum est asino praebere lactucas, cum ei sufficiant cardui,' loquens de vulgo, cui rudia et vilia et imperfecta sapientiae sufficiunt alimenta. Nec oportet margaritas spargi inter porcos; nam rerum majestatem minuit, qui vulgat mystica: nec manent secreta, quorum turba est conscia, sicut in libro Gemmarum edocetur. Atque Aristoteles in libro secretorum [1] dicit, 'se fore fractorem sigilli coelestis, si secreta naturae vulgaret.' Et propter hoc sapientes licet darent in scriptis radices arcanorum sapientiae, tamen ramos et fructus vulgo philosophantium non dederunt. Nam vel omiserunt scribere, vel per sermones figurativos et multis modis, de quibus non est ad praesens dicendum, occultaverunt. Quoniam secundum sententiam Aristotelis libro secretorum et Socratis magistri sui, secreta scientiarum non scribimus in pellibus caprarum et ovium, ita quod a vulgo valeant aperiri. Sapientissimi enim et maxime experti multotiens maximam difficultatem in libris reperiunt antiquorum. Et cum philosophus dividat probabile primo libro [2] Topicorum, separat vulgus a sapientibus: Nam dicit probabile esse quod videtur, vel pluribus, vel sapientibus; sub omnibus enim comprehenduntur vulgus et sapientes communiter, ergo per plures vulgus designatur; quia de consortio sapientum non est ipsum vulgus insensatum. Et hoc accidit ei non solum propter sui propriam stultitiam, sed quia in pluribus caput

[1] This work was perhaps the most widely diffused of the apocryphal writings current in the middle ages on Aristotle and Alexander. It was translated from the Arabic by a certain Philip, who dedicates the translation to Guido of Valence, archbishop of Tripoli, with the remark that the treatise was addressed by Aristotle to Alexander, who wished the philosopher to accompany him on his Eastern expedition. It is sometimes spoken of as *liber de regimine principum*, also as *liber decem scientiarum*. Bacon refers to it frequently, and wrote a copious commentary on it, which is preserved among the Tanner MSS. in the Bodleian (116).

[2] *Top.* lib. i. cap. 2, § 7.

recipit languidum et infirmum, quod erroribus et proclive est et auctor imperfectionis cujus nutu ducitur in omnem eventum. Et ideo vulgus imperitum nunquam ad perfectionem sapientiae potest ascendere, nam nescit uti rebus dignissimis: quas si aliquando casu contingat omnia in malum convertit, et ideo justo Dei consilio negatae sunt ei viae perfectionis, et optime secum agitur quando permittitur non errare. Nam suum nomen ostendit omnia quae praedicta sunt, et apud omnes auctores vocatur vulgus imperitum vel insensatum. Imperitia vero in errore et in imperfectione consistit, et ideo vulgo familiaris est error et imperfectio. Nam multi sunt vocati, pauci vero electi ad veritatis divinae receptionem, et similiter philosophicae. Quare Philosophus dicit secundo Topicorum, quod sentiendum est ut pauci, licet loquendum sit ut plures; quia stultitiam vulgi aliquando simulare prudentia summa aestimatur, praecipue cum est in furore suo. Ex his omnibus colligitur malitia et stultitia horum trium et damnum humani generis infinitum, et ideo suspecta sunt in omni causa, nec est eis confidendum: et maxime vulgi sensus est negligendus propter dictas speciales rationes, non quia aliquando cadant supra vera, sed quia ut in pluribus falsis implicantur. Et rarissime exemplum et consuetudo perfectionem habent, vulgus autem nunquam pertingit ad eam, ut prius explanatum est.

CAPITULUM V.

We are warned against them by Scripture and by the wisdom of the ancients.

Munimen vero ad defensionem contra haec habere non possumus nisi mandata et consilia Dei ac scripturae suae et Juris Canonici, sanctorum et philosophorum et omnium sapientum antiquorum, sequamur. Et si his mandatis et consiliis adhaereamus, non possumus errare, nec debemus in aliquo reprobari. In praedictis igitur per philosophum principaliter allegatum est ostendere horum pravitatem et stultitiam, propter quas ipsa possemus et debemus evitare, sed propter causas assignatas de inducendis philosophorum testimoniis silui fere in aliis; nunc autem possum invenire consimilia testimonia, in quibus consilium vel mandatum de obviando istis expressius habetur. Contra igitur sensum vulgi teneamus mandatum Exodi, 'Non sequeris turbam ad faciendum malum; nec in judicio plurimorum adquiesces sententiae, ut a vero devies.' Et consilium Tullii in fine[1] de quaestionibus impleamus dicentis, 'Tu autem etsi in oculis sis multitudinis, tamen in ejus judicio stare noli, nec quod ipsa putet id putare pulcherrimum.' Et Seneca libro de copia verborum invitat nos et consolatur recedentes a via multitudinis, dicens, 'Nondum felix es, si nondum turba te deriserit.' Et contra consuetudinem impleamus consilium Cypriani, 'Consuetudo sine veritate vetustas erroris est, propter quod relicto errore, sequamur veritatem.' Et Augustinus praecipit quod 'veritate manifesta, cedat consuetudo veritati, quia consuetudinem veritas et ratio semper excludit.' Et Isidorus, 'Usus auctoritati cedat; pravum usum lex et ratio vincat.' Et ideo Tullius de immortalitate animae laudans et extollens eos, qui fugiunt consuetudinem dicit 'magni animi est cogitationem a consuetudine revocare.' Nomen autem auctoritatis favorabile est. Et ideo majores nostri venerandi sunt, sive habeant auctoritatem veram sive apparentem, quae est auctoritas ducum vulgi. Et contra usum apparentis auctoritatis non solum propria sunt consilia et mandata secundum vias Dei et sanctorum et philosophorum et omnium sapientum, sed quicquid contra vitia humanae fragilitatis apud auctores veros dici potest, commune est eis qui auctoritate abutuntur praesumpta. Et ideo si consilia et mandata habemus contra defectus verorum auctorum multo magis contra abutentes. Sed quia auctores veridici, ut sancti et philosophi principales, juverunt nos in consideratione veritatis, ideo regratiandum est eis, sicut Aristoteles primo Metaphysicae regratiatur suis praedecessoribus, et in fine Elenchorum[2] de inventis multis ipsemet cupit habere grates. Qui vero prima principia jecerunt, eis non solum regratiandum, sed quasi totum cum quadam reverentia ascribendum. Secundum quod Seneca vult tertio[3] naturalium, 'si quid a posterioribus inventum est, nihilominus referendum

[1] The word *secundi* seems omitted. The passage is in *Tusc. Disp.* ii. 26.

[2] *De Soph. Elench.* xxxiv. § 9.

[3] The passage is not in the third, but in the sixth book, cap. 5. It is not very accurately quoted.

antiquis, quia magni animi fuit rerum latebras primitus dimovere, et plurimum ad inveniendum contulit, qui speravit posse reperire, et quamvis propter humanam fragilitatem in multis defecerunt, tamen excusandi sunt.' Nam in libro memorato ait Seneca, 'cum excusatione veteres audiendi sunt.'

CAPITULUM VI.

The wisest of men are liable to error.

Sed quoniam propter peccatum originale, et propria cujuslibet peccata partes imaginationis laesae sunt, (nam ratio caecutiens est, memoria fragilis, ac voluntas depravata; et verum et bonum uno modo sunt, falsum autem cuilibet vero oppositum et malum cuilibet bono contrarium infinitis modis variantur; quia ut philosophus secundo Ethicorum exemplificat, facile est diverti a signo dato propter multiformem declinationis possibilitatem, sed veritas et virtus sunt in puncto; infinitae etiam sunt veritates, virtutesque, atque innumerabiles gradus in qualibet veritate et virtute,) manifestum est quod mens humana non sufficit dare quod necessarium est in omnibus, nec potest in singulis vitare falsum nec malum. Et ideo ad auctorum dicta verorum potest convenienter addi et corrigi in quampluribus. Et hoc egregie docet Seneca in libro quaestionum naturalium; quoniam libro tertio[1] dicit, 'opiniones veteres parum exactas esse; et rudes circa verum adhuc errabant; nova erant omnia primo tentantibus, postea eadem illa limata sunt; nulla res consummata est, dum incipit.' Et in quarto[2] libro dicit, 'veniet tempus, quo ista quae latent nunc, in lucem dies extrahet, et longioris aevi diligentia. Ad inquisitiones tantorum una aetas non sufficit. Multa venientis aevi populus ignota nobis sciet, et veniet tempus, quo posteri nostri tam aperta nos nescisse mirentur.' Et ideo dicit in prologo majoris voluminis, quod nihil est perfectum in humanis inventionibus et infert, quanto juniores tanto perspicaciores, quia juniores posteriores successione temporum ingrediuntur labores priorum. Et cum jam per causam et auctoritatem verificatum est quod volo, nunc tertio

[1] *loc. cit.* vi. 5.

[2] *Nat. Quaest.* vii. 25.

per effectum probabo. Nam semper posteriores addiderunt ad opera priorum, et multa correxerunt, et plura mutaverunt, sicut maxime per Aristotelem patet, qui omnes sententias praecedentium discussit. Et etiam Avicenna et Averroes plura de dictis ejus correxerunt. Ad haec reprehenditur de mundi aeternitate, quam nimis inexpressam reliquit; nec mirum, cum ipsemet dicit se non omnia scivisse. Nam quadraturam circuli se ignorasse confitetur, quod his diebus scitur veraciter. Et cum ignoravit hoc, multo magis majora. Ac Avicenna, dux et princeps philosophiae post eum, ut dicit commentator super causam de Iride in libro Metaphysicae Aristotelis, et opera in totam philosophiam ab eo digesta, sicut ab Aristotele, hoc manifestant, dixit se materiam Iridis ignorasse, sicut praecedens commentator fatetur. Et in tertio physicorum se ignorasse unum ex decem praedicamentis, viz. praedicamentum habitus, dicere non veretur. Et proculdubio in libro de philosophia vulgata errores et falsa continentur, ut in undecimo Metaphysicae ponitur error de mundi productione, in quo dicitur quod Deus propter infinitam unitatem quam habet, et ne recipiat varietatem dispositionum, non potest creare nisi unum, scilicet, angelum primum, qui creavit secundum cum coelo primo, et secundus tertium cum coelo secundo, et ultra. Et cum in undecimo ponit omne peccatum habere fines suae purgationis in alia vita, et animas peccatrices redire ad gloriam, manifeste errat, et sic in multis. Et Averroes, maximus post eos, in multis redarguit Avicennam; et sic sapientes nostri eum in pluribus corrigunt et non immerito, quia proculdubio erravit in multis locis, quamvis in aliis optime dixit. Et si isti majores erraverunt, multo fortius juniores. Quoniam autem in errores inciderunt, multo magis defecerunt in necessariis, superflua et inutilia cumulantes, dubia et obscura et perplexa spargentes; et haec omnia in libris eorum manifesta sunt, et per effectum in nobis probantur. Nam tanta difficultate videndi veritatem premimur et vacillamus, quod fere quilibet philosophorum contradicit alii, ita quod vix in una vanissima quaestione vel in uno vilissimo sophismate vel una operatione sapientiae, sicut in medicina, et chirurgia, et aliis operationibus secularium, unus cum alio concordat.

Sed non solum philosophi immo sancti aliquid humanum in hac parte passi sunt. Nam ipsimet retractaverunt dictorum suorum quamplurima. Unde Augustinus, qui major in inquisitione veritatum occultarum reputatur, fecit librum de retractatione eorum, quae non bene praedixerat. Et Hieronymus super Isaiam, et alii plures suam non verentur retractare sententiam. Nam celeritate dictandi in transferendo pluries se confitetur deceptum fuisse, et multis aliis modis; et sic omnes alii doctores fecerunt. Sancti etiam ipsi mutuo suas correxerunt positiones et sibi invicem fortiter resistebant. Etiam Paulus Petro resistebat, sicut ipsemet confitetur. Et Augustinus Hieronymi sententias reprehendit, et Hieronymus Augustino in pluribus contradicit. Haec exempla eorum manifesta sunt, ut inferius exponetur, et posteriores priorum dicta correxerunt. Nam Origenem maximum doctorem secundum omnes in multis posteriores reprobant, quia inter caetera posuit errorem Avicennae de animabus peccatricibus, quod nulla finaliter damnabitur. Et cum multi doctores sancti et famosi exposuerunt Israel, ut dicatur *vir videns Deum*, venit Hieronymus super Genesim et probat falsam esse expositionem [1] per rationes irrefragabiles, sicut inferius exponetur. Unde dicit, 'quamvis auctoritatis sint, et eorum umbra nos opprimat qui Israel virum videntem Deum interpretati sunt, tamen magis consentimus Deo vel angelo qui hoc nomen imposuit, quam auctoritati alicujus eloquentiae secularis.' Et doctores catholici in studiis solemnibus constituti nunc temporis in publicis multa mutaverunt, quae sancti dixerunt, eos pie exponentes ut possunt, salva veritate.

CAPITULUM VII.

Plato, Aristotle, Augustine, Boetius, preferred truth to authority.

Quoniam igitur haec ita se habent, non oportet nos adhaerere omnibus quae audimus et legimus, sed examinare debemus districtissime sententias majorum, ut addamus quae eis defuerunt, et corrigamus quae errata sunt, cum omni

[1] Nevertheless in Jerome's Commentaries on Isaiah and on Amos (vol. iv. p. 27, and vol. vi. p. 1032 of Migne's ed.) the interpretation *videns Deum* is accepted.

tamen modestia et excusatione. Et ad hanc audaciam erigi possumus, non solum propter necessitatem, ne deficiamus vel erremus, sed per exempla et auctoritates, ut in nullo simus reprehensibiles de praesumptione. Nam Plato dicit, 'Amicus est Socrates, magister meus, sed magis est amica veritas.' Et Aristoteles dicit, 'se magis velle consentire veritati, quam amicitiae Platonis, doctoris nostri.' Haec ex vita Aristotelis et primo Ethicorum, et libro secretorum, manifesta sunt. Et Seneca dicit libro de quatuor virtutibus cardinalibus, 'Non te moveat dicentis auctoritas; non quis, sed quid.' Et Boetius libro de disciplina scholarum, 'stultum est magistratus orationibus omnino confidere, nam primo credendum est, donec videatur quid sentiat; postea est fingendum eundem in dicendo errasse, si forte reperire queat discipulus, quod expresse objiciat sedulitati magistrali.' Et Augustinus dicit ad Hieronymum, 'quod solos auctores scripturae sacrae vult credere in scribendo non errasse, sed in scripturis aliorum, quantumcunque sanctitate et doctrina polleant, non vult verum putare, nisi per canonem et alios auctores vel per rationes sufficientes possint probare quod dicunt.' Et ad Vincentium dicit 'negare non possum, nec debeo, sicut in ipsis majoribus, ita multa esse in tam multis opusculis meis, quae possunt justo judicio et nulla temeritate culpari[1].' Et in prologo libri tertii de Trinitate dicit, 'sic meis libris nisi certum intellexeris, noli firmum tenere.' Item ad Fortunatianum; 'neque quorumlibet disputationes quamvis catholicorum et laudatorum hominum veluti scripturas canonicas habere debemus, ut nobis non liceat, salva honorificentia quae illis debetur hominibus, aliquid in eorum scriptis improbare atque respuere, si forte invenerimus quod aliter senserint quam veritas habeat, divino adjutorio vel ab aliis intellectum vel a nobis. Talis ego sum in scriptis aliorum, quales volo esse intellectores meorum[2].' Si igitur propter necessitatem vitandae falsitatis et consequendi perfectiorem statum sapientiae possumus, et debemus, et consulimur per sanctos perfectos et philo-

[1] Aug. *de Anima*, lib. iv. cap. I.

[2] Aug. *Ep.* cxlviii. § 15, vol. ii. p. 628, Migne's Ed. The correspondent was not Fortunatus the Manichaean, as Jebb has it.

sophos dignos, ut eorum dicta mutemus loco et tempore et ad eorum sententias addamus ; multo fortius licentiamur ad hoc et cogimur in eis, quae vulgi sunt et eorum qui vulgo praesunt imperito, praecipue cum capita multitudinis istius temporis non attingunt ad dignitatem auctoritatis sanctorum et philosophorum magnorum nec sapientum antiquorum, quorum aliquos nostro tempore conspeximus.

Capitulum VIII.

Remedium vero contra haec tria non est, nisi ut tota virtute auctores validos fragilibus, consuetudini rationem, sensibus vulgi sententias sanctorum aut sapientum reponamus, et non confidamus in argumento triplicato, scilicet, hoc exemplificatum est, vel consuetum, aut vulgatum, igitur tenendum. Patet enim ex praedictis secundum sententias sanctorum et omnium sapientum, quod longe magis sequitur oppositum conclusionis ex eisdem praemissis. Et licet totus mundus sit his causis erroris occupatus, tamen audiamus libenter contraria consuetudini vulgatae. Nam hoc est magnum remedium horum malorum, secundum quod Averroes dicit in fine secundi physicorum, quod mala consuetudo auferri potest per consuetudinem audiendi contraria. Nam multum operatur in opinione, quae est per consuetudinem, audire res extraneas, quod confirmat per effectum, dicens, quod ideo opinio vulgi est fortior quam fides philosophantium, quia vulgus non consuevit audire diversa, sed philosophantes multa audiunt. Propter quod igitur Vestra Sapientia non miretur [1], si contra consuetudinem vulgi et exempla vulgata laborem. Nam haec est via sola perveniendi ad considerationem veritatis et perfectionis.

Their example we should imitate.

Capitulum IX.

Non solum quidem hae causae generales sunt malorum in studio et in vita, sed quarta nequior his tribus est, similiter communis in omni statu, et apud quamlibet personam regnare

The fourth cause of error, false conceit of

[1] J. adds the words 'nec indignetur auctoritati,' which are not very intelligible. O. omits them.

knowledge, is by far the most pernicious.

demonstratur. Conjunxi quidem praedictas tres causas, et propter hoc quod sapientes eas saepius conjungunt et separavi hanc ab illis propter malitiam principalem. Haec enim est singularis fera, quae depascit et destruit omnem rationem, quae est apparentis sapientiae desiderium, quo fertur omnis homo. Nam quamcunque parum sciamus et licet vile, idem tamen extollimus; celebramus etiam multa quae ignoramus ubi possumus occultare ignorantiam, et scienter ostendimus, ut de nihilo gloriemur. Et quicquid nescimus, ubi scientiam ostentare non valemus, negligimus, reprobamus, reprehendimus et adnihilamus. ne videamur aliquid ignorare, quatenus saltem mundo muliebri et fuco meretricio nostram ignorantiam infami remedio coloremus, unde utilissima et maxima et omni decore plena et sua proprietate certissima a nobis per hanc viam et ab aliis relegemus. Haec vero pestis propter malitiam absolutam quam habet recepit cumulum suae pravitatis in eo, qui est fons et origo causarum trium praedictarum. Nam propter zelum nimium sensus proprii et excusandi ignorantiam statim exoritur fragilis auctoritatis praesumptio, qua nitimur propria extollere, et reprehendere aliena. Deinde cum omnis homo diligat opera sua, ut dicit Aristoteles, nostra libenter trahimus in consuetudinem. Et cum nemo sibi soli errat, sed sententiam suam spargere gaudet in proximos, ut dicit Seneca libro secundo epistolarum, nostris adinventionibus occupamus alios et eas in vulgum quantum possumus dilatamus. Necesse est vero causas universales hic praemitti, ad hoc ut error vitetur et veritas relucescat. Nam in morbo spirituali evenit sicut in morbo corporali. Medici enim per signa causas proprias et particulares morbi cognoscunt; sed tam hanc quam illas praecedit causarum universalium notitia, quas a communibus naturalium medicus habet scire; quia dicit philosophus libro de Sensu et Sensato[1], ubi terminant principia philosophiae naturalis incipiunt principia medicinae. Similiter igitur in cura ignorantiae et erroris, ut veritas sana inducatur, ante ipsius propositi exhibitionem requiritur ut

[1] *De Sensu*, cap. I *τῶν περὶ φύσεως οἱ πλεῖστοι καὶ τῶν ἰατρῶν οἱ φιλοσοφωτέρως τὴν τέχνην μετιόντες, οἱ μὲν τελευτῶσιν εἰς τὰ περὶ ἰατρικῆς, οἱ δ' ἐκ τῶν περὶ φύσεως ἄρχονται περὶ τῆς ἰατρικῆς.*

signa et causae particulares ostendantur; sed prae omnibus exiguntur causae universales, sine quibus nec signa aliquid ostendunt nec causae particulares demonstrant. Nata enim nobis est via cognoscendi ab universalibus ad particularia, sicut dicit philosophus in principio Physicorum. Nam, ignoratis communibus, ignorantur quae post[1] communia relinquuntur.

Haec autem causa quarta multum invaluit ab antiquo, sicut nunc improba invenitur quoque in theologicis, sicut in philosophia manifesto, per experientiam et exempla. Moyses enim simplicissimus recepit sapientiam legis a Deo, contra quem Pharao et Aegyptii, et populus Hebraeorum, et omnes nationes murmurabant, ita quod vix plebs Dei electa hanc sapientiam recipere voluit, et tamen praevaluit lex contra adversarios, qui neglexerunt et impedierunt sapientiam quam didicerunt. Similiter Dominus Jesus Christus omni simplicitate et sine plica falsitatis incedens, et apostoli simplicissimi intulerunt sapientiam mundo, quibus satis contradictum erat per ignorantiam solam tantae novitatis, et tandem, licet cum summa difficultate, sacra veritas est recepta. Deinde sancti doctores cum profluvia expositionum legis divinae voluerunt dare, et magno impetu aquarum sapientiae ecclesiam irrigare, diu reputabantur haeretici, et compositores falsitatum. Nam sicut prologi beati Hieronymi in bibliam et alia ejus opera probant, ipse vocabatur corruptor scripturae et falsarius et haeresium seminator, et in tempore suo succubuit, nec potuit sua opera in publico promovere; sed tandem post mortem suam veritas suae translationis claruit et sua expositio, et per omnes ecclesias dilatatae sunt, ita ut nullum vestigium translationis antiquae, scilicet LXX interpretum, qua prius usa fuerat ecclesia, valeat reperiri. Dum etiam beatissimus Papa Gregorius auctoritate functus est, ejus libris non fuit contradictum: sed post mortem famosi in ecclesia egerunt ad hoc, ut comburerentur, et per miraculum Dei pulcherrimum fuerunt salvati, et apparuit mundo sapientia cum sua veritate et in demonstratione plenissima. Et similiter cucurrit impedimentum veritatis apud omnes sacrae scripturae doctores; nam renovantes studium semper receperunt contradictionem

Illustrations from life of Moses, Jerome, Aristotle, Avicenna.

[1] Not postea, as in J.

et impedimenta, et tamen veritas invalescebat et invalescet usque ad dies Antichristi.

Similiter de philosophia. Aristoteles enim voluit contradicere prioribus et multa renovare, qui licet sapientissimus habuit tamen repulsam, et occultationes suae sapientiae visus est usque fere ad haec tempora recepisse. Nam primus Avicenna revocavit philosophiam Aristotelis apud Arabes in lucem plenam. Vulgus enim philosophantium ipsum ignoravit. Pauci enim et modicum philosophiae Aristotelis attigerunt ante tempus Avicennae, qui diu post tempus Machometi Imperatoris philosophatus est. Avicenna vero praecipuus Aristotelis expositor et maximus imitator multas rebellitates passus est ab aliis. Nam Averroes major post eos et alii condemnaverunt Avicennam ultra modum; sed his temporibus gratiam sapientum obtinuit quicquid dicit Averroes, qui etiam diu neglectus fuit et repudiatus ac reprobatus a sapientibus famosis in studio, donec paulatim patuit ejus sapientia satis digna, licet in aliquibus dixit minus bene. Scimus enim quod temporibus nostris Parisiis diu fuit contradictum naturali philosophiae et metaphysicae Aristotelis per Avicennae et Averrois expositores, et ob densam ignorantiam fuerunt libri eorum excommunicati, et utentes eis per tempora satis longa. Cum igitur haec ita se habent, et nos moderni approbamus viros praedictos tam philosophos quam sanctos; et scimus quod omnis additio et cumulatio sapientiae quas dederunt, sunt dignae omni favore, licet in multis aliis diminuti sint, et in pluribus superflui, et in quibusdam corrigendi, et in aliquibus explanandi, manifestum nobis est quod illi, qui per aetates singulas impediverunt documenta veritatis et utilitatis quae oblata fuerunt eis per viros praedictos, nimis erraverunt, et vitiosi plurimum fuerunt in hac parte; sed hoc fecerunt propter scientiae extollentiam et propter ignorantiam. Ergo in nobis ipsis debemus facere idem argumentum, ut cum nos respuamus et vituperemus quae ignoramus, clamemus hoc esse propter ignorantiae nostrae defensionem, et ut illa modica, quae scimus, sublimius attollamus. Permittamus igitur labores introduci congaudentes veritati, quia proculdubio licet cum difficultate veritas semper praevalebit, donec Antichristus et ejus appareant praecursores. Nam semper bonitas Dei est parata sapientiae donum augmentare per subsequentes, et subsequentium sententias in melius transformare.

CAPITULUM X.

Caeterum duo sunt hic discutienda, videlicet, praesumptae scientiae ostentatio, et excusatio ignorantiae infructuosa. Propter primum debemus advertere diligenter, quod cum infinitae sunt veritates Domini et creaturarum, atque in qualibet sunt gradus innumerabiles, oportet quod pauca sciantur a quolibet, et ideo de multitudine scitorum non oportet quicquam gloriari. Et cum intellectus noster se habet ad maxima, sicut oculus vespertilionis ad lumen solis, ut prius secundum philosophiam allegatum est, oportet quod sint parva quae veraciter sciamus; nam pro certo ubi intellectus noster de facili comprehendit, illud valde modicum est et vile, et quanto cum majori difficultate intellexit, tanto est nobilius quod adquiritur. Sed tamen omne id super quod potest intellectus noster, ut intelligat et sciat, oportet quod sit indignum respectu eorum, ad quae in principio credenda sua debilitate obligetur, sicut sunt divinae veritates et multa secreta naturae et artis complentis naturam, de quibus nulla ratio humana dari potest in principio; sed oportet quod per experientiam illuminationis interioris a Deo recipiat intellectum, videlicet in sacris veritatibus gratiae et gloriae, et per experientiam sensibilem in arcanis naturae et artis expergefactus inveniat rationem.

Two heads of this error: exaggeration of knowledge: palliation of ignorance.

Et adhuc minora sunt longe quae credimus, quam quae ignoramus, sicut sunt secreta Dei, et arcana vitae aeternae, quae utcunque vidit Apostolus, ad tertium coelum raptus, nesciens utrum in corpore vel extra corpus, quae tanta sunt, ut non liceat homini loqui de illis. Et similiter in rebus creatis; nam ob magnam intellectus nostri difficultatem certum est, antequam videatur Deus facie ad faciem nunquam homo sciet aliquid in fine certitudinis. Et si per infinita seculorum secula viveret in hac mortalitate, nunquam ad

Perfect knowledge unattainable by man.

perfectionem sapientiae in multitudine scibilium et certitudine pertingeret. Nam nullus est ita sapiens in rerum naturis, qui sciret certificare de veritatibus omnibus quae sunt circa naturam et proprietates unius muscae, nec sciret dare causas proprias coloris ejus, et quare tot pedes haberet, et non plures neque pauciores, nec rationem reddere de membris ejus. Est igitur homo impossibilis ad perfectam sapientiam in hac vita, et ad perfectionem veritatis nimis est difficilis, et pronus[1] et proclivis ad falsa et vana quaecunque; quapropter non est homini gloriandum de sapientia, nec debet aliquis magnificare et extollere quae scit. Pauca enim sunt et vilia respectu eorum quae non intelligit sed credit, et longe pauciora respectu eorum quae ignorat. Et quoniam respectu eorum quae scit homo, restant infinita quae ignorat, et sine omni comparatione majora et meliora et pulchriora: insanus est qui de sapientia se extollit, et maxime insanit qui ostentat et tanquam portentum suam scientiam nititur divulgare.

The simple often surpass the learned in wisdom.

Praeterea quis audet de sapientia gloriari, qui totam medullam, quam unus quantumcunque studiosus addiscit per viginti vel quadraginta annos et cum maximis expensis et laboribus gravissimis, valet uni puero docili certo scripto et verbo sufficienter ostendere per annum unum, vel in minori tempore? Nam hoc probavi in puero praesenti, qui in paupertate magna et modicam habens instructionem, quantitatem anni vix ponens in addiscendo, novit tot et tanta quae omnes mirantur qui eum cognoscunt. Nam securus dico quod licet aliqui sciunt plus de philosophia et linguis, et in diversis diversi ipsum excedunt, non tamen inter Latinos sunt qui eum ex omni parte transcedunt, et ipse singulis eorum est in aliquibus par, et in quibusdam singulos excedit. Nec est aliquis inter Latinos quin multa bona valeat ab isto puero auscultare. Nec aliquis tam sapiens est cui non sit necessarius multis modis. Quamvis omnia didicit meo consilio et regimine et adjutorio, et multa ipsum docui verbo et scripto, tamen me senem in multis transcendit propter meliores radices quas recepit, ex quibus potest fructus salubres expectare, ad quos ego nunquam pertingam. Quare igitur gloriabor de

[1] sic O. J. has multum parvus.

scientia? Non dico, quidam sapientes et experti potuerunt per suam virtutem propriam facilius et citius cernere multa secreta sapientiae, quam hic puer per seipsum, quia non est expertus vires suas, nec percipit quantum novit nec quid potest facere juxta fundamenta sibi tradita, sed sicut in radicibus excedit alios, ut dictum est, sic si sano et efficaci consilio juxta fontalem plenitudinem quam habet dirigeretur, nullus seniorum consequeretur eum in sapientialium profluviis rivorum. Et quoniam sapientes se sentiunt magis vacuos quam insipientes, vel qui nolunt de ipsis confiteri, ideo videmus homines quanto sapientiores sunt, tanto humilius se inclinare ad doctrinam alterius suscipiendam, nec dedignantur simplicitatem docentis, sed ad rusticos, et ad vetulas, et pueros se humiliant. Quoniam simplices et idiotae aestimati sciunt multoties magna quae latent sapientes, sicut Aristoteles docet de somno et vigilia[1], secundo libro. Nam cum simplicibus est sermocinatio Dei secundum scripturam, et experientia reddit nos certos in hac parte, quoniam plura secreta sapientiae semper inventa sunt apud simplices et neglectos quam apud famosos in vulgo. Quia homines famosi in eis occupantur, quae vulgantur, et haec non possunt esse magnalia, sicut patuit in prioribus; plura etiam utilia et digna sine comparatione didici ab hominibus detentis magna simplicitate, nec nominatis in studio, quam ab omnibus doctoribus meis famosis. Proposui igitur Sapientiae Vestrae hoc exemplum, et transmisi per eum, non solum propter duas causas superius annotatas, sed in argumentum perfectum, ut nullus glorietur de sapientia, nec despiciat simplices qui sciunt proponere ea quae famosis hominibus in scientia Deus non concessit, et sciunt renovare et revelare multa secreta, quae sapientes vulgati nondum perceperunt.

CAPITULUM XI.

Ignorance most foul when it

Secundum vitium, quod hic reperitur, est, quod ignorantia retinet locum persuasionis, sed cum veritas impugnatur, nefas

[1] The reference seems to be to *De divin. per somnum*, cap. 2 πάνυ γὰρ εὐτελεῖς ἄνθρωποι προορατικοί εἰσι καὶ εὐθυόνειροι, and again, τὸ πάθος τοῦτο συμβαίνει τοῖς τυχοῦσι, καὶ οὐ τοῖς φρονιμωτάτοις.

usurps the place of knowledge.

est et puritas malitiae, deinde ignorantiae turpitudo crescit magis et clarius revelatur. Crescit quidem, quia nititur a se et ab aliis excludere sapientiam; apertius quidem revelatur, quoniam coram Deo et hominibus[1] veraciter innotescit hoc secundum judicium omnis sapientis; ergo cum judex teneatur habere scientiam causae, non habet homo ignorans auctoritatem judicandi de his quorum habet ignorantiam. Et ideo si de illis affirmet vel neget, ejus judicio stari non debet, immo ex hoc vehementius resistendum, quod sententia qualiscunque feratur ex ignorantia ea auctoritatem non habet: Unde si verum diceret, verisimile non esset. Nam, ut dicit Seneca libro de virtutibus cardinalibus, nullam auctoritatem habet sententia, ubi qui damnandus est damnat. Quapropter sive sapientes apud vulgum, seu secundum veritatem, sive bonus, sive sanctus affirmet vel reprobet quod ignoret, et maxime in excusatione suae ignorantiae, vel ostentatione sapientiae apparentis, approbari non debet ex hac parte, sed negligi et contradici, quamvis ex aliis fuerit magnifice collaudandus.

CAPITULUM XII.

Knowledge of great value is now neglected: as e. g. Mathematics.

Hanc vero causam malorum nostrorum cum aliis tribus ideo specialiter introduxi, ut sciamus nunc, sicut in retroactis temporibus, multa quae sunt utilissima et omnino necessaria, studio absolute considerato et quatuor modis relate praetactis, negari negligi et ex sola ignorantia reprobari. Et pro infinitis latius explicandis posterius in singulis partibus sapientiae, volo nunc aliqua extra praemittere grossiora. Cum enim linguarum cognitio et Mathematicae est maxime necessaria studio Latinorum, ut tactum est superius, et exponetur loco opportuno, et fuit praecise in usu sanctorum et omnium sapientum antiquorum, nos moderni negligimus, adnihilamus, et reprobamus, quia ista et eorum utilitatem nescimus. Deinde, si aliqui sapientes et sancti alia neglexerunt aut humana fragilitate devicti aut ex causa rationabili, nos praesentis temporis obstinate et pertinaciter negligimus et reprobamus, fortificantes nostram ignorantiam propter hoc quod sancti et

[1] sic O. J., omnibus sanctis.

sapientes neglexerunt, non volentes considerare quod in omni homine est multa imperfectio sapientiae, tam in sanctis quam in sapientibus, ut prius evidenter et multipliciter probatum est tam per eorum exempla et auctoritates, quam per rationem et experientiae certitudinem. Praeterea non solum volumus humanam fragilitatem considerare, sed etiam [1] causas rationabiles, quas sancti et sapientes multi habuerunt, quare pro loco et tempore multa utilia vitaverunt propter abusum hominum in eis, quia ea convertebant ad majoris impedimenta utilitatis et salutis. Ne igitur nos simus causa erroris nostri et fiat magnum sapientiae impedimentum ex eo quod vias sanctorum et sapientum non intelligimus, ut expedit, possumus auctoritate sanctorum et sapientum antiquorum multipliciter prius assignata considerare pia mente et animo reverenti, propter veritatis dignitatem quae omnibus antefertur, si sancti et sapientes aliqua quae humanam imperfectionem important protulerunt in quibus seu affirmatis seu negatis non oportet quod nos imitemur ex fronte.

Wise men of old sought to amend and correct error.

Scimus quidem quod non solum dederunt nobis consilium et licentiam hoc faciendi, sed conspicimus quod ipsi multa posuerunt magna auctoritate, quae postea majori humilitate retractaverunt; et ideo latuit in eis magna imperfectio prioribus temporibus. Quod si vixissent usque nunc, multo plura correxissent et mutassent. Cujus signum est sicut et argumentum principalis intenti, quod doctores posteriores quamplurima de sententiis sanctorum mutaverunt, et pie et reverenter interpretati sunt in sensum quem eorum litera non praetendit. Praeterea sancti ad invicem fortiter contendebant et mutuas positiones acriter mordebant, et reprobabant ut taedeat nos conspicere, et supra modum miremur; quod evidens est in epistolis beatorum Augustini et Hieronymi, et multis aliis. Cum enim Hieronymus comparavit se bovi lasso, qui fortius pedem figit, propter hoc quod jam in studio sacro senuerit, et Augustinus fuit junior eo, quamvis episcopus, respondit monacho pontifex quod bos senex fortius figit pedes non animi vigore sed corporis senectute. Et cum Augustinus multa quaereret ab Hieronymo, dicit Hieronymus, 'Diversas

[1] MSS. have nec, but etiam seems required.

tu nominas quaestiones, sed ego sentio meorum opusculorum reprehensiones continuas. Praetermitto salutationes, quibus meum demulces caput; taceo de blanditiis, quibus reprehensionem meam niteris consolari, ut ad ipsas causas veniam; observari legis ceremonias non potest esse indifferens, sed aut bonum aut malum; tu vero bonum, ego malum assero; dum aliud vitas, ad aliud devolveris; dum enim metuis Porphyrium blasphemantem[1] Ebionis haeretici laqueos incurris.' Et talia innumerabilia colliguntur ex libris sanctorum, qui tam in rebus de quibus est contentio quam in modo reprehensionis, multum in scientiis humanae fragilitatis ostendunt, qua affirmabant quod non debebant. Sed constat, non ex certa scientia hoc fecerunt; ergo ex apparenti, et aestimata laboraverunt in hac parte.

Capitulum XIII.

Errors of ancients due to ignorance of foreign languages.

Caeterum non solum ex hujus mortalitatis imperfectione multa reprobant, quibus nos non oportet obstinate inhaerere, immo magis ad eorum honores pie et reverenter interpretari secundum leges veritatis; multa etiam et maxima neglexerunt ex causis certis. Una est, quia non fuerunt translata in linguam latinam, nec ab aliquo Latinorum composita, et ideo non fuit mirum si illorum non aestimabant valorem. Platonis enim libros doctores omnes assumebant in manibus, quia translati fuerunt; sed libri Aristotelis non fuerunt tunc temporis translati. Nam Augustinus fuit primus translator Aristotelis et expositor, sed in minimo et in primo libellorum suorum, videlicet in praedicamentis: nec fuit philosophia Aristotelis tunc temporis Graecis philosophis nota, nec Arabicis, sicut prius tactum est. Et ideo sancti, sicut et alii, neglexerunt philosophiam Aristotelis, laudabant Platonem. Et quia intellexerunt quod Aristoteles persecutus est sententias Platonicas, Aristotelem in multis reprobant, et dicunt rationem haereses congregasse; sicut Augustinus dicit in libro de Civitate Dei ipsum, adhuc magistro suo Platone

[1] blasphemantem om. in J. The passage occurs in *Ep.* cxii. § 16, vol. i. p. 927 of Migne's ed.

vivente, multos in suam haeresim congregasse. Sed tamen omnium philosophantium testimonio Plato nullam comparationem respectu Aristotelis noscitur habuisse. Si igitur sancti philosophiam ejus vidissent, pro certo ea usi essent, et altius extulissent, quia veritatem non negassent manifestam, nec maxima pro minimis declinassent. Nam Augustinus ipse transtulit librum Praedicamentorum de Graeco in Latinum pro filio suo, et exposuit diligenter, plus laudans Aristotelem de hoc nihilo, quam nos pro magna parte suae sapientiae. Quoniam in principio dicit, 'cum omnis scientia et disciplina non nisi oratione tractetur, nullus tamen, o fili mi, in quovis genere pollens inventus est, qui de ipsius orationis vellet origine principiove tractere; idcirco miranda est Aristotelis diligentia, qui disserendi de omnibus cupidus ab ipsius coepit examine, quam sciret et praetermissam a cunctis[1] et omnibus necessariam.' Et in fine dicit, 'haec sunt, fili carissime, quae jugi labore assecuti ad utilitatem tuam de Graeco in Latinum vertimus, scilicet, ut ex his quoque bonam frugem studii a nobis suscipias[2].' Et Alcuinus de expositoribus sacrae scripturae unus, et magister Caroli Magni, illam translationem Augustini de Praedicamentis mire laudavit, et metrico prologo decoravit in his verbis,

'Continet iste decem naturae verba libellus,
Quae jam verba tenent rerum ratione stupenda
Omne quod in nostrum poterit decurrere sensum;
Qui legit, ingenium veterum mirabile laudet,
Atque suum studeat tali exercere labore.
Nunc Augustino placuit transferre magistro
De veterum gazis Graecorum clave Latina,
Quem tibi, Rex magne, Sophiae sectator, amator,
Munere qui tali gaudes, modo mitto legendum.'

Boetius quidem fuit longe post sanctos doctores, qui primus incepit libros Aristotelis plures transferre. Et ipse aliqua logicalia et pauca de aliis transtulit in Latinum. Nec adhuc medietatem, nec partem meliorem habemus. Nam Aristoteles quidem fuit diu ignotus et philosophantibus, nedum aliis, et vulgo Latinorum. Caeterum sancti grammaticalia logicalia

[1] multis, J. [2] Aug. *Categor.* cap. 1 et 22.

et rhetorica, et communia metaphysicae multum efferunt, et abundanter in sacris utuntur. Unde Augustinus in libro de doctrina christiana, 2°, 3°, 4° docet ista applicari ad divina, et in aliis locis: nec non et sancti caeteri idem volunt. Sed de aliis parum et raro loquuntur, imo multam negligunt et negligi edocent aliquando, sicut per Ambrosium patet super epistolam ad Colossenses, et per Hieronymum super illam ad Titum, et per Rabanum de pressuris ecclesiasticis, ac etiam in locis aliis pluribus. Sed constat omnibus philosophantibus et theologis scientias has nullius valoris esse respectu caeterarum nec alicujus dignitatis. Constat omnibus[1] si sancti habuissent usum scientiarum philosophiae magnarum, nunquam cineres philosophicos in tantum extulissent, et ad sacros usus convertissent, quanto enim sanctae scientiae[2] meliores sunt et majores, tanto sunt ad divina aptiores. Sed quia ad manus eorum non devenerunt libri nisi grammatici, logici, rhetorici et de communibus metaphysicae, ideo his se juverunt secundum gratiam eis datam; et quicquid poterant de his laudabiliter extrahere, converterunt copiosius ad legem Dei, ut in expositionibus eorum et tractatibus singulariter manifestum est, et hoc suo loco planius exponetur[3].

CAPITULUM XIV.

Why the early Church neglected philosophy.

Deinde considerandum est diligenter, quod quamvis multa habuissent de majoribus scientiis, non fuit tempus utendi eis nisi in duobus casibus, scilicet, astronomia pro calendario, et musica pro officio divino. Patet enim per historias, quod Eusebius Caesariensis, et beatus Cyrillus, et Victorius et Dionysius Abbas Romanus, cujus doctrinam nunc sequitur ecclesia per leges astronomiae, et alii ex mandato apostolico laboraverunt in hac parte; sed aliae scientiae majores fuerunt neglectae, et praecipue istae quae judicia et opera sapientiae magnifica noscuntur continere. Et hujus causa fuit quintuplex. Nam philosophia ante Christi adventum dedit leges mundo,

[1] Quapropter, J. [2] sanctis, J.

[3] The last part of the sentence is omitted in O. and D. Like many other passages in the first three parts of the work, it has been supplied from Jul.

praeterquam populo Hebreorum, tam de cultu divino quam de moribus et legibus justitiae et pacis inter cives, et belli contra adversarios. Et quoniam hujusmodi leges fuerunt datae, quantum potuit humana ratio, ut Aristoteles vult in fine Ethicorum, ubi transit ad librum legum dicens, 'dicamus nunc quantum possibile est philosophiae in rebus humanis,' &c. principes mundi noluerunt recipere legem Christi, quae fuit supra humanam rationem, et ipsi fuerunt regulati secundum philosophos, et ideo philosophia impedivit ingressum fidei, et tardavit in hoc, quod mundus ea deductus legi celsiori horruit consentire.

Caeterum non solum hoc modo philosophia tardavit fidem Christi, sed jure legum suarum de reipublicae defensione ab omni contrario sensu visa est per judicia futurorum et ostensionem rerum occultarum praesentium, et per opera mirabilia supra naturae et artis communiter operantium potestatem, contendere cum fidei praedicatoribus, quorum proprium fuit non per naturam et artem, sed per Dei virtutem, prophetiam de futuris eructare, occulta producere in lucem, miracula suscitare. Nam quod philosophiae potestas valeat magnifica peragere, quae vulgus non solum laicorum sed etiam clericorum duceret pro miraculis, sequentia declarabunt. Necnon rectores rerumpublicarum ubique per consilia philosophorum suas leges zelantium persecutionis et mortis judicia sanctis Dei graviter intulerunt. Insuper ars magica per totum orbem invalescens, occupans homines in omni superstitione et fraude religionis, quamvis fuerat philosophis odiosa, et ab omnibus debellata, ut certificabitur suo loco, tamen sancti primitivi invenientes mundum occupatum in utraque pro eodem artificio utramque reputabant, quoniam ambae fidei fructum impediebant multis modis. Nam sicut magi Pharaonis Moysi resistebant et populum Aegypti mandato Dei inobedientem faciebant, sic fuit in principio ecclesiae per artis magicae violentiam. Quae cum in eundem effectum, scilicet contra opus fidei, cum philosophia concordabat, totum ejus vituperium in philosophiam redundabat.

Praeterea Deo placuit a principio ecclesiae quod nullum testimonium humanum ei daretur, sed ut veritas fidei tanto vigore mundo radiaret, ut probaretur solum Deo auctore promulgari per testes ab ejus imperio destinatos. His igitur de causis philosophia fuit ab ecclesia in principio et sanctis Dei non solum neglecta, sed eis odiosa ; non tamen propter aliquod quod in ea continetur contrarium veritati. Nam licet imperfecta sit respectu professionis Christianae, tamen ejus potestas non est sectae Christi dissona, immo totaliter ad eam disposita, et ei utilissima et omnino necessaria, sicut omnes credunt, et certificabitur evidenter. Non igitur propter aliquid malum philosophiae ecclesia Dei neglexit et reprobavit eam a principio, sed propter abusores ejus, qui noluerunt eam suo fini, qui est veritas Christiana, copulare. Et propter hanc causam Ecclesia primitiva non fuit sollicita de translatione magnarum scientiarum philosophiae, et ideo sancti doctores Latini copiam philosophicorum non habent, et configurantes se principiis Ecclesiae neglexerunt multa dignissima, sicut in principio propter causas supradictas neglecta fuerunt, non propter aliquid falsum vel indignum quod in philosophia reperiri possit, ut certius suo loco per ipsos Dei sanctos patebit. Nam ostendetur quod sancti Patriarchae et Prophetae a principio mundi omnes scientias receperunt a Deo, quibus illam magnam vitae longitudinem dedit, ut possent experiri quid eis fuerat revelatum, quatenus, fide Christi introducta et evacuata artis magicae fraudulentia, potestas philosophiae ad divina utiliter rapiatur.

CAPITULUM XV.

The example not to be followed by later theologians.

Sicut vero doctores sacri magnificas scientias philosophiae non habuerunt in usu, sic nec posteriores, scilicet Gratianus, Magister sententiarum et historiarum, Hugo de Sancto Victore, et Ricardus de eodem. Nam non fuerunt eorum temporibus translatae, nec in usu Latinorum, et ideo neglexerunt eas, nec dignas sacris mysteriis sciverunt judicare, sed humano sensu respuebant quorum usum non habebant, et in multis obloquuntur, sumentes nihilominus occasionem ex hoc, quod sancti doctores prius easdem neglexerunt ; sed non attendebant causas sanctorum, scilicet, quod non sunt translatae in eorum tempore, et etiam quia ecclesia eas jubere transferri propter causas quinque prius tactas neglexit. Moderni vero doctores vulgi, licet multa de philosophia sint translata, tamen non habent eorum usum, cum et in parvis et vilibus delectati duos libros logicae meliores negligunt, quorum unus translatus est cum commentario Alpharabii[1] super illum, et alterius expositio per Averroem facta sine textu Aristotelis est translata. Et longe magis caetera, quae minorem obtinent dignitatem, sicut novem scientias mathematicae, et sex scientias magnas naturales, quae multas alias scientias sub se comprehendunt, atque morales quatuor partes negligunt dignissimas; et suae ignorantiae quaerunt miserabile solatium per Gratianum et caeteros magistros authenticos, qui non habuerunt notitiam partium philosophiae. Nam post Christum sancti non sunt usi dignitate philosophiae, at non propter hoc, quod ipsa sit sacris contraria sententiis vel indigna, cum ad theologica absolute intelligenda, et respectu ecclesiae Dei et reipublicae fidelium ac conversionis infidelium utiliter et magnifice possit adjuvare, sicut certificabitur suo tempore. Et tanto miserabilius est quod multitudo studentium modernorum magnas negligit scientias, cum tamen fuerunt introductae post Gratianum. Aliqui adhuc vivunt qui in studiis eas perlegerunt.

CAPITULUM XVI.

No sudden change called for. Reform must begin in the highest place.

Quamvis autem istas causas malorum omnium universales persecutus sum, et vellem omnia reduci ad auctoritatem solidam, et sensum sapientum et expertorum, qui pauci sunt, non tamen credat Serenitas Vestra, quod ego intendo Vestriae clementiam Sanctitatis excitare, ut auctores fragiles et ipsam multitudinem Majestas Papalis violenter invadat; nec quod ego indignus sub umbra Gloriae Vestrae suscitem aliquam super facto studii molestiam; sed ut mensa Domini ferculis sapientalibus cumulata, ego pauperculus micas mihi colligam decidentes. Poterit enim Vestrae Potentiae magnitudo sibi et successoribus suis providere de totius sapientiae compendiosa plenitudine non solum absolute habenda, sed quatuor modis

[1] See note to p. 100.

praedictis comparata. Deinde cum Vestrae Paternitatis discretio planiorem de his certitudinem reportaverit, poterit auctoritatis vestrae judicium studiosis et sapientibus de facili persuadere, ut quod vulgus studentium capere non potest, cupidi sapientiae se gaudeant obtinere; insuper quantum sufficit multitudini spes promittit. Nam Hieronymus dicit super Isaiam, 'Multitudo, accepta veritate, de facili mutat sententiam.' Et hoc verum est nisi quando captiosis malesanis retractatur. Nam licet vulgus de se sit proclive ad malum, et quia saepius invenitur caput languidum, tamen nisi qui praesit impediat, satis facile est ad bonum imperfectum, quia instabilis est, et semel mota modum servare non potest, et ideo de facili quantum est de se vertitur ad contraria secundum regimen praesidentis; quoniam omni vento doctrinae flectitur, velut arundo, et quod principi ejus placet legis habet vigorem. Nos enim hoc videmus in omni congregatione hominum, quod secundum arbitrium capitis membra moventur. Nam si qui praeest bonum negligit, subditi obdormiunt; si ad malum excitati, in idem currunt cum furore; si ad bonum, similiter sine discretione festinant. Et si vias perfectionis monet[1], tunc olfacit a longe multitudo, sed gustare non potest nec ab ea debet requiri, ut superius ostensum est. Quod si non est temporis vestri omnia apud vulgum consummare; poterit Vestra Magnificentia locare fundamenta, fontes eruere, radices figere, ut Vestrae Sanctitatis successores quod feliciter inceptum fuerit valeant facilius adimplere.

[1] movet, J.

PARS SECUNDA

HUJUS PERSUASIONIS.

CAPITULUM I.

RELEGATIS igitur quatuor causis totius ignorantiae humanae generalibus, volo in hac secunda distinctione unam sapientiam esse perfectam ostendere, et hanc in sacris literis contineri; de cujus radicibus omnis veritas eruitur. Dico igitur, quod est una scientia dominatrix aliarum, ut theologia, cui reliquae penitus sunt necessariae, et sine quibus ad effectum pervenire non potest; virtutem in suum jus vindicat, ad cujus nutum et imperium caeterae jacent; una tamen est sapientia perfecta, quae in sacra scriptura totaliter continetur per jus canonicum et philosophiam explicanda, et expositio veritatis divinae per illas scientias habetur. Nam ipsa cum eis velut in palmam explicatur, et tamen totam sapientiam in pugnum colligit per seipsum. Quoniam ab uno Deo data est tota sapientia et uni mundo, et propter unum finem. Quapropter haec sapientia ex sua triplici comparatione unitatem sortietur. Caeterum via salutis una licet gradus multi; sed sapientia est via in salutem. Omnis enim consideratio hominis, quae non est salutaris, est plena caecitate, et ad finalem inferni deducit caliginem; propter quod multi sapientes famosi[1] damnati sunt, quia veram sapientiam non habuerunt, sed

Theology the mistress science.

[1] In his commentary on the *Secretum Secretorum*, Bacon, alluding to the legend that Aristotle ascended to heaven in a column of fire, remarks: 'Haec est sententia philosophorum paganorum. Sed nobis Christianis non est licitum hoc sentire nec firmare; quod nisi habuisset fidem Christi revelatam ei, ac fuisset instructus a prophetis, salvari non potuit. Nescimus tamen quod Deus fecit ei et aliis dignis prophetis et philosophis quibus dedit magnalia sapientiae.' [Tanner MSS. 116, *Sec. Secret.* cap. i.]

apparentem et falsam, unde se aestimantes sapientes stulti facti sunt secundum scripturam. Augustinus loquens de sacra scriptura dicit libro secundo de doctrina Christiana, 'si verum est, hic invenitur; si contrarium, damnatur.' Et vult quod ubicunque invenerit Christianus, Domini sui intelligat veritatem esse. Veritas Jesu Christi est sapientia sacrae scripturae. Ergo non alibi veritas est, nisi quae in illa continetur scientia. Et Ambrosius super epistolam ad Colossenses dicit, 'Omnis ratio supernae scientiae et terrenae creaturae in eo est, qui est caput et auctor, ut qui hunc novit, nihil ultra quaerat, quia hic est perfecta virtus et sapientia. Quicquid alibi quaeritur, hic perfecte invenitur.' Cum ergo sacra scriptura dat nobis hanc sapientiam, manifestum est quod hic omnis veritas sit conclusa; si aliqua est sapientia huic contraria, erit erronea, nec habebit nisi nomen sapientiae; sed diversitas, quam non faciat alibi, contrarietatem hic tamen inducit, sicut patet per evangelicam auctoritatem, 'Qui non est mecum, contra me est.' Sic de hac sapientia verum est, ut quod illi annexum non est contra illam esse probetur, et ideo Christiano abhorrendum.

CAPITULUM II.

In theology all studies (including the Canon Law) are rooted.

Haec autem manifestius patent consideranti divisionem scientiarum. Nam si nitamur separare scientias ab invicem, non possumus dicere theologiam. Sub una enim parte philosophiae, scilicet morali quam Aristoteles civilem nominavit, continetur jus civile, ut inferius innotescet. Canonicum[1] vero jus a scripturis sacris nominatur, non ab aliis, sicut ipsum nomen demonstrat; quae scripturae canonicae dicuntur libri Veteris Testamenti, sicut Decretorum parte prima distinctione

[1] The first text-book on Canon Law was the Decretum of Gratian, a monk of Bologna, completed probably in 1143. In 1234 Gregory IX published five books of Decretals, to which a sixth book was subsequently added by Boniface at the end of the century. Bacon, giving way to the strong prejudice felt by himself and other theologians against the Civil Law, maintains that the Canon Law is founded in the main on Scripture. But in reality the Canon Law is based on the Civil Law. 'Everything in the Canon Law was Roman which was not of directly Christian or Jewish origin.' (Rashdall, *Mediaeval Universities*, i. p. 133.)

nona habetur, aut canones nuncupantur. Nam *Canon* Graece, *Regula* Latine dicitur. Et tam jus canonicum quam jus divinum regularem modum vivendi reddere comprobatur. Caeterum jus canonicum totaliter fundatur super auctoritate scripturae et expositore. Nam aut pro constitutionibus allegantur auctores sacrae scripturae, ut Augustinus et alii, aut summi pontifices pro suis statutis indicant auctoritates et exempla Novi et Veteris Testamenti; et ideo hoc jus non est nisi explicatio voluntatis Dei in scriptura. Item jus canonicum vocatur ecclesiasticum, quo regitur in spiritualibus Dei ecclesia, tam in capite quam in membris. Sed nihil aliud sonat scriptura nisi hoc regimen. Praeterea jus naturale continetur in sacra scriptura sicut docetur in principio decreti manifeste. Jura canonica non possunt esse aliena a jure divino, imo de fontibus illius debent derivari, et jus commune est divinum vel humanum; divinum est quod spiritu Dei allatum est mundo in sua scriptura; humanum quando sensu hominis est adinventum. Sed constat ecclesiam Dei regi jure canonico. Quapropter idem jus est divinum, de thesauro sacrae scripturae eruendum. Et hoc manifestum est consideranti partes juris canonici. Nam vel ordinat gradus ecclesiasticorum officiorum, vel sacra Dei determinat, vel forum conscientiae discutit, vel causas ecclesiasticas discindit. Sed horum omnium radices et ipsa stipes erecta apud sacram scripturam reperiuntur. Rami vero penes expositores ejusdem, ut in canone folia, flores, fructus salutiferi capiantur. Nam sermones canonici suavis ornatus foliis comparantur secundum scripturam. Flores autem et fructus sunt segetum aurei palmites et uvarum maturitio. Et ideo jus canonicum sine potestate scripturae in uno corpore continetur, sicut unius arboris corpus ex radicibus et stipite, ramis, floribus, et fructibus constituitur.

CAPITULUM III.

Quod autem philosophia[1] non sit aliena a Dei sapientia, sed in ipsa conclusa, manifestandum est. Si enim a philosophis

St. Augustine has maintained

[1] philosophiae, J.

that theology includes philosophy.

tanquam injustis possessoribus rapere debent Christiani utilia quae in libris eorum continentur, sicut dicit Augustinus, patet quod philosophia est condigna sacrae veritati. Et iterum in libro, scilicet de doctrina Christi, dicit quod philosophorum aurum et argentum non ipsi instituerunt, sed de communibus quasi metallis divinae providentiae, quae ubique est effusa, eruitur; quod praefiguratum fuisse dicit, sicut Aegyptii fecerant vasa atque ornamenta de auro vel argento et vestem quam ille populus exiens de Aegypto sibi potius tanquam ad usum meliorem vindicavit, sic doctrinae gentilium liberales disciplinas usui veritatis aptiores et morum praecepta utilissima continent, deque Deo ipso colendo multa inveniuntur apud eos. Ulterius hoc explicat, dicens, in omnibus humanis tractatibus, quae sunt moralia, vel historialia vel artificialia, naturalia, logicalia vel grammaticalia, sunt nobis necessaria. Nam pro moralibus dicit vestem, quae illorum est hominum quidem institutio sed tamen accommodata humanae societati qua in hac vita carere non possumus, in usum converti christianum debet. De historialibus dicit, historia gentilium plurimum nos adjuvat ad sanctos libros intelligendos. De aliis vero considerationibus tam artificialibus quam naturalibus dicit, Artium autem caeterarum, quibus adfabricatur domus, et hujusmodi, medicinae, vel agriculturae, vel quorum omnis effectus actio est, ut saltationum, cursionum, &c., harum autem cognitio usurpanda est ad judicandum, ne omnino nesciamus quid scriptura velit insinuare, cum de his artibus aliquas facit figuras. Pro omnibus naturalibus dicit, Benignam sane operam faceret pro sacra scriptura, qui proprietates temporum et locorum, lapidum, et caeterarum rerum inanimatarum, plantarum et animalium colligeret. Et pro logicalibus dicit, Nam de eis pro theologia possunt quaedam necessaria colligi et condigna, sed non video, ut ait, utrum hoc possit sine eis. Atque in libro secundo de ordine disciplinae dicit, Ad sacram scientiam nullus debet accedere sine scientia logicali. De mathematicis dicit Cassiodorus, Geometriam, arithmeticam, astronomiam, musicam cum sollicita mente revolvimus, sensum acuunt, limamque ignorantiae detergunt, et ad illam divinam contemplationem, Deo largiente, perducunt; quas merito sancti patres legendas persuadent, quoniam ex magna parte appetitus a carnalibus rebus extrahitur, et faciunt desiderare quae solo corde possumus respicere. Et Augustinus doctos multos sanctos fuisse commemorat, cum quaerit, Nonne aspicimus quanto auro subsarcinatus exierit de Aegypto Cyprianus doctor suavissimus et martyr beatissimus, quanto Victorinus et alii multi innumerabiles Graeci, et Moyses quidem eruditus omni sapientia Aegyptiorum?

Capitulum IV.

Corroboration by other authorities.

Non solum autem beatus Augustinus, sed et alii sancti idem asserunt. Nam Hieronymus ad magnum oratorem dicit, Si scripturas sacras legeres, quis nesciat et in Moyse ac prophetarum libris quaedam assumpta ex gentilium libris? Et inducit ad hujus probationem prophetas ipsos et omnes doctores famosos a principio ecclesiae, qui philosophorum doctrinis fidem Christi persuaserunt principibus et infidelibus, ac roboraverunt multipliciter. Et Beda super librum Regum dicit, quod liberalium scientiarum sapientiam quasi suam sumere licitum est Christianis ac divinam, alioquin Moysen et Daniel sapientia Aegyptiorum et Chaldaeorum non paterentur erudiri. Iterum de factura templi dicit Solomonem cum suis servis signare Christum, et Hiram cum suis significare philosophos et sapientes gentiles, ut templum Dei, hoc est ecclesia, non solum sapientia apostolica, sed philosophorum construeretur; quia gentiles ab errore conversi atque ad veritatem evangelii transformati, melius ipsos gentium errores noverant, et quo certius noverunt, eo artificiosius hos expugnare atque evacuare didicerunt. Paulus evangelium quod per revelationem didicerat, melius novit; sed Dionysius melius revincere poterat falsa dogmata, quorum cum erroribus argumenta a puero noverat; et ideo Salomon dicit, Scis enim quod non est in populo meo vir, qui noverat ligna caedere, sicut Sidonii. Haec et hujusmodi multa allegat venerabilis Beda. Sed et Paulus Apostolus Epimenidis poetae usus est versiculo, scribens ad Titum, 'Cretenses semper mendaces, malae bestiae, ventres pigri.' In alia quoque epistola Menandri ponit senarium, 'Corrumpunt bonos mores confabulationes pessimae.'

CAPITULUM V.

The principle which stirs our intellectual powers is without us, not within.

Causae autem, quare sancti affirmant quod quaerimus, et figuratum fuisse declarant, possunt assignari; primo propter hoc quod ubicunque veritas invenitur, Christi judicatur secundum sententias et auctoritates superius allegatas. Idcirco quamvis aliquo modo veritas philosophiae dicatur esse eorum; ad hanc tamen habendam primo lux divina influxit in animos eorum, et eosdem superillustravit; 'Illuminat enim omnem hominem venientem in hunc mundum,' sicut dicit scriptura; cui sententiae philosophi ipsi concordant. Nam ponunt intellectum agentem et possibilem [1]; anima vero humana dicitur ab eis possibilis, quia de se est impotens ad scientias et virtutes, et eas recipit aliunde. Intellectus agens dicitur, qui

[1] This passage deals with one of the most important of mediaeval controversies in which Aquinas was at issue with Averroes and his Arabian predecessors. These maintained the existence of an universal reason, of which individuals were more or less the partakers. Aquinas refutes this view in *S. T.*, Pars 1, Quaest. lxxvi. Art. 2: 'Utrum intellectivum principium multiplicetur secundum multiplicationem corporum.' See also his systematic treatise, *De Unitate Intellectus*. It will be seen that Bacon in this passage appears to side with the Arabians against St. Thomas. Cf. also the corresponding passage in *Opus Tertium*, cap. 23. But in his more elaborate treatment of the subject in *Communia Naturalium* (unpublished), Pars iv. cap. 14, he is careful to distinguish his doctrine from that of Averroes. That there should be a divine light pervading the world, was one thing: that the intellect of mankind should be one and the same substance, was quite another thesis which cut at the root of morality. Renan (Averroes, p. 71 et seq.) shows that, though the doctrine attacked by St. Thomas is commonly identified with Averroes, it had been held by Alkindi, Alfarabi, and Avicenna centuries before. It may be noted, in passing, that Jebb's error in writing repeatedly *interius* for *intellectus* has gone far to make this passage unintelligible.

Aristotle can hardly be called as a witness for the Arabian side of the controversy, though his distinction between intellect and the other physical faculties is of course emphatic enough. Cf. *De Anima*, iii. 4, § 5 *τὸ μὲν γὰρ αἰσθητικὸν οὐκ ἄνευ σώματος, ὁ δὲ (νοῦς) χωριστός*. Again, ii. 2, § 9 *τοῦτο μόνον ἐνδέχεται χωρίζεσθαι, καθάπερ τὸ ἀΐδιον τοῦ φθαρτοῦ*. The passage of the translation of which Bacon complains is *De Anima*, iii. 5 *ἐπεὶ δ' ὥσπερ ἐν ἁπάσῃ τῇ φύσει ἐστί τι τὸ μὲν ὕλη ἑκάστῳ γένει (τοῦτο δ' ὃ πάντα δυνάμει ἐκεῖνα), ἕτερον δὲ τὸ αἴτιον καὶ ποιητικὸν τῷ ποιεῖν πάντα, οἷον ἡ τέχνη πρὸς τὴν ὕλην πέπονθεν, ἀνάγκη καὶ ἐν τῇ ψυχῇ ὑπάρχειν ταύτας τὰς διαφοράς*.

influit in animas nostras illuminans ad scientiam et virtutem; quia licet intellectus possibilis possit dici agens ab actu intelligendi, tamen sumendo intellectum agentem ut ipsi sumunt, vocatur influens et illuminans possibilem ad cognitionem veritatis. Et sic intellectus agens secundum majores philosophos non est pars animae, sed est substantia intellectiva alia et separata per essentiam ab intellectu possibili; et quia istud est necessarium ad propositi persuasionem, ut ostendatur quod philosophia sit per influentiam divinae illuminationis, volo illud efficaciter probare; praecipue cum magnus error invaserit vulgus in hac parte, necnon multitudinem magnam theologorum, quoniam qualis hic est in philosophia, talis in theologia esse probatur. Dicit enim Alpharabius in libro de intellectu et intellecto, 'Quod intellectus agens, quem nominavit Aristoteles in tertio tractatu suo de anima, non est in materia, sed est substantia separata.' Et Avicenna quinto de anima et decimo metaphysices idem docet, necnon ipse philosophus dicit, 'Quod intellectus agens est separatus a possibili et immixtus.' Item vult quod intellectus agens sit incorruptibilis secundum esse et substantiam, quoniam dicit ipsum differre a possibili penes incorruptionem, sed possibilis est incorruptibilis secundum substantiam, et corruptibilis secundum esse, propter separationem ejus. Ergo agens secundum esse et substantiam erit incorruptibilis; quapropter non erit pars animae, quoniam tunc secundum esse suum in corpore corrumpetur, quando separetur; et dicit, quod se habet ad possibilem, sicut artifex ad materiam, et sicut lux solis ad colores. Artifex enim est extra materiam in quam agit, et separatus ab ea per essentiam; similiter lux solis expellens tenebras a corporibus separata est ab eis per essentiam, et advenit aliunde. Dicit etiam, intellectus agens scit omnia et [est] semper in actu, quod nec animae nec angelo convenit, sed soli Deo; item a digniore parte magis habet res denominari, ergo magis dicetur sciens per agentem, quam ignorans per possibilem, ante inventionem et doctrinam. Item Aristoteles [1] dicit quod

[1] Aristotle puts forward this view tentatively before he had come to a definite conclusion as to the distinction between reason and other faculties. *De Anima*, ii. 1, § 12 *Ἄδηλον εἰ οὕτως ἐντελέχεια τοῦ σώματος ἡ ψυχὴ ὥσπερ πλωτὴρ πλοίου.*

intellectus est in corpore, sicut nauta in navi quantum ad hoc, quod non est alligatus alicui parti, sicut nec nauta navi; sed nauta non est profectio, sed motor tantum.

Cum igitur haec sententia sit consona veritati, ut textus philosophi evidenter praetendit, et expositores declarant, ne aliquis cavillator a latere insurgat, allegans idem quo vulgus decipitur, dico quod Aristoteli imponuntur ista verba, 'Quoniam in omni natura est aliquid quod agat et aliud quod patiatur, ista est in anima,' immo respondeo quod multoties falso translatum est et obscurum. Nam cum tertio caeli et mundi dicatur, quod circulus et figura orbicularis replent locum [1], istud est falsum; ut sciunt experti in naturalibus et geometricis, sicut Averroes demonstrat ibidem. Et quod tertio Meteorologicorum dicitur de iride est falsum etiam. Nam experientia docet, quod quandocunque luna sit plena et pluat, nec ipsa sit nubibus cooperta, accidit iris. Et sunt multa alia falso translata, cujus causa patebit ex tertia parte hujus operis, et plura obscura, in quibus quilibet alii potest contradicere. Et in hoc loco accidit utrumque vitium vel saltem secundum, quod probo per ipsum Aristotelem. Nam ipse dicit secundo physicorum, quod materia non coincidit cum aliis causis in [2] eodem secundum numerum, ergo in nulla natura sunt simul agens et materia, igitur nec in anima. Si igitur ad literam teneatur textus male translatus, tunc omnino falsus est et contra Aristotelem alibi, et contradicit sibi tantus auctor; et qualitercunque contingat verbum suum in secundo Physicorum est verum, et ab omnibus concessum; ergo sermo suus tertio de anima est falso translatus, et indiget expositione. Nam nihil aliud intendit, nisi quod in anima, videlicet, in operatione requiruntur duo, scilicet, agens et materia; sicut in omni natura, id est operatione naturae duo exiguntur, scilicet efficiens et materia. Et idem est verum, sed agens semper est aliud a materia et extra

[1] In *De Caelo*, iii. 8, § 1, Aristotle distinctly says that only three plane figures can form a continuous surface, the triangle, square, and hexagon; and only two solid figures can occupy space, the pyramid and cube.

The reference to the lunar rainbow is *Meteor.* iii. 2, § 9. It fully bears out Bacon's view as to faulty translation. What Aristotle really said was that during an experience of fifty years he had only twice observed the lunar rainbow.

[2] sic, O. cum eodem, J.

eam secundum substantiam, licet operetur in ea. Caeterum possumus aliter hunc locum consolari. Nam Aristoteles quarto[1] Physicorum dicit, quod octo modi sunt essendi in actu, quorum unus est ut movens in moto, quia movens seu agens est secundum virtutem suam in materia sua, licet non secundum substantiam. Et sic est in omni natura in qua operatur, et ita in anima; et sic nullo modo sequitur quod intellectus agens sit pars animae, ut vulgus fingit; et haec sententia est tota fidelis et a sanctis confirmata. Et Augustinus dicit in soliloquiis et alibi, 'Quod soli Deo est anima rationalis subjecta in illuminationibus et influentiis omnibus principalibus.' Et quamvis angeli purgent mentes nostras et illuminent et excitent multis modis, et sunt ad animas nostras tanquam stellae respectu oculi corporalis, tamen Augustinus ascribat Deo influentiam principalem; soli influentia luminis cadentis per fenestram ascribitur, et angelus aperienti fenestram comparatur, secundum Augustinum. Et quod plus est, vult in pluribus locis quod non cognoscimus aliquam veritatem nisi in veritate increata et in regulis aeternis. Cum igitur Deus illuminaverit animas eorum in percipiendis veritatibus philosophiae, manifestum est quod eorum labor non est alienus a sapientia divina.

CAPITULUM VI.

Tertia causa, propter quam sapientia philosophiae reducitur ad divinam, est quia non solum mentes eorum illustravit Deus ad notitiam sapientiae adquirendam, sed ab eo ipsam habuerunt et eam illis revelavit. Et Augustinus dicit super Johannem, eis praestitit Deus sapientiam. Aristoteles in libro secretorum asserit manifeste totam philosophiam fuisse a Deo datam et revelatam; et unus de maximis philosophis, scilicet Tullius in quaestionibus Tusculanis quaerit, 'Philosophia quid est nisi donum, ut ego credo, inventum Dei?' Unde et dicit, 'quod nec poeta grave plenumque carmen sine coelesti aliquo instinctu effundit.' Et Augustinus octavo de Civitate Dei docet et approbat quod Socrates pater philosophorum firma-

Philosophy is a revelation

[1] *Nat. Auscult.* iv. 3, § 1.

vit, quod non potest homo causas rerum scire, nisi in luce divina, et per donum ejus. Et quilibet potest per se experiri quod nihil primo ab homine invenitur quod sit de potestate philosophiae. Et pono de minimo exemplum; quoniam licet universalia Porphyrii sunt apud eum sufficienter explicata per logicam, metaphysicam, et naturalem philosophiam sufficientissime expositam, tamen non est homo ita bene studiosus, quin oportet ut doctores habeat et per longa tempora audiat et studeat, antequam sciat totam veritatem universalium. Et nullus vix ante mortem cognoscit; quod patet propter discordiam omnium [1], quia aliqui ponunt ea solum in anima, aliqui extra, aliqui medio modo. Si igitur talis ignorantia est horum, multo magis per se nunquam perveniet homo ad veritatem philosophorum. Quapropter veritatem horum est necesse a principio fuisse homini revelatam. Et cum puerilis revelatio est necessaria, multo fortius in tota sapientia philosophiae, quod et a Deo est, et ille dedit et revelavit, et ideo oportet quod suae sapientiae sit conformis.

CAPITULUM VII.

needed for the acquisition of divine truth.

Caeterum totius philosophiae decursus, consistit in eo, ut per cognitionem suae creaturae cognoscatur creator, cui propter reverentiam majestatis et beneficium creationis et conservationis et futurae felicitatis serviatur in cultu honorifico et morum pulchritudine et legum utilium honestate; ut in pace et honestate vivant homines in hac vita. Philosophia enim speculativa decurrit usque ad cognitionem creatoris per creaturas. Et moralis philosophia morum honestatem, leges justas, et cultum Dei statuit, et persuadet de futura felicitate utiliter et magnifice secundum quod possibile est philosophiae. Haec sunt certa discurrentibus per omnes partes philosophiae

[1] Bacon's attitude in the controversy on Universals is not easy to define with precision. 'Universale,' he says (*Communia Naturalium*), 'non est nisi convenientia plurium individuorum.' The universal *ante rem* he entirely rejected. 'Individuum est prius secundum naturam.' See Charles's monograph, pp. 239 and 383–9. A property or attribute common to several individuals ('natura aliqua communis solis individuis') would seem to be his definition of it. But Bacon attached less importance to the controversy than his contemporaries.

principales, sicut sequentia docebunt. Cum igitur haec sint omnino necessaria christianis, et omnino consona sapientiae Dei, manifestum est quod philosophia necessaria est legi divinae et fidelibus in ea gloriantibus.

CAPITULUM VIII.

Spiritual truth underlies physical.

Item omnes sancti et sapientes antiqui in suis expositionibus sensum literalem colligunt ex naturis rerum et proprietatibus earum, ut per convenientes adaptationes et similitudines eliciant spirituales sensus : quod declarat Augustinus libro de doctrina Christi secundo, ponens exemplum de verbo Domini dicentis, 'Estote prudentes sicut serpentes, et simplices sicut columbae.' Nam voluit Dominus per hoc, ut ad similitudinem serpentis totum corpus exponentis pro defensione apostoli et apostolici viri se et sua darent pro Christo capite suo et pro fide sua. Et propter hoc omnis creatura in se vel in suo simili, vel in universali vel in particulari, a summis coelorum usque ad terminos eorum ponitur in scriptura, ut sicut Deus fecit creaturas et scripturam, sic voluit ipsas res factas ponere in scriptura ad intellectum ipsius tam sensus literalis quam spiritualis. Sed tota philosophiae intentio non est nisi rerum naturas et proprietates evolvere, quapropter totius philosophiae potestas in sacris literis continetur; et hoc maxime patet, quia longe certius ac melius et verius accipit scriptura creaturas, quam labor philosophicus sciat eruere. Quod pro infinitis exemplis pateat ad praesens in Iride. Philosophus Aristoteles suis obscuritatibus nos perturbat ut nec aliquid quod dignum sit valeamus per eum intelligere. Nec mirum, cum Avicenna dux et princeps philosophiae fateatur se naturam Iridis ignorasse; et causa hujus est, quia philosophi causam Iridis finalem ignoraverunt; et ignorato fine ignorantur ea quae sunt ad finem; quia finis imponit necessitatem iis quae ad finem ordinantur, ut Aristoteles vult secundo Physicorum. Causa vero finalis Iridis est dissipatio humiditatis aqueae, sicut patet ex libro Geneseos; unde semper in apparitione Iridis est nubium resolutio in stillicidia infinita, ut consumantur aqueae humiditates tam in aere, quam in mari et terra, quia una pars Iridis cadit in sphaeras aquae et terrae. Consumptio vero aquae non potest esse per Iridem, nisi propter radios solis facientis eam. Nam per varias reflectiones et fractiones congregantur radii infiniti, et congregatio radiorum est causa resolutionis et consumptionis aquarum, et ideo Iris generatur per reflectiones multiplices. Non enim possunt radii congregari, nisi per fractionem et reflectionem, ut postea patebit. Ex scriptura igitur Geneseos cum dicitur 'Ponam arcum meum in nubibus coeli, ut non sit amplius diluvium super terram,' accipitur causa finalis ipsius Iridis. Ex quo investigari potest causa efficiens, et modus generandi Iridem, qui modus non fuit notus philosophis sufficienter secundum quod libri eorum manifestant nobis. Et ita est de omni creatura. Impossibile enim est quod homo sciret veritatem creaturae ultimam secundum quam accipitur in scriptura, nisi fuerit specialiter a Deo illustratus. Nam creaturae accipiuntur ibi propter veritates gratiae et gloriae eliciendas, quas philosophi nescierunt, et ideo ad potestatem ultimam sapientiae creaturarum non venerunt, sicut sacra scriptura eam in suis continet visceribus. Unde tota philosophia jacet in sensu literali sacris mysteriis gratiae et gloriae decorata, tamquam quibusdam picturis et coloribus nobilissimis redimita.

CAPITULUM IX.

Philosophy revealed to the patriarchs.

Distinctio finalis[1] hujus partis, in qua ad confirmationem omnium praedictorum et dicendorum ostenditur quod tota sapientia revelata est primo sanctis; et ostenditur propositum in universali.

Et hoc ultimo confirmari potest per hoc, quod eisdem personis data est philosophiae plenitudo quibus et lex Dei, scilicet, sanctis patriarchis et prophetis a principio mundi, Et non solum est necessarium propter articulum qui hic trac-

[1] In the MSS. this is called Distinctio tertia, but as no first and second Distinctions are noted, I have substituted the word 'finalis.' This concluding portion of Part II, which is of great importance, seems to have been added by Bacon as an appendix. The MS. has no division into chapters; I have supplied this.

tatur, sed propter totum negotium sapientiae certificandum. Nam impossibile fuit homini ad magnalia scientiarum et artium devenire per se, sed oportet quod habuerit revelationem, qua probata nihil debet apud nos dubitari de arcanis sapientiae repertis apud auctores. Sed nullum capitulum sapientiale est tanti laboris sicut est certificatio hujus rei, eo quod est magnum fundamentum totius comprehensionis humanae. Atqui contrarietates et dubia multipliciter intercurrerunt, et oportet auctores et volumina abundantius revolvi quam pro aliquo alio articulo, qui in toto sapientiae studio valeat reperiri. Dico igitur quod eisdem personis a Deo data est philosophiae potestas quibus et sacra scriptura, videlicet, sanctis ab initio, ut sic appareat una sapientia esse completa et omnibus necessaria. Soli enim patriarchae et prophetae fuerunt veri philosophi qui omnia sciverunt, non solum legem Dei, sed omnes partes philosophiae. Hoc enim nostra scriptura satis nobis ostendit, quae Joseph erudivisse principes Pharaonis et senes Aegypti prudentiam docet; et Moysen fuisse peritum in omni sapientia Aegyptiorum. Et Bezaleel et Eliab hoc demonstrant, qui omni intellectu et sapientia rerum naturalium fuerunt illustrati; uno enim flatu Spiritus Sanctus eos illuminavit et docuit totam potestatem naturae in rebus metallicis et caeteris mineralibus. Sed et Solomon sapientior omnibus praecedentibus et subsequentibus secundum testimonium scripturae plenam obtinuit philosophiae potestatem. Et Josephus primo antiquitatum libro, capite secundo, dicit, quod cum filii Adae per Seth fuerunt viri religiosi et ab ipso Deo dilecti, Deus dedit eis sexcentos annos vivere propter gloriosas partes philosophiae in quibus studuerunt, ut, quod Deus eis revelavit, possent experiri per vitae longitudinem; et addit, quod Noe et filii ejus docuerunt Chaldaeos partes philosophiae, et quod Abraham intravit Aegyptum et docuit Aegyptios. Et postea in octavo libro, quod nullam naturam inexaminatam Solomon praeteriit, sed de omnibus philosophatus est, et disciplinam proprietatum earum evidenter exposuit, et tangit quomodo descendens ad singula composuerit quatuor millia librorum et quinque.

Et maximus Aristoteles ipsa veritate coactus dicit in libro

By these the thinkers of Greece were instructed.

secretorum; 'Omnem sapientiam Deus revelavit suis prophetis et justis et quibusdam aliis, quos praeelegit et illustravit spiritu divinae sapientiae, et dotavit eos dotibus scientiae. Ab istis enim sequentes philosophi philosophiae principium et originem habuerunt et scripserunt artium et scientiarum principia et secreta, quia in scriptis eorum nihil falsum nihil reprobandum invenitur, sed a sapientibus approbatum.' Et Averroes dicit super partem coeli et mundi, 'quod in tempore antiquorum ante Aristotelem et alios philosophos fuit philosophia completa, ad cujus completionem Aristoteles suo tempore aspirabat.' Et apud Albumazar in majori introductorio et alibi, et penes aliquos habetur multipliciter, quod Noe et filii ejus multiplicaverunt philosophiam; et praecipue Sem praevaluit in hac parte. Deinde post istos fuerunt viri [1] . . . nomine vulgato. Omnes philosophi et poetae majores et juniores fuerunt post Noe et filios suos, et Abraham. Nam et Aristoteles et omnes consentiunt in hoc, quod primi philosophantes [2] fuerunt Chaldaei et Aegyptii, unde adhaeret sententiis patrum Chaldaeorum in undecimo Metaphysicae. Quia licet Noe et filii ejus docuerunt Chaldaeos, antequam Abraham docuit Aegyptios, tamen non fuit studium more scholastico ita cito institutum, sed paulatim crevit ordo ejus et exercitium.

Quatenus omnis igitur dubitatio tollatur in hac parte, videamus decursum et seriem infidelium philosophorum, et poetarum, et omnium sollicitantium de studio sapientiali, et percipiemus quod post Abraham et decessores suos, quibus a Deo sapientia revelata est, inventi sunt singuli qui aliquem titulum adepti sunt. Nam quantumcunque volumus strictius computare Zoroastres invenit artes Magicas, secundum Augustinum vicesimo primo de civitate Dei; et secundum omnes auctores hoc vulgatum est; sed hic fuit Cham filius Noe, ut Clemens, libro suo, et magister historiarum, et speculum historiale conscribunt. Deinde Io, quae postea Isis dicta est, dedit literas Aegyptiis, ut Augustinus dicit libro de civitate Dei octavo

[1] There is a hiatus here in O. & D. It may have been supplied by Jul. But fire has rendered the passage illegible.

[2] There is a reference of this kind in a fragment of an apocryphal work attributed to Aristotle entitled Μαγικός, quoted by Diog. *Laert.* i. 1.

decimo. Ante hujus tempora non fuit secundum Augustinum sapientiae studium literis et scriptis pertractatum, quamvis doctrinis Abraham instructi fuerunt. Et Isis, ut Augustinus ait, dicitur fuisse filia Machi, qui fuit primus rex Argivorum, qui regnavit primo anno Jacob et Esau nepotum Abrahae, sicut Augustinus et historiae confitentur. Quanquam et alii voluerunt, quod Isis veniret de Aethiopia in Aegyptum, et eis literas dedit, et multa beneficia contulit, sicut recitat Augustinus. Sed tamen ante tempus Machi non fuit, ut in ordine regum Aegypti in chronicis reperitur. Eodem tempore fuit Minerva aetate virginali apparens, multorum, ut ait Augustinus libro praedicto, inventrix, quae Pallas dicitur, et apud poetas Dea sapientiae nuncupatur, et Athena vocatur, atque Tritonia, ut dicit Augustinus. Et Isidorus hunc locum esse in Africa, qui Trito vocatur, recitat octavo libro etymologiarum, et Plinius quinto libro, a quo Pallas dicitur Tritonia et fuit tempore diluvii Ogygis regis, quod illi ascribitur, quia in Achaia accidit tempore ejus, qui secundum Augustinum, Eusebium et Hieronymum, et Solinum libro de mirabilibus mundi, fuit tempore Phoronei filii Machi. Regnavit autem Machus quinquaginta annis, et Phoroneus filius ejus sexaginta, cujus tempore facta est repromissio Jacob, sicut patri suo, ut dicit Augustinus. Et ideo Ogyges fuit tempore Jacob; unde Solinus dicit, diluvium primum in Achaia fuisse tempore Ogygis et Jacob patriarchae. Quod diluvium fuit ante diluvium Deucalionis per 600 annos, ut idem narrat Solinus. Nam ut Hieronymus et Eusebius narrant, regnante Cecrope primo rege Atheniensium, sub quo Moyses eduxit filios Israel de Aegypto, fuit Deucalionis diluvium.

Et sub Phoroneo moralis philosophia incepit apud infideles. Nam Augustinus dicit, quod sub legum et judiciorum institutis Graecia clarior facta est[1]. Sed post fuerunt mores et jura vivendi; quod patet per inhibitionem sanguinis et licentiam de usu carnium post diluvium, et de emptione et venditione apud Abraham pro spelunca; atque ex sanctitate Abrahae et patrum suorum leges honestas et sacras vivendi concludit ab eis fuisse edoctos. Et cum minus utiles scientias perfecerunt, non debuit tantorum virorum sapientia scientiam

[1] *De Civitate Dei*, lib. xviii. cap. 3.

morum utilissimam negligere. Deinde primus inter viros titulo majoris sapientiae doctor fuisse perhibetur, quia optimus sapientiae investigator fuit, Prometheus[1], quem poetae ferunt de luto formasse homines, cujus frater, ut dicit Augustinus, fuit Atlas magnus astrologus; unde occasionem, ut Augustinus refert, fabula invenit, quod eum portare coelum finxerit, quamvis mons ejus nomine nuncupatur cujus altitudine potius coeli portatio vulgo videatur, qui in extremis Africae maritimis prope Gades Herculis attollitur velut in caelum. Sed priores fuerunt filii Noe et Abraham qui fuerunt periti astronomi, ut Josephus narrat et Isidorus tertio libro, et Clemens libro primo. Nam hi secundum Augustinum floruerunt, quando Moyses natus est. Et Isidorus concordat libro quinto dicens, quod Atlas fuit sub servitute filiorum Israel. Atlas vero, ut dicit Augustinus, fuit avus maternus Hermetis Mercurii majoris, qui magnarum artium peritus floruit et eas hominibus tradidit, propter quod eum tanquam Deum post mortem venerati sunt. Et hic, ut dicit Augustinus octavo decimo libro, fuit tempore quo Moyses eduxit filios Israel; cujus nepos fuit Hermes Mercurius, qui ad doctrinam alterius est dictus Trismegistus, qui famosus fuit philosophus Aegypti, maxime in moralibus, sicut Augustinus docet octavo de Civitate Dei. Et hic scripsit ad Asclepium, sicut patet in libro de divinitate, qui satis habetur, cujus Asclepii avus fuit Aesculapius primus medicinae auctor apud infideles. Sed tamen Isidorus dicit in tertio libro etymologiarum, quod Apollo fuit pater Aesculapii, qui primus inter philosophos infideles dicitur docuisse artem medicinae. Nam et patri ascribitur medicina quantum ad prima documenta; sed filio magis, qui hanc artem applicavit et certiori modo docuit. Nam Apollo per carmina et hujusmodi remedia processit, Aesculapius per veritatem experientiae, Isidorus dicit, et creditur esse Apollo magnus, qui a poetis fingitur esse inter Deos et dare responsa in templo Apollinis in Delphis, unde vocatur Apollo Delphicus. Et tamen ante istos fuit inaestimabilis gloria medicinae, secundum quod Aristoteles tangit in libro de regimine vitae[2], quam Adae et

Prometheus.

Aesculapius.

[1] *De Civitate Dei,* lib. xviii. cap. 8.

[2] Another title for the *Secretum Secretorum.*

Enoch magis ascribit quam sequentibus philosophis. Et cum medicina magis sit necessaria homini quam multae aliae scientiae, non est dubium quin filii Adae et Noe illam invenerunt, quibus sapientiae plenitudo data fuit, et quibus concessum est tam diu vivere propter studium sapientiae perficiendum.

Capitulum X.

Post hoc, tempore Othonielis Judicis Israel, regnavit Cadmus Thebaeus, qui primus dedit literas Graecis, ut in chronicis Cluniacensibus edocetur. Et Beda in libro temporum juniori et caeteri concordant, quod sub Aoth judice Amphion musicus floruit, qui Aoth fuit proximus post Othonielem. Et sub Barach fuit alius Apollo philosophus secundum Chronicam Cluniacensem, auctor medicinae, temporaneus Herculi secundo, cujus facta celebrantur; sicut dicit Augustinus decimo octavo libro de Civitate Dei[1]. Qui Hercules in tempore Abimelech judicis Trojam devastavit, et pilas suas in India statuit, et in Gadibus columnas erexit, et dolorem morbi non ferens seipsum tempore Jepthae judicis cremavit, ut per Augustinum decimo octavo libro, et dicta chronica confirmatur. De hoc Hercule secundo propter hoc narravi, quia alius fuit Hercules prope tempus Mercurii majoris, qui parum post eum fuit, ut narrat Augustinus, et post eum fuit tertius, qui certamen Olympiacum constituit, quod intermissum filius ejus instauravit post excidium Trojae anno 408, ut Solinus scribit. Unde multi decepti fuerunt, unum esse Herculem aestimantes, qui omnia fecerit, quae de pluribus scripta sunt.

Cadmus.

Hercules.

Similiter erratum est de hoc philosopho Apolline. Nam omnes, ut dicit Augustinus, aestimant ipsum fuisse illum, qui pro Deo in Delos insula colebatur, tanquam unus et idem esset; cujus contrarium ostenditur multis testimoniis. Nam ille Apollo, qui in templis dabat testimonia, invenitur saltem respondisse quando primo facta est civitas Athenarum, ut Athena, quae est Minerva, pro Dea coleretur; et hic philosophus non potest esse, qui pro Deo Delphico colebatur. Sed

Apollo.

[1] omitted in J.

iste, de quo dicit Augustinus, fuit filius Latonae, cujus soror Diana. Et Isidorus tertio libro idem dicit. Similiter non videtur esse ille, de quo Hieronymus scribit in epistola ad Paulinum, quae bibliis praeponitur Latinorum; nam ille Hiarcum invenit in aureo throno sedentem et docentem, qui Hiarcus dicitur esse Abrachis[1] astronomus, qui post mortem Alexandri magni fuit, sicut docet Ptolemaeus in Almagesti. Et ideo secundum hoc tres fuerunt Apollines, sicut Hercules. Deinde sub Gideone fuerunt Orpheus et Linus, secundum quod Beda refert. Et hi, scilicet Amphion, Orpheus, et Linus, suo tempore dicti sunt poetae theologi, secundum quod Augustinus dicit[2], eo quod Diis carmina faciebant; secundum autem Solinum Nicostrates mater Evandri regis Romani dicta est a vaticinio Carmentis, quae in Capitolino monte Romae habitavit, et Latinis primo literas dedit. Et haec, ut Beda refert, fuit tempore Jair judicis Israel; secundum tamen chronicam Cluniacensem fuit tempore Judicis post Jair, transactis septemdecim annis. Sed de hoc non est cura quantum ad praesentem intentionem.

Orpheus.

Erythraean Sibyl.

Propter Sibyllas vero et maxime Erythraeam, quae omnes praedictos et praedictas philosophantes infideles longe supergressa est, oportet etiam nos aliqua certificare. Nam Augustinus refert octavo decimo de Civitate Dei, quod multi auctores scripserunt eam fuisse tempore Trojani belli, et alii voluerunt eam fuisse tempore Romuli, et Achaz vel Ezechiae regis Judae. Et excidium Trojae fuit ante Romulum per quadringentos triginta annos. Nam Solinus probat Romam fuisse conditam Olympiade septima quadringentesimo tricesimo tertio anno post bellum Trojanum, sicut docet evidenter per Herculem et Picum filium suum et per alios. Et secundum Augustinum, octavo decimo de Civitate Dei, vult quod Troja capta sit judicante Hebraeos Abdon. Deinde Hesiodus philosophus successit Homero ante Romam conditam, ut ait Tullius in quaestionibus Tusculanis. Et postea Archilochus, regnante Romulo, sicut ibi describitur, et tempore Achaz vel Ezechiae regis Judae. Et

[1] Abrachis or Abraxis was the Arabian spelling of Hipparchus.

[2] *De Civitate Dei*, lib. xviii. cap. 14.

similiter regnaverunt Numitor et ejus nepos Romulus, et tunc cessavit regnum et nomen Albanorum, et vocati sunt Romani reges. Et rex tunc erat in Judaea Achaz, vel sicut alii putant, Ezechias; et sub eodem Romulo Thales Milesius fuisse perhibetur, qui fuit unus de septem sapientibus, et primus secundum Augustinum[1].

CAPITULUM XI.

Seven wise men of Greece.

Nam post poetas theologos crevit sapientia, et auctores sapientiae vocati sunt Sophi, i.e. sapientes; secundum tamen Bedam in libro temporum, et secundum Isidorum quinto Etymologiarum, Thales fuit sub Josia, qui rerum naturas scrutatus est, et fuit astrologus. Tempore quo populus Hebraeorum, ut Augustinus refert, ductus est in captivitatem, alius de septem sapientibus apparuit, scilicet, Pittacus nomine, et alii quinque fuerunt tempore captivitatis, quorum nomina sunt haec, Solon Atheniensis, Chilon Lacedaemonius, Periander Corinthius, Cleobulus Lydius, Bias Pierius. De his Solon dedit leges Atheniensibus, ad quas transferendas decem viros populus Romanus misit, et vocantur leges duodecim tabularum, sicut scribit Isidorus quinto libro. Aliud vero genus hominum sapientiae deditum post eos exortum est in lingua Graeca, quae tamen vocatur Italica, scilicet ex ea parte, quae Italia dicebatur, antiquitus Magna Graecia, et hi studuerunt in Italia licet Graeci, etiam in lingua Graeca. Et isti non voluerunt se vocari sapientes sed amatores sapientiae, quorum princeps fuit Pythagoras Samius a Samo insula; a quo cum quaereretur, Quis esset, respondit, philosophus, i.e. amator sapientiae: sicut dicit Augustinus[2] octavo de Civitate Dei. Sed octavo decimo libro dicit, quod Pythagoras apparuit eo tempore quo Judaeorum soluta est captivitas, et secundum Tullium in libro primo quaestionum Tusculanarum, Tarquinio Superbo regnante Romanis, qui fuit septimus a Romulo, et ultimus rex Romanorum. Postquam consules exorti sunt, venit in Italiam Pythagoras, et illam Magnam Graeciam tenuit

Pythagoras, and the Italian school.

[1] *De Civitate Dei*, lib. xviii. cap. 24.

[2] Ibid. lib. viii. cap. 2. Cf. lib. xviii. cap. 25.

cum honore, cum disciplina, cum auctoritate, et postea sic viguit Pythagoreorum nomen, ut nulli alii docti viderentur. Et Tarquinius, ut scribit Beda, tempore Cyri regis Persarum, qui laxavit captivitatem Judaeorum, incepit regnare. Ac regnavit tempore Cambysis filii ejus, et duorum fratrum magorum, et Darii, in cujus anno secundo templum aedificatum est. Et tunc clarus Pythagoras, ut dicit Beda, habebatur, et Zorobabel. Aggaeus, Zacharias, et Malachias prophetae claruerunt. Pythagoras quidem edoctus fuit a Pherecide Syro, ut dicit Tullius libro praedicto, qui Pherecides primus animas hominum posuit immortales, cujus tempora non certificantur nisi per tempus Pythagorae discipuli sui ; quamvis et Isidorus libro primo dicat, quod Pherecides scripsit historias tempore Esdrae, qui potuit forte esse versus finem vitae ipsius Pherecidis et in juventute Esdrae. Nam a tempore, quo dictus Pythagoras dicitur floruisse, fluxerunt triginta sex anni quibus regnavit Darius, et decem quibus Xerxes, et septem menses quibus Arthabas, et sex anni quibus Artaxerxes Longimanus, antequam Esdras ascendit de Babylonia in Jerusalem. Nam septimo anno regni ejus, primo die mensis primi, Esdras secundum Scripturas et chronicas profectus est.

CAPITULUM XII.

The Ionic school leading to Socrates, Plato, Aristotle.

Haec autem duo genera philosophantium, scilicet Ionicum et Italicum, ramificati sunt per multas sectas et varios successores usque ad doctrinam Aristotelis, qui correxit et mutavit omnium praecedentium positiones, et philosophiam perficere conatus est. Successerunt vero Pythagoras, Archytas Tarentinus, et Timaeus, inter alios maxime nominati. Sed praecipui philosophi, ut Socrates et Plato et Aristoteles, non descenderunt ex hac linea, immo vero Ionici et veri Graeci fuerunt; quorum primus fuit Thales Milesius. Quomodo autem huic caeteri successerunt, ostendit Augustinus octavo[1] libro de civitate Dei. Nam post Thaletem fuit primus Anaximander ejus discipulus, cujus successor fuit Anaximenes,

[1] Cap. 2.

et hi duo fuerunt tempore Judaicae captivitatis. Augustinus et alii similiter concordant in hoc. Anaxagoras vero et Diogenes Anaximenis auditores fuerunt, et eidem successerunt sub Dario Hydaspis, cujus anno secundo templum coepit aedificari. Anaxagorae, ut dicit Augustinus, successit Archelaus ejus discipulus, cujus auditor fuit Democritus secundum Isidorum octavo libro. Et Socrates, secundum Augustinum octavo libro, Archelai fuisse discipulus perhibetur. Socrates autem secundum Bedam natus est sub Arthaba, qui Persis regnavit mensibus septem, cui in idem regnum successit Artaxerxes Longimanus, in cujus anno septimo Esdras descendit de Babylone, et ideo simul fuerunt Esdras et Socrates. Sed prior natu fuit Esdras, sicut ex nunc dictis claret. Et ideo dicit Augustinus decimo octavo libro de Civitate Dei, quod post Esdram fuit Socrates, i.e. posterior natu. Nam quando floruit Esdras apud regem Persarum et Judaeos, tunc Socrates exortus est. Hic Socrates dicitur pater philosophorum magnorum, quoniam Platonis et Aristotelis magister fuit, a quibus omnes sectae philosophantium descenderunt. Plato quidem, secundum Bedam in tractatu majori de temporibus, natus est sub Sogdiano, qui mensibus septem regnavit, cui successit Darius cognomine Nothus, quanquam sub eodem Dario Beda in eodem tractatu de temporibus scribat natum esse Platonem. Sed in illo tractatu tempus Sogdiani, quia modicum fuit, computat sub regno Darii. Nam continuat eum cum Artaxerxe Longimano. Nascente vero Platone, Hippocrates medicus, ut dicit Beda, habetur insignis, et hoc tempore Empedocles et Parmenides inventi sunt; sed Plato Socratica primo addiscens, et ea quae Graeca fuerunt, Aegyptum petiit, ad Archytam Tarentinum, et Timaeum laboriosissime peragravit, ut dicit Hieronymus ad Paulinum. Et contra Rufinum scribit Hieronymus, quod Plato post Academiam et innumerabiles discipulos sentiens multum deesse suae doctrinae, venit ad Magnam Graeciam, ibique ab Archyta Tarentino eruditus, elegantiam et leporem cum hujusmodi miscuit discipulis. Et iste Plato omnibus philosophis antefertur secundum sanctos, quoniam ejus libri ad eorum manus devenerunt, et quia sententias de Deo pulchras, et de moribus et de vita futura multa conscripsit quae sacrae Dei sapientiae multum concordant, ut in morali Philosophia explanabo; et ab hoc aestimaverunt multi Catholici viri quod audiverat Jeremiam prophetam in Aegypto. Nam Aegyptum petiit propter sapientiam, et a barbaris sacerdotibus instructus est, ut scribit Tullius libro Academicorum quinto. Sed tamen Augustinus dicit quod non fuit tempore Jeremiae. Nam Jeremias ut dicit nono de Civitate Dei, primo prophetavit tempore quarti regis a Romulo qui vocatus est Ancus Martius[1], et in tempore quinti regis scilicet Tarquinii Prisci. Sed Plato tunc non fuit; immo post tempus Jeremiae fere per annos centum, ut dicit Augustinus octavo libro, natus est Plato. Sed ut alii aestimabant invenit LXX interpretes a quibus instrueretur, sicut Augustinus dicit octavo libro. Et Tullius libro de Senectute dicit quod Plato mortuus est LXXXI° anno vitae suae, id est in fine Artaxerxis qui Ochus dicebatur, ut scribit Beda.

CAPITULUM XIII.

Aristotle, Avicenna, Averrhoes.

Ante[2] vero mortem Socratis natus est Aristoteles, quoniam per tres annos auditor ejus fuit, sicut in vita Aristotelis legitur. Et secundum Bedam natus est sub Artaxerxe, qui successit Dario Notho. Et in decimo septimo anno vitae suae fuit auditor Socratis, et ipsum per tres annos audivit, et post mortem Socratis factus est auditor Platonis secundum Bedam, et ipsum audivit, viginti annis, ut in vita sua legitur. Et post mortem Platonis vixit quadraginta tres annis, unde in universo non vixit nisi sexaginta sex annis, sicut ex dictis patet. Et hoc similiter patet in libro Censorini de die natali, quoniam ipse Censorinus refert contra passionem mortalem per tres annos eum magnitudine animi magis

[1] The remaining part of this paragraph is omitted in Jebb's ed.

[2] What follows, as far as the last sentence but one in cap. 15, is omitted by O. from this second part, and interpolated in Part III between Tertio and Quarto; as also is the last paragraph of Part II. Further, the order in which these paragraphs are placed in Part III is not the same as in J. O. proceeds: 'Hoc Domino Alexandro notum; et multis aliis potest hoc idem adhuc etiam ostendi per proprietates duas Metaphysicae,' &c.

quam medicinae virtute luctatum fuisse. Hic Aristoteles magister Alexandri magni effectus duo millia hominum misit per mundi regiones, ut naturas rerum exquirerent, sicut Plinius narrat in Naturalibus octavo libro et mille libros composuit, ut in ejus vita legitur. Hic enim praecedentium philosophorum errores evacuavit, et augmentavit philosophiam aspirans ad ejus complementum quod habuerint antiqui patriarchae, quamvis non potuit singula perficere. Nam posteriores ipsum in aliquibus correxerunt, et multa ad ejus opera addiderunt, et adhuc addentur usque ad finem mundi; quia nihil est perfectum in humanis inventionibus, ut in prioribus est expositum. Hunc natura firmavit, ut dicit Averroes in tertio de Anima, ut ultimam perfectionem hominis inveniret. Hic omnium philosophorum magnorum testimonio praefertur philosophis, et philosophiae ascribendum est id quod ipse affirmavit; unde nunc temporis autonomatice Philosophus nominatur, in auctoritate philosophiae, sicut Paulus in doctrina sapientiae sacrae apostoli nomine intelligitur[1]. Quievit autem et siluit philosophia Aristotelis, pro majori parte aut propter occultationem exemplarium et raritatem, aut propter difficultatem, aut propter invidiam, aut propter guerras Orientis[2], usque post tempora Mahometi, quando Avicenna et Averroes et caeteri revocaverunt philosophiam Aristotelis in lucem plenam expositionis. Et licet alia Logicalia et quaedam alia translata fuerunt per Boetium de Graeco, tamen tempore Michael Scoti, qui annis Domini 1230 transactis apparuit deferens librorum Aristotelis partes aliquas de Naturalibus et Metaphysicis[3] cum expositoribus authenticis, magnificata est philosophia Aristotelis apud Latinos. Sed respectu multitudinis et magnitudinis suae sapientiae in mille tractatibus comprehensae, valde[4] modicum adhuc in linguam Latinam est translatum, et minus est in usu vulgi studentium. Avicenna quidem praecipuus imitator et expositor Aristotelis, et complens philosophiam secundum quod ei fuerit possibile, triplex volumen condidit philosophiae, ut ipse dicit in prologo libri Sufficientiae; unum

[1] Om. in Jebb.
[2] Om. in Jebb.
[3] Jebb has Mathematicis.
[4] vel, Jebb.

vulgatum juxta communes sententias philosophorum Peripateticorum, qui sunt de secta Aristotelis; aliud vero secundum puram veritatem Philosophiae, quae non timet ictus lancearum contradicentium, ut ipse asserit; tertium vero fuit cum termino vitae suae, in quo exposuit priora[1], secretiora naturae et artis recolligens. Sed de his voluminibus duo non sunt translata; primum autem secundum aliquas partes habent Latini, quod vocatur Assephae, i. e. liber sufficientiae. Post hunc venit Averroes, homo solidae sapientiae, corrigens dicta priorum et addens multa, quamvis corrigendus sit in aliquibus, et in multis complendus[2].

Capitulum XIV.

It is for Christian theologians to carry on their work.

Ex his sequitur necessario, quod nos Christiani debemus uti philosophia in divinis, et in philosophicis multa assumere theologica, ut appareat quod una sit sapientia in utraque relucens. Quam necessitatem voluero certificare, non solum propter unitatem sapientiae, sed propter quod inferius tangam, oportet nos in philosophia revolvere sententias fidei et theologiae magnificas quas reperimus in libris philosophorum et in partibus philosophiae; ut non sit mirum quod in philosophia tangam sacratissimas veritates, quoniam philosophis Deus concessit multas sapientiae suae veritates. Oportet igitur ut trahatur philosophiae potestas ad sacram veritatem quantum possumus; nam valor philosophiae aliter non lucescit[3]. Nam philosophia secundum se considerata nullius utilitatis est. Philosophi vero infideles damnati sunt, 'et cum cognoverunt Dominum, non sicut Dominum glorificaverunt et ideo stultificati sunt et evanuerunt a cogitationibus suis.' Et ideo philosophia non potest aliquid dignitatis habere, nisi quantum de ea requirit Dei sapientia. Totum enim residuum

[1] This is the reading of Jul.

[2] Jul. here adds a passage which fire has rendered illegible. The succeeding paragraph is headed, 'Quarta Distinctio: in qua ostenditur necessitas utendi philosophia in divinis, non solum propter unitatem sapientiae perfectae sed quia in sequentibus partibus hujus persuasionis adductae sunt veritates divinae quae de pluribus philosophorum extrahuntur.'

[3] 'nam . . . lucescit,' supplied from Jul.

est erroneum et inane, et propter hoc dicit Alpharabius in libro de scientiis, quod sicut puer indoctus se habet ad hominem sapientissimum in philosophia sic homo talis ad sapientiam Dei. Praeterea semper crescere potest in hac vita studium sapientiae, quia nihil est perfectum in humanis inventionibus. Quapropter antiquorum defectus deberemus nos posteriores supplere, quia introivimus in labores eorum, per quos, nisi simus asini, possumus ad meliora excitari; quia miserrimum est semper uti inventis, et nunquam inveniendis, ut dicit Boetius, et probatum est efficaciter superius suo loco. Item Christiani debent ad suam professionem quae sapientia Dei est caetera pertractare, et vias philosophorum infidelium complere; non solum quia posteriores sumus, et debemus addere ad eorum opera, sed ut cogamus sapientiam philosophorum nostrae deservire. Nam hoc philosophi infideles faciunt ipsa veritate coacti in quantum eis datum est; nam totam philosophiam reducunt ad divina; ut ex libris Avicennae in Metaphysicae et Moralibus, et per Alpharabium et Senecam et Tullium, et per Aristotelem patet; nam omnia reducunt ad Deum, sicut exercitus ad principem, inferentes de angelis et aliis multis, quoniam principales articuli fidei reperiuntur in eis.

They had many glimpses of Christian truth

Nam ut in Moralibus exponetur, Deum esse docent, et quod sit unus in esse, infinitae potentiae et sapientiae et bonitatis, trinis in personis, Pater, et Filius et Spiritus Sanctus, qui omnia creavit ex nihilo; et de Domino Jesu Christo et de beata virgine multa tangunt. Similiter de Antichristo, atque de angelis et custodia hominum per eos[1], necnon de resurrectione mortuorum, et de judicio futuro, et de vita futurae felicitatis quam Deus promisit obedientibus sibi, et de miseria futura, quam proponit inferre his qui mandata ejus non observant. Scribuntque innumerabilia de morum honestate, de legum gloria, de legislatore qui debet accipere legem a Deo per revelationem, qui sit mediator Dei et hominum, et vicarius Dei in terra, et Dominus terreni mundi, de quo cum probatum fuerit quod recepit legem a Deo, ei credendum sit in omnibus, exclusa omni dubitatione et haesitatione, qui debet totum genus humanum ordinare in cultu Dei et legibus jus-

[1] 'per eos,' om. in O.

titiae et pacis, et in virtutum exercitio propter reverentiam Dei ac futuram felicitatem; et quod idolorum cultura destrui debet. Haec et his similia[1] habuerunt philosophi. In libris enim eorum hujusmodi reperimus, sicut probatio certa docebit in sequentibus, et quilibet potest experiri qui vult libros philosophorum perlegere; et negare non possumus quin scripta sunt ab eis, undecunque hujusmodi receperunt. Nec mirandum est quod philosophi talia scribant; nam omnes philosophi fuerunt post patriarchas et prophetas, sicut prius de hoc facta est consideratio in suo loco; et legerunt libros prophetarum et patriarcharum qui sunt in sacro textu, et similiter alios libros, quos fecerunt tangentes Christi mysteria ut in libro Enoch et in testamento patriarcharum in libro Esdrae, 3°, 4°, 5°, et in multis aliis libris de quorum aliquibus sit mentio in sacro textu ut de libris Nathan, Samuelis, et Abdon prophetarum. In hujusmodi enim libris tanguntur expresse articuli fidei, et longe expressius quam in Canone Scripturae. Nam praeter caeteros libros liber de testamentis patriarcharum ostendit omnia, quae de Christo impleta sunt. Quilibet enim patriarcha in morte praedicavit filiis suis et tribui suae, et praedixit eis ea quae de Christo tenenda sunt, sicut manifestum est ex libro suo. Et hi libri licet non sint in Canone, tamen sancti et sapientes Graeci et Latini usi sunt eis a principio Ecclesiae. Nam beatus Judas de hoc Enoch accepit auctoritatem, et Augustinus decimo de Civitate Dei, multum fundatur super illum librum, ut ostendatur quod primo fuit sapientia apud sanctos quam apud philosophos, et ait quod jam propter nimiam antiquitatem ille liber non est in auctoritate, quam propter aliquid aliud. De libris autem aliis manifestum est quod in usu sanctorum et sapientum antiquorum sunt propter hoc quod planas veritates de Christo continere noscimus. Philosophi igitur curiosi et diligentes in studio sapientiae peragrarunt regiones diversas, ut sapientiam inquirerent, et libros sanctorum perlegerunt, et didicerunt ab Hebraeis multa. Nam Avicenna in radicibus moralis philosophiae

[1] After 'similia' O. has 'nunquam,' which is omitted in Jul., and which spoils the sense.

recitat verba Esaiae de vita aeterna, dicens illam esse quam oculus non vidit, nec auris audivit, et recitat, Eleemosyna tollit peccatum, sicut propheta veritatis dicit, scilicet Tobias. Et Augustinus vult decimo octavo de Civitate Dei, quod Plato legerat librum Geneseos, propter creationem mundi quam posuit similem ei quae ibi describitur. Et quod legis librum legit, videlicet Exodi, propter nomen Dei quod ibi ponitur, scilicet, Ego sum qui sum. Nam hoc usus est Plato, et alibi non potuit invenire, ut dicit Augustinus. Et praeter sacros libros prophetales composuerunt libros philosophiae; immo totam philosophiam his perfecerunt. Et quod philosophi non habuerunt nisi ab eis, ostensum est in praecedentibus evidenter, et quia una est sapientia, quae sufficit humano generi; ideo sancti in libris philosophicis miscuerunt divina multa cum aliis, quantum potuit philosophia recipere. Et propterea propter istos libros philosophicos sanctorum multa perceperunt philosophi de divinis veritatibus.

CAPITULUM XV.

Praeterea cum philosophi fuerunt dediti veritatibus et omni vitae bonitati, contemnentes divitias, delicias, et honores, aspirantes ad futuram felicitatem quantum potuit humana fragilitas, immo victores effecti humanae naturae, sicut Hieronymus scribit de Diogene in libro contra Jovinianum, non est mirum, si Deus, qui in his minoribus illuminavit, daret eis alia lumina veritatum majorum. Et si non principaliter propter eos, tamen propter nos, ut eorum persuasionibus mundus disponeretur ad fidem. Et ad hoc facit quod Sibyllae multae inventae sunt, scilicet decem; sicut omnes sancti concordant, et Augustinus decimo octavo de Civitate Dei [1], et Isidorus libro Etymologicorum septimo. Necnon historiae et philosophiae et poetae concordant universaliter in his Sibyllis. Sed certum est eas recitasse divina, et ea quae de Christo habentur et de judicio futuro, et hujusmodi. Ergo multo magis probabile est quod philosophi sapientissimi et optimi a Deo receperunt hujusmodi veritates. Quod vero

which in the poems of the Sibyls were even more explicit.

[1] *De Civitate Dei*, lib. xviii. cap. 23.

Sibyllae locutae sunt praeclare veritates divinas, manifestum est per sanctos et alios, et sufficit recitare quod Augustinus dicit decimo octavo de Civitate Dei [1]. 'Dixerunt igitur istae mulieres hujusmodi sermones; dabunt Deo alapas manibus incestis, imputato ore exspuent venenatos sputus; dabit vero ad verbera simpliciter suum dorsum, colaphos accipiens tacebit, et corona spinea coronabitur. Ad cibum autem fel, et ad potum acetum dederunt. Insipiens gens, Dominum tuum non intellexisti ludentem mortalium mentibus, sed spinis coronasti, et horridum fel miscuisti. Templi vero velum scinditur, et medio die nox erit tenebrosa tribus horis, et morte morietur tribus diebus, somno suscepto.' Et iterum metrice dixit Sibylla,

> 'Judicii signum, tellus sudore madescet,
> Ex coelo rex adveniet per secla futurus,
> Scilicet in carne praesens ut judicet orbem.
> Unde Deum cernent incredulus atque fidelis,
> Celsum cum sanctis ejus jam termino in ipso.
> Sic animae cum carne aderunt, quas judicet ipse.
> Exuret terras ignis pontumque polumque;
> Sanctorum sed enim cunctae lux libera carni
> Tradetur, sontes aeternum flamma cremabit.
> Actus occultos retegens tunc quisque loquetur
> Secreta, atque Deus reserabit pectora luci.
> Eripitur solis jubar et chorus interit astris:
> Solvetur coelum, lunaris splendor abibit.
> Dejiciet colles, valles extollet ab imo.
> Sic pariter fontes torrentur, fluminaque igni,
> Tartareumque Chaos monstrabit terra dehiscens,
> Excidet e coelis ignisque et sulphuris amnis [2].'

Si igitur mulierculae fragiles hujusmodi dixerunt, longe magis credendum philosophos sapientissimos hujusmodi gustasse veritates. Et Augustinus vult decimo octavo de Civitate Dei, alios percepisse Dei veritatem, quam illi qui de linea Abraham usque ad Christum et deinceps descenderunt. Nam

[1] Quoted by St. Augustine from Lactantius.

[2] This passage from the Sibylline poem is given most incorrectly in the MSS. I have corrected it from *De Civitate Dei*, lib. xviii. cap. 23.

Job scivit resurrectionem et Dei veritates. Et in chronicis Eusebii legitur, quod Irene et Constantino imperantibus fuit cadaver effossum, in quo inveniebatur scriptura haec, 'Credo in Christum; sub Irene et Constantino iterum videbit me Sol.' Et nunc tempore Domini Alexandri Papae quarti Saracenus in Borea mundum contemnens, vacans in lege sua Deo et virtuti et contemplationi alterius vitae, recepit visitationem angelicam et consilium ut converteretur ad fidem Christi, et baptizatus est. Hoc Domino Alexandro notum est et multis aliis, et adhuc recolunt quamplures.

CAPITULUM XVI.

Philosophy recognizing its own imperfection leads to the higher truths of theology.

Potest hoc idem adhuc etiam ostendi per proprietates duas metaphysicae. Nam haec scientia est de illis, quae omnibus rebus et scientiis conveniunt, et ideo ostendit numerum scientiarum, et quod oportet esse aliam scientiam ultra philosophiam, cujus proprietates tangit in universali; licet in particulari non posset eam assignare. Scit enim philosophia suas imperfectiones, et quod deficit a plena cognitione eorum quae maxime sunt cognoscenda, sicut Aristoteles docet in Metaphysica; et Avicenna similiter, ut tactum est superius. Et propter hoc devenit philosophia ad inveniendam scientiam altiorem, quam dicit esse scientiam divinorum, quam Theologiam perfectam vocant philosophi, et ideo philosophia elevat se ad scientiam divinorum. Item solliciti fuerunt philosophi super omnia inquirere sectam in qua esset salus hominis, et dant modos probandi hic praeclaros, sicut ex moralibus manifestum est. Et invenerunt certitudinaliter, quod aliqua debet esse secta fidelis et sufficiens mundo, cujus proprietates assignant, quae non possunt reperiri nisi in secta Christi, ut probatur suis locis, et ostenditur quod de bonitate Dei est de necessitate humana quod sciatur haec secta fidelis. Sed non potest hoc probari infidelibus per legem Christi, nec per auctores sacros, quia ex lege disputationis possunt negari omnia quae in lege Christi sunt, sicut Christiani negant ea quae in aliis legibus continentur. Et etiam quia Christum negant, non est mirum, si auctores Christianorum negent.

Persuasio autem fidei necessaria est; sed non potest hoc esse nisi duobus modis, aut per miracula quae sunt supra fideles et infideles, de quibus nullus potest praesumere; aut per viam communem fidelibus et infidelibus[1]; sed hoc non est nisi per philosophiam. Ergo philosophia habet dare probationes fidei Christianae. Articuli vero hujus fidei sunt principia propria theologiae; ergo philosophia habet descendere ad probationes principiorum theologiae, licet minus profunde quam ad principia aliarum scientiarum. Et hoc modo supponatur ex hac ratione, donec veniatur ad probationem sectarum. Nam ibi ostendetur quod moralis philosophia efficacius theologia[2] deservit in hac parte, et ideo licet secundum veritatem hujusmodi sunt theologica, nihilominus tamen sunt philosophica, sed propter theologiam.

Capitulum XVII.

Moral philosophy is the connecting link.

Praeterea tota philosophia speculativa ordinatur in finem suum, qui est philosophia moralis. Et quia finis imponit necessitatem eis, quae sunt ad finem ut Aristoteles dicit secundo Physicorum, ideo philosophia speculativa semper aspirat ad finem suum, et erigit se ad eum, et quaerit vias utiles in ipsum, et propter hoc potest philosophia speculativa praeparare principia moralis philosophiae. Sic igitur se habent duae partes sapientiae apud infideles philosophos: sed apud Christianos philosophantes scientia moralis proprie et perfecte est theologia, quae super majorem philosophiam infidelium addit fidem Christi, et veritates quae sunt proprie divinae[3]. Et hic finis habet suam speculationem praecedentem, sicut moralis philosophia infidelium habet suam; quae igitur est proportio finis ad finem, est proportio speculationis ad speculationem: sed finis ut lex Christiana supra legem philosophorum addit articulos fidei expressos, per quos complet legem moralis philosophiae[4], ut fiat una lex completa. Nam lex Christi leges et mores philosophiae sumit et assumit, ut certum est per sanctos, et in usu theologiae et ecclesiae. Ergo

[1] Thus in Jul. Jebb's reading here is confused and inaccurate.
[2] Sic Jul. Jebb's reading, 'in homine,' is unintelligible.
[3] This passage has been corrected from Jul.
[4] This is the reading of Jul.

speculatio Christianorum praecedens legem suam debet super speculationem alterius legis addere ea quae valent ad legem Christi docendam et probandam, ut surgat una speculatio completa, cujus initium erit speculativa philosophorum infidelium, et complementum ejus erit superinductum theologiae, et secundum proprietatem legis Christianae. Et ideo philosophia apud Christianos debet sapere multum de divinis, plus quam apud philosophos infideles, et propter hoc debent philosophi considerare philosophiam, ac si modo esset de novo inventa, ut eam facerent aptam fini suo. Et ideo debent multa addi in philosophia Christianorum; quae philosophi infideles scire non potuerant. Et hujusmodi sunt rationes exsurgentes in nobis ex fide et auctoribus legis et sanctorum qui sapiunt philosophiam; et possunt esse communia philosophiae completae et theologiae. Et haec cognoscuntur per hoc quod debent esse communia fidelibus et infidelibus, ut sint ita nota cum proferuntur et probantur, quod negari non possunt a sapientibus et instructis in philosophia infidelium. Nam philosophi infideles multa ignorant in praesenti[1] de divinis quae proponerentur eis, ut probarentur per principia philosophia completae, hoc est, per vivacitates rationis quae sumunt originem a philosophia infidelium; licet complementum a fide Christi reciperent sine contradictione, et gauderent de proposita sibi veritate, quia avidi sunt sapientiae et magis studiosi quam Christiani. Non tamen dico, quod aliquid de spiritualibus articulis fidei Christianae reciperetur in probatione, sed multae veritates sunt communes rationales, quas omnis sapiens de facili reciperet ab alio, quamvis secundum se ignoraret.

Non igitur mirentur philosophantes, si habeant elevare philosophiam ad divina et ad theologiae veritates et sanctorum auctoritates, et uti eis abundanter cum fuerit opportunum[2], et probare eas cum necesse est, et per illas alias probare; quoniam proculdubio philosophia et theologia communicant in multis. Et sancti non solum loquuntur theologice, sed philosophice, et philosophica multipliciter introducunt. Et ideo Christiani, philosophiam volentes complere, debent in

[1] Sic O.; J. has particulari.

[2] Sic Jul.; O. has optimum.

suis tractatibus non solum dicta philosophorum de divinis veritatibus colligere, sed longe ulterius progredi, usquequo potestas philosophiae totius compleatur. Et propter hoc complens philosophiam per hujusmodi veritates non debet dici theologicus nec transcendere metas philosophiae; quoniam ista quae sunt communia philosophiae et theologiae potest secure tractare et ea quae communiter habent recipi a fidelibus et infidelibus. Et talia multa sunt praeter dicta philosophorum infidelium, quae tanquam propria infra limites philosophiae debet recte philosophans colligere, ubicunque ea invenit, et tanquam sua habet congregare, sive in libris sanctorum, sive philosophorum, sive in sacra scriptura, sive in historiis, sive alibi. Nullus enim auctor est quin praeter principalem intentionem aliqua incidenter recitet quae sunt alibi magis propria; et hujus causa est annexio scientiarum, quia quaelibet ab alia quodam modo dependet; sed omnis, qui debito modo tractat, debet quae sunt propria ei assignare, et quae necessaria et suae competentia dignitati, et ideo ubicunque ea inveniat velut sua cognoscere, et tanquam propria habet rapere, et in locis propriis collocare. Propter quod philosophans Christianus potest multas auctoritates et rationes et sententias quamplurimas de scriptis aliis, quoque de libris philosophorum infidelium adunare, dummodo sint propria philosophiae, vel communia ei et theologiae, et quae communiter habent fideles et infideles reperire. Et nisi hoc fiat, non perficietur, sed multum ei derogabitur. Et non solum debet hoc fieri propter complementum philosophiae, sed propter conscientiam Christianam, quae habet omnem veritatem ducere ad divinam, ut ei subjiciatur et famuletur. Atque propter hoc, philosophia infidelium est penitus nociva, et nihil valet secundum se considerata, nam philosophia secundum se ducit ad caecitatem infernalem et ideo oportet quod secundum se sit tenebrae et caligo.

Capitulum XVIII.

Summary. His consideratis [1], patet intentum principale; et manifestum est quod omnes philosophi infideles et Poetae et Sibyllae et

[1] From this point to the end of Part II is omitted here in O. and is transposed to Part III. Vide note on p. 54.

quicunque sapientiae sunt dediti, inventi sunt post philosophos veros et fideles, qui fuerunt filii Seth et Noe cum filiis suis, quibus Deus dedit vivere sexcentos annos propter studium sapientiae complendum, ut dicit Josephus primo Antiquitatum, asserens quod in minore tempore non potuerant complere philosophiam praecipue propter astronomiam in qua est major difficultas, eo quod a caelestibus homines mortales multum distant; sed Deus eis revelavit omnia, et dedit eis vitae longitudinem, ut philosophiam per experientias complerent. Sed propter malitiam hominum qui abusi sunt viis sapientiae, ut Nimroth, et Zoroaster, et Atlas, et Prometheus, et Mercurius aut Trismegistus, et Aesculapius, et Apollo, et alii qui colebantur sicut Dii propter sapientiam, Deus obscuravit cor multitudinis, et recidit paulatim usus philosophiae usquequo iterum Salomon eam revocavit et perfecit omnino, sicut Josephus docet octavo Antiquitatum. Et iterum propter peccata hominum evanuit studium sapientiae, donec Thales Milesius resumpsit eam, et ejus successores dilataverunt, usquequo Aristoteles consummavit, quantum fuit possibile juxta idem tempus. Sed isti ab Hebraeis didicerunt omnia, sicut Aristoteles dicit in libro secretorum, quoniam philosophi infideles, ut Nimroth et alii, fuerunt post Seth, Noe, Sem, et Abraham; et post Salomonem, qui secundo perfecit eam, fuerunt reliqui philosophi infideles ut Thales et Pythagoras, Socrates et Plato, et Aristoteles. Ideo manifestum est, quod philosophiae perfectio fuit primo data sanctis, Patriarchis et Prophetis, quibus lex Dei similiter fuit ab uno et eodem Deo revelata; quod non fuisset factum, nisi philosophia omnino esset sanctis Dei et legi sacrae conformis et utilis ac necessaria propter intellectum legis et excusationem et defensionem; insuper ut fiat ejus persuasio, et probetur et communicetur et dilatetur; nam omnibus his modis necessaria est, sicut discurrendo per partes singulas philosophiae apparebit. Et ideo philosophia non est nisi sapientiae divinae explicatio per doctrinam et opus, et propter hoc una est sapientia perfecta, quae sacris literis continetur.

PARS TERTIA

HUJUS PERSUASIONIS.

DE UTILITATE GRAMMATICAE [1].

Knowledge implies study of the languages in which knowledge is recorded.

Declarato igitur, quod una est sapientia perfecta, quae sacris literis continetur per jus canonicum et philosophiam, qua mundus habet regi, nec alia requiritur scientia pro utilitate generis humani, nunc volo descendere ad ea hujus sapientiae magnifica, quae maxime valent exponi. Et sunt quinque, sine quibus nec divina nec humana sciri possunt, quorum certa cognitio reddit nos faciles ad omnia cognoscenda. Et primum est Grammatica in linguis alienis exposita, ex quibus emanavit sapientia Latinorum. Impossibile enim est, quod Latini perveniant ad ea quae necessaria sunt in divinis et humanis, nisi notitiam habeant aliarum linguarum, nec perficietur eis sapientia absolute, nec relate ad ecclesiam Dei et reliqua tria praenominata. Quod volo nunc declarare, et primo respectu scientiae absolutae. Nam totus textus sacer a Graeco et Hebraeo transfusus est, et philosophia ab his et Arabico deducta est; sed impossibile est quod proprietas unius linguae servetur in alia. Nam et idiomata ejusdem linguae variantur apud diversos, sicut patet de lingua Gallicana, quae apud Gallicos et Picardos et Normannos et Burgundos multiplici variatur idiomate. Et quod proprie dicitur in idiomate Picardorum horrescit apud Burgundos, immo apud Gallicos viciniores: quanto igitur magis accidet hoc apud linguas diversas? Quapropter, quod bene factum est in una lingua,

[1] I have kept this title, which is given in all the MSS. But it must always be remembered that it is not grammar in the ordinary acceptation of the word, but knowledge of certain foreign languages, of which Bacon is urging the importance.

non est possibile ut transferatur in aliam secundum ejus proprietatem quam habuerit in priori.

Unde Hieronymus, in epistola de optimo genere interpretandi, sic dicit, 'Si ad verbum interpretor, absurdum resonat.' Quod si cuiquam videatur linguae gratiam interpretatione non mutari, Homerum exprimat in Latinum ad verbum. Si quis autem eundem in sua lingua per se interpretetur, videbit ordinem ridiculosum, et poetam eloquentissimum vix loquentem. Quicunque enim aliquam scientiam ut logicam vel aliam quamcunque bene sciat, eam, etsi nitatur in linguam convertere maternam, videbit non solum in sententiis sed in verbis deficere. Et ideo nullus Latinus sapientiam sacrae scripturae et philosophiae poterit ut oportet intelligere, nisi intelligat linguas a quibus sunt translatae.

Many words have no Latin equivalent.

Et secundo considerandum est quod interpretes non habuerunt vocabula in Latino pro scientiis transferendis, quia non fuerunt primo compositae in lingua Latina. Et propter hoc posuerunt infinita de linguis alienis, quae sicut nec intelliguntur ab eis qui linguas ignorant, sic nec recte proferuntur nec scribuntur ut decet; atque, quod vile est, propter ignorantiam linguae Latinae posuerunt Hispanicam, et alias linguas maternas, quasi infinities pro Latino. Nam pro mille millibus exemplis unum ponatur de libro vegetabilium Aristotelis, ubi dicit, 'Belenum in Perside pernitiosissimum, sed transplantatum Jerusalem fit comestibile.' Hoc vocabulum non est scientiale laico Hispanicorum. Nam jusquiamus vel semen cassilaginis est nomen ejus in Latino. Quae sicut multa alia prius ab Hispanis scholaribus derisus cum non intelligebam quod legebam, ipsis vocabula linguae maternae scientibus, tandem didici ab eisdem.

A translator should know (1) his subject, (2) the two languages he is handling.

Tertio, oportet quod interpres optime sciat scientiam quam vult transferre, et duas linguas a quibus et in quas transferat. Solus Boethius primus interpres novit plenarie potestatem linguarum. Et solus dominus Robertus, dictus Grossum Caput, novit scientias[1]. Alii quidem mendici translatores

[1] Of this bold reformer of ecclesiastical abuses, and champion of civil liberties, Bacon is never tired of speaking. He held the see of Lincoln from 1235 till his death in 1253. 'Quietis nescius, multis adversans, quam plurimisque ei

defecerunt multum tam in scientiis quam in linguis; quod ostendit ipsorum translatio. Nam tanta est perversitas et horribilis difficultas, maxime in libris Aristotelis translatis [1],

adversantibus,' is Matthew Paris' description of him. Visiting Pope Innocent IV in 1250 to protest against the abuses of the Hospitallers and Templars, and finding that his opponents had deafened the papal ear with bribes, 'O gold, gold,' he cried, 'how great is thy power, especially in the Court of Rome!' In the last year of his life, he wrote to the Pope, plainly telling him that his vow of obedience to the Church forbade compliance with his extortionate demands for money. 'Filialiter et obedienter non obedio, contradico et rebello.' The Pope could not contain himself for anger. 'Who is this mad old man,' he cried, 'who dares to sit in judgement on our actions?' But Cardinal Aegidius and others restrained his wrath. 'If we are to tell the truth,' they said, 'the things he writes are true. He is a Catholic man of most holy life, more religious and more excellent than we. He is held for a great philosopher, deeply learned in Latin and Greek literature, zealous for justice, a teacher in theological schools, a preacher to the people, a lover of chastity, an uprooter of simony.' (Matthew Paris, ad ann. 1253.)

It would seem probable that his scientific works were written before his elevation to the bishopric. Among them are, *Libellus de physicis lineis angulis et figuris per quas omnes actiones naturales complentur*; a treatise *De natura locorum*; and a treatise *De artibus liberalibus*. These were printed in Venice, 1514. Among the Harleian MSS. are two, as yet unedited: 7402, a treatise on the Calendar; and 4350, on the Celestial Sphere. Of Greek, according to Bacon, his knowledge was but slight; but he introduced several Greek scholars into England, one of whom, Nicolas, translated the apocryphal Testaments of the twelve patriarchs, brought from Athens by John of Basingstoke. (Cf. Matthew Paris, ad ann. 1252). It seems probable that Bacon in the early part of his career profited much from Grosstête's teaching. The language of the *Libellus de physicis lineis*, as to the propagation of force, and as to the laws of reflection and refraction, bears a remarkable resemblance to that of Bacon.

[1] Cantor, in his *History of Mathematics*, speaking of the school of translation set up at Toledo in the twelfth century under the direction of Raymund, the archbishop of that city, by Dominic Gundisalvi and John of Seville, remarks: 'Their labours were conducted in a circuitous fashion which had its consequences. The Arabic was first translated into Castilian, and from this the Latin version was made. Bearing in mind that the Arabic text was taken from the Greek by men whose powers of translation were not wholly beyond suspicion, we may imagine what sort of Aristotelian philosophy reached the mediaeval student after three repetitions of bungling.' Cantor, vol. i. p. 684. Jourdain, in his work *Recherches critiques sur l'âge et l'origine des traductions latines d'Aristote* (Nouvelle édition, 1843), has supplied specimens of these translations which enable us to form some judgement of their value; since he distinguishes those made directly from the Greek text from those made from Arabic versions. The latter are not so inferior to the former as might have been expected; probably because the Arab scholars of the tenth and eleventh centuries knew more Greek than the European scholars of the twelfth and

quod nullus potest eos intelligere. Sed quilibet alii contradicit, et multiplex reperitur falsitas, ut patet ex collatione diversorum interpretum et textuum diversarum linguarum.

Et similiter in textu sacro inveniuntur falsa, et male translata quamplurima. Nam Hieronymus probat translationem LXX interpretum et Theodotionis et Aquilae multas habuisse falsitates, quae fuerunt vulgatae per totam Ecclesiam. Et omnes stabant maxime pro translatione LXX, sicut pro vita, et reputabatur Hieronymus falsarius et corruptor scripturarum, donec paulatim claruit veritas Hebraica per solum Hieronymum in Latinum conversa. Ne tamen nimia novitae deterreret Latinos, ideo, ut ipse scribit, aliquando coaptavit se LXX interpretibus et aliquando Theodotioni, aliquando Aquilae, et ideo multa dimisit, et propter hoc remanserunt plura falsa. Nam ut Augustinus probat de doctrina Christi libro secundo male translatum est quod habetur in libro Sapientiae, 'Spuria vitulamina non dabunt radices altas.' Nam debent esse spuriae plantationes, ut Augustinus probat per Graecum. Et tamen Hieronymus dimisit hoc, sicut alia, propter pacem Ecclesiae et doctorum. Atque scitur manifeste, quod Hieronymus humanum aliquid passus aliquando in translatione sua oberravit, sicut ipsemet pluries confitetur. Nam quod decimum nonum male transtulerat Isaiae, resumit in saginali quinto dicens, 'in eo quo transtulimus *incurvantem* et *lascivientem* nos autem verbum Hebraicum *acmon* dum celeriter scripto vertimus, ambiguitate decepti *refraenantem* diximus.' Et aliud quod in eodem capitulo male transtulerat revolvit dicens, 'Melius reor proprium errorem reprehendere, quam dum erubesco imperitiam confiteri, in errore persistere. In eo vero quo transtuli, "et erit terra Juda Aegypto in festivitatem," in Hebraico legitur *agga*, quod interpretari potest etiam festivitas, unde *aggus* in festivum vertitur, et *timor*, quod significantius Aquila transtulit *gerosin*, cum aliquis pavidus et timens circumfert oculos, et adveniens formidat initium; ergo, si voluerimus in bonam partem accipere quod recordatio Judaeae Aegypto sit gaudii recte festivitas dicitur; sin autem, ut

thirteenth; although sometimes the Arabic translation was made from an intermediate Syriac version.

arbitror, in timore pro festivitate vertitur in formidinem vel pavorem.'

Parts of Scripture and of the Fathers still untranslated.

Quarta causa est et ratio hujus rei, quod quamplurima adhuc desunt Latinis tam philosophica quam theologica. Nam vidi duos libros Machabaeorum in Graeco, viz. tertium, et quartum, et Scriptura facit mentionem de libris Samuel et Nathan et Gad videntis et aliorum, quos non habemus. Atque cum tota certificatio historiae sacrae sit a Josepho in Antiquitatum libris, et omnes sancti expositionum suarum radices accipiant a libris illis, necesse est Latinis ut habeant illum librum incorruptum; sed probatum est quod codices Latini omnino sunt corrupti in omnibus locis, in quibus vis historiae consistit; ita ut textus ille sibi contradicat ubique, quod non est vitium tanti auctoris; igitur ex translatione mala hoc accidit et corruptione ejus per Latinos. Nec est remedium nisi de novo transferantur vel ad singulas radices corrigantur. Similiter libri doctorum magnorum, ut beatorum Dionysii, Basilii, Johannis Damasceni, et aliorum multorum deficiunt; quorum tamen aliquos dominus Robertus praefatus episcopus vertit in Latinum, et alii quosdam alios ante eum; cujus opus est valde gratum theologis. Et si libri istorum translati essent, non solum augmentaretur sapientia Latinorum, sed haberet Ecclesia fortia adjutoria contra Graecorum haereses et schismata, quoniam per sanctorum eorum sententias, quibus non possunt contradicere, convincerentur.

So too of Avicenna and of Aristotle.

Similiter fere omnia secreta philosophiae adhuc jacent in linguis alienis. Nam solum quaedam communia et vilia ut in pluribus translata sunt; et de hujusmodi etiam multa desunt. Nam libri integri omittuntur in mathematicis et in naturalibus et in logicalibus et aliis, praeter magna secreta scientiarum, et artium, et naturae arcana quae nondum sunt translata; ut est secunda philosophia Avicennae, quam vocant orientalem, quae traditur secundum puritatem philosophiae in se, nec timet ictus contradicentium lancearum; et tertia quae fuit contermina [1] vitae suae, in qua experientias secretas congregavit, sicut ipse in prologo primae philosophiae suae annotavit. Et similiter cum Aristoteles complevit octo partes naturalis

[2] *Sic*, Jul. O. has conscientia.

philosophiae principales, quae multas sub se continet scientias, de prima parte non habemus omnia, de aliis vero quasi nihil. Et eodem modo de metaphysica, quae sunt novem, cum ipse compleverit eas, nihil habemus quod de metaphysica ulla dignitate vocari potest propter defectus multiplices et praegrandes. De mathematicis vero, cum sint quinque scientiae magnae, non habemus nisi primam, et parum de secunda. Etiam de logica[1] deficit liber melior inter omnes alios, et alius post eum in bonitate secundus male translatus est, nec potest sciri, nec adhuc in usu vulgi est, quia nuper venit ad Latinos, et cum defectu translationis et squalore. Nec est mirum si dico istos libros logicae meliores; nam oportet esse quatuor argumenta veridica; duo enim movent intellectum speculativum seu rationem, scilicet dialecticum per debilem habitum et initialem, qui est opinio, ut disponamur ad scientiam, quae est habitus completus et finalis, in quo quiescet mens speculando veritatem. Et hic habitus non adquiritur per demonstrationem. Sed cum voluntas seu intellectus practicus sit nobilior quam speculativus, et virtus cum felicitate excellit in infinitum scientiam nudam, et nobis est magis necessaria sine comparatione, necesse est ut habeamus argumenta ad exercitandum per intellectum practicum, praecipue cum magis simus infirmi in hac parte quam in speculatione. Libenter enim gustamus de ligno scientiae boni et mali; sed difficiles sumus ad lignum vitae, ut virtutem dignitatem amplectamur propter futuram felicitatem. Quapropter oportet quod habeat intellectus practicus sua adjutoria ut excitetur per propria argumenta, sicut speculativus per sua, et ideo necesse fuit ut traderetur de his argumentis quibus moralis philosophia et theologia utuntur abundanter. Nam sicut speculativae scientiae gaudent argumentis speculativis opinionis et scientiae nudae, sic practicae scientiae, ut theologia, et moralis philosophia practica, considerant argumenta, quibus ad praxim, i.e. ad opus bonum excitemur, et flectamur ad amorem felicitatis aeternae. Et hic sunt duo modi flectendi nos; unus est qui promovet animam ad credendum et consentiendum et commiserandum, et ad com-

The practical part of his Logic, viz. his Poetic and Rhetoric, is still wanting.

[1] The inclusion of Rhetoric and Poetic in Logic is restated in Part iv. ch. 2, and also in *Op. Tertium*, cap. 75.

placendum, et eorum actus, et ad contraria cum necessitate. Et hoc argumentum vocatur rhetoricum, et est respectu intellectus practici, sicut argumentum dialecticum ad intellectum speculativum. Et hic habitus qui flectit nos ad amorem boni operis habetur per argumentum poeticum; quia poetae, ut Horatius et alii Graeci et Latini, vitia prosequuntur, et virtutes magnificant, ut alliciantur homines ad honorem et odium peccati. Nam, ut ille dicit,

'Aut prodesse volunt, aut delectare poetae.
Omne tulit punctum qui miscuit utile dulci.'

Non enim parum prodest civibus, qui delectat in moribus; oportet enim non solum docere, sed delectare et promovere. Unde tam poeta quam orator debet haec tria facere, ut doçendo reddat auditores dociles, per delectationem faciat attentos, et promovendo seu flectendo cogat in opus. Et haec argumenta in salutiferis rebus sunt fortissima, in puris speculativis impotentia, sicut demonstratio efficacissima est in speculationibus nudis, sed impotens est omnino in practicis, et in his quae pertinent ad salutem, secundum quod Aristoteles dicit, primo moralis philosophiae, quod perprimum est mathematicum uti argumento rhetorico, et rhetorem demonstrationem experiri, quoniam, ut dicit secundo, haec scientia non est contemplationis gratia, sed ut boni fiamus. Aristoteles igitur fecit libros de his argumentis, et Alpharabius in libro de scientiis affirmat duas partes logicae debere constitui de his duobus argumentis, quia sola logica deberet docere cujusmodi sunt argumenta, et qualiter componantur propter usum omnium aliarum scientiarum. Et tunc logica speculativis scientiis per argumenta, videlicet duo, quae sunt dialecticum et demonstrativum, moralibus autem ministrat practica argumenta. Et quia theologia et jus canonicum mores et leges et jura determinant, ideo haec duo argumenta sunt eis necessaria. Etiam quamvis Latini nondum habent scientiam horum argumentorum secundum artis logicae traditionem, tamen necessaria sunt multis modis. Qualiter autem componantur haec argumenta, non est ad praesens dicendum. Sed in hoc opere, quod Vestra Beatitudo postulavit dicitur explicari. Nihil tamen de scientiis speculativis utilius est propter fidem probandam infidelibus ut flectantur ad credendum et amorem fidei Christianae. Et similiter ut artificialiter praedicemus omnibus quibus praedicatio necessaria est, et sic de aliis utilibus persuasionibus ad salutem. Magnum autem adjutorium habemus per Augustinum in tertio et quarto de doctrina Christi, et per libros Tullianos, et per libros Senecae et epistolas, quae possunt colligi in lingua Latina de his argumentis, quamvis ipsa textura Aristotelis nobis deficiat.

The fathers make much use of words of foreign origin.

Quinta ratio est ad hoc, quoniam ex sensu sunt compositae et expositae, et ideo cum scientiae fuerunt traditae Latinis a linguis alienis, omnes sancti et philosophi Latini, qui exponunt scientias, usi sunt caeteris copiose, et multiplicant nobis vocabula Graeca et Hebraea et Chaldaea et Arabica, praeter ea quae in textibus continentur. Et nos sumus filii et successores sanctorum et sapientum usque ad haec ultima tempora. Nam vidimus aliquos de antiquis qui multum laboraverunt, sicut fuit dominus Robertus praefatus translator, Episcopus Lincolniensis, et dominus Thomas venerabilis antistes sancti David nuper defunctus, et frater Adam de Marisco, et magister Hermannus translator[1] et quidam alii sapientes. Sed quoniam eos non imitamur, ideo ultra illud quod credi potest deficimus a potestate scientiarum, quia expositiones authenticas non possumus intelligere, et per eas nec intellectum scientiarum possumus obtinere. Et pro infinitis pono duo exempla. Hieronymus dicit in prologo Daniel, quod Daniel et Esdras scribuntur libris Hebraicis, sed Chaldaeo sermone, et una pericope Jeremiae. Hanc autem pericopen Jeremiae omnes theologi dicunt esse Threnos Jeremiae: quia pericope idem est quod pars parva seu particula; sed omnes Hebraei sciunt quod scribimus *Threni* literis Hebraicis et Hebraeo sermone. Deinde possumus hanc pericopen sumere decimo capitulo Jeremiae, ubi dicitur sic, 'Ergo dicetis eis, Dii qui coelum et terram non fecerunt pereant de terra, de his quae sub coelo sunt.' Nam hoc tantum in Jeremia habet sermonem Chaldaeum, sicut omnes Hebraei sciunt literati. Et certum est quod Hebraei et Chaldaei eandem habent linguam, sed diversum idioma, sicut

[1] Omitted in O.

Gallicus et Picardus. Idioma enim est proprietas linguae apud aliquam nationem determinatam, unde Hebraeus dicit *Eloim* pro Deo vel Diis; Chaldaeus dicit *Eloa*, pro coelo vel coelis. Pro *non*, Hebraeus dicit *lo*, Chaldaeus dicit *la*, et sic in aliis.

The Hebrew alphabet.

Tamen quod haec pericope scribatur sermone hic Hebraeo et Chaldeo, ponetur alphabetum[1] Hebraeum, ut facilius valeat intelligi quaestio proposita, et primo scribuntur figurae Hebraicae; secundo in linea superiori ponuntur nomina; et supremo assignantur litterae quae literis Hebraicis correspondent; ut literarum Hebraicarum sciamus virtutes et potestatem sonorum, secundum quod quaedam sunt vocales et quaedam consonantes.

י	ט	ח	ז	ו	ה	ד	ג	ב	א
iot	teth	heth	zain	vaf	he	dalet	gimel	beth	aleph
i	t	h	z	v	e	d	g	b	a

ף	פ	ע	ס	נ	מ	ל	ך	כ
pe	pe	ain	samech	nun	mem	lamet	chaf	chaf
p	p		s	n	m	l	ch	ch

ת	ש	ר	ק	צ
taf	shin	rish	koph	tsadik
t	sh	r	k	ts

Sunt autem sex vocales, viz. aleph, ain, he, heth, iot, vau; reliquae sunt consonantes: he et heth aspirantur, ut he in principio, heth non solum in primo sed in fine, et generatur in gutture, he in ore; aleph similiter in ore; ain in gutture; sed considerandum quod solum iot habet unum sonum sicut i nostrum, et sit consonans et vocalis sicut apud nos j. Vau vero, ut dicit Hieronymus in Hebraicis quaestionibus, habet duplicem sonum, viz. v nostrum et o: reliqua vero quatuor habent sonum quinque vocalium nostrarum, viz. a, e, i, o, et u; sicut patet per Hieronymum in libro interpretationum. Et

[1] Neither this Hebrew alphabet nor the Greek alphabet which follows will be found in Jebb's edition, which has moreover many other omissions, supplied here from Jul This is the more strange as Jebb undoubtedly had this MS. before him, and made use of it in the first part. He had the advantage of using it a year or two before it was so damaged by fire as to render much of it illegible.

hanc diversitatem sonorum signant per puncta et tractus. Nam si sub aleph trahatur linea sine puncto sic, אַ, vel cum puncto, אָ sonatur a. Si vero duo puncta fiant jacentia sub aleph e transverso אֵ, vel duo stantia אְ, vel tria in modum trianguli אֶ, vel quinque puncta hoc modo אֱ, sonatur e. Si vero tria puncta iaceant sub aleph ex obliquo descendentia sic אֻ sonatur u. Si vero unus punctus ponatur sub litera אִ sonatur i. Si vero unus punctus fiat supra sonatur o, sic, אֹ. Et ita est de ain, et he, et heth, quae habent hos quinque sonos per istorum signorum diversitatem. Et cum vaf sonatur v potest esse signum trium punctorum ut dictum est sic וֻ, vel potest poni unus punctus sic וֹ.

Ideo oportet quod ad consonantes ponantur haec signa, ut sciatur sonus vocalis syllabicandus cum consonante: ut si volo designare ba, be, bi, bo, bu, scribam sic: בָ בֵ בִ בֹ בֻ. Et habent alia signa per quae designant sonos consonantium aliquando fortificari aliquando remitti. Unde quando tractus ponitur super literam tunc remittitur: quando punctus in ventre ponitur tunc fortificatur. Ut quando super Dalet ponitur tractus sic דֿ tunc debilem sonum reddit ut nostri z, ut cum dico, *adamas*. Quando punctus in ventre ejus collocatur, sic, דּ, tunc fortiter sonat, ut cum dico, *dabo*. . . .

Manifestus ergo et vilis est error omnium in hac parte propter ignorantiam harum linguarum.

The Greek alphabet.

Aliud exemplum accipiam de Graeco et multa exempla Graeca addentur in sequentibus. Sed volo hic ponere alphabetum Graecum cum diphthongis quibus scribunt; multo enim evidentius per hoc patebit quae dicenda sunt.

a	b	g	d	e	z	i	th	i
alpha	vita	gemma	delta	e pente	zita	ita	thita	iota
α	β	γ	δ	ϵ	ζ	H	θ	ι

k	l	m	n	x	o	p	r	s	t
kappa	labda	mi	ni	xi	o micron	pi	ro	sima	taf
κ	λ	μ	N	ξ	ο	π	ρ	σ	τ

y	ph	ch	ps	
ypsilo	phi	chi	psi	o mega
υ	φ	χ	ψ	ω

Sunt autem septem vocales quantum ad figuras diversas, quum habent triplex i et duplex o; sed quatuor tantum habent quantum ad sonum principalem, videlicet, a, e, i, o. Diphthongus apud Graecos est conjunctio duarum vocalium. Sonus unius vocalis habetur ut vocalis cum consonante. Et finales literae in diphthongis sunt iota et ipsilo. Potest igitur ipsilo consequi alpha sic, *αυ*, et tunc sonat quantum a cum v consonante, quia sonus aliquantulum similis est sono ipsius a cum f, et ideo vulgariter exemplificamus quod sonat af. Et potest consequi e, sic *ευ*, et tunc sonat quantum e vocalis cum v consonante, quasi ef, ut dictum de alpha et ipsilo. Vel potest consequi iota, sic *ιυ*, et sonat quasi if, ut dictum de aliis. Vel ipsilo potest consequi o micron, sic *ου*, et tunc sonat u vocalem. Et sic solum habent Graeci sonum hujus vocalis u. Quum autem iota consequitur alpha sic, *αι*, tunc sonat e, quasi e. Quum e, sic *ει*, tunc sonat i per iota: si *ο*, sic *οι*, tunc sonat y per ypsilo. Et hi octo diphthongi vocantur proprii. Et alii tres dicuntur improprii; et fiunt per subscriptionem hujus literae iota ad alpha, ita, et o mega, sic, *ᾳ*, *ῃ*, *ῳ*. Aliquando ponitur iota in linea, sicut in aliis diphthongis sic *αι*, *ηι*, *ωι*. Sed remanet sonus literae principalis, scilicet ejus cui subscribitur iota. Nam cum subscribitur *α*, quae est alpha, sonat a: quum vero *η*, quae est ita, sonat ita: si vero *ω*, quae est o mega, sonat solum o mega. Et his tribus diphthongis utuntur Graeci semper in dativo casu primae declinationis.

... Et[1] licet exemplificare in praesenti loco de Jacob qui cum obviaret Esau fratri suo veniens de Mesopotamia et distaret dicit, Vidi faciem tuam quasi faciem Dei vidissem. Quaerit Augustinus, qualiter poterat homo sanctus hominem reprobatum comparare Deo? et solvit quod multipliciter Deus in scriptura accipitur pro vero Deo, aliquando aliter. Et hoc multis modis: sed ut LXX interpretes designaverunt quod non loquebatur de vero Deo ideo apposuerunt articulum Graecum ad nomen dei. Nam hoc est de proprietate articuli ut veritatem rei designet. Sed hoc non apparet in Latino,

[1] There is evidently a hiatus here. The subject considered is now the use of the article in Greek.

quia Latini non habent articulum. Nam satis innotescit in Gallico. Unde cum dicitur Parisius *Li reis vent*, iste articulus li designat proprium et verum regem talis loci, quasi regis Franciae. Et non sufficeret hoc ut denotaret adventum regis Angliae. Nullus enim diceret de rege Angliae veniente Parisius, *Li reis vent*, sed adjungeret aliud dicens, *Li reis de Engletere vent*. Et ideo articulus solus sufficit ad veritatem et proprietatem rei de qua est sermo designandam. Propter quod Augustinus dicit quod Graecum hoc sit πρόσωπον θεοῦ, quod in Graeco sonat prosopon theu; cum articulo enim non sic, prosopon tu theu, πρόσωπον τοῦ θεοῦ. Prosopon hic significat vultum vel faciem. Theu est genitivus casus hujus nominis theos quod est Deus, et tu est articulus genitivi. Magna ergo necessitas est ut Latini sciant linguas propter dicta sanctorum et caeterorum sapientum.

The text of the Bible has become extremely corrupt.

Sexta ratio est propter errorum falsitatum infinitarum correctionem in textu tam theologiae quam philosophiae, non solum in litera, sed in sensu. Quod autem correctio sit necessaria, probo per corruptionis magnitudinem. Et quoniam violentius et periculosius erratur in textu Dei quam textu philosophiae, ideo convertam linguarum potestatem ad corruptionem textus sacri, ut pateat necessitas earum, propter [1] corruptionem infinitam exemplaris vulgati quod est Parisiense. Et Deus novit quod nihil tam valida indigens correptione potest Apostolicae Sedi praesentari sicut haec corruptio infinita. Nam litera ubique in exemplari vulgato falsa est, et si litera sit falsa vel dubia, tunc sensus literalis et spiritualis falsitatem et dubitationem ineffabilem continebit, quod volo nunc ostendere sine contradictione possibili. Nam Augustinus contra Faustum dicit, 'Si discordia in Latinis codicibus est, recurrendum est ad antiquos et plures. Nam antiqui praeponendi sunt novis, et plures paucioribus prae-

[1] From here to the end of the following sentence, supplied from Jul. The corruption of the Biblical text is fully discussed in the *Opus Minus*, under the head of the fifth sin of Theology (Brewer, pp. 330–349). See also the valuable Memoir of Abbé Martin, *La Vulgate Latine au xiiie siècle d'après Roger Bacon.* Paris: Maisonneuve, 1888. The *exemplar Parisiense* appears to have been a text very carelessly compiled about 1230, and probably very carelessly copied, to meet the demands of the multitude of students who were flocking to Paris.

feruntur.' Sed omnes antiquae Bibliae quae jacent in monasteriis, quae non sunt adhuc glossatae nec tactae, habent veritatem translationis, quam sacrosancta a principio recepit Romana Ecclesia, et jussit per omnes Ecclesias divulgari.

Especially the Paris text.

Sed hae in infinitum contradicunt exemplari Parisiensi: igitur hoc exemplar magna indiget correctione per antiqua. Caeterum Augustinus ibidem dicit, 'Quod si dubitatio adhuc remaneat in antiquis bibliis, recurrendum est ad linguas, scilicet Hebraicam et Graecam'; et hoc dicit secundo de doctrina Christiana, et ostendit in exemplis. Et Hieronymus hoc dicit ad Frecellam sororem et super Zachariam, et omnes sancti concordant; sed antiquis bibliis concordant linguae Graecorum et Hebraeorum contra exemplar Parisiense; igitur oportet quod corrigatur. Caeterum Hieronymus dicit ad Damascenum in hoc casu, 'Ubi est diversitas, non est veritas nota.' Sed illi qui nituntur cum omni veritate quantum possint corrigere textum sunt duo ordines Praedicatorum et Minorum. Jam de correctione formaverunt varias scripturas, et plus quam una biblia contineat; contendunt ad invicem, et contradicunt infinities, et non solum ordines ad invicem, sed utriusque ordinis fratres sibi invicem contrariantes plus quam ordines totales; nam omnis dominus alii contradicit, et in eadem correctores sibi invicem succedentes mutuas eradunt positiones cum infinito scandalo et confusione. Unde cum ad viginti annos praedicatores redegerunt correctionem in scripturis, jam venerunt alii, et novam ordinaverunt correctionem, quae continet plus medietate unius bibliae; quantum vix ponatur in tanta scriptura quantum Novum continet Testamentum. Et quia vident se errasse in antiqua correctione, jam fecerunt statuta quod nullus ei adhaereat; et tamen secunda correctio propter horribilem sui quantitatem simul cum veritatibus multis habet sine comparatione plures falsitates quam prima correctio.

Illustrations.

Quod autem dixi in universali, potest patere in exemplis; nam infinities accidit corruptio additione, subtractione, conjunctione, divisione orationis, dictionis, syllabae, literae, diphthongi, aspirationis notae: et non solum litera, sed sensus literalis et spiritualis mutantur; et non solum cadunt haec vitia circa unam orationem, sed circa multas, immo penes folia quamplurima; et de singulis unum ponam exemplum vel duo. Nam multi prologi superflui ponuntur in textu, cum non sint prologi textus in quibus redditur ratio translationis librorum quibus praeponuntur, sed sunt epistolae familiaribus missae, ut epistola Hieronymi ad Paulinum, quae in capite bibliae reputatur prologus et vocatur a vulgo, quae tamen in libro epistolarum Jeronymi continetur; vel respectu prologi in commentarios et in originalia non in textum, sicut idem quod praemittitur ante librum Ecclesiastis. Nam patet quod prologus est ibi originalis, et patet ex sententia. Et sic est de multis aliis, quae non sunt in bibliis antiquitatis. De una oratione superflua est exemplum Deuteronomii 27. 'Maledictus qui dormit cum uxore proximi sui, et dicet omnis populus Amen;' quoniam nec antiqui codices, nec Hebraeus, nec Graecus habent versum hunc. De superfluitate dictionis horribile est ac nefandum octavo Genesis, cum dicitur quod 'Corvus ad arcam non est reversus,' et Hebraei et Hieronymus in originali habent affirmativam. Et accepta est negatio a paucis temporibus de alia translatione, scilicet LXX interpretum, cujus falsitatem Hieronymus ostendit locis infinitis, et jam a tempore Isidori et antea evacuata est. Nam ipse dicit in libro de officiis quod generaliter omnes ecclesiae Latinae utuntur translatione Hieronymi, pro eo quod veracior sit in sententiis et clarior in verbis; excepto quod propter nimium usum psallendi in ecclesia solius psalterii translatio scilicet LXX interpretum remansit. Sed antiquitus Romana Ecclesia jussit translationem hanc ubicunque haberi. Augustinus et alii et ipsemet Hieronymus tempore suo usi sunt sicut Ecclesia translatione antiqua. Et ideo Augustinum quum recitat textum hunc decimo sexto de Civitate Dei et exponit oportuit quod uteretur translatione quae fuit vulgata et recepta apud Latinos, nec potuit aliud facere. Omnis vero glossator qui infixit glossas super textum accepit auctoritatem Augustini de Civitate Dei et eam posuit infra textum, sed non mutavit eam nec intulit negationem . . . inter caeteros hoc fecit. Et sic vulgatus est error horribilis cum contradictorium pro contradictione ponatur. Nam videtur in philosophia quod ejusdem libri est aliquando duplex et triplex translatio; et una habet diversum vel aliquando contrarium alteri. Sed nullus est qui ausus est translationem unam miscere cum alia.

Quod autem ecclesiastici habent in legendo negationem hoc est de corruptione exemplaris apud studentes ad ecclesiasticos derivatum, et de syllabae mutatione, et per consequens totius dictionis. Et exemplum mirabile est de Joseph, qui dicitur in exemplari vulgato venditus fuisse triginta argenteis propter exemplum Domini, sed secundum antiquos codices et Hebraeum, et Graecum, et Arabicum, et Hieronymum in originali, et Josephum in antiquitatum libro, debent esse viginti non triginta. Et similiter in Psalterio ad syllabae mutationem mutatur tota dictio cum infinito errore, cum dicitur, 'Sitivit anima mea ad Deum fontem vivum.' Nam cum ecclesia in solo Psalterio utatur translatione LXX interpretum, Hieronymus correxit hanc translationem bis, et posuit *fortem* ubi ponimus *fontem* per errorem propter similitudinem dictionis, et propter hoc quod in praecedenti versu fit mentio de fonte; sed ut dixi Hieronymus correxit *fortem*, et ita est in Hebraeis bibliis et in psalteriis antiquis monasticis. Nam hoc diligenter inspexi; et omnino certum est quod non est hic error vilissimus propter similitudines praedictas.

De literae mutatione est exemplum notabile primo Judicum, cum dicitur in monte Ares, quod interpretatur testaceo, ut penultima litera sit e non i, sed communiter habetur testacio per i, ut sit nominativus casus, et idem quod testificatio a *teste*; sed si deberet esse ablativus, derivatur a *testa.* Nam in omnibus antiquis bibliis est *testaceo* per e, et in Graeco, et in Hebraeo, ubi habetur hares, Hieronymus transtulit *testam*, vel aliquid derivatum a *testa*; nam *hares* in Hebraeo *testam* vel aliquid praedictorum significat in Latino. Unde Hieronymus in sexto libro super Isaiam exponens idem verbum decimo sexto capite, 'His qui laetantur super muros cocti lateris,' dicit hares testam sive coctum laterem significat. Et in decimo octavo libro super idem vicesimo quarto Isaiae, 'Erubescet luna,' dicit quod hares testam sive coctum vel ariditatem sonat. Quod vero tricesimo primo et tricesimo secundo Jeremiae confunduntur haec nomina Ananeel et

Anameel per errorem, ut in litera indifferenter ponitur m in penultima, est error magnus in mutatione unius literae. Nam Hieronymus dicit in originali, quod Ananeel per n scriptum est turris, per m est filius patruelis Jeremiae, et sic invenitur in Hebraeo. De aspirationis nota exemplum primo ad Thessalonicenses, cum dicitur, ad cujus *ore,* ut sit ablativus casus hujus nominis *os*, *oris*, et non genitivus hujus nominis *hora*, *horae*; scribitur enim in ablativo casu, et glossatur non a sancto, sed a magistro sententiarum, qui glossavit epistolas; sed sicut defecit multipliciter in expositione propter ignorantiam Graeci, ita fecit hic. Quum procul[1] dubio in Graeco, a quo sumptus est genitivus hujus nominis *hora*, invenitur *horas*, et aspiratur tam apud Graecum quam apud Latinum. Scilicet *Os*, *oris*, non aspiratur. Hoc enim verbum, *hora*, est Graecum, licet Latino nomine declinatur, sicut Domina: sed Graecus declinat sic, *hora*, *horas*, *hora*, *horam*, *hora*. Unde nominativus et dativus, et vocativus, similes sunt: accusativus in am, genitivus in as; ablativum non habent Graeci. Et hoc in Graeco est *horas*, sicut ego legi diligenter et omnibus possim probare qui sciunt Graecum, et in Graeco invenitur aspiratio. Haec exempla volui assumere ut quae probant quod necesse est linguas sciri propter textus Latini corruptionem tam in theologia quam in philosophia. Quomodo vere de corruptione plene probo et in speciali per omnes corruptiones bibliae, in aliud tempus differtur propter rei magnitudinem quae potest Vestrae Sanctitati praesentari, sed non nunc ut sufficiat, sed magis per alium in sequentibus explicabo.

Even when the text is correct it is wrongly interpreted.

Septima ratio est, quia necesse est ut Latini sciant linguas specialiter propter sensus falsitatem, etsi litera esset verissima. Nam tam in theologia quam in philosophia necessariae sunt interpretationes et praecipue in textu sacro, et in textu medicinae et scientiarum secretarum; quae nimis occultantur propter ignorantiam interpretationum. Nam medici confusi sunt propter malas interpretationes, quas vocant synonyma; non est autem eis possibile uti medicinis authenticis propter errorum istorum synonymorum ignorantiam; et ideo accidit

[1] The following passage, down to the end of the paragraph, is omitted in O. and D., and has been supplied from Jul.

in manibus eorum infinitum periculum. Eodem modo est in textu sacro; nam summa difficultas, quae est apud ipsum sciendum est propter varietatem et obscuritatem infinitarum interpretationum, et in exemplo familiari pro infinitis aliis apparet. Vulgus enim hoc nomen Israel pro patriarcha interpretatur virum videntem Deum. Et praevaluit hoc in usu usque ad tempus Hieronymi, et etiam usque quo sua translatio et sua expositio jussae sunt per omnes ecclesias divulgari. Sed ipse dicit in originali quantae grandis auctoritatis sunt, et eorum verbis nos opprimit qui Israel virum videntem Deum interpretati sunt. Nos tamen magis consentimus Domino vel Angelo qui illi hoc nomen imposuit quam auctoritati alicujus saecularis eloquentiae. Et ideo probat egregie quod affirmat. Nam illi qui sic interpretati sunt crediderunt quod hoc vocabulum significet idem conjunctum vel divisum, sicut respublica apud nos. Sed hoc non est generaliter verum, immo in pluribus habet instantiam in omni lingua. Nam apud Hebraeos *Is* est vir, *Ra* videns, *El* Deus; et ideo crediderunt multi quod hoc nomen patriarchae habet resolutionem in illa tria. Sed Hieronymus reprobat per multa argumenta; quatuor enim possunt sumi e dictis suis a parte vocis, et quatuor vel quinque a parte rei. Nam in illis tribus nominibus aliae literae sunt et plures quam in nomine patriarchae, et aliter ordine et syllabicatae reperiuntur. Ex hoc ergo triplici argumento sumto penes literas concluditur per Hieronymum quod id significari non potest hinc inde; cum potentia significationis ejusdem sumitur propter vocis identitatem, sed vocem et literas nimis variari, quum in nomine patriarchae sunt hae quinque literae per ordinem: Iod, Sin, Resh, Aleph, Lamet, sicut ipsum Hebraeum hic positum declarat יִשְׂרָאֵל, Israel. Sed in hoc triplici vocabulo hae octo literae habent hunc ordinem, scilicet, Aleph, Iod, Sin, Resh, Aleph, He, Aleph, Lamet, ut hic Hebraeum ostendit, אִישְׂרָאהָאֵל. Et quarto argui potest explicatione. Nam sicut puncta ostendunt nomen proprium non retinet apud Hebraeum sonum praecisum illorum vocabulorum. Nam secundum majorem quasi Iserael sonatur in quatuor syllabis, tamen ibi vocabulorum sonus in solis tribus syllabis coarctatur, quoniam punctum sub litera sonat i, et duo puncta sonant e, et linea cum puncto sub ea sonat a. Sed argumenta fortiora trahuntur ex sensu vocis secundum Hieronymum. Et hoc ostendit ipsum Hebraeum hic scriptum hoc modo:

וַיֹּאמֶר לֹא יַעֲקֹב יֵאָמֵר עוֹד שִׁמְךָ כִּי אִם־יִשְׂרָאֵל כִּי־שָׂרִיתָ עִם־אֱלֹהִים וְעִם־אֲנָשִׁים וַתּוּכָל׃

Et textus Graecus habet sic: 'quia invaluisti cum Domino [et cum hominibus valebis[1]].'

Nam secundum Hieronymum et per textum Hebraeum, et Graecum, et Latinum, et per Josephum patet quod Israel non debet dici vir videns Deum, sed principalis vel princeps cum Deo, quoniam in Hebraeo ad literam est sic: 'Et dixit Deus non vocabitur nomen tuum a modo Jacob, sed Israel; quoniam principalis vel princeps fuisti cum Deo, et cum hominibus poteris principari.' Et ideo dicit Hieronymus quod sensus est, 'Non vocabitur nomen tuum supplantator, hoc est Jacob, sed vocabitur nomen tuum princeps cum Deo, hoc est Israel. Quoniam ego princeps sum, sic tu, qui mecum luctari potuisti, princeps vocaberis. Si autem mecum pugnare invaluisti, quanto magis cum hominibus; hoc est, cum Esau, quem formidare non debes[2]?' Et quoque Latinum habet: 'Quoniam si contra Dominum fuisti, quanto magis contra homines praevalebis?' Et Josephus primo antiquitatum libro, Israel ideo appellatum dicit, quia contra Angelum steterit. Omnia igitur haec, scilicet principari cum Deo, et invalescere, et fortem esse, et stare cum Deo ut patet reducuntur ad eundem sensum, sed diversis vocabulis interpretationum, quorum nullum de virtute significationis suae potest elicere visionem Dei. Et ideo vera interpretatio est *princeps cum Deo*. Et adhuc confirmat hoc Hieronymus per argumentum derivationis; nam Sarith et ab Israel nomine derivatur, et principem sonat. Unde Sara uxor Abrahae princeps dicitur,

[1] The words in brackets are omitted in the MSS.

[2] See Jerome's *Liber Hebraicarum Quaestionum in Genesim*. Jerome adds: 'Quamvis igitur grandis auctoritatis sint, et eloquentiae ipsorum umbra nos opprimat qui Israel *virum* sive *mentem videntem deum* transtulerunt, nos magis Scripturae et Angeli qui ipsum Israel vocavit auctoritate ducimur quam cujuslibet eloquentiae saecularis.'

ut dicit Hieronymus super septimum decimum capitulum Geneseos; quapropter si vulgus vel aliqui antiqui, ut Eusebius Caesariensis in libro nominum Hebraeorum quem Hieronymus in Latinum vertit, et alii, famosa abutentes interpretatione dicunt *Israel* interpretari per *virum videntem Deum*, dicamus cum Hieronymo[1]. Illud vero quod in libro Nominum interpretatur Israel *Vir videns Deum* omnium pene sermone decretum non tam vero quam violenter interpretatum videtur. Et igitur per Eusebium in libro Nominum quem Hieronymus transtulit in Latinum et per Ambrosium et alios forsitan sanctos allegare si quis concedat quod recta hujus vocabuli, *Israel*, expositio sit *Vir videns Deum*, dicendum est quod locuti sunt secundam vulgatam expositionem, antequam veritas fuerit patefacta quam postea beatus Hieronymus vera et perfecta interpretatione Latinis revelavit, sicut in ejus libris continetur et in glosa etiam habetur.

Et si forsan dicitur quod consuetudo vulgi theologorum modernorum hanc interpretationem frequentet, patet responsio per supradicta secundum Augustinum, et Cyprianum, et Isidorum, et alios et per varias declarationes[2]. Nam secundum eos manifestatae veritati cedat consuetudo, et relicto errore vulgi sequamur veritatem. Et quod ex mera ignorantia venit non debet allegari, sicut accidit in proposito, et praecipue contra auctorem et doctorem sacrum non licet contraire, nisi pro se rationes sufficientes et auctoritates allegat. Et ad omnem affirmationem poterit quilibet Hebraicae peritos consulere, et inveniet sententiam Hieronymi ratam et inconcussam[3]. Summa vero necessitas remediorum falsitatis requiritur in his interpretationibus propter formam Hebraei sermonis; nam in interpretationibus vulgatis quae in fine bibliae ponuntur, sunt infinitae occasiones errorum: propter hoc erramus quod unum vocabulum aestimatur simplex secundum normam Latinorum, quod est multiplex apud Hebraeos: et abundantius erratur quod tali vocabulo dantur variae interpretationes tanquam

[1] The remainder of the paragraph is omitted in O. and is supplied from Jul.

[2] et . . . declarationes om. in O.

[3] This sentence omitted in O.

ejusdem sint vocabuli Hebraei, cum tamen quaelibet sit diversi, eo quod vocabulum Hebraeum apud nos male consideratum in scriptura una habet diversas literas apud Hebraeos, penes quos recepit diversas interpretationes, secundum quod Hieronymus ponit exemplum in epistola de mansionibus. Nam oportet si *ar* scribatur per *aleph*, significat *lumen*; si per *ain*, *inimicum*, si per *heth*, *foramen*; per *he*, *montem*; dicit igitur quod vicesimo Numerorum quidam interpretati sunt his modis quatuor; sed opiniones tres destruit, quia in Hebraeo scribitur hic per *he*, et ideo solum *montem* in hoc loco designat. Sed in praedicando et legendo theologi recurrunt ad omnes quatuor expositiones in hoc vocabulo, et sic alias, igitur multipliciter errant.

Latin grammar has been formed on the model of Greek and Hebrew.

Est ultima ratio scientialis de necessitate linguarum, quod Grammatica in lingua Latinorum tracta est a Graeco et Hebraeo. Nam litteras accepimus a Graecis, ut docet Priscianus, et [1] totam rationem tractandi partes omnes Priscianus accepit a Graecis et miscet Graecum in multa abundantia per omnes libros suos. Et ipsa vocabula linguae Latinae, et tam theologica quam philosophica, ab alienis linguis pro parte maxima sunt transfusa, quorum aliqua suspicantur Latini esse alterius linguae; et de aliquibus non considerant quod ab aliena lingua descendunt. Multa vero aestimantur quod sint penitus Latina, cum tamen sint Graeca, vel Hebraea, vel Chaldaea, seu Arabica, in quibus tam in pronunciatione quam in scriptura et sensu accidit multiplex error Latinorum; nec est modicum errare in vocabulis, quia per consequens errabitur in orationibus, deinde in argumentis, tandem in his quae aestimantur concludi [2]. Nam Aristoteles dicit quod 'qui nominum ignari sunt, saepe paralogizantur.' Et primum et principale fundamentum doctrinae ponit Boethius in certa et integra cognitione terminorum, sicut docet in libro de disciplina scholarium, atque nos experimur hoc in singulis scientiis. Nam principalis difficultas [3] et utilitas est, quod homo sciat intelligere vocabula quae dicuntur in scientia et prudenter

[1] The following six lines supplied by Jul., om. in O.

[2] The last six words om. in O.

[3] difficultas et om. in O.

sine errore proferre; quando veraciter scit hoc, potest per scriptum perficere sine ulteriori doctrina si sit diligens in studio. Nam textus scientiarum sunt ei plani, quando noverit proprie ac certe intelligere et interpretari; et sine difficultate potest quemlibet sapientem intelligere, et cum quolibet sufficienter conferre, et a quolibet si necesse est edoceri. Et Aristoteles dicit in primo coeli et mundi, quod parvus error in principio est magnus in principiatis; qui enim in fundamentis errat, necesse est ei in errore totum aedificium cumulare.

Aestimamus igitur linguam nostram Latinis dictionibus esse compositam, et pauca esse vocabula aliarum linguarum, cum tamen quae communiter utuntur sunt de linguis alienis, ut domus, scyphus, clericus, laicus, diabolus, Satanas, ego, pater, mater, ambo, leo, ago, malum [1], et sic de infinitis, quae vix in magno volumine possunt congregari; praecipue si scrutemur vocabula singularum scientiarum, et maxime theologiae et medicinae; quo volumine nihil esset utilius, si vocabulorum omnium recta scriptura ac pronunciatio debita cum fideli derivatione et recta interpretatione probarentur [2]. Sed nunc in his quatuor erratur in magnum totius detrimentum sapientiae, quod paucis exemplis potest intelligi. Nos enim non consideramus ordinem linguarum, nec quod prior lingua non recepit interpretationem posterioris, nec quod diversae linguae in eo quo diversae sunt non se mutuo exponunt: sed quod dicit Hieronymus, et maxime prior ex posteriori non potest originem habere, ut certum est omni homini rationem habenti; unde Graecum non oritur ex Latino, nec Hebraeum ex Graeco, et non debet Hebraeum capere etymologiam ex Graeco, nec Graecum ex Latino: unde Hieronymus dicit contra quosdam in dicto loco memorato, quod Sara non Graecam sed Hebraicam debet habere rationem, Hebraea enim est. Et saepius dicit quod Lenaeus a ληήνη, id est, *lacus*, dicitur, non a *lenio*, quia Graecum non potest Latinam etymologiam recipere; sed quod hoc facimus graviter et indifferenter. Et nos contra; nam dicimus quod *amen*, licet sit Hebraeum, dicitur ab *a*, quod est *sine*, et *mene* Graeco, quod est defectus. Et cum

[1] The last nine examples om. in O.

[2] J.'s reading, praeliarentur, is unintelligible. Jul. has probarentur.

parasceue sit Graecum, dicimus quod derivatur a *paro*, *paras*, et *coena*, *coenae*, quae sunt Latina. Et[1] dicunt quod *dogma* dicitur a *doceo*, et sic de infinitis quae omnia falsa sunt. Et non solum vulgus Latinorum sed auctores in his oberrant, et Hugo et ejus sectatores qui aestimant *jubileum* a *jubilo* derivari, cum tamen *jubileum* debet esse Hebraeum. *Jubilo* est Latinum; sed non debet dici *Jubileum* ut litera *i* sit in secunda syllaba sicut in *Jubilo*. Debet *i* esse *e* litera ut dicatur *jubeleus*, sicut vult Isidorus et Papias, et omnes libri antiqui sic habent. Nam dicitur a *jobel* quod est Hebraeum. Ita aestimamus quod multa vocabula quae sunt in usu Latinorum debent exponi per alias linguas.

Many words of Latin origin wrongly thought to be foreign.

Assueti autem in hoc credimus quod longe plura, quam veritas sit, capiunt etymologiam aliunde. Nam sola illa vocabula, quae oriuntur et derivantur ex Graeco et Hebraeo, debent habere interpretationes per linguas illas. Ea enim, quae pure Latina sunt, non possunt habere expositionem nisi per vocabula Latina. Nam purum Latinum est omnino diversum ab omni lingua, et ideo non potest habere interpretationem aliunde: sed Latini non hoc considerant: imo indifferenter pura Latina per alias linguas interpretantur. Unde multis modis hoc verbum *coelum* quod est pure Latinum Graece interpretantur dicentes quod *coelum* dicitur quasi *casa helios*, i.e. domus solis, nam sol dicitur helios; sed incongrue dicunt et falso. Debent enim dicere, *Casa heliu*, cum helios sit nominativi casus, heliu genitivi. Deinde falso dicitur. Nam sicut Varro peritissimus Latinorum et Plinius in prologo plene confirmat, coelum dicitur a *coelo*, *coelas*, quod est *sculpo*, *is*, quia stellis sculptum est et ornatum. Quod etiam est lege scripturae vocabulorum. Nam *coelo*, *coelas*, pro *sculpo*, *sculpis*, scribitur per diphthongum *ae* in omnibus libris antiquis. Et sic hoc verbum *coelum* apud omnes codices antiquos scribitur per diphthongum eandem. Et ideo derivatur a coelo quod est sculpo. Et ex hoc sequitur quod non derivatur a *celo*, *celas* quod est *occulto*, *occultas*, sicut illi qui huic nomini dant etymologiam absurdam dicentes sic dici quod occultatur et elongatur a nobis, vili errore sicut

[1] The following passage, to the end of the paragraph, supplied from Jul.

priores decepti. Similiter hoc nomen *ave*, quod est pure Latinum, Graece exponunt, dicentes quod dicitur ab *a*, quod est *sine*, et *ve*, quasi *sine ve*: sed hoc fieri non debet, quia hoc vocabulum non sumitur a Graeco vocabulo cognatae significationis. Nam *chaere* in Graeco signat *ave* in Latino, sed haec duo non concordant. Hic ergo est unus modus quo in infinitis Graecis vocabulis errat Latinus.

Greek diphthongs misunderstood.

Alius modus est, quod in Graecis vocabulis non intelligimus eorum scripturam, quam habent multipliciter variatam; quia vocabula consimilia in sono distinguunt in significato: unde habent triplex et duplex o, et duplex t, ac p et c; et habent undecim diphthongos, et multa alia, ut sic varietatem suorum vocabulorum in significando designent. Nam *cenos* quod est inanis, a quo cenodoxia, i. e. inanis gloria, de quo Deuteronomi septimo, per e breve scribitur. Et *cenos* quod est novus, a quo encenia, i. e. innovationes, ut nova festa et dedicationes, de quo Johannes decimo, et scribitur per ae diphthongum, sic *caenos*. *Cenos* vero quod communis, a quo *cenobium* et *epicenium*, scribitur per oi diphthongum, quam Latinus proferat e, sed deberet proferri i, ut diceret *cinos*, unde ab hoc dicitur *cinomia*, quod est secundum Hieronymum in correctione Psalterii, communis vel omnimoda musca. Unde Papias dicit quod scribitur per diphthongum in prima syllaba, sic *coinomia*. Et hoc manifestum est in Graeco Psalterio. Et *cynos*, canis, qui scribitur per *y* Graecum, unde *cynomia*, i.e. musca canina, de qua Exodi octavo, et *xenos* per *x*, quod est peregrinus, a quo *xenia*, quae sunt munera seu dona, de quibus liber Machabaeorum, et secundus Ecclesiastici. Et *schenos* per *sche* est *funis*, a quo *schenobates*, qui graditur in fune et super funem. *Scena* est *umbra*, vel *tabernaculum*, a quo *scenopegia*, i. e. fixio tabernaculi, et *scenofactoria* ars est in qua Paulus Apostolus laborabat. Cum igitur derivativa istorum vocabulorum et composita sic variantur in significatis, licet sint similia in sermone et sono, manifestum est quod non est possibile evadere in sensu literali sine errore, qui non advertit scripturam hujusmodi. Unde magni viri et famosi expositores aliquando decepti sunt, sicut Rabanus, qui dicit quod *scenofactoria ars* docet facere *funes*, quia aestimabat quod *schenos*, quod est *funis*, esset idem a quo nomen derivatur. Sed Beda docet contrarium volens quod a scena derivatur, et hoc manifestum est per scripturam vocabuli. Scilicet[1] ignoratione Actuum in Graeco textu scribitur vocabulum penultima syllaba sine aspiratione et per vocalem quae vocatur *ita*, quod est i longum, et sic scribitur *scena* pro *tabernaculo*; sed *schenos* pro *fune* scribitur per oe diphthongum et per aspirationem. Et sic contentio est inter doctores de *cinomia* et de caeteris praedictis. Unde de *xeniis* credit vulgus quod nihil sit, et corrigunt in textu Latino dicentes exenia. Sed in Bibliis antiquis non est sic, nec in Graeco; nec potest sic dici secundum Graecam grammaticam quia oporteret quod *ex* praepositio Graeca poneretur primo quod non est possibile, quia vocabulum incipit per consonantem sicut patet per grammaticam Graecam. Per hunc modum accidit error quasi in infinitis vocabulis.

Differences of gender and of termination in Greek and Latin.

Tertius modus est, quod licet Latini multum communicant cum Graecis, tamen in aliquibus differant, quod non observatur ut oportet. Nam cum dicat Priscianus et omnes Latini sciunt, quod nomen arboris apud Latinos est foeminini generis et terminatur in us, et nomen fructus est neutrius, et terminatur in um; ut *pomus*, *pomum*; *pyrus*, *pyrum*; et sic de aliis, aestimatur quod hoc sit intelligendum de omnibus vocabulis, quae sunt in usu Latinorum, ut de *malo* et de *amygdalo* et aliis. Nam regula Latinorum est solum intelligenda de Latinis dictionibus, non de Graecis nec aliis. Et quod hoc sit verum patet primo quod Latinus dat regulas de Latinis, et non pertinet ad eum ordinare regulas de linguis alienis. Deinde Priscianus dicit quod omne Graecum cadens in usum Latinorum retinet genus suum quod habuit apud Graecos. Et ideo cum malum pro arbore sit Graecum et neutrius generis, est sic apud usum Latinorum. Et ideo tam pro fructu quam pro arbore est ejusdem generis et ejusdem terminationis. Et hoc probatur per Virgilium qui dicit in Georgicis *mala insita* arbores quae inservantur in fructus. Et super hoc Servius commentator qui fuit major quam Priscianus, cujus auctoritate saepius utitur, dicit, quod hoc, *omne nomen arboris est foeminini generis*,

[1] This sentence, and the three following, have been corrected from Jul.

intelligendum est de Latinis, non de Graecis. Et certum est quod *malum* est Graecum, licet secundum morem Latinorum aliquantulum sit aliter prolatum. Nulla dictio apud Graecos terminatur in m literam, sed in n; et Latini omnes consueverunt terminare dictiones suas in m, ut *scamnum*, *lignum*, *pomum*, et hujusmodi. Idem multoties Latinus mutat aliquam vocalem in vocabulo Graeco, ut ubi dicit Graecus *grammaticos*, Latinus dicit *grammaticus*, et sic multipliciter; et sic est hic. Nam Graecus dicit *melon* pro arbore et fructu, Latinus mutat e in a, sicut n in m, et dicit *malum*. Sed ista mutatio non mutat vocabulum secundum substantiam et secundum radicem, quia acceptum est a Graeco, licet aliter prolatum, et hoc omnes auctores testantur. Caeterum per textum Latinorum in antiquis libris, tam de theologia, quam de philosophia, invenitur semper *malum* pro arbore. Nam in primo Joel invenitur communiter apud omnes Biblias malum pro arbore, etiam usque correctores dimiserunt illud in novis Bibliis: et quarto Canticorum ubi dicitur, '*sicut malum inter ligna sylvarum*'; sic exponit Beda in originali, et duodecimo. Ecclesiastis est *amygdalum* et *malogranatum* in singulari et *malogranata* in plurali, quod non fieret, si *malum* non esset neutrius generis. Mutantur igitur hujusmodi vocabula secundum formam Latinorum. Et praecipue mirum est quod in aliquo correctores dimittunt antiquam literam et in alio abradunt, quod est omnino contra rationem.

Mistakes in prosody due to neglect of these differences.

Similiter in pronunciatione literarum Graecarum multum erratur, propter hoc, quod Latini volunt formam suam servare in Graecis dictionibus; et in hoc peccatur maxime, cum omnes Poetae et omnes antiqui Latini proferebant secundum primam institutionem. Sed nos moderni violavimus hoc multis modis contra usum omnium antiquorum; verbi gratia, cum Priscianus dicit, quod nomina possessiva desinentia in *nus* longantur et acuuntur in penultima, ut Bovinus, Latinus, equinus, intelligenda est regula de Latinis dictionibus, non de Graecis, propter aliquas rationes tactas prius. Et ideo cum Adamantinum, Byssinum, Chrystallinum, Hyacinthinum, Bombycinum, Onychinum, Amethystinum, Smaragdinum, et hujus modi sunt Graeca, debent breviari in penultima, sicut Graeci faciunt. Praeterea nec ista sunt possessiva. Nam duae tantum sunt terminationes possessivorum apud Graecos, scilicet in cos, ut Grammaticos; et in nios, ut Uranios, i. e. coelestis. Caeterum omnes Poetae Latini breviant penultimam, et ideo non est poetica licentia quia communiter fit ab omnibus et ubicunque: quod enim raro fit et ex causa, licentiae poeticae ascribendum est, sed non quod fit semper et communiter. Unde Juvenalis, '*amethystina convenit illi.*' Et idem dicit, '*Grandia tolluntur crystallina*'; penultimam corripiendo, sicut omnes faciunt, et nullus facit contrarium: igitur non est poetica licentia sed ex lege naturali. Et cum secundo Regum septimo decimo capitulo habeatur, '*siccaret ptisanas,*' expositio famosa vocabulorum Bibliae, cui omnes adhaerent, nititur probare quod media sit producta; et auctor illius expositionis defendit se per versum Horatii,

'Tu cessas, agedum sume hoc ptisanarium orizae.'

Sed error est, nam sicut per omnes auctores probatur, nunquam abscinditur in metro, nisi una syllaba in fine dictionis; et ideo sic debet scandi, 'Ptisanari' orizae,' ut haec syllaba *sa* brevietur, et haec syllaba *na* longetur. Et hoc patet aliter, quia in omnibus derivativus *a* ante *rium* longatur; ut contrarium, *armarium* et hujusmodi infinita quae observantur in hac scansione, sed non modo vulgato cum dicitur *ptisanar' orizae* ut duae syllabae auferantur, quia ibi breviatur haec syllaba *na*, ut patet. Ergo oportet quod media hujus dictionis *ptisana* sit brevis. Praeterea erratur in scriptura, nam in novis Bibliis habetur *tipsanas* quod nihil est, et debet p anteponi, sicut in hoc nomine Ptolemaeus. Et in hoc modo erratur infinities in aliis vocabulis, et tam violenter mutamus veras accidentium causas et regulas, quod non est remedium per magistros. Quoniam consuetudo cogit omnes male proferre, ut in uno patet exemplo pro mille millibus. Butyrum habet penultimam correptam apud auctores; unde Statius,

'Lac tenerum cum melle bibit, butyrumque comedit.'

Et Macer in libro herbarum:

'Cum butyro modicoque oleo decocta tumorem.'

Et Graecus sic breviat. Atque componentia ipsum requirunt hoc. Nam componitur de *tyros* et *bos*, et *tyros* est breve in prima syllaba et est lacticinium, quod a bove venit. Sed longe sunt majores errores apud multos, et ignorantia veritatis apud omnes circa accentus. Sed major disputatio requiritur quam praesens scriptura concedit[1].

Importance to the Church of linguistic studies,

[2] Cum jam manifestavi quomodo cognitio linguarum sit necessaria Latinis propter studium sapientiae absolutum nunc volo declarare quomodo oportet eum haberi propter sapientiam comparatam ad Dei Ecclesiam et rem publicam fidelium et confusionem infidelium et eorum reprobationem qui converti non possunt. Nam quadrupliciter in eis necessaria est ecclesiae, primo videlicet propter officium divinum, eo quod Graecis et Hebraeis et Chaldaeis utuntur in officio sicut in Scriptura.

(1) for explanation of liturgy,

Et plura accipimus quorum Scriptura non facit usum, ut *agios*, *atheos*, *athanatos*, *iskiros*, *ymas*, *eleison*, et hujusmodi. Cum ergo ignoramus scripturam et pronuntiationem rectam et sensum multum deficimus a veritate et devotione psallendi. Nam loquimur sicut pica et psittacus et talia bruta animalia quae voces emittunt humanas, sed nec recte proferuntur nec intelliguntur quae dicuntur. Quum enim dicimus *alleluia* infinities in anno, deceret multum et expediret ut omnes per totam ecclesiam psallentes scirent quid sint duo vocabula, scilicet *allelu* et *ia*. Nam *allelu* significat idem quod *Laudate*, et *ia* denotat *Dominum*, quum est unum de decem nominibus Dei, sicut Hieronymus scribit ad Marcellum; et praecipue significat *invisibilem*, et Deus est maxime invisibilis. Verum non quodcunque invisibile, sed Deum designat. Et cum in omni Missa dicimus Osanna, haec dictio est composita ex corrupto et integro. Nam ut Hieronymus dicit ad Damasum Papam, *Os* est idem quod Salvifica, et *anna* est indicatio deprecantis, secundum quod per aleph scribitur syllaba prima; unde significat idem quod, *salva deprecor*. Nam aliter scribitur prima syllaba per e literam, et tunc significat conjunctionem quod Latinus sermo non habet. Et cum gloriosam Virginem

[1] The subject of accentuation is treated with some fullness in the Greek grammar by Bacon preserved in Corpus Christi College, Oxford; and also in *Compendium Studii*, cap. xi (Brewer, pp. 508–514).

[2] All that follows, to the end of Part III, is omitted in Jebb's edition, and has been restored here from Jul.

salutamus, dicentes, Ave Maria gratia plena Dominus tecum, multum esset necessarium ad intellectum veracem et intellectum ut quilibet litteratus sciret sensum vocabuli; et praecipue cum multi aestimantes scire in hoc oberrent. Mirum quidem vocabulum est *Maron*, et significat *dominum* a quo venit *Maria*, et idem est quod dominatrix, ut dicit Hieronymus in interpretationibus. Et hoc valde competit beatissimae Virgini, quae dominatur supra omnem immunditiam peccati expellendam a nobis et diabolicae fraudis et nequitiae, quia ipsa est terribilis peccato et de moribus sicut castrorum acies ordinata. Haec vero interpretatio est peritissima et sine calumnia. Dicit autem Hieronymus quod multi aestimaverunt interpretari . . . . quod ipse non recipit et dicit quod debet dici Stella Maris, vel amarum mare secundum Hebraicam interpretationem. Et vere dicitur Stella Maris ut nos dirigat ad portum salutis, et Amarum Mare quia in omni paupertate et amaritudine temporali vixit in hoc mundo, et ipsius animam pertransivit gladio in morte Filii, ut sit nobis exemplum omnis patientiae et consolatrix in omnibus adversitatibus hujus mundi. Necesse est ergo nobis in omnibus psalmodiis et obsecrationibus nostris ut sciamus recte proponere et intelligere quaecunque et juxta verborum proprietatem devote nostras petitiones sonare, ut quod recte et devote petimus, Dei et sanctorum pietatem et merita ecclesiae consequamur. Sed hoc non possumus facere sine notitia vocabulorum alterius linguae. Et ideo multum expedit et necessarium est ut hoc sciamus.

(2) for explanation of sacraments and consecrations,

Secunda causa est quod Ecclesiae necessaria est cognitio linguarum propter sacramenta et consecrationes. Nam intentio necessaria est sacramento, ut theologi sciunt. Et intentionem praecedit intellectus et notitia rei faciendae. Et ideo per omnem modum expediret Ecclesiae ut sacerdotes et praelati omnia vocabula sacrificiorum et sacramentorum et consecrationum scirent recte proferre et intelligere, sicut a principio sacri et summi pontifices et omnes sancti patres et institutores ordinum ecclesiasticorum constituerunt et sciverunt qualiter in verbis et sensibus mysteria Dei consisterent. Unde incipiendo a primis ut ab exorcismis et catharizationibus, et sic per baptismum et omnia sacramenta discurrendo, non solum decens sed expediens et necessaria est ut ab eis qui ministrant sacramenta sciretur recta pronuntiatio et debitus intellectus, quatenus in nullo derogaretur sacramento. Sed modo per universam Ecclesiam innumerabiles proferunt verba instituta ab Ecclesia et nesciunt quod dicunt, nec verborum servant rectam pronuntiationem, quod esse non potest sine injuria sacramenti. Utinam fiat cum plena efficacia effectus sacramentalis! Et cum Ecclesia statuit hoc ex certa notitia, et omnes patres antiqui sciverunt rectam pronuntiationem et sensum vocabulorum secundum quod competebat sacramentis, nos nullam habemus excusationem; sed turpis et vilis ignorantia est nulla tergiversatione excusanda. Et quum in consecrationibus ecclesiarum cuspide baculi pastoralis fuerint factae literae alterius linguae secundum ordinem alphabeti, certum est quod paucissimi faciunt figuras debitas secundum quod a sanctis patribus et Ecclesia fuerint institutae, propter ignorantiam characterarum alterius linguae. Et praecipue in hoc erratur quod tres figurae sunt quae nullo modo scribi deberent in Graeco alphabeto. Nam procul dubio figurae quae vocantur episemon, koppa, sanpi non sunt de alphabeto Graecorum, nec Graecis inservierunt in ordine literarum; sed sunt figurae et notae numerorum. Modo vero Latini non considerant quod Graeci numerant per literas alphabeti, et quod ad complendam computationem interserunt tres figuras prius nominatas, scilicet has ϛ, ϟ, ϡ. Sed hoc faciunt quum numerant, non quasi nominetur figura pro literis et scriptis. Unde in scribendo nunquam utuntur his tribus figuris nec ponunt eas in ordine alphabeti. Sed Ecclesia instituit quod literae solae alphabeti scriberentur in consecratione ecclesiae, et deceret uti literis non notis numerorum. Quapropter valde indignum est quod per universam Ecclesiam fiat hujus erronea scriptura.

Et vile est quod haec nomina IHC, XPC, scribuntur per literas Graecas et aestimatur quod sunt Latinae, aut nescitur cujus modi sunt Graecae. Nam procul dubio in hoc nomine IHC prima est iota quae valet i nostrum; secunda est ita, quae valet e longum. Tertia est sima quae valet s nostrum.

Et in hoc nomine XPC, prima est chi quae valet ch aspiratum. Secunda est ro quae valet r nostrum. Tertia est sima.

Tertia vero causa est de notitia linguarum Ecclesiae Dei necessaria. Nam multi Graeci et Chaldaei et Armeni et Syrii et Arabes et aliarum linguarum nationes subjiciuntur Ecclesiae Latinorum, cum quibus multa habet ordinare et illis varia mandare. Sed non possunt haec recte pertractari nec ut oportet utiliter nisi Latini sciant linguas earum. Cujus signum est quod omnes dictae nationes vacillant fide et moribus, et ordines ecclesiae salutares per sincerum non recipiunt in lingua materna. Unde accidit quod apud tales nationes sunt mali Christiani et ecclesia non regitur ut oportet.

(3) for due regulation of foreign Churches,

Quarta causa est propter totius Ecclesiae doctrinam a principio usque in finem dierum. Nam dicit Dominus, Iota unum aut unum apex non peribit a lege donec omnia fiant. Et ideo docetur publice in libro de senibus Scripturarum quod singulae literae alphabeti Graeci figurabant super populum antiquum, et oportet numerum centenariorum annorum quibus decurrebat status illius gentis juxta singulas aetates et saecula [computare] secundum spirituales vires et potestates literarum. Et deinde . . . . ecclesiae Latinae per necessitates literarum Latinarum. Et consimilis est consideratio super ecclesia Graeca per literas sui alphabeti. Et in hujus consideratione mirabili tempora ponuntur secundum omnes status Ecclesiae usque in finem, et per quot centenarios annos durabit quaelibet immutatio quae accidet Ecclesiae in decursu suo, cui si prophetias et testimonia digna necteremus possemus per Dei gratiam praesentire utiliter ea quae Ecclesia recipiet tam in prosperis quam in adversis. Et ideo nil utilius esset hujus veritatis literarum consideratione et aliis considerationibus similibus. Nam ad certificationem harum rerum multae materiae requiruntur, quarum saltem una non ignobilis est per literas linguarum diversarum. Et nequeo satis admirari cum videatur inexpertis habere debile fundamentum, scilicet literas alphabeti quae sunt prima puerorum rudimenta, sed secundum documentum apostoli munera sunt magis necessaria et majori honore circumdanda. Et sicut Deus elegit infirma ut fortia

(4) for Christian doctrine,

quaecumque confundat, ita in rebus quas reputamus minimas posuit majestas majora quam possit intelligere mens humana. Et sic est in his literis triplicis alphabeti. Unde non sine maxima causa in epitaphio Domini scriptum est Hebraice Graece Latine, ut doceremur quod Ecclesia cruce Domini redempta habeat considerare veritates literarum triplicis alphabeti; praecipue cum Ecclesia incepit in Hebraeis et profecit in Graecis et consummata est in Latinis.

(5) for commerce with foreign nations.

Quinto multum est necessaria reipublicae Latinorum dirigendae cognitio linguarum propter tria. Unum est commeatio utilitatum necessariarum et mercatis et negotiis sine quibus Latini esse non possunt, quia medicina et omnia pretiosa recipiuntur ab aliis nationibus. Et inde oritur magnum malum Latinis et fraus eis infertur infinita quia linguas ignorant alienas, licet per interpretationes eloquantur. Nam hae raro sufficiunt.

. . . .

Ex his ergo quae circa linguas dicta sunt patens est quod Latini magnum habent sapientiae detrimentum propter linguarum ignorantiam. Unde ex hac parte gloriari non possunt de sapientia; immo multum inglorii et cum vario sapientiae damno languent.

PARS QUARTA

HUJUS PERSUASIONIS.

In qua ostenditur potestas mathematicae in scientiis, et rebus, et occupationibus hujus mundi.

DISTINCTIO PRIMA,

HABENS CAPITULA TRIA.

CAPITULUM I.

The key to the greater sciences is mathematic.

Manifestato quod multae praeclarae radices sapientiae dependent ex potestate linguarum, per quas est introitus in sapientiam Latinorum, nunc volo revolvere fundamenta ejusdem sapientiae penes scientias magnas, in quibus est specialis potestas respectu caeterarum scientiarum et rerum hujus mundi. Et sunt quatuor scientiae magnae, sine quibus caeterae scientiae sciri non possunt, nec rerum notitia haberi: quibus scitis, potest quilibet gloriose proficere in sapientiae potestate sine difficultate et labore, non solum in scientiis humanis, sed divina. Et cujuslibet istarum tangetur virtus non solum propter sapientiam absolute, sed respectu caeterorum praedictorum. Et harum scientiarum porta et clavis est mathematica, quam sancti a principio mundi invenerunt, ut ostendam, et quae semper fuit in usu omnium sanctorum et sapientum prae omnibus aliis scientiis. Cujus negligentia jam per triginta vel quadraginta annos destruxit totum studium Latinorum. Quoniam qui ignorat eam non potest scire caeteras scientias nec res hujus mundi, ut probabo. Et, quod pejus est, homines eam ignorantes non percipiunt suam ignorantiam, et ideo remedium non quaerunt. Ac per contrarium hujus scientiae notitia praeparat animum et elevat ad omnium certificatam cognitionem, ut si radices sapientiae datas circa illam cognoscat, et eas radices recte applicet ad caeterarum scientiarum et rerum cognitiones, tunc omnia sequentia poterit scire sine errore et sine dubitatione, ac de facili et potenter. Sine his enim nec praecedentia nec consequentia sciri possunt; unde perficiunt priora et regulant, sicut finis ea quae sunt ad finem, et disponunt et aperiunt viam ad sequentia. Ad quod nunc intendo innuere per auctoritatem et rationem; et primo in scientiis humanis et rebus istius mundi, deinde in divina, et ultimo prout ad Ecclesiam et caetera tria comparantur.

CAPITULUM II.

In quo probatur per auctoritatem, quod omnis scientia requirit mathematicam.

Authorities for this view. Boethius.

Per auctoritatem quidem sic procedo. Dicit Boetius[1] in secundo prologo Arithmeticae, quod 'mathematicae quatuor partibus si careat inquisitor, verum minime invenire possit.' Et iterum 'Sine hac quidem speculatione veritatis nulli recte sapiendum est.' Et adhuc dicit 'Qui spernit has semitas sapientiae, ei denuncio non recte philosophandum.' Et iterum, 'Constat quisquis haec praetermiserit, omnis sapientiae perdidisse doctrinam.' Quod etiam omnium virorum authenticorum sententia confirmat dicens, 'Inter omnes priscae auctoritatis viros, qui Pythagora duce puriore mentis ratione viguerunt, constare manifestum est, haud quemquam in philosophiae disciplinis ad cumulum perfectionis evadere, nisi cui talis prudentiae nobilitas quodam quasi quadrivio investigatur.' Et in particulari ostenditur per Ptolemaeum et ipsum Boetium[2]. Cum enim

[1] Boethius, as appears from a letter addressed to him by Theodoric which is quoted by Cassiodorus, made Latin translations of many of the Greek mathematicians; Euclid, Nicomachus, Archimedes, Ptolemy, &c. These, however, have disappeared; the two books *De Institutione Arithmetica*, and the five books *De Musica*, with a few doubtful fragments of his *Geometry*, alone remain. The expression quadrivium, as applied to the four sciences of arithmetic, music, geometry, and astronomy, seems to have originated with Boethius.

[2] In the introduction to the *Almagest* Ptolemy remarks that mathematic holds an intermediate place between the invisible and incomprehensible object of

sint modi tres philosophiae essentiales, ut dicit Aristoteles in sexto Metaphysicae, mathematicus, naturalis, et divinus, non parum valet mathematicus ad reliquorum duorum modorum scientiae comprehensionem, ut docet Ptolemaeus in capitulo primo Almagesti quod et ipse ibidem ostendit. Et cum divinus sit dupliciter, ut patet ex primo Metaphysicae, scilicet Philosophia prima, quae Deum esse ostendit, cujus proprietates excelsas investigat, et civilis scientia quae cultum divinum statuit, multaque de eo secundum possibilitatem hominis exponit, ad utramque istarum multum valere mathematicam idem Ptolemeus asserit et declarat. Unde Boetius in fine arithmeticae mathematicas medietates asserit in rebus civilibus inveniri. Dicit enim quod 'arithmetica medietas reipublicae comparatur quae a paucis regitur, idcirco quod in minoribus ejus terminis major proportio fit, musicam vero medietatem optimatum dicit esse rempublicam. eo quod in majoribus terminis major proportionalitas invenitur. Geometrica medietas popularis quodammodo exaequatae civitatis est: namque vel in minoribus vel in majoribus aequali omnium proportionalitate componuntur. Est enim inter omnes paritas quaedam medietatis aequum jus in proportionibus conservantis.' Et quod sine his respublica regi non potest, Aristoteles et ejus expositores in moralibus in pluribus locis docent. De his vero medietatibus exponetur, quando ad divinas veritates applicabuntur. Cum vero omnes modi Philosophiae essentiales, qui sunt plures quam quadraginta scientiae ad invicem distinctae, reducantur ad hos tres, sufficit nunc per auctoritates dictas persuasum esse valorem mathematicae respectu modorum philosophiae essentialium.

Ptolemy.

Modi autem philosophiae accidentales sunt grammatica et logica. Et quod sine mathematica non possunt sciri scientiae istae patet per Alpharabium in libro de scientiis. Nam etsi grammatica pueris ministrat ea quae vocis sunt et proprietates

Alpharabius proves dependence of grammar and of logic on mathematic.

Theology, and the shifting phenomena of nature. Τὰ μὲν ἄλλα δύο γένη τοῦ θεωρητικοῦ μᾶλλον ἄν τις εἰκασίαν ἢ κατάληψιν εἴποι· τὸ μὲν θεολογικὸν διὰ τὸ παντελῶς ἀφανὲς αὐτοῦ καὶ ἀνεπίληπτον, τὸ δὲ φυσικὸν διὰ τὸ τῆς ὕλης ἄστατον καὶ ἄδηλον· ὡς διὰ τοῦτο μηδέποτε ἂν ἐλπίσαι περὶ αὐτῶν ὁμονοῆσαι τοὺς φιλοσοφοῦντας· μόνον δὲ τὸ μαθηματικόν, εἴ τις ἐξεταστικῶς αὐτῷ προσέρχοιτο βεβαίαν καὶ ἀμετάπιστον τοῖς μεταχειρομένοις τὴν εἴδησιν παράσχοι. He indicates also the value of mathematical study in elevating and consolidating character.

ejus in prosa, et metro, et rhythmo, nihilominus tamen hoc facit pueriliter, et per viam narrationis, non per causas, nec per rationes. Nam alterius scientiae est dare causas horum, scilicet illius, quae vocum naturam plenarie habet considerare, et haec sola est musica, cujus species et partes multae sunt. Nam una est prosaica, et altera est metrica et tertia est rhythmica, et quarta est melica in cantu. Et praeter has habet plures. Et prosaica docet causas omnium elevationum vocum in prosa, secundum accentuum differentias et secundum cola et commata et periodos et hujusmodi [1]. Et metrica docet omnes rationes et causas pedum et metrorum. Et rhythmica de omni modulatione et proportione suavi rhythmorum docet, quia omnia ista sunt quaedam genera cantus, licet non sic ut in cantu usuali. Nam accentus dicitur quasi accantus, de *accino*, *accinis*. Unde ad musicam pertinent sicut docet Cassiodorus in musica, et Censorinus in libris de accentu, et sic de aliis. Hoc autem testantur auctores musicae, et libri de illa scientia. Et his concordat Alpharabius in libro de divisione scientiarum. Ergo grammatica dependet causaliter ex musica.

Eodem modo logica. Nam finis logicae est compositio argumentorum quae movent intellectum practicum ad fidem et amorem virtutis et felicitatis futurae, ut prius ostensum est, quae argumenta traduntur in libris Aristotelis de his argumentis, ut declaratum est. Sed haec argumenta debent esse in fine pulchritudinis, ut rapiatur animus hominis ad salutiferas veritates subito et sine praevisione, ut docetur in illis libris. Et Alpharabius [2] hoc docet maxime de poetico, cujus sermones

[1] For a fuller discussion of punctuation, metre, and rhythm, see *Opus Tertium*, ch. 62 and 63.

[2] Abu Nasr Muhammed ibn Muhammed ben Tarchân ben Auzelag el Fârâbî, known in the Western world as Alpharabius, was born A. D. 870 in the town of Farab (afterwards called Otrar) in Turkestan, and is one of many instances proving that Mahommedan learning in the Middle Ages was not limited to the Arab race. He was ignorant of Arabic till he came to Bagdad for the purposes of study. Having acquired it he became one of the most zealous students of Aristotle, devoting himself specially to the *Physics* and the *De Anima*. The latter part of his life was spent in Damascus, where he died in 950 A. D. Of his original works the most important was *Liber de scientiis earumque numero partibus et praestantia*. This work was translated by Dominicus Gundisalvi, at

debent esse sublimes et decori, et ideo cum ornatu prosaico, et metrico, et rhythmico insigniti, secundum quod competit loco et tempori et personis et materiae de qua sit persuasio. Et sic docuit Aristoteles in libro suo de poetico argumento, quem non ausus fuit interpres Hermannus transferre in Latinum propter metrorum difficultatem, quam non intellexit, ut ipse dicit in prologo commentarii Averrois super illum librum. Et ideo finis logicae pendet ex musica. Sed finis est nobilissimum in re, et imponit necessitatem eis quae sunt ad finem, ut Aristoteles dicit in secundo Physicae; nec habent utilitatem suam ea quae naturaliter ordinantur ad finem, nisi quando ad finem suum comparantur, ut patet in singulis. Et ideo tota utilitas logicae nascitur ex comparatione logicalium omnium ad hujusmodi argumenta, et ideo cum dependeant ex musicalibus necesse est logicam mendicare potestatem musicae. Et haec

Toledo, in the twelfth century. In the following century Hermannus Alemannus translated Aristotle's *Rhetoric* from the Arabic, and with it certain commentaries of Alpharabius. 'Omnia enim,' he says, 'in glosa super hunc librum exquisite Alfarabius pertractavit.' Hermann observes, however, that 'Alfarabius multa exempla Graeca propter ipsorum obscuritatem pertransiens derelinquit.' The Arabic version of Aristotle's *Poetics* seems to have been made not by Alfarabius but by Averroes. Hermann remarks on the extreme difficulty of rendering this version into Latin, 'propter disconvenientiam modi metrificandi in Graeco cum modo metrificandi in Arabico, et propter vocabulorum obscuritatem, et plures alias causas.'

Alpharabius is one of the authors (Aristotle, Avicenna, and Algazel being the others) from whom David the Jew compiled the work *De Causis*, of which Albertus Magnus gives a long description, and which is cited both by Bacon and by Thomas Aquinas. See Jourdain, pp. 112, 138–145, 184–5; also Wuestenfeld, *Geschichte der Arabischen Aerzte und Naturforscher* (Göttingen, 1840).

The short treatise, *De Scientiis*, is interesting for its comprehensive and encyclopaedic survey of Science. It consists of five chapters. 1. De Scientia Linguae et partibus ejus. 2. De Scientia Logicae et partibus ejus. 3. De Scientiis doctrinalibus, quae sunt Geometria, Arithmetica, Scientia de Aspectibus, Scientia stellarum doctrinalis, Scientia Musicae, Scientia de Ponderibus, Scientia de Ingeniis. 4. De Scientia naturali et partibus ejus; et Scientia divina et partibus ejus. 5. De Scientia civili et partibus ejus, et de Scientia judicandi, et de Scientia eloquendi.

Like Bacon, Alfarabi includes rhetoric and poetic as parts of Logic. 'Rhetorica movet animum auditoris et inclinat ad illud quod voluerit, ut credatur id quod dicit et generet in eo cognitionem proximam certitudini. Proprium est autem Poeticae sermonibus sensum facere imaginari pulcrum aliquid et dubium quod non est, ita ut auditor credat et aliquid abhorreat vel appetat . . . quoniam Imaginatio plus operatur in homine quam scientia vel cognitio.'

omnia sunt secundum sententiam Alpharabii in libro de scientiis, et patent similiter per Aristotelem et Averroem in libris suis, licet Latini horum usum non habeant. Sed non solum dependet cognitio logicae a mathematica propter suum finem, sed propter medium et cor ejus, quod est liber posteriorum, nam ille liber docet artem demonstrandi. Sed nec principia demonstrationis, nec conclusiones, nec ipsa tota potest cognosci, nec manifestari nisi in mathematicis rebus, quia ibi solum est demonstratio vera et potens, ut omnes sciunt et exponetur post. Quapropter necesse est logicam a mathematicis dependere.

Without mathematic the Categories are unintelligible.

Item propter suum principium, non solum propter medium et finem. Nam liber praedicamentorum est primus liber logicae secundum Aristotelem. Sed constat praedicamentum quantitatis cognosci non posse sine mathematica. Nam sola mathematica constituitur de quantitate cognoscenda. Quantitati vero annexa sunt praedicamenta de quando et ubi. Nam quando attinet tempori, et ubi oritur ex loco. Praedicamentum habitus non potest cognosci sine praedicamento ubi, ut docet Averroes in quinto Metaphysicae. Major vero pars praedicamenti qualitatis continet passiones et proprietates quantitatum, quia omnia quae sunt in quarto genere qualitatis vocantur qualitates in quantitatibus Et omnes passiones earum quae absolute debentur eis sunt qualitates, de quibus magna pars geometriae et arithmeticae constituuntur, sicut sunt rectum et curvum, et caetera quae lineae debentur, et triangulatio et omnis reliqua figuratio, quae superficiei et corpori assignantur; et primum incompositum in numeris, ut docet Aristoteles quinto Metaphysicae et caeterae passiones numerorum absolutae. Quicquid autem dignum est consideratione in praedicamento relationis est proprietas quantitatis, ut sunt proportiones et proportionalitates, et medietates geometricae, et arithmeticae, et musicae, et species majoris inaequalitatis et minoris. Substantiae vero spirituales non cognoscuntur per philosophiam nisi per corporales, et maxime supercoelestes, secundum quod Aristoteles docet undecimo Metaphysicae. Nec inferiora cognoscuntur nisi per superiora, quia coelestia sunt causae inferiorum. Sed coelestia non cognoscuntur nisi per quantitatem, sicut patet ex astronomia. Et ideo omnia praedicamenta dependent ex cognitione quantitatis, de qua est mathematica, et ideo virtus tota logicae dependet ex mathematica.

CAPITULUM III.

In quo probatur per rationem quod omnis scientia requirit mathematicam.

Mathematical examples used to illustrate other sciences.

Quod vero per auctoritatem de tota mathematica ostensum est, potest nunc per rationem similiter ostendi. Et primo, quia aliae scientiae utuntur exemplis mathematicis, sed exempla ponuntur propter evidentiam rerum de quibus scientiae constituuntur; quare ignoratis exemplis, ignorantur ea propter quorum intelligentiam adducuntur. Cum enim alteratio in naturalibus non inveniatur sine augmento et diminutione quibuscunque nec haec sine alteratione: non potuit Aristoteles ad purum manifestare differentiam inter augmentum et alterationem per aliquod exemplum naturale, quia semper concomitantur se aliquo modo; propter quod posuit exemplum mathematicum in quadrangulo quod addito gnomone crevit, et non alteratur. Quod exemplum ante vigesimam secundam propositionem sexti libri elementorum non potest intelligi. In illa enim sexti probatur, quod quadrangulus minor est omnino similis majori. Et ideo non alteratur minor, cum fiat major de minore per gnomonis additionem.

Comprehension of mathematical truth innate.

Secundo, quia mathematicarum rerum cognitio est quasi nobis innata. Socrati enim interroganti geometrica a puero pusione, ut recitat Tullius primo Tusculanarum quaestionum ita respondebat, quasi geometriam didicisset. Et hoc saepe expertum est in multis; quod non accidit in aliis scientiis, ut ex sequentibus magis erit manifestum. Quapropter cum sit quasi innata, et tanquam praecedens inventionem et doctrinam, seu saltem minus indigens eis quam aliae scientiae, prima erit inter scientias et praecedens alias, disponens nos ad eas; quoniam quae innata sunt vel prope disponunt ad acquisita.

Tertio, quia haec scientia prius est inter omnes partes

Earliest discovered.

philosophiae inventa. A principio enim humani generis haec primo inventa est. Quoniam ante diluvium et post, per filios Adae, et per Noe et filios ejus, sicut ex prologo compositionis astrolabii secundum Ptolemaeum, et ex Albumazar in majori introductorio astronomiae, et ex primo antiquitatum libro manifestum est, et hoc quantum ad omnes ejus partes, scilicet geometriam, arithmeticam, musicam, astronomiam. Illud autem non contigisset, nisi quia haec scientia est prior aliis, et naturaliter eas praecedens. Quare manifestum est, quod haec debet primo sciri, ut per eam promoveamur ad omnes scientias posteriores.

Easiest of apprehension.

Quarto, quia nobis est nata via a facilibus ad difficilia. Sed haec scientia est facillima. Quod manifestum est in eo, quod non refugit intellectum alicujus. Laici enim et omnino illiterati figurare et computare sciunt, et cantare, et haec sunt opera mathematicae. Sed primo incipiendum est ab his quae sint communia laicis et literatis ; et non solum damnosum est clericis, sed omnino turpe et vile, quod ipsi ignorant quod laici utiliter et pulchre sciunt. Quinto videmus quod clerici licet rudissimi mathematicalia possunt scire, quamvis ad alias scientias non valeant attingere. Insuper semel et bis audiendo plus homo de ea potest cognoscere certo et veraciter sine errore, quam decies de aliis partibus philosophiae, ut patet experienti. Sexto, quoniam nata est nobis via ab his quae conveniunt statui et ingenio puerili, quia pueri a nobis notioribus et primo addiscendis incipiunt. Sed hujusmodi est mathematica, quoniam primo docentur canere, et eodem modo possunt capere modum figurandi et numerandi, et longe facilius et necesse esset eis scire de numeris ante cantum ; quia in proportionibus numerorum tota ratio numeri exemplariter explicatur, sicut auctores musicae docent, tam in ecclesiastica musica, quam in philosophica. Sed ratio numerorum a figuris dependet, quia numeri lineares, et superficiales, et corporales, et quadrati, et cubici, et pentagoni et hexagoni, et caeteri, a lineis et figuris et angulis cognoscuntur. Expertum enim est, quod pueri melius et citius addiscunt mathematicalia, ut manifestum est in cantu, et etiam per experientiam scimus, quod pueri melius addiscunt et capiunt mathematicalia, quam

Accessible to the dullest. The first step in teaching.

alias partes philosophiae. Nam Aristoteles dicit in sexto Ethicorum quod juvenes possunt cito scire mathematicalia, non sic naturalia, nec metaphysicalia, nec moralia. Quare disponendus est animus prius per haec quam per alia.

It leads from things known to us to things known to nature.

Septimo, ubi non sunt eadem nobis nota et naturae, nata est nobis via a notioribus nobis ad notiora naturae. Sive simpliciter et facilius scimus ea, quae nobis notiora sunt, et cum magna difficultate devenimus in ea, quae sunt notiora naturae. Et nota naturae sunt male et imperfecte nobis cognita, quia intellectus noster se habet ad ea, quae sunt sic manifesta naturae, sicut oculus vespertilionis ad lucem solis, ut vult Aristoteles secundo Metaphysicae; sicut sunt maxime Deus et angeli, et vita futura et coelestia, et aliae creaturae nobiliores aliis. quia quanto sunt nobiliores, tanto sunt nobis minus notae. Et haec vocantur nota naturae. et simpliciter. Ergo per oppositum ubi eadem sunt nota nobis et naturae, multum proficimus circa nota naturae, et omnia quae ibi sunt, et ad ea possumus attingere ut sciamus ea perfecte. Sed in mathematica tantum, ut dicit Averroes [1] primo physicae et septimo metaphysico et super tertio coeli et mundi, sunt eadem nobis nota et naturae sive simpliciter. Ergo sicut in mathematica ad ea quae sunt nobis nota complete attingimus, sic ad ea quae sunt nota naturae et simpliciter. Quare ad intima illius scientiae possumus simpliciter attingere. Cum ergo hoc non valeamus in aliis, manifestum est quod haec est magis nota. Quapropter ab ea sumenda est origo nostrae cognitionis.

It is the type of certainty.

Item octavo, quia omne dubium fit notum per certum, et omnis error evacuatur per solidam veritatem. Sed in mathematica possumus devenire ad plenam veritatem sine errore, et ad omnium certitudinem sine dubitatione: quoniam in ea convenit haberi demonstrationem per causam propriam et necessariam. Et demonstratio facit cognosci veritatem. Et similiter in ea contingit haberi ad omnia exemplum sensibile, et experientiam sensibilem figurando et numerando, ut omnia

[1] In his commentary on the first book of *De Natur. Auscult.*, Averroes says, 'Illa quae sunt cognita apud nos in rebus naturalibus non sunt illa quae sunt cognita simpliciter, id est, naturaliter. Quod est contrarium in mathematicis: illa enim quae sunt cognita in illis simpliciter, et sunt causae priores in esse, sunt cognita apud nos.'

ad sensum manifestentur: propter quod non potest esse dubitatio in ea. Sed in aliis scientiis excluso mathematicae beneficio, tot sunt dubitationes, tot opiniones, tot errores a parte hominis, ut non possint explicari, ut manifestum est, quoniam demonstratio per causam propriam et necessariam non est in eis ex propria potestate, eo quod in naturalibus propter generationem et corruptionem propriarum causarum, sicut effectuum, non est necessitas. In metaphysicis non potest fieri demonstratio nisi per effectum. Quoniam inveniuntur spiritualia per corporales effectus et creator per creaturam, sicut patet in illa scientia. In moralibus non possunt esse ex propriis demonstrationes, ut Aristoteles docet. Et similiter nec in logicalibus nec grammaticalibus, ut planum est, possunt esse demonstrationes potissimae propter debilitatem materiae de qua sunt illae scientiae. Et ideo in sola mathematica sunt demonstrationes potissimae per causam necessariam. Et ideo solum ibi potest homo ex potestate illius scientiae devenire ad veritatem. Similiter in aliis scientiis sunt dubitationes, et opiniones, et contrarietates a parte nostra, ut vix concordetur in una vilissima quaestione, nec in uno sophismate; non enim sunt in eis ex sua proprietate experientiae figurationum et numerationum, per quas omnia certificari debent. Et ideo in sola mathematica est certitudo sine dubitatione.

Quare patet quod si in aliis scientiis debemus venire in certitudinem sine dubitatione et ad veritatem sine errore, oportet ut fundamenta cognitionis in mathematica ponamus; quatenus per eam dispositi possumus pertingere ad certitudinem aliarum scientiarum, et ad veritatem per exclusionem erroris. Et haec ratio potest per simile magis manifestari, et principale etiam propositum nonum Euclidis. Sicut enim cognitio conclusionis se habet ad cognitionem praemissarum, ut si sit in eis error et dubitatio, non possit veritas haberi per eas de conclusione, nec certitudo, quia dubium non certificatur per dubium, nec verum per falsum probatur, licet possit syllogizari ex falsis, syllogismo inferente non probante; sic est de scientiis totalibus, quod illae in quibus sunt dubitationes vehementes et multiplices, atque opiniones et errores, dico saltem a parte nostra, oportet quod hujusmodi dubitationes et falsitates evacuentur per aliquam scientiam nobis certam, et in qua nec dubitamus nec erramus. Cum enim conclusiones et principia propria eis sint partes totalium scientiarum, sicut pars se habet ad partem, ut conclusio ad praemissas, sic scientia ad scientiam, ut scilicet scientia, quae est plena dubitationibus et opinionibus respersa atque obscuritatibus, non valeat certificari, nec manifestari, nec verificari, nisi per aliam scientiam notam et verificatam, et nobis certam et planam, sicut est de conclusione per praemissas. Sed sola mathematica, ut prius habitum est, manet nobis certa et verificata in fine certitudinis et verificationis. Quapropter per hanc oportet omnes alias scientias sciri et certificari.

Consideration of the subject-matter of mathematics leads us to the same conclusion,

Et quoniam jam per proprietatem istius scientiae ostensum est, quod mathematica est prior aliis, et eis utilis et necessaria, nunc ostenditur hoc per rationes sumptas a parte sui subjecti. Et primo sic, quia nobis est via nata a sensu ad intellectum, quoniam deficiente sensu deficit scientia quae est secundum illum sensum[1], ut dicitur primo Posteriorum, quoniam secundum quod proficit sensus, proficit humanus intellectus. Sed quantitas est maxime sensibilis, quia est sensibile commune, et ab aliis sensibus sentitur, et nihil potest sentiri sine quantitate quapropter maxime potest intellectus proficere circa quantitatem. Secundo, quia ipse actus intelligendi secundum se ipsum non perficitur sine quantitate continua, quia dicit Aristoteles in libro de Memoria[2] et Reminiscentia quod omnis intellectus noster est cum continuo et tempore. Unde quanta et corpora intelligimus intuitu intellectus, quia species eorum apud intellectum sunt. Incorporeorum autem species non sic recipiuntur intellectu nostro; aut si fiant in eo, secundum quod Avicenna dicit tertio Metaphysicorum, non tamen hoc percipimus propter occupationem fortiorem intellectus nostri circa corpora et quanta. Et ideo per viam argumentationis et admirationis corporalium et quantorum investigamus rerum incorporalium notitiam, sicut Aristoteles facit in libro

[1] Φανερὸν δὲ καὶ ὅτι εἴ τις αἴσθησις ἐκλέλοιπεν ἀνάγκη καὶ ἐπιστήμην τινὰ ἐκλελοιπέναι. *Analyt. Post.* i. 18.

[2] Οὐκ ἐνδέχεται νοεῖν οὐδὲν ἄνευ τοῦ συνεχοῦς, οὐδ' ἄνευ χρόνου τὰ μὴ ἐν χρόνῳ ὄντα. *De Memoria*, cap. 1.

undecimo Metaphysicorum. Quapropter proficiet maxime intellectus circa ipsam quantitatem, eo quod quanta et corpora in quantum hujusmodi approprianturintellectui humano secundum statum communem intelligendi. Unumquodque est propter quod et illud magis.

in which wise men of all ages concur.

Ad omnem autem confirmationem potest ratio ultima sumi ex experientia sapientum; nam omnes sapientes antiqui laborarunt in mathematica, ut omnia scirent, sicut nos vidimus de aliquibus nostri temporis, et audivimus de aliis, qui per mathematicam, quam bene sciverunt, omnem scientiam cognoverunt. Inventi enim sunt viri famosissimi, ut Episcopus Robertus Lincolniensis, et Frater Adam de Marisco, et multi alii, qui per potestatem mathematicae sciverunt causas omnium explicare, et tam humana quam divina sufficienter exponere. Hujus autem rei certitudo patet in scriptis illorum virorum, ut de impressionibus, sicut de iride et de cometis, et de generatione caloris, et locorum mundi investigatione, et de coelestibus et aliis, quibus tam theologia quam philosophia utitur. Quapropter manifestum est quod mathematica est omnino necessaria et utilis aliis scientiis.

Hae rationes sunt universales, sed in particulari contingit hoc ostendi descendendo ad omnes partes philosophiae, quomodo per applicationem mathematicae sciuntur omnia. Et hoc nihil aliud est, quam ostendere scientias alias non debere sciri per argumenta dialectica et sophistica quae introducuntur communiter, sed per demonstrationes mathematicas descendentes in veritates et opera aliarum scientiarum et regulantes eas, sine quibus nec possunt intelligi, nec manifestari, nec doceri, nec disci. Si quis vero in particulari descenderet applicando mathematicae potestatem ad singulas scientias, viderit quod nihil in eis posset sciri magnificum sine mathematica. Sed hoc nihil aliud esset, nisi constituere tractatus certos de omnibus scientiis, et per vias mathematicae verificare omnia quae scientiis caeteris sunt necessaria. Sed hoc non est praesentis speculationis.

DISTINCTIO SECUNDA.

In qua ostenditur, quod res hujus mundi requirunt mathematicam, habens capitula tria.

CAPITULUM I.

In primo docetur in universali quod coelestia et inferiora requirunt mathematicam.

In the science of the heavens the need of mathematics is obvious.

Quod de scientiis jam ostensum est, potest de rebus manifestari. Nam impossibile est res hujus mundi sciri, nisi sciatur mathematica. De coelestibus enim certum est omnibus, quia duae scientiae magnae mathematicae sunt de eis, scilicet astrologia speculativa, et astrologia practica. Prima speculatur quantitates omnium quae sunt in coelestibus, et omnia quae ad quantitatem reducuntur, tam discretam quam continuam quantitatem. Nam numerum coelorum et stellarum, quarum quantitas potest per instrumenta comprehendi, certificat, et figuras omnium, et magnitudines et altitudines a terra ac spissitudines et numerum ac magnitudinem ac parvitatem, ortum et occasum signorum stellarum, et motum tam coelorum quam stellarum, et quantitates et varietates eclipsium. Item descendit ad quantitatem et figuram habitabilis, et omnium partium ejus magnarum, quae vocantur climata, et ostendit diversitatem horizontium et dierum ac noctium secundum singula climata. Haec ergo determinantur hic, et multa eis annexa. Practica vero descendit ad hoc, ut ad omnem horam sciamus loca planetarum et stellarum, et aspectus et compositiones earum et omnia quae in coelestibus renovantur, atque descendit ad ea quae fiunt in aere, cujusmodi sunt cometae, et irides, et caetera ibi renovata, ut sciamus loca eorum, et altitudines, et magnitudines, et figuras, et multa quae oportet considerare in his. Et haec omnia fiunt per instrumenta ad haec idonea, et per tabulas, et per canones, id est, regulas ad haec certificanda inventas, quatenus via paretur ad judicia, quae fieri possunt secundum potestatem philosophiae, non solum in naturalibus, sed in his quae sumunt inclinationem ex natura, et gratis sequuntur coelestem dispositionem; et non solum ad judicia praesentium praeteritorum et futurorum, sed ad opera miranda, ut omnia prospera hujus mundi promoveantur, et adversa reprimantur, utiliter ac magnifice. Nec sunt haec dubia. Nam patriarchae et prophetae a principio mundi certificaverunt haec, sicut caetera. Et Aristoteles renovavit certificationem antiquorum, et produxit in lucem. Et omnes sapientes in rebus magnis in hoc concordant, et experientia docet. Sed de his expositio fiet suo loco.

Terrestrial things are governed by celestial.

Planum ergo est, quod coelestia sciuntur per mathematicam, et quod praeparatur per eam via ad haec inferiora. Quod autem haec inferiora non possunt cognosci sine mathematica, patet primo per hoc, quod non scimus res nisi per causas, si proprie accipiatur scientia, sicut Aristoteles dicit. Sed coelestia sunt causae inferiorum. Ergo non scientur haec inferiora, nisi sciantur coelestia, et illa sine mathematica sciri non possunt. Ergo horum inferiorum scientia dependet ex eadem. Secundo possumus videre ex propriis, quod nihil horum inferiorum nec superiorum sciri potest sine mathematicae potestate. Nam omnis res naturalis producitur in esse per efficiens et materiam in quam operatur, nam haec duo concurrunt primo. Agens enim per suam virtutem movet et transmutat materiam, ut fiat res. Sed virtus efficientis et materiae sciri non potest sine magna mathematicae potestate, sicut nec ipsi effectus producti. Sunt ergo haec tria, efficiens, materia et effectus. Et in coelestibus fit mutua influentia virtutum, ut lucis et aliarum, et est in eis alteratio, licet non ad corruptionem. Et sic potest ostendi, quod nihil in rebus sciri potest sine geometriae potestate. Habemus ex hoc argumento, quod similiter aliae partes mathematicae sunt necessariae: qua ratione enim illa, et aliae: et proculdubio longe magis, quia nobiliores sunt. Si ergo propositum ostendatur in geometricis, non est necesse in hac persuasione de aliis fieri sermonem.

The laws of force in both revealed by mathematic.

Primo ergo ostendo propositum geometricae a parte

efficientis. Omne enim efficiens agit per suam virtutem quam facit in materiam subjectam, ut lux solis facit suam virtutem in aere, quae est lumen diffusum per totum mundum a luce solari. Et haec virtus vocatur similitudo, et imago, et species [1] et multis nominibus, et hanc facit tam substantia quam accidens, et tam spiritualis quam corporalis. Et substantia plus quam accidens, et spiritualis plus quam corporalis. Et haec species facit omnem operationem hujus mundi; nam operatur in sensum, in intellectum, et in totam mundi materiam per rerum generationem, quia unum et idem fit ab agente naturali in quodcunque operetur, quia non habet deliberationem; et ideo quicquid ei occurrat facit idem. Sed si in sensum et intellectum agat, fit species, ut omnes sciunt. Ergo in contrarium, et in materiam fit species. Et in his quae habent rationem et intellectum, licet multa faciant secundum deliberationem et electionem voluntatis, tamen haec operatio, quae est generatio speciei, est naturalis in eis sicut in aliis. Unde substantia animae multiplicat suam virtutem in corpore et extra corpus, et quodlibet corpus extra se facit suam virtutem, et angeli movent mundum per hujusmodi virtutes. Sed Deus facit virtutes de nihilo, quas multiplicat in rebus; agentia creata non sic, sed alio modo de quo non est ad praesens curandum. Hujusmodi ergo virtutes agentium in hoc mundo faciunt omnem operationem. Sed duo sunt modo attendenda circa ista: unum est ipsa multiplicatio speciei et virtutis a loco suae generationis; et aliud est operatio varia in hoc mundo propter rerum generationem et corruptionem. Secundum sciri non potest sine primo. Et ideo oportet primo ipsam multiplicationem describi.

What is meant by species.

CAPITULUM II.

In quo canones multiplicationis virtutum agentium secundum lineas et angulos explicantur.

[1] All that is said in this fourth part of the *Opus Majus* on the subject of *species* must be studied in connexion with the much fuller treatment of the subject in the treatise *De Multiplicatione Specierum*.

Force in a homogeneous medium propagated in straight lines.

Omnis autem multiplicatio vel est secundum lineas, vel angulos, vel figuras. Dum vero species in medio raritatis unius incedit, ut in toto coelo, et in toto igne, et in toto aere, vel in tota aqua, semper tenet vias rectas, quia Aristoteles dicit quinto Metaphysicae quod natura operatur breviori modo quo potest, et linea recta est omnium brevissima. Quod etiam patet per vicesimum primi elementorum Euclidis dicentis, in omni triangulo duo latera sunt tertio longiora.

The law of refraction.

Sed cum corpus secundum est alterius raritatis et densitatis, ita ut non sit omnino densum, sed permutat aliquo modo transitum speciei, sicut aqua, quae est quodam modo rara, et quodam modo densa, et crystallus similiter et vitrum, et hujusmodi per quorum media possumus videre, tunc species aut venit perpendiculariter super corpus secundum, et adhuc incedit per lineam rectam sicut prius; aut si non cadat perpendiculariter, tunc de necessitate mutat incessum rectum, et facit angulum in introitu corporis secundi. Et hujus declinatio ab incessu recto vocatur fractio radii et speciei[1]. Et haec causa est, quia perpendicularis fortior est et brevior, et ideo natura operatur meliori modo super eam, sicut docent geometricae demonstrationes, de quibus postea fiet mentio magis suo loco. Sed haec fractio est duplex, quoniam si corpus secundum est densius, prout accidit descendendo a coelo in haec inferiora, tunc omnes virtutes stellarum quae non cadunt perpendiculariter super globum elementorum, franguntur inter incessum rectum et perpendicularem ducendam a loco fractionis. Et si corpus secundum est subtilius sicut est ascendendo ab aqua superius, tunc inter fractionem et perpendicularem ducendam a loco fractionis

[1] Refraction is described for the first time in the optical work of Ptolemy, the Greek text of which is lost, but of which an Arabic version was translated into Latin in the twelfth century by the admiral Eugenio, who served under Roger, king of Sicily. This translation has been recently edited (1885) by Gilberto Govi of Turin. The account of refraction will be found in the fifth section of the work (p. 142 et seq.). Ptolemy did more than state the fact of refraction. He measured the amount of deviation of the refracted ray for different angles of incidence, in the three media of air, water, and glass. As he shows himself to be aware of the error caused by refraction in astronomical observations (p. 151), it is strange that no mention of it should have been made in the *Almagest.*

cadit incessus rectus. Et haec est admiranda varietas in operatione naturae, sed nec mirum, cum infinita mirabilia fiunt per naturam secundum leges istarum fractionum; et per artificium juvans naturam possunt fieri ea quae mundus capere non potest, ut in scientia perspectiva explicabo. Sed cogetur per haec temporibus Antichristi ad ea quae ipse volet pro magna parte.

Artificial convergence of rays.

Quod autem haec sint vera auctores docent, et omnes periti sciunt, atque instrumenta possunt ad haec fieri, ut sensibiliter hujusmodi incessus videamus; sed usquequo habeamus instrumenta, possumus per effectum naturalem hoc probare sine contradictione, ut docet haec figura. Accipiatur ergo dimidia sphaera crystalli vel vas vitreum, cujus inferius sit rotundum, plenum aqua. Cum ergo a centro solis veniunt radii per aerem ad corpus crystalli, vel vitri, quod est densius aere, illi qui non sunt perpendiculares super corpus tale (et hi sunt qui non vadunt in centrum ejus, ut ex geometricis planum est), franguntur inter incessum rectum et perpendicularem ducendam a loco fractionis, ut est radius *a c*, qui cum transiverit per totum corpus vasis, venit ad aerem subtiliorem et non perpendiculariter. Ergo oportet ut sic vadat, ut incessus rectus sit inter ipsum et perpendicularem ducendam a loco fractionis, et ideo non ibit in *c*, sed declinat ad *f*, super principalem perpendicularem, quae venit a sole, ut est *Of* radius. Et eodem modo est ex alia parte, quod per duplicem fractionem *h f* concurret in eodem puncto *f*, in quo radius *O f*. Sed infiniti radii exeunt a sole super hoc corpus; ergo infiniti congregabuntur in eodem puncto per duplicem fractionem. Sed congregatio luminum est causa caloris. Ergo ibi fiet calor comburens. Et hoc est verum, ut patet ad sensum; nam si apponatur combustibile,

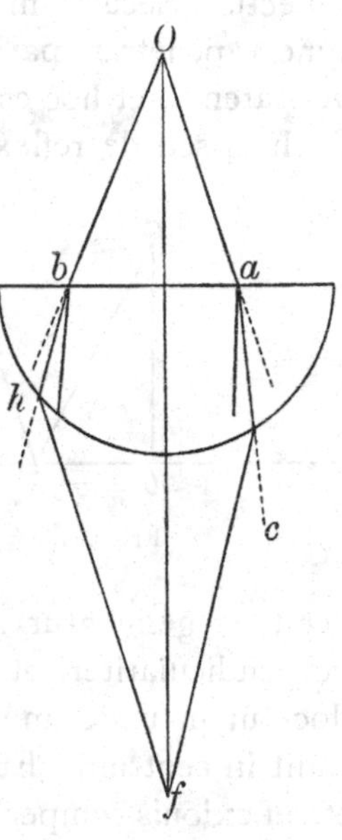

FIG. I.

ut lana, vel bombyx, vel pannus, comburetur. Cum ergo hic sit combustio, et hoc fieri non potest nisi per congregationem radiorum, et radii non possunt congregari nisi per duplicem fractionem, quia una non sufficeret, nec tertia requiritur, ergo oportet ponere hanc varietatem fractionem, quod est mirabile in oculis sapientum. Nam unde est, quod natura sic operatur? Certe nihil est jucundum naturae, vel voluntati, nisi quod reficit varietas; sed causae occultae sunt. Nec oportet causas modo investigare, cum per experientiam certissimam istud miraculum sciamus, et in sequentibus aliae experientiae subjungentur.

Law of reflection.

Quando vero secundum corpus est ita densum, quod nullo modo permittet transitum speciei, dico de sensibili transitu, quantum ad judicium visus humani, tunc dicimus speciem reflecti. Secundum tamen Aristotelem et Boetium visus lyncis penetrat parietes. Ergo species transit secundum veritatem [1], et hoc est verum: sed visus hominis non judicat de hoc, sed de reflexione, quae necessario fit. Nam propter difficultatem transitus per densum, cum in aere a quo venit inveniat viam facilem, multiplicat se copiosius in partem a qua venit. Et potest esse primo in generali duobus modis; nam aut cadit perpendiculariter super densum, et tunc redit in se omnino per eandem viam a qua venit, et geminatur radius in eodem loco, ut *a b* radius cadit perpendiculariter, et hoc est in planis ad angulos rectos, ut docetur in undecimo geometriae, sicut in sphaericis, quando cadit in centrum. Et causa hujus est; quia anguli incidentiae et reflexionis semper sunt aequales, ut multiplex demonstratio docet, et auctores omnes supponunt, et instrumenta ad hoc facta edocent ad oculum. Sed non sunt nisi duo anguli recti ex

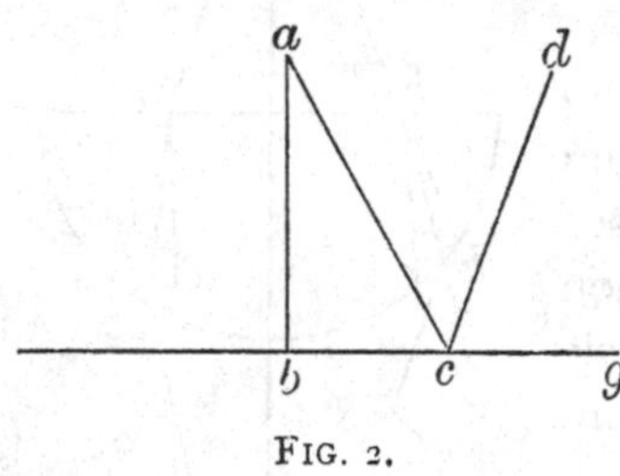

FIG. 2.

[1] Recent discoveries give much interest to this expression. Here, as in the case afterwards to be noticed of the time occupied in the passage of light, Bacon conceived that processes might take place of which human sense was not keen enough to take cognizance. See *De Mult. Specierum*, Pars II, cap. 5.

casu *a b* ad densum. Ergo per eosdem redibit radius reflexus, et ideo in eodem loco. Sed linea *a c* quae cadit ad obliquos angulos, et non perpendiculariter, non redit in se ipsam, sed usque ad *d*, propter aequalitatem angulorum incidentiae et reflexionis. Quandocunque vero cadit radius ad angulos obliquos, tunc angulus acutus vocatur angulus incidentiae; et ab illo obtuso angulo separatur per lineam reflexam angulus aequalis angulo incidentiae, qui angulus continetur inter lineam reflexam et densum, ut est angulus *d c g*, et hic vocatur angulus reflexionis, quem oportet necessario aequari angulo acuto a parte altera, et hoc ad visum probamus in speculis. Nam non possumus videre res, nisi oculus sit in termino reflexionis, ut si oculus sit in *d*, videbit; et si non non videbit per illum radium reflexum. Et haec nota sunt, atque experientiae satis dabuntur de hoc in sequentibus.

Artificial convergence of reflected rays.

Possunt autem congregari radii infiniti per reflexionem. sicut per multiplicationem, ut fiant combustiones validae. Sed a plana superficie non possunt radii congregari in unum quia unus vadit ad unum locum et alius ad alium. Nec a convexo speculo; sed a concavo sphaerico. columnari. et pyramidali, annulari, et ovali, et sic de aliis. Si ergo speculum concavum sphaericum ad solem ponatur, concurrunt radii infiniti in punctum unum per reflexionem. Et ideo oportet, ut speculo concavo ad solem posito ignis accendatur, sicut dicitur ultimo proposito libri de speculis, et ibidem demonstratur. Sed instrumentum ad hoc factum esset pulchrum valde, et tunc videretur ad oculum, sicut prius dictum est de fractione. Unde si fieret speculum de chalybe bono, vel de argento, facilius accideret combustio; sed una combustio non fit per omnes radios cadentes in speculo, sed per solos illos qui cadunt in circumferentiam unius circuli circa axem speculi, quia omnes, qui cadunt in una circumferentia, cadunt ad angulos aequales, et ideo reflectuntur ad punctum unum in axe, quia anguli reflexionum sunt aequales, et qui cadunt in alia circulatione, redeunt ad aliud punctum, et qui in tertia ad tertium, et sic de infinitis circulis imaginandis circa axem speculi: oportet enim quod ad puncta diversa vadant radii cadentes in diversis circumferentiis, propter hoc quod non cadunt ad angulos aequales. Et illi qui cadunt in minori circulo altius reflectuntur, et qui in maximo ad punctum infimum, scilicet ad polum sphaerae, seu ad extremitatem axis, reflectuntur. Sed nec natura nec ars contentae sunt hujusmodi combustione, imo volunt sic figurare corpora, ut omnes radii cadentes in totam superficiem speculi concurrant in punctum unum. Et sic adhuc ut in omni distantia quam volumus; et hoc est ultimum quod geometriae valet facere potestas. Nam hoc speculum potenter combureret omne quod posset objici. Et credendum est quod Antichristus his utetur, ut civitates et castra et exercitus comburat. Quoniam si modica congregatio radiorum per fractionem vel speculum concavum comburit sensibiliter, quanto ergo magis in infinitum, quando radii infinities infiniti congregabuntur per hoc speculum? Aestimant sapientes hoc esse necessarium. Et auctor in libro de speculis comburentibus docet hoc instrumentum fieri, sed gratis in illo libro occultavit multum de artificio, et dicit quod in alio libro posuit residuum, quod non est translatum apud Latinos. Sed sunt Latini qui, mala gratia illius auctoris occultantis perfectionem suae sapientiae, devenerunt ad hoc magnificum naturae secretum, quia ille auctor multum excitat peritos in sapientia ut residuum perficiant, et docet quod debet esse quasi annularis figurae, vel ovalis, ut, si amputarentur coni unius ovi, fieret annularis figura, si vero unus conus remanet, fit ovalis. Tali vero figura artificialiter facta secundum quod competit, oportet quod omnes radii cadentes in totam superficiem speculi cadant ad angulos aequales, et ideo reflectantur ad consimiles, et propter hoc in punctum unum. Elaboratur autem circa hoc speculum faciendum a peritissimo Latinorum [1], et Vestrae Magnificentiae gloria poterit praecipere, ut compleatur, cum vobis fuerit annotatus. Haec autem triplex multiplicatio secundum lineas

[1] This was Peter Peregrinus of Maricourt. See *Opus Tertium*, cap. 13. Reference to *De Mult. Specierum* (Pars II, cap. 7) will show that Bacon was aware of the properties of a surface produced by the rotation on its axis of a conic section. Vitello, Bacon's contemporary, in the ninth book of his *Optics*, after proving geometrically that parallel rays falling on a parabolic mirror are reflected to the same point, explains how such mirrors are constructed (lib. ix. prop. 44).

dicitur esse principalis propter hoc, quod ab ipso agente venit.

Sed quarta est magis mundo necessaria, quamvis vocetur accidentalis multiplicatio. Nam lumen accidentale vocatur respectu lucis principalis venientis a re, quoniam hoc non venit ab agente, sed a multiplicationibus principalibus, ut in domo cadit per fenestram multiplicatio principalis a sole, sed in angulo domus venit a radio fenestrae lux accidentalis. Non possent autem corpora mortalium semper exponi speciebus principalibus sine sui corruptione, et ideo temperavit Deus omnia per hujusmodi species accidentales.

Diffused rays.

Quinta vero est aliena ab istis, nam non tenet leges communes naturae, sed sibi vindicat privilegium speciale. Et haec multiplicatio non fit nisi in medio animato, ut in nervis sensuum: nam species sequitur tortuositatem nervi, et non curat de incessu recto, et hoc fit per virtutem animae regulantis incessum speciei, secundum quod opera rei animatae requirunt. De hac in perspectivis veritatibus aliquid dicetur. Quatuor primae sunt communes rebus mundi inanimatis, secundum quas natura delectatur operari; quinta ad sensum noscitur pertinere.

Propagation of rays through nerves.

CAPITULUM III.

In quo datur multiplicatio secundum figuras.

Deinde considerandum est, quo modo fiat multiplicatio secundum figuras. Et oportet quod multiplicatio fiat sphaerice. Nam agens multiplicat se aequabiliter in omnem partem, et secundum omnes diametros, et omnes differentias positionis, quae sunt sursum, deorsum, ante, retro, dextrorsum, sinistrorsum. Ergo undique exeunt lineae in omnem partem ab agente tanquam a centro; sed lineae undique exeuntes ab uno loco non possunt terminari, nisi ad superficiem concavam sphaerae. Et hoc patet, quia oculus non videt nisi per speciem venientem, sed si infiniti oculi ponerentur undique, omnes viderent eandem rem; ergo per infinitas lineas exit species: sed infinitae non terminantur nisi ad superficiem sphaericam. Si vero dicatur,

From the agent force proceeds in every direction.

quod lumen intrans per magnum foramen triangulare vel alterius polygoniae figurae, non cadit sphaerice, sed quando intrat per parvum foramen; dicendum est, quod latera parvi foraminis parum distant, et ideo lux in parva distantia potest figuram suam recuperare, sed quando transit per figuram magnam, non potest ita de facili, sed in aliqua distantia sufficienti, si obstacula amoverentur. Quod patet per xiv et xv primi elementorum Euclidis, ut ostendit figura.

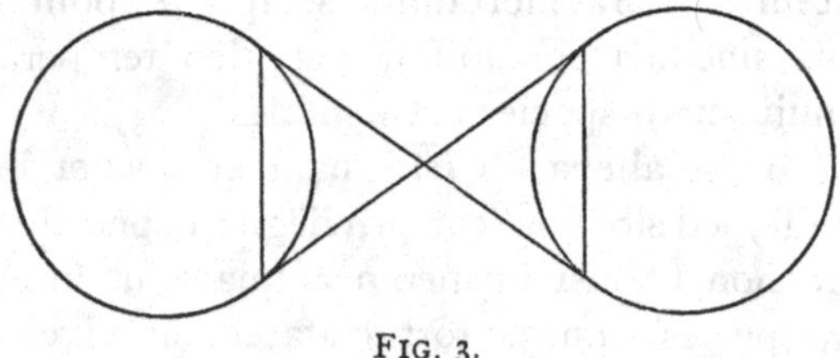

FIG. 3.

Nam trahantur radii ab intersectione quantum est ab intersectione ad solem, oportet per dictas propositiones ut bases triangulorum sint aequales. Sed illae bases sunt diametri luminum. Ergo oportet ad minus, ut diameter speciei sit aequalis diametro solis in aliqua distantia, et per consequens multiplicatio erit sphaerica aequalis, et potest variari secundum diversitatem distantiae, sed semper sphaerice. Nec est instantia de luce ignis, quae ascendit in figura pyramidali: quia haec non est multiplicatio ex propria natura lucis, sed est propter motum corporis ipsius ignis, cujus accidens est lux, et accidens fertur secundum motum sui subjecti, sicut lux solis in sole. Pyramidaliter vero ignem necesse est ascendere, quoniam partes interiores semper elongantur a frigido circumstanti, et ideo minus impediuntur et citius expediunt se quam exteriores, et propter hoc altius ascendunt, et caeterae quanto iis propinquiores, tanto citius sese expediunt, et applicant se interioribus, aliquantulum deficientes ab intimarum altitudine, et sic gradatim per ordinem remotiores minus exaltantur, quia magis impediuntur a contrario circumstanti; et ideo oportet quod pyramis enascatur. Sed in sphaera possunt omnes figurae regulares inscribi, ut patet ex xiii libro elementorum Euclidis, inter quas una est pyramis.

Et licet jam secundum rationem inscriptionis geometricae

non possunt figurae irregulares inscribi, nec figurae rotundae; possunt tamen omnes figurae protrahi, et signari in sphaera. Et ideo non solum in sphaerica multiplicatione inveniemus pyramides lateratas, quarum proprium est inscribi in sphaera, sed pyramides rotundas [1], quae signari possunt et figurari in sphaerica multiplicatione. Et haec est figura, quam specialiter elegit natura in omni multiplicatione et actione, et non quamcunque pyramidem, sed illam cujus basis est superficies agentis, et cujus conus cadit in aliquod punctum patientis, quia sic potest a tota superficie agentis species venire ad singula puncta patientis per singulas pyramides et infinitas, ut patet in figura. Nam a quolibet puncto patientis fiunt radii infiniti, et ideo possunt combinari infinities, ut fiant pyramides rotundae infinitae, quarum omnium est una basis, scilicet, superficies totius agentis; et ad quamlibet partem patientis venit unus conus unius pyramidis, ut virtus veniat a toto agente ad quodlibet punctum patientis, et non ab aliqua parte determinata, quatenus virtus completa perveniat et tota, non partialis et imperfecta, ut fiat actio completa, quia natura facit secundum quod melius est.

Cones of rays impinge on each point of the surface acted on.

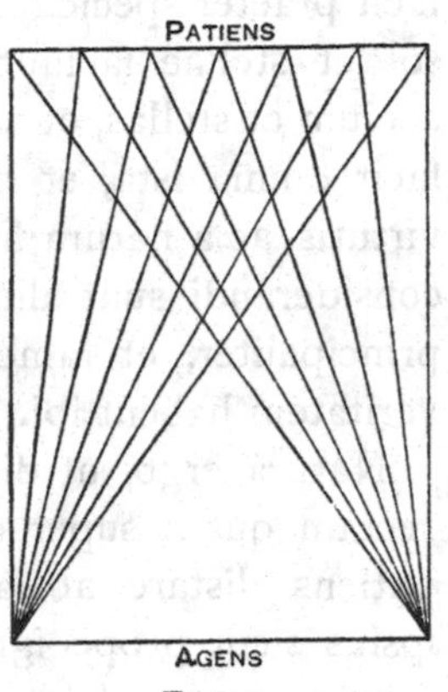

FIG. 4.

DISTINCTIO TERTIA.

In qua declaratur varietas actionis naturalis per geometriam, habens tria capita.

CAPITULUM I.

His consideratis circa multiplicationem, sunt aliqua consideranda circa actionem ulteriorem. Nam lux per viam

Indirect action of force.

1 This is Bacon's expression for a cone. He uses the word *conus* to denote the apex of the cone.

multiplicationis suae facit speciem luminosam, et haec actio dicitur univoca, quia effectus est univocus, et unigenius, et conformis agenti. Sed alia est multiplicatio aequivoca, ut lux generat calorem, calor putrefactionem, putrefactio mortem, et vinum inebriat, et sic de omni agente, quod multos effectus facit praeter speciem suam et virtutem sibi univocam. Et sic sol et stellae faciunt omnia hic inferius, et angeli movent coelum et stellas, et anima corpus suum: virtus tamen agentis facit omnia ista, et haec est completa actio agentis et suae virtutis, ac a natura finaliter desiderata. De hac ergo actione considerandi sunt aliqui canones seu regulae, et propter eam principaliter, et tamen locum habent in actione univoca, et veritatem habent ibi.

Force most potent when directed vertically.

Natura ergo, ut dictum est, fortius operatur super lineam rectam quam super curvam, quia brevior est, et minus facit patiens distare ab agente, et ideo plus capit de virtute ipsius sicut prope ignem aliquis plus calefit quam remotior. Caeterum melius est aequale quam inaequale, ut dicit Boetius in practica geometriae. Sed in linea recta est aequalitas. Item omnis virtus unita est fortioris operationis, sicut dicitur in libro de causis. Sed uniformitas et unitas major est in linea recta, sicut dicit Aristoteles quinto Metaphysicae. Nam in curva est angulus, qui facit dispersionem et difformitatem, et repugnat unitati. Quapropter natura operatur fortius super lineam rectam, quam super fractam vel reflexam. Sed linea recta, quae cadit ad angulos aequales, et perpendiculariter, sive in planis, sive in sphaericis, illa est super quam natura eligit operari tum propter aequalitatem et majorem uniformitatem, tum propter brevitatem. Nam per xix propositum primi elementorum Euclidis, in omni triangulo majori angulo majus latus opponitur. Sed ex xvii primi ejusdem libri, major angulus in triangulo est rectus, scilicet, *a c b*. Ergo ei opponitur maximum latus scilicet, *a b*. Sed illa cadit non perpendiculariter. Ergo perpendicularis *b c* est brevior: quare virtus veniens super eam operabitur

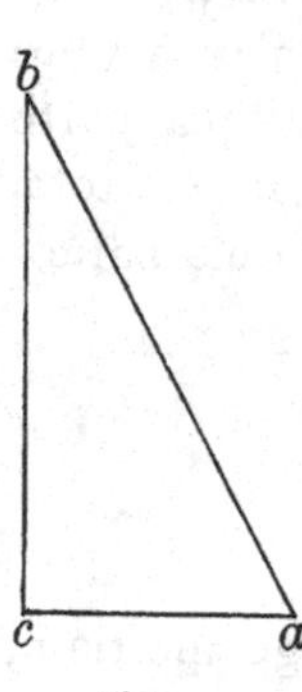

FIG. 5.

fortius. Perpendicularis autem fortitudo non solum patet per demonstrationem, sed per experientiam, unde lapis cadens ab alto perpendiculariter fortius scindit ; et si homo cadit ab alto magis laeditur : nam qui hominem cadentem ab alto pelleret ab incessu perpendiculari, dummodo esset prope terram, non laederetur ; quod si non fieret, moritur ex casu perpendiculari, et totus conquassatur.

Et considerandum, quod si a diversis punctis agentis veniant ad eandem rem radii, unus perpendicularis super agens, et alius non perpendicularis super ipsum, semper erit perpendicularis brevior. Ut posito, quod agens sit *a c*, tunc *a b* est brevior quam *c b*, et ideo fortior quantum est ex ratione hac. Et licet *c b* cadet super *e d* patiens ad angulos rectos, et *a b* non, ut sic *c b* fortificetur ex hujusmodi causa, nihilominus non attinget ad fortitudinem *a b*, quia brevior est, et quia *b* punctus *e d* similiter punctus est alterius lineae ut *f g*, ad quam *a b* est perpendicularis, et sic *b* habet rationem patientis principaliter[1]. Et maxime super corpus sphaericum fit actio completissima, cum virtus non cadat ibi ad aliquem angulum, sed transeat superficiem et corpus sphaericum, et nulla est diversitas vel difformitas et plenitudo actionis ibi invenitur. Nisi enim signentur circuli in sphaera non est ibi angulus ; sed cum signantur tunc anguli lineae perpendicularis erunt aequales, sicut in planis, et complebitur actio. Sed tamen non sunt recti sed obtusi ut patet geometrae.

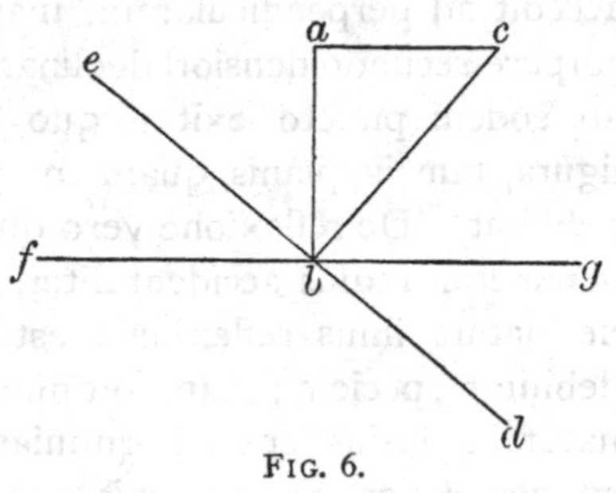

FIG. 6.

Deinde sciendum est, quod natura operatur super lineas fractas fortius, quam super reflexas, quia fractio est in partem incessus recti, sed reflexio in contrarium vadit, et contra incessum naturalem quem appetit natura virtutis incedentis in continuum et directum nisi impediatur, et ideo reflexio multum debilitat speciem et virtutem, et magis quam fractio. Sed hoc

Refracted rays stronger than reflected.

[1] This sentence and the first part of the following are omitted in J.

intelligendum est de fractione et reflexione secundum proprietatem incessus in eis. Si tamen consideremus, quod est reflexio in eodem medio, et fractio in diversis, oportet quod duplex medium magis impediat quam unum, et hoc saltem quando reflexio fit in medio subtili, et fiat fractio in secundo densiori, ut est in vase vitreo. Nam si congregentur radii a speculo comburente, et post perspicuum comburens, oportet quod major sit combustio, sicut inferius suis locis explicabitur. Fractio vero quae est in corpore secundo densiori minus debilitat, quam ea quae est in corpore secundo subtiliori. Nam incessus perpendicularis est fortissimus, et ideo quae magis accedit ad perpendicularem, magis est fortis. Sed fractio in corpore secundo densiori declinat versus perpendicularem, quae ab eodem puncto exit a quo fractio, ut patet superius in figura, tam in planis, quam in sphaericis, et ideo nulla minus debilitat. De reflexione vero quae est ad angulos rectos, licet geminetur radius accidentaliter, et sic fit fortior actio, tamen de natura illius reflexionis est, quod per se loquendo plus debilitat speciem ; nam omnino est in contrarium conatus naturalis ipsius speciei, quoniam per eandem lineam redit species super quam venit. Sed quando est ad angulos obliquos, non est omnino in contrariam partem, sed a latere, et ideo non tantum debilitat haec reflexio, sicut alia. Quantum est de natura reflexionis dico, sed propter geminationem virtutis in eodem loco, et propter aequalitatem angulorum, et conditiones perpendicularis, fortior est actio.

Of convergent rays the greater number oblique.

Et tamen considerandum hic, quod per casum radiorum ad angulos obliquos possunt plures radii congregari per intersectionem, quam per radios cadentes ad angulos rectos, non solum ex proprietate speculorum, ut dictum est, sed propter radios occurrentes sibi infinities ex lege incidentiae et reflexionis ad angulos obliquos, sicut accidit in aere, quando propter casum hujusmodi et reflexionem intersecant se radii in quolibet puncto infinities, et fit calor. Nam pauci sunt incidentes perpendiculariter super aliquam rem, quia non nisi ab uno puncto agentis cadit unus perpendicularis ad unum punctum patientis, et ideo sunt pauci reflexi. Sed infiniti non perpendiculares exeunt a quolibet puncto agentis, et infiniti reflexi sunt eis respondentes. Deinde per casum perpendicularem tantum duo conjunguntur in eodem loco aeris, scilicet, incidens et reflexus ejus compar. Sed per casum ad angulos obliquos incidentes infiniti se intersecant in quolibet puncto aeris. Et similiter incidentes penetrant reflexos non sibi compares, et reflexi reflexos infinities. Nam ad omne punctum terrae incidunt radii infiniti, et ab eodem infiniti reflectuntur, et ideo fortior operatio nascitur sic per accidens ex incidentibus et reflexis ad angulos obliquos, quam ad rectos. Ars vero potest juvare naturam in formatione actionis; nam potest sic figurare specula, ut fiat congregatio virtutum magna per specula concava, et maxime per ovalia, sicut dictum est. Sed virtus principalis, scilicet recta fracta et reflexa, est fortior accidentali, quia non venit accidentalis ab agente, sed a specie agentis, et est species speciei, propter quod debilior est.

CAPITULUM II.

In quo consideratur fortitudo actionis secundum figuras.

Et cum pyramis, ut dictum est, requiratur ad actionem naturae, considerandum est quod conus brevioris pyramidis fortius operatur, tum quia minus distat ab agente, tum quia radii conterminales circa conum pyramidis brevioris magis vicinantur, et vicinia radiorum ac congregatio fortius operatur; et hoc patet in figura. Nam per xvii primi elementorum Euclidis, omnes anguli circa punctum unum in superficie non valent nisi quatuor rectos; ergo quatuor anguli apud conum pyramidis brevioris valent reliquos quatuor apud conum longioris. Sed per xxi ejusdem, angulus in cono pyramidis brevioris est major quam angulus in cono pyramidis longioris, scilicet *a* est major quam *c*; et per xv ejusdem, anguli contrapositi sunt aequales, scilicet *a* et *b*, item *c* et *d*; ergo *c* et *d* simul sumpti sunt minores quam *a* et *b* simul conjuncti; ergo cum quatuor

Short cones of rays more potent than long.

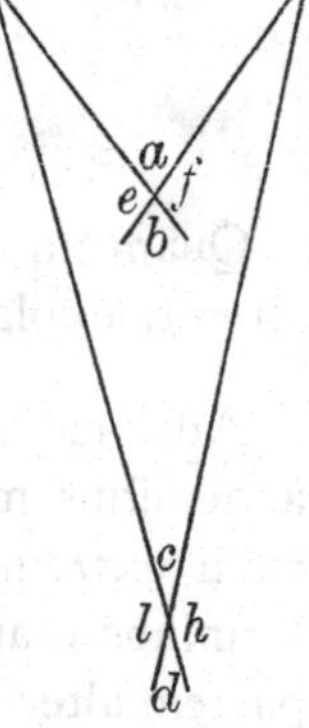

FIG. 7.

simul sumpti aequantur aliis quatuor simul sumptis per xiii, tunc oportet quod *h* et *l* sint majores quam *f* et *e*. Quapropter radii qui continent *e* magis vicinantur quam radii continentes *h*. Et eodem modo radii continentes *f* propinquiores sunt quam radii continentes *l*, et sic de infinitis radiis qui conterminales sunt in pyramide breviori, oportet quod omnes magis vicinentur quam radii qui simul terminantur in cono pyramidis longioris. Sed vicinia virtutum est causa fortioris actionis. Sed tamen cum incessus perpendicularium sit fortissimus, et omnis accessus ad perpendiculares fortior est recessu, tunc radii pyramidis longioris cum magis accedant ad perpendiculares *a c b d*, erunt fortiores. Item tot radii veniunt ad conum pyramidis longioris, sicut ad conum brevioris, quoniam infiniti utrobique. Sed conus pyramidis longioris acutiorem habet angulum per xxi. Ergo radii ejus magis uniuntur. Ergo fortius comburent. Et dicendum est, quod hae rationes ad utramque partem demonstrationes sunt, sed fortiores sunt ad primam partem, et ideo praevalent. Unde quantum possunt rationes ultimae, tantum concludunt, sed aliae potentiores sunt et efficacius operantur.

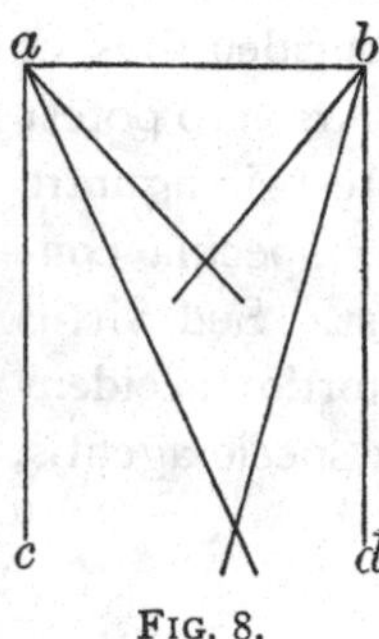

FIG. 8.

CAPITULUM III.

Quantum alteretur de patiente, et quantum de agente alteret, explanans.

Mutual action of large and small spheres.

Ad haec subjungendum est quod in corporibus sphaericis aequalibus medietas cujuslibet recipit virtutem alterius, quia radii extremi contingunt corpora illa, et ideo transeunt per terminos diametri, et non attingit aliquis radius ad aliquam partem alterius medietatis. Sed corpus minus recepit virtutem majoris in majori sua portione, propter hoc quod radii extremi corporis majoris non aequidistant semper, sed concurrunt, et possunt amplecti plus medietate minoris. Nam diameter corporis majoris est major diametro corporis minoris; et ideo possunt radii exeuntes a terminis diametri corporis majoris transire ultra diametrum corporis minoris, ut majorem ejus portionem comprehendant ; et e contrario corporis majoris

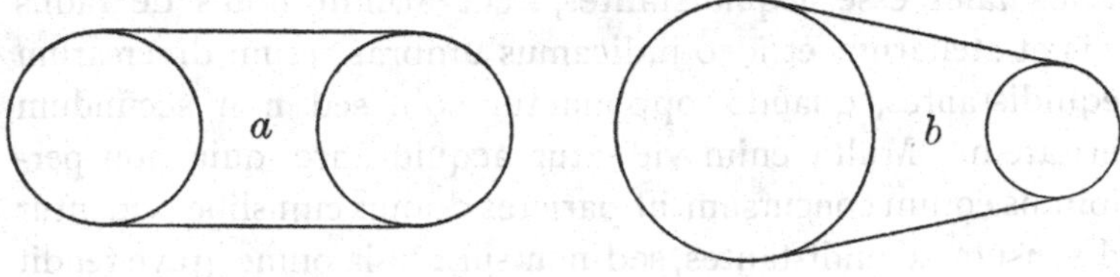

FIG. 9.

minor portio recipit virtutem a minori corpore. quia diameter minoris corporis non aequatur diametro majoris, sed alicui chordae minoris portionis corporis majoris. Et nullus punctus sphaerici corporis potest facere virtutem a se, nisi ad spatium quod separatur ab eo per lineam quae contingat illud corpus in puncto illo a quo fit virtus ; quoniam ab *a* puncto non potest aliqua linea cadere inter lineam contingentem et corpus sphaericum, ut demonstratur in xv tertii elementorum Euclidis. Et ideo spatium, quod est in angulo contingentiae, et sic per totam retro lineam contingentem versus corpus, non recipiet virtutem ab *a* puncto, sed totum spatium ultra lineam contingentiae, in quo sunt puncta *bc de*, recipiet virtutem. Et omnes radii qui exeunt a superficie corporis sphaerici, quorum directio cadit in centrum corporis illius, sunt perpendiculares super ipsum et tales undique exeunt infiniti ut patet in figura.

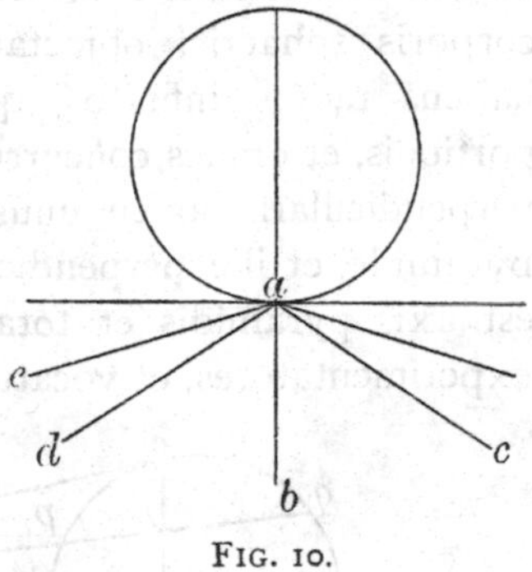

FIG. 10.

Et tum ab eodem puncto superficiei corporis, a quo fit radius perpendicularis, super idem corpus fiunt radii infiniti, ut patet ad *a* punctum, et sic de omnibus ; sed unus solus, scilicet *a b*, radius est perpendicularis super illud corpus, quia ille solus, si protrahatur in continuum et directum, vadit in centrum corporis, et ideo ille est fortissimus, et habet longe plus de virtute.

Convergent rays from distant objects appear parallel.

Et isti radii perpendiculares, eo quod concurrunt in centrum corporis, non sunt aequidistantes ; potest tamen oculus esse ita longe a corpore, quod non percipiat concursum, et judicabit radios tales esse aequidistantes, sicut accidit nobis de radiis solis et stellarum, et ideo judicamus umbras rerum diversarum aequidistantes, quando opponuntur soli, sed non secundum veritatem. Multa enim videntur aequidistare, quia non percipimus eorum concursum, ut parietes domus cujuslibet videntur ad sensum aequidistantes, sed non sunt quia omne grave tendit ad centrum naturaliter, et ideo rueret domus, si essent omnino aequidistantes. Et circuli meridiani diversarum civitatum videntur esse aequidistantes, et lineae meridiei, quia non percipimus concursum, et tamen concurrunt in polo mundi.

The central ray of each cone is most operative.

Est etiam sciendum, quod radii qui cadunt in centrum corporis sphaerici a quo veniunt, sunt illi per quos judicamus stellas per foramina instrumentorum. Unde astronomus et perspectivus, qui experiuntur hujusmodi, utuntur istis radiis quia sensuales sunt et fortes. Quamvis enim a portione aliqua corporis sphaerici objecta rei patienti veniat una pyramis habens radios infinitos, qui fiunt a singulis punctis illius portionis, et omnes concurrunt in conum pyramidis cum radio perpendiculari, tamen unus solus est perpendicularis in una pyramide, et ille perpendicularis dominatur in fortitudine, et est axis pyramidis et tota pyramis ab eo nominatur apud experimentatores, et vocatur radius corporis agentis, ut patet

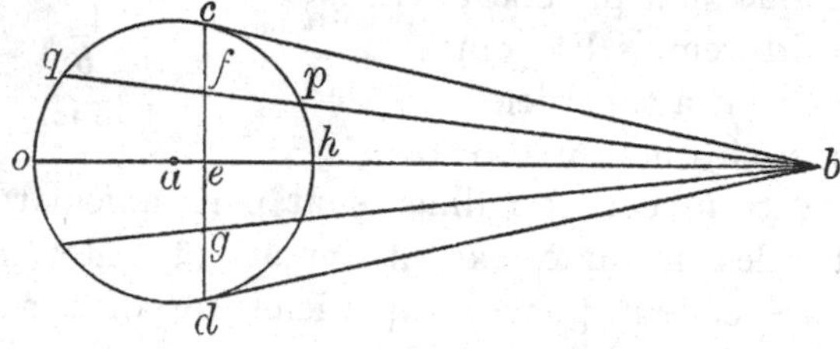

FIG. II.

in figura. Nam sit *a* centrum, et *d c* sit portio solis objecta terrae, et *b* conus pyramidis cadentis in terram, planum est quod *e b* radius cadit in centrum solis, et nulli alii qui sunt de corpore pyramidis, quamvis sint infiniti. Nam *g b* declinat a centro solis, ut patet, et sic de omnibus aliis *f b*, *c b*. Et ideo est perpendicularis super corpus solis, et axis pyramidis, et ideo est fortior, et plus habet de virtute, quoniam virtus venit secundum hunc radium a tota profunditate solis, quod non accidit in aliis. Nam diameter *h o* est longior quam *p q*, et quam omnes lineae cadentes in circulo a latere diametri, et ideo plus capit de substantia solis, ideo plus habet de virtute. Et *h b* linea est brevior quam *c b*, et omnes aliae quae a portione solis descendunt in terram, quapropter plus habet de virtute secundum praedicta. Et haec, quae nunc dicta sunt, patent ex viii tertii elementorum Euclidis.

DISTINCTIO QUARTA.

CAPITULUM I.

In quo canones dicti applicantur ad lucem stellarum.

His principiis et hujusmodi datis per vias geometriae, potest homo verificare omnem actionem naturae, quia omnis veritas circa operationem agentis in medium, vel in materiam generabilem, vel in coelestia, et in totam mundi machinam, sumit ortum mediate vel immediate ex jam dictis, et quibusdam similibus, quia non potui omnia in hac persuasione ponere, quae opus majus requirit. Et quod dico manifestare volo per aliqua exempla in diversis rebus mundi, et incipiam a superioribus. Aristoteles vero dicit in primo Meteorologicorum [1] quod omnes stellae habent lucem a sole; et hoc patet per eclipsin lunae: nam quando terra interponitur inter solem et lunam, ipsa eclipsatur, et quando non, tunc illuminatur; et ideo similiter esset de aliis, si essent in tali situ in quo luna. Sed non sunt, nam conus pyramidis umbrae non attingit nisi usque ad orbem Mercurii, et ideo sola luna potest cadere in umbra terrae. Et tamen stellae inferiores eclipsant superiores, sicut Aristoteles vult secundo Coeli et Mundi, quando cadunt [2]

Illustrations.

Lunar eclipses.

Meteor. i. 8, § 6. Aristotle explains that the earth's shadow reaches to the moon, but not much beyond; and that the distance of the stars from the earth is many times as great as that of the sun: wherefore, *ἀνάγκη πάντα τὸν ἥλιον τὰ ἄστρα περιορᾶν, καὶ μηθενὶ τὴν γῆν ἀντιφράττειν αὐτῶν.*

[2] In *Meteor.* i. 6, § 11, Aristotle speaks of an occultation of one of the stars in Gemini by Jupiter.

inferiores inter solem et superiores, et hoc bene accidit, sed non est ita notabile, sicut de luna. Sed quoniam, ut patet ex praedictis, rara et perspicua permittunt transitum speciei, ut aer et species oculi et stellarum transeunt per orbem ignis, et per medium omnium orbium septem planetarum, necesse est quod sint rara et perspicua, et quod non terminant visum. Ergo non sunt densa. Ergo nec visibilia, quia solum est visibile, ut docet Avicenna tertio de Anima, quod potest visum terminare, et verum est hoc. Sed si non sunt visibilia, non sunt lucida, quia lucidum est visibile. Et loquor de lucido quod habet lucem fixam, et non transeuntem ac propriam quae potest multiplicare a se radios, ut stella et ignis; non loquor de lucido quod recipit lucem transeuntem sicut aer, quem Aristoteles vocat lucidum[1] in secundo de Anima: sed hoc est aequivoce. Quapropter errant, qui aestimant sphaeram ignis lucere naturaliter, sicut hic inferius, et praecipue cum magis sit rarus quam aer, et ideo minus visibilis, et propter hoc minus aptus luci, quia densitas est causa illuminationis, ut dicit Averroes secundo Coeli et Mundi, et libro de Substantia Orbis. Et similiter turpius erratur a vulgo, quando ponit orbes stellarum lucere, praecipue cum dicant quoddam falsum, et imponunt Averroi illud. Nam dicunt, quod stella non differt ab orbe nisi per majorem aggregationem et minorem lucis. Sed Averroes non dicit hoc, sed contrarium docet et probat: bene enim dicit totum hunc sermonem praeter casum genitivum ultimum, qui est *lucis*, loco cujus dicit *perspicui coelestis*. Et quia fere omnia vocabula sermonis vulgati et ipsius Averrois sunt eadem, et aestimant perspicuum et lucidum esse idem, propter verbum Aristotelis secundo de Anima, ubi accipitur lucidum aequivoce, imponunt Averroi quod velit orbem lucere propria luce et fixa, ut stella, licet minus. Sed dicit contrarium, volens quod propter fortitudinem actionis quam habet stella in hoc mundo, oportet quod multum habeat de substantia coeli congregata in suo corpore, et ideo perspicuum coeleste, quod dispergitur in partibus orbis,

[1] *De Anima*, ii. 7, § 2 Ἔστι δή τι διαφανές· διαφανὲς δὲ λέγω ὃ ἔστι μὲν ὁρατόν, οὐ καθ' αὑτὸ δ' ὁρατὸν ἀλλὰ δι' ἀλλότριον χρῶμα· τοιοῦτον δ' ἐστὶν ἀὴρ καὶ ὕδωρ . . . φῶς δ' ἐστὶν ἡ τούτου ἐνέργεια τοῦ διαφανοῦς ᾗ διαφανές.

condensatur in corpore stellae, ut habeat fortem virtutem in alteratione mundi. Et ideo docet ibi, et in libro de substantia orbis, quod stella solà lucet, et nulla pars orbis. Et hic error vulgatur apud philosophantes, et apud omnes Theologos, quando loquuntur, et scribunt de stellis.

Quoniam vero in libro de proprietatibus elementorum dicitur, quod sol est sicut candela, et stellae sicut specula, aestimat totum vulgus studentium quod lumen quod venit ad nos de luna et stellis sit lux solis reflexa a superficiebus earum. Sed hoc est impossibile propter aequalitatem angulorum incidentiae et reflexionis, ut patet in figura. Nam ut prius visum est, si ita esset, tunc oportet angulum *a* incidentiae, et angulum *b* reflexionis aequari. Ergo *b c* radius non vadit nisi in partem determinatam terrae et non ubique, et sic de toto lumine quod venit ad superficiem lunae. Totum enim est quasi unus radius, et cadit ad angulos inaequales, et reflectitur in partem determinatam. Ergo si veniret ad terram, non illuminaret luna nisi partem Horizontis aliquam determinatam; sed nos videmus, quod illuminat totum hemisphaerium sicut sol. Ergo lux illa quae venit a luna et a stellis non est reflexa. Et Averroes secundo coeli et mundi hac demonstratione utitur, et sua confirmat auctoritate, quod non est lumen solis reflexum a superficie stellarum, quod ad nos descendit ab eis, sed proprium lumen et innatum, eductum tamen de potentia materiae in corpore stellae per virtutem solis venientis ad stellam, quae virtus alterat et transmutat stellam, et facit lumen in ea; et quando habet lumen naturaliter genitum in ea, sicut sol habet lucem creatam, tunc potest multiplicare lucem a se undique sicut sol. Et tunc concedendum est quod lumen solis reflectitur a superficie lunae, sed non venit ad terram, sed ad aliam partem mundi

Is the light of the moon and stars reflected from the sun?

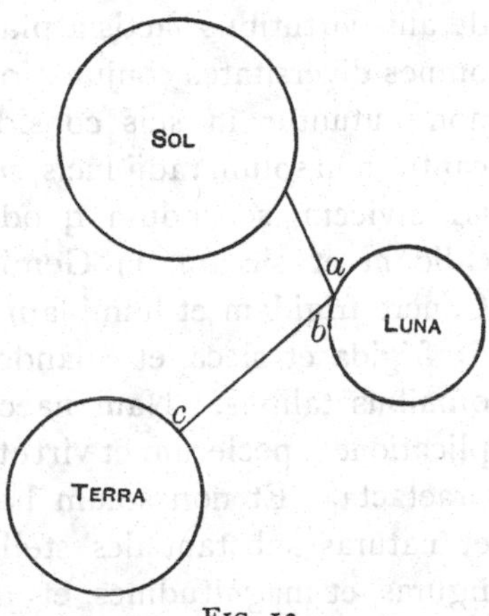

FIG. 12.

declinat in coelestibus secundum aequalitatem angulorum incidentiae et reflexionis.

More than half of the earth illuminated by the sun.

Et per supradicta patet, quantum illuminatur de terra et de stellis per solem. Nam majores earum portiones semper illuminantur, quia minores sunt sole. Sol enim est centies septuagies fere major tota terra, sicut ostendit Ptolemaeus in quinto Almagesti [1], et sic possunt omnia quae pertinent ad illuminationes et protractiones radiorum in coelestibus verificari, ut in diversa figuratione lunae secundum aetates, et quare in eclipsibus appareat rubea et pallida, et tota ratione eclipsis; nam hoc est propter lumen principale, quod a radiis principalibus venit infra umbram, et quodammodo sic imperfecte illuminatur per radios accidentales. Et non solum de his, sed de aliis virtutibus factis a planetis et stellis in alias secundum omnes diversitates conjunctionum et aspectuum, quibus astronomi utuntur in suis considerationibus, in quibus multiplicantur non solum radii lucis, sed virtutes substantiales stellarum ad invicem, secundum quod dicimus lunam in Ariete esse calidam et siccam, in Geminis calidam et humidam, et in Cancro frigidam et humidam: et quando conjungitur Saturno fit frigida et sicca, et quando Iovi calida et humida, et sic de omnibus talibus. Nam haec omnia certificantur per multiplicationes specierum et virtutum determinatas juxta principia praetacta. Et non solum haec, quae pertinent ad qualitates et naturas substantiales stellarum, sed ea quae pertinent ad figuras, et magnitudines, et altitudines, et numerum coelorum et stellarum, et ad hujusmodi consimilia.

Capitulum II.

In quo canones supradicti applicantur ad totum mundum.

In space are more media than one, as refraction proves.

Et non solum haec de coelestibus verificantur per dicta, sed de elementis et toto mundo. Nam cum philosophi ante Aristotelem posuerunt omnia esse unum corpus mundi, potest destrui hoc per leges fractionis. Nam si quis per instrumenta quibus experimur ea quae sunt in coelestibus, cujusmodi

[1] Ptol. *Syntaxis*, i. 16.

vocantur armillae vel alia, accipiat locum alicujus stellae circa aequinoctialem in ortu suo, et deinde accipiat locum ejusdem quando venit ad lineam meridiei, distare sensibiliter inveniet eam in loco meridiei plus a polo mundi septentrionali, quam quando fuit in ortu. Ergo visus videt stellam diversis modis in illis diversis temporibus: nam si videret eodem modo tunc stellam in eodem loco inveniret semper. Sed quando stella est in linea meridiei, tunc stella accedit ad zenith capitis aspicientis, qui est punctus in coelo suprapositus capiti, quapropter cadunt radii in visum perpendiculariter, et in centrum mundi, et ideo non franguntur, et propter hoc videt visus stellam per rectas lineas in suo loco vero. Ergo quando visus errat in ortu stellae non videbit per lineas perpendiculares, quia multum distat stella a zenith capitis, et ideo radii cadunt ad angulos obliquos, quapropter franguntur, et ideo visus tunc videt per lineas fractas, et errat in loco stellae. Sic autem Ptolemaeus[1] in libro quinto de opticis docet considerare, et Alhazen in libro septimo et ego consideravi in instrumentis hoc idem, et certum est. Quoniam ergo fractio radiorum cadit in hoc mundo, planum est quod plura corpora sunt in mundo. Et prima fractio invenitur secundum considerationem dictam in superficie ignis immediate sub orbe coelesti, scilicet sub sphaera lunae: quapropter sphaera ignis est diversa a sphaera coeli; cum tamen Platonici et Augustinus, et multi autores antiqui tangant Platonice quod ignis et coelum sunt unius naturae. Sed hoc est impossibile propter demonstrationem dictam, et propter alias demonstrationes naturales quas Aristoteles in libro Coeli et Mundi affert, quibus nullus modernus modo contradicit: nam trita est haec veritas in naturalibus. Haec tamen demonstratio vulgo naturalium est ignota, nam Aristoteles non tangit eam, nec ejus expositores. Quoniam autem non est fractio in sphaera aeris, ut iidem autores docent, et certitudo experientiae, ideo multum turbantur sapientes, an sphaera aeris et ignis sint duae vel una. Videtur enim per autores praedictos, et propter privationem fractionis, quod una sit superficies aeris et ignis, et unum corpus. Sed hoc est impossibile, quia Aristoteles dicit tertio Coeli et Mundi,

Ray passing from lunar sphere to sphere of fire is refracted.

[1] Ptol. *Opticae*, v. p. 151, ed. Govi. Cf. Alhazen, vii. 15.

quod aer est gravis[1] in sphaera sua, et sequitur naturaliter superficiem aquae non ignis; nam si ignis ascenderet in sphaera sua, aer non sequeretur, ut dicit, quia cum aqua descendit, aer sequitur ejus superficiem, ut videmus ad oculum. Quapropter non erunt unum corpus aer et ignis; et tota dubitatio solvitur per legem fractionis. Nam tria requiruntur ad hoc ut sit fractio, scilicet ut corpus secundum habeat superficiem distinctam a primo, et quod sit alterius raritatis, scilicet magis rarum vel minus, et quod radii cadant ad angulos obliquos. Quod si aliquid istorum deficiat, non est fractio possibilis. Propter primum enim non est fractio in eodem corpore, licet habeat partem unam rariorem et aliam minus raram, sicut aer est rarior superius quam inferius. Et propter secundum non est fractio in orbibus coelestibus, quia sunt ejusdem raritatis, quantum ad sensum nostrum. Et propter hoc idem non est fractio in sphaera aeris, quia aer gradatim subtiliatur usquequo in sua parte suprema aequetur subtilitati ignis in parte inferiori, et ideo non est ibi fractio. Quoniam autem fractio haec est inter incessum rectum et perpendicularem ducendam a loco fractionis, ut docent hi autores, et ipsa experientia, ideo sequitur quod corpus secundum est densius priori et ideo corpus sub orbe coelesti est densius quam orbis coeli. Quapropter oportet ponere plenam diversitatem corporum mundi penes coelum et elementum. Et quando habuerimus haec, tunc per radios, et pyramides luminosas stellarum venientium ad instrumenta certificabimus omnia quae sunt in coelestibus, scilicet numerum coelorum, et stellarum magnitudinem et spissitudinem, et omnia quae sunt in coelis.

CAPITULUM III.

In quo per multiplicationes praedictas investigantur complexiones locorum mundi circa polos ejus.

Laws of climate. Polar

Posthaec descendemus ad sphaeras elementorum et investigabimus omnes complexiones eorum secundum singulas partes

[1] *De Coelo*, iv. 4, § 5 ἐν τῇ αὑτοῦ γὰρ χώρᾳ πάντα βάρος ἔχει πλὴν πυρός, καὶ ὁ ἀήρ.

habitationis, et inveniemus quod loca sub polis mundi sunt naturaliter inhabitabilia propter frigus. Nec obstant nobis rationes superius annotatae pro pyramidibus longioribus, nam sicut ibi dictum est, demonstrationes pro pyramidibus brevioribus vincunt in hac parte, quamvis omnis homo qui nesciret hujusmodi radices multiplicationis lucis vinceretur per pyramides longiores. Sed verum est, quod coni pyramidum venientium ad loca sub polis nimis distant a sole, et ideo debiles sunt, nec possunt nisi elevare vapores ab aqua et terra in aerem, et non possunt eos consumere; et ideo oportet quod aer sit densus et caliginosus et congelatus per vapores ubique et semper, ut ibi plantae et animalia non possint vivere, sicut docet Ptolemaeus in libro de dispositione sphaerae. Sed si cum pyramidibus longioribus addamus autoritatem Aristotelis secundo de vegetabilibus dicentis, 'quod apud eos[1] ubi prolongantur dies,' &c. et hoc est, ut docet commentator super illud verbum, ubi est dimidius annus dies, et dimidius annus nox, non sunt animalia nec plantae, quia calor combussit materiam eorum; videbitur quod illa loca sub polis sint inhabitabilia propter calorem, non propter frigus. Nam proculdubio ibi est dimidius annus dies, et tunc videretur quod calor nimis abundaret, dum sol ingreditur primum gradum Arietis, usquequo veniat ad primum gradum Librae, et hoc per dimidium annum. Et adhuc additur quod diluculum matutinum incipit per mensem et dimidium ante ortum solis super terram, et crepusculum vespertinum durat per mensem et dimidium ultra occasum solis, ita ut lumen solis appareat super terram ante ortum et occasum per totum tempus dictum, sicut nobis in aestate post occasum et in aurora apparet lux solis; quare per longitudinem crepusculi et aurorae multum augmentabitur calor, quia non habent noctem profundam nisi per tres menses anni, ut certum est per omnes autores et singulos sapientes. Item sol nunquam recedit ab eis, nisi per maximam solis declinationem, quae fere est 24[2] gradus, quia aequinoctialis est

Temperature low, notwithstanding length of day in summer.

[1] The reference apparently is to the apocryphal work, *De Plantis*, lib. ii. 7 Ἐὰν ὁ ἥλιος μακρότητα τῇ ἡμέρᾳ προσάγῃ ἐν τῇ κινήσει αὐτοῦ καὶ κατακυριεύῃ τῆς ὑγρότητος οὐκ ἔχει τὸ φυτὸν δυνάμεις φύλλα καὶ καρποὺς προάγειν.

[2] It may be noted, once for all, that in the oldest MSS. of the *Opus Majus*,

Horizon eorum. Sed a nobis recedit sol per duplicatam declinationem, scilicet fere per 48 gradus. Ergo ex hujusmodi appropinquatione augmentabitur calor. Quapropter concluditur secundum veritatem, quod loca sunt ibi inhabitabilia propter calorem. Sed cum Plinius in naturalibus et Martianus in descriptione regionum mundi invenerunt per experientiam certam, quod regiones sub polis sunt temperatissimae in hoc mundo, sicut ipsi dicunt, et allegant experientiam hominum qui ibi fuerunt, non possumus negare quin ibi sint regiones temperatissimae ; et quis concordabit tantam contrarietatem? Certe nullus, nisi optime sciat principia multiplicationis et actionis specierum. Persuasionem ergo do ad hoc, quod naturaliter secundum dispositionem loci respectu coeli et solis oportet quod locus sit inhabitabilis propter frigus; et sic currunt auctoritates permultae.

Temperature modified by mountain ranges.

Sed propter accidentalem dispositionem loci, simul cum aliquibus causis naturalibus, potest ibi aliqua regio esse combusta, et aliqua temperata. Nam propter rationes de longioribus pyramidibus, et propter longitudinem diei, et propter magnitudinem crepusculi et aurorae, et propter hoc, quod sol non recedit ab eis nisi per maximam solis declinationem, si adjungamus causam accidentalem cum his, forsan inveniemus quod quaerimus. Nam proculdubio, secundum quod Plinius et Martianus[1] et alii docent, montes maximi sunt ad ubera aquilonis, ut montes Riphaei et Hyperborei et alii, quorum altitudo immensa est, propter quam possunt prohibere frigus aquilonis, sicut accidit in montibus Italiae apud loca quae sunt inter solem et montes. Et simul cum hoc additur, quod montes inveniuntur lapidei, et alii coagulati in crystallum et salem, sicut videmus in pluribus locis mundi, qui montes habent superficies magis politas et aequales, propter quod ab eis potest fieri major et melior reflexio quam a montibus asperis. Nam a superficie polita

those which are contemporary with Bacon, Arabic numerals are generally used; invariably so where the numbers are large.

[1] For Pliny's account of the enviable life of the Hyperboreans see *Hist. Natural.* iv. 12. Martianus Capella was a writer of the fifth century. His work, much read in the Middle Ages, was a scientific encyclopaedia bearing the fanciful title, *De nuptiis philologiae et Mercurii de septem artibus liberalibus.*

et aequali et leni fit sensibilis reflexio, ut patet in speculis: et hoc est, quia partes concordant in unam actionem, et non dissipatur species, sed integra redit sicut venit; sed propter inaequalitatem superficiei corporis asperi nulla pars concordat cum alia, sed elevatior primo reflectit, et depressior secundario, et sic tota species dissipatur nec venit integra, propter quod non videmus per corpora aspera sed polita. Bonitas ergo reflexionis, quae potest inveniri in locis aliquibus circa polos propter polituram superficierum montium, valet cooperari ad calorem generandum simul cum altitudine montium. Et ulterius considerandum est, quod montes habent varias figuras, nam aliquis potest habere figuram ad modum speculorum comburentium, et alius ad modum speculorum sphaericorum vel columnarium vel pyramidalium, et ubi figura speculi comburentis invenietur cum caeteris causis caloris, necesse est ibi esse combustionem validam, ita ut nihil ibi vivere possit, et sic intelligendus est Aristoteles cum ejus commentatore. Ubi autem temperantur causae caloris et frigoris, et penes altitudinem montium, et penes lenitatem superficierum, et penes figuram, necesse est quod sit locus temperatus, et sic intelligendi sunt Plinius et Martianus et caeteri experimentatores.

Capitulum IV.

In quo investigatur complexio locorum, qui sunt in medio mundi.

Climate of temperate and tropical zones.

Mundi vero loca in quibus sumus per totam habitationem usque versus finem tertii climatis sunt tolerabilis caloris. Hierusalem quidem est in tertio climate, sed ultra sub Tropico Cancri incipit torrida zona, et locus malae habitationis. Et propter hoc, quod via solis est inter duos Tropicos, aestimat vulgus quod totus ille locus est combustus et quod nihil ibi sit temperatum; et ideo aestimant, quod locus sub aequinoctiali circulo sit maxime combustus, quia sol super capita habitantium bis transit in anno, et non attingit ad Tropica nisi semel, ut in solstitio aestivali venit ad Tropicum Cancri, et in solstitio hyemali accedit ad Tropicum Capricorni. Sed eundo de uno ad aliud transit bis super aequinoctialem circulum, scilicet in principio Arietis et Librae, et hoc est in principio veris et autumni, propter quod aestimatur a multis, quod locus sub aequinoctiali sit maxime combustus, et ad hoc faciunt rationes secundum multiplicationes praedictas. Nam ibi saltem bis in anno sunt pyramides breviores et radii cadentes ad angulos rectos ac perpendiculares, et per consequens non franguntur; redeunt in se ut geminentur radii, et ideo videtur prima facie quod tota fortitudo actionis naturalis ibi concurrit longe plus quam alibi. Sed constat nobis quod sub Tropicis sunt Aethiopes combusti. Ergo videtur quod locus sub aequinoctiali circulo sit combustissimus, sicut vulgus ponit. Sed proculdubio Ptolemaeus vult libro praedicto, quod locus ille sit temperatus respectu Tropicorum. Atque Avicenna docet primo de Animalibus, et primo artis medicinae, quod locus ille est temperatissimus. Et propter hoc theologi ponunt his diebus, quod ibi sit paradisus, et ideo vulgus indoctum errat in hoc loco. Et tenebimus sententiam Ptolemaei ad minus, quicquid sit de opinione Avicennae et theologorum, quamvis leges multiplicationis specierum hic concludant quantum possunt. Sed causae fortiores sunt in contrarium, quas Avicenna egregie assignat, videlicet quod declinatio solis magna est ibi, et aequidistantes sunt distinctae et distant multum. Nam cum tota declinatio solis sit fere 24 gradus de Coluro, fere 12 debentur signis aequinoctialibus, scilicet Arieti et Virgini, Librae et Piscibus, et octo fere debentur Tauro, Leoni, Scorpioni, Aquario, et ut fere quatuor debentur Geminis, Cancro, Sagittario, Capricorno. Et ideo quando sol est in signis ubi declinatio est fere 4 graduum, fertur propter confusionem aequidistantium super eandem regionem per 40 dies, et comburit ut in Tropicis. Sed minus in aliis ubi declinatio est octo graduum, minime vero ubi est 12, et hoc est apud aequinoctialem.

Effect of solar eccentricity.

Alia causa est propter aequinoctium, quia aer tantum temperatur de nocte, ut non possit esse calor superfluus de die, et propter hoc ibi est aequalitas aeris, et nobiles complexionum dispositiones, ut docet Avicenna. Nec est dubium quin locus sit temperatus, sed an sit temperatissimus, non percipio adhuc. Et ideo non est certificatum an paradisus debeat ibi esse. Quoniam si eccentricitas solis sit prout mathematici ponunt, impossibile est pure temperatum esse sub aequinoctiali. Nam una pars eccentrici, quae vocatur oppositum augis descendit ad terram per quinque partes semidiametri eccentrici magis quam reliqua quae dicitur aux[1]. Et ideo quando sol venit ad oppositum augis comburit terram omnino, ut nihil ibi vivere possit, tum propter appropinquationem, tum propter casum radiorum ad angulos rectos, tum propter confusionem aequidistantium in quibus moratur super eandem regionem ut comburat eam; et hoc est quando sol est in Sagittario et Capricorno et Scorpione. Nam oppositum augis est in Sagittario, et in locis propinquis ei comburit similiter et distemperat. Et ideo sol distemperabit Libram et Arietem. Nam jam transivit medietatem eccentrici, quando venit ad initium Librae, et Libra satis appropinquavit ad oppositum augis, ut intemperiem caloris possit causare, sicut patet in figura. Sed constat, quod paradisus habet plenum temperamentum. Horum autem certificatio non est praesentis persuasionis.

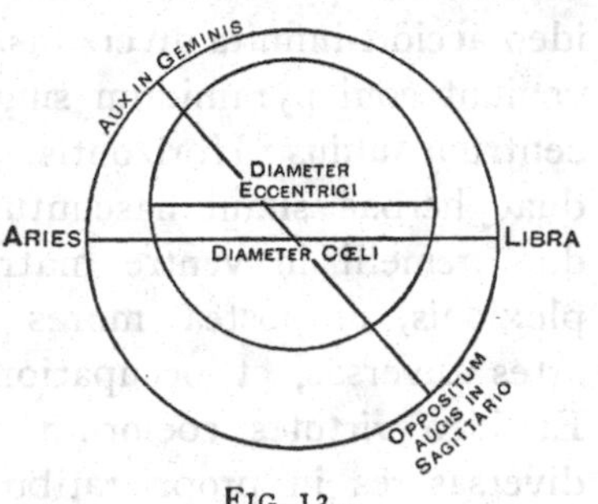

FIG. 13.

CAPITULUM V.

Et ideo[2] redeo ad propositum dicens quod complexiones locorum mundi inveniri non possunt, nisi homo sciat leges Influence of climate on temperament.

[1] *Aux* is the Arabian form of the word *apsis*: used together with the related expression *oppositum augis* to denote the two points of an orbit which are at the greatest or least distance from the centre of motion. In modern astronomy the aphelion or apogee of the earth's or the moon's orbit would be called *summa apsis*, or *aux*; the perihelion or perigee would be *ima apsis*, or *oppositum augis*. It need hardly be mentioned that in the Hipparchian or Ptolemaic theory the sun's orbit was supposed to be eccentric to the circle of the heavens.

[2] I follow O. in making this the beginning of a new chapter.

multiplicationum praesignatas, quoniam nec falsum vitabit, nec verum poterit confirmare. Sed locus est principium generationis, quemadmodum et pater, ut dicit Porphyrius. Et nos videmus, quod omnia variantur secundum loca mundi diversa non solum in naturalibus, sed homines in moribus; quoniam alios mores habent Aethiopes, alios Hispani, alios Romani, et alios Gallici. Nam et Picardi, qui sunt veris Gallicis vicini, habent tantam diversitatem in moribus et in lingua ut non sine admiratione possit esse unde sit tanta diversitas locorum propinquorum.

Sed quoniam res hujus mundi in diversis locis constitutae, quantumcunque propinquae sunt, recipiunt conos diversarum pyramidum venientium a toto coelo objecto eis, ideo accidit infinita diversitas. Nam ad singula puncta terrae veniunt coni pyramidum singularum, et quilibet punctus est centrum unius Horizontis novi. Et ideo videmus, quod duae herbae simul nascuntur de terra sine medio, et ideo duo gemelli in ventre matris sortiuntur diversitatem complexionis, ut postea mores habeant diversos, et sequantur artes diversas, et occupationes difformes per totam vitam. Et ideo virtutes coelorum et stellarum producunt ubique diversas res in proprietatibus et naturis, et in rebus generatis secundum propagationem. Et non solum multiplicatio virtutis coelestis operatur, sed patris et matris, quoniam descinduntur virtutes in seminibus, ut docent medici. Et praecipue ab anima matris continuatur multiplicatio virtutis et speciei super foetum usque in complementum generationis et nativitatis. Et cum puer in nativitate exponitur aeri novo tanquam alteri mundo, tunc recipit conos pyramidum coelestium secundum singulas partes, et sic recipit impressiones novas, quas nunquam dimittit, quia quod nova testa capit, inveterata sapit. Et tunc confirmatur complexio radicalis, quae semper manet usque ad finem vitae, licet complexio currens mutetur tota die. Et ad hanc radicalem complexionem sequuntur inclinationes ad mores et ad scientias et ad linguas, et ad quaecunque artificia et negotia, et ad omnem diversitatem quam videmus in omnibus. Et si coeli dispositio sit mala in conceptione et nativitate pueri, tunc coni pyramidum laedunt complexionem, et per consequens inclinatur homo ad malos mores et artes perversas, secundum diversitatem coelestis constellationis: et si constellatio bona est, tunc complexio est bona, et sequitur inclinatio ad mores bonos et scientias utiles: et si mediocritas accidat in coelesti constellatione, tunc homo mediocris est in omnibus quantum est ex naturali dispositione, licet poterit se mutare per libertatem arbitrii, et per gratiam Dei, et per tentationem diaboli et per bonum aut malum consilium, maxime a juventute. Et nunc quae dixi in universali, possunt confirmari si tempus esset per exempla in singulis rebus. Sed jam de combustione facta per crystallum et vitrum et specula concava et alia quae ad oculum patent, ostenditur nobis sensibiliter, quod fractiones et reflexiones et hujusmodi multiplicationes possunt effectus naturales producere. Atque vulgatum est quod irides et cometae et plures aliae impressiones inflammatae in aere, et circuli circa solem et lunam, fiunt per hujusmodi multiplicationes radiorum, cum aliam causam non possunt habere. Et sic de omnibus aliis, licet non sit in singulis evidens, quia non omnium agentium species sunt visibiles.

CAPITULUM VI.

In quo datur causa fluxus et refluxus maris per radios.

Et nunc ponam unum exemplum omnibus occultum, et pono ubi minus videtur quod multiplicatio secundum lineas et angulos determinatos requiratur, ut in re quam difficillima, quae tamen per multiplicationem redditur satis plana; et est de fluxu et refluxu maris. Alpetragius vero in libro suo de motibus coelestibus aestimat, omnia corpora mundi praeter terram moveri motu coeli primi, et hoc verum est: sed secundum quod magis elongantur tardius moventur, et cum majori impedimento. Unde aqua tardius et irregularius movetur in sphaera sua, quam alia corpora mundi. Addit ergo iste, quod hic motus facit fluxum et refluxum; sed non placet hic, quia fluxus et refluxus sunt determinati et certi, et

The tides.

currunt sicut luna variatur in partibus coeli. Sed motus aquae a motu coeli est confusus, et inordinatus, et irregularis propter hoc, quod virtus coeli primi nimis elongatur ab ejus origine, quando est in aqua, et ideo praevalet virtus aquae propria, scilicet sua gravitas, quia nititur quiescere in loco suo, propter quod non potest hic motus esse ita regularis et distinctus temporibus certis penes accessus et refluxus ut nos videmus in mari. Et ideo Albumazar in majori introductorio astronomiae determinat omnes differentias fluxus et refluxus, et narrat quod accidunt omni die et nocte secundum quod luna est in diversis partibus sui circuli et respectu solis. Sed non dicit nobis causam, nisi quod luna est causa, et quod quando luna est in uno loco tunc est fluxus, quando in alio tunc est refluxus. Propter quod considerandum est, quod quando luna ascendit super mare alicujus regionis, ejus radii cadunt ad angulos obliquos, ut quilibet qui novit casum angulorum potest hoc scire. Et quia cadunt ad angulos tales, oportet quod sint debilis virtutis, ut prius ostensum est. Et ideo solum possunt elevare vapores a fundo maris, et ampullas tumentes, et ingurgitantes aquas maris, ut expellantur a canalibus suis, quos vapores non possunt radii ad aerem extrahere nec consumere propter debilitatem suam; et ideo oportet ut aqua fluat a sedibus suis, donec durat hujusmodi ebullitio vaporum. Sed cum luna accedit ad medium coeli, cadunt magis et magis radii ejus ad angulos rectos, et fortificantur super corpus maris, ac extrahunt vapores ad aerem et consumunt, unde debilitatur fluxus paulatim, secundum quod luna appropinquat lineae meridiei; et quando venit ad illam lineam sunt vapores castigati et consumpti, ita ut statim dum luna descendit ad aliam quartam coeli incipiat refluxus, quia cessante causa cessat effectus. Et pono exemplum sensibile ad istud. Nam in pulmento posito super ignem, ignis in principio resolvit vapores, et facit eos exire orificium vasis: quando vero fortificatus est et continuatur consumit vapores, et liquor residet in fundo vasis. Nam universaliter calor debilis resolvit vapores et non consumit, calor enim solis fortis consumit, unde in nocte, et in mane, et in vespere fit resolutio vaporum major quam in meridie, et tunc consumuntur. Et similiter

Flood tide when moon is approaching meridian.

calor naturalis in principio post cibum et potum resolvit vapores fortiter et facit somnum. Sed quando invaluit diu super cibum consumit eos, et homo evigilat. Nam vapores resoluti obstruunt vias spirituum et calorum venientium ad sensus a corde et capite, et ideo oportet ut quiescant sensus, et hoc est dormire. Et sic de omni virtute agente, cujus est proprium resolvere et consumere. Primo resolvit dum debilis est, et postea fortificata consumit.

Et dicit Albumazar et alii concordant, quod luna facit effectum consimilem in quartis oppositis. Nam dum luna est in quarta inter oriens et meridiem fit fluxus in illa quarta et in quarta opposita, quae est inter occidens et angulum terrae, et dum resolvit quando est in quarta inter angulum medii coeli et angulum occidentis. Similiter refluit tunc mare ab angulo terrae usque ad angulum orientis. Sed non dant causam hujus rei; solum dicentes quod luna effectum in quartis oppositis habet consimilem. Sed quomodo operabitur luna ubi non est? et constat, per medium terrae non transeunt ejus radii. Ad hoc autem juvat nos multiplicatio reflexa. Nam proculdubio coelum stellatum aut nonum est densum per totum, nam visus noster stat ad alterum illorum, sed non terminatur visus nisi per densum, ad quorum alterum multiplicati radii lunae existentes in quarta una reflectuntur ad quartam oppositam, et sic virtus lunae est directa in quarta una, et ejus reflexio in eodem tempore in quarta opposita. Solis autem virtus est nimis potens, et ideo ad quoscunque angulos cadat potest consumere vapores cito, quos resolvit. Aliarum vero stellarum virtus est debilis propter earum elongationem a terra, quamvis multo sint majores quam luna sit, et propter debilitatem non faciunt hujusmodi operationem, quamvis ad angulos diversos cadant earum radii ut lunae. Nec tamen hic intelligo, quod sit ebullitio cum fervore caloris ut in liquore super ignem, sed similitudo est quantum ad resolutionem vaporum quam facit virtus lunae, aut per radios lucis, aut per radios suae naturae substantialis, aut per utrosque. Quando enim de nocte per radios stellarum, et in mane et in vespere per solem fit resolutio vaporum de terra et aqua, non propter hoc ebulliunt cum fervore, et tamen in

Explanation of flood tide when moon is in opposite quarter.

maxima copia elevantur, et sunt nubes infinitae. Et ideo non oportet ut aquae maris ferveant in fluxu.

CAPITULUM VII.

In quo multiplicatio virtutum comparatur ad sanitatem et infirmitatem corporis humani.

Influence of stellar rays on health and disease.

Mira ergo potestas est hujus multiplicationis cum omnia fiant secundum leges ejus et occulta et manifesta. Et horum notitia non solum necessaria est in scientialibus, sed utilitates magnas nobis praestat in corpore et in anima, si diligenter investigemus, nam valde utile est scire has leges in conservatione sanitatis. Quoniam cum non possumus vitare omnes incidentias specierum et virtutum rerum malarum et nocivarum sanitati, nec possumus semper aptare corpora nostra ad pleniores casus virtutum venientium a rebus salutiferis, semper tamen debemus esse solliciti ut rerum nocivarum virtutes non capiamus principales, scilicet rectas fractas et reflexas, sed accidentales si possumus, et si non possumus vitare omnes principales, saltem vitemus rectas, et si non omnes rectas, declinemus casum ad angulos rectos, et si non possumus saltem caveamus pyramidis breviores. Et hae considerationes habent locum, quando homo exponitur coelestibus impressionibus nocivis, ut soli in aestate, et lunae de nocte, quae exhaurit corpora. Unde multi mortui sunt non caventes sibi a radiis lunae. Et praecipue quando homo exponitur radiis Saturni et Martis, quoniam isti duo inducunt laesionem magnam et corruptionem in rebus, ut experientia docet. Similiter quando homo recipit species a locis corruptis et immundis, et quando multiplicantur species leprosorum et infirmorum et maxime illorum qui habent morbos contagiosos, nec non eorum qui habent malas complexiones, et praecipue mulieris menstruatae; quoniam si ipsa aspiciat speculum novum, apparet nubes sanguinea in speculo ex violentia menstrui inficientis, ut Aristoteles dicit in libro de somnis et vigilia secundo[1], et de serpentibus

[1] *De Insomniis*, cap. 2 ὅταν τῶν καταμηνίων ταῖς γυναιξὶ γινομένων ἀμβλέψωσιν εἰς τὸ κάτοπτρον γίνεται τὸ ἐπιπολῆς τοῦ ἐνόπτρου οἷον νεφέλη αἱματώδης.

et aliis rebus venenosis. Et maxime considerandum est istud, quando homines et animalia nascuntur, et habent desiderium nocendi et animum malignandi. Nam ad hoc reducitur fascinatio, unde quod habet virtutis capit ex hac causa, quoniam proculdubio tunc fit fortior impressio, eo quod natura obedit cogitationibus animae et desiderio, et excitatur ad fortiorem operationem, sicut Avicenna docet octavo de Animalibus et quarto de Anima per exempla et experientias varias, et certum est hoc. Unde Solinus narrat in libro de mirabilibus mundi [1], quod in quadam regione aquilonari sunt mulieres habentes geminas pupillas in oculis, quae cum irascuntur interficiunt homines solo visu, de quibus dicit Ovidius, nocet pupilla duplex. Et maxime cavendum est ne partes nobiliores, sicut oculi et facies, exponantur hujusmodi speciebus: nam vidi medicum excaecari dum intendebat curae habentis infirmitatem oculorum, propter multiplicationem speciei venientis ab oculis patientis. Oportet enim in istis nocivis uti magnis cautelis, et maxime quando malum est grave vel intolerabile. Sicut Alexander doctrina Aristotelis ut historiae narrant, basilisci speciem venenosam positi super murum civitatis ad interficiendum exercitum per corpora magna polita retorsit in eandem civitatem, ut per proprium destrueretur venenum. Et per contrarium aptatio corporis ad species rerum salubrium recipiendas, quanto efficacius potest homo, multum est utilis per omnem modum tam sanis quam infirmis.

The evil eye.

Capitulum VIII.

De infinitate Materiae [2].

Quoniam autem diutius tenui persuasionem ut ostenderem, quomodo in rebus mundi a parte suorum efficientium et

Matter is not of one kind.

[1] Solinus probably lived in the third century A.D. His work, *Collectanea Rerum Memorabilium*, is in the main a recast of Pliny. It is from his work rather than from that of Pliny, that such compilers as Isidore and Martianus Capella derived their information.

[2] Cf. *Opus Tertium*, cap. 38, where this subject is fully treated. It is discussed at still greater length in that part of Bacon's encyclopaedic work, entitled *Communia Naturalium*; copious extracts from which are given by Charles, pp. 369–389.

generantium nihil potest sciri sine geometrica potestate, nunc volo breviter pertransire a parte materiae illorum, ostendendo quod necesse est verificare materiam mundi per demonstrationes in lineis geometricis explicandas, si potenter volumus errores infinitos evacuare. Multitudo vero philosophantium non solum in forma propria philosophiae, sed in usu theologiae dicit et asserit, quod una est materia numero in omnibus rebus et quod solum est diversitas a parte formarum.

[1] Et arguit ad hoc quod si excludamus per intellectum a materia coeli et lapidis formas eorum non est assignare per quod differant quod actus a forma dividit, ut dicit Aristoteles septimo Metaphysicae. Et in primo Physicorum [2] dicit quod omnia sunt unum in materia sed differunt in forma: et in secundo Metaphysicae dicit quod nihil est in materia distinctum. Et si materia plurificaretur sicut forma, quod materia erit communis et praedicabilis de pluribus sicut forma. Et allegant quod omne universale naturam formae habet. Nam partes definitionis ut dicit Aristoteles septimo Metaphysicae sunt formae [3]; sed partes definitionis sunt per genus et differentiam et componunt speciem; ergo species est forma tantum sicut genus et differentia. Et constat quod reliqua duo universalia Porphiriana, id est proprium et accidens, sunt purae formae, quia sunt de predicamentis accidentium. Ergo omne universale est forma; et ideo materia non habebit rationem universalis sed singularis erit et una in numero omnibus. Et per hujusmodi fundamenta et auctoritates male translatas nituntur persuadere. Sed hic est error infinitus. Nullus enim major est in veritatibus speculativis, quia hoc dato impossibile est servare rerum generationem, et ignorabitur totus decursus naturae. Sed quod plus est, si hic error discutiatur, invenietur valde propinquus haeresi, aut omnino haereticus, quo nihil magis sit profanum, quia sequitur necessario quod materia sit Deus et creator. Quod volo ad

[1] From here to nituntur persuadere omitted in J. It has been restored from O.

[2] ἡ δ' ὕλη ἄγνωστος καθ' αὑτήν. *Met.* vi. 10, § 13. Cf. *Natural. Auscult.* ii. 1, §§ 11, 12. The passage in *Natural. Auscult.* i. 4, § 1, appears to refer to the opinion of others.

[3] Ἀλλὰ τοῦ λόγου μέρη τὰ τοῦ εἴδους μόνον ἐστίν. *Met.* vi. 10, § 12; cf. 12, §§ 4–9; also vii. 2, § 8.

praesens ostendere solum per rationes geometricas ut promisi, quamquam rationes naturales et metaphysicae sunt copiosae et efficaces, de quibus alias grandis sermo potest fieri. Et necesse est propter sensum vulgi, qui violentus est ubique.

If so, it would possess infinite power, and be equivalent to Deity.

Dico ergo quod si materia potest esse una numero in duobus, eadem ratione in tribus et in infinitis. Nam consimile argumentum facit Aristoteles in capite de vacuo in quarto physicae, dicens[1], quod si aliquid unum et idem potest esse in duobus, tunc potest esse in tribus, et sic in infinitis; sed quod potest esse in infinitis habet infinitam potentiam. Ergo materia est infinitae potentiae quare et essentiae infinitae, ut probabitur, et ideo erit Deus. Sed consequentiae dictae non possunt contradicere, scilicet quod sequatur quod materia eadem possit esse in infinitis si essent, et ideo concedunt ei potentiam infinitam sed negant essentiae infinitatem, quia nihil habet essentiam infinitam nisi Deus. Et cum arctantur in hac parte circa infinitatem potentiae, incipiunt verbis vacillare, dicentes quod potentia materiae est infinita, sicut potentia continui. Sed istud nihil est, quia haec potentia non est respectu actus infiniti, nec respectu infinitorum simul et in actu considerandorum, et ideo haec potentia non est actu infinita, sed potentia tantum. Sed potentia infinita in tantum datur materiae secundum dictam positionem, ut si essent actu infinita, posset materia eadem numero esse actualiter in illis infinitis. Ergo actualiter et intensive habet potentiam infinitam. Caeterum potentia continui non est similis potentiae divinae: quia potentia divina non habet arctationem et limitationem, quantum ad actualem existentiam in pluribus et in infinitis. Ad quam non se extendit potentia continui, sed materiae conceditur illa eadem existentia in pluribus, et sic in infinitis, per auctoritatem et demonstrationem Aristotelis. Quapropter potentia materiae non est similis potentiae continui sed divinae.

Caeterum aliter vacillant in hac parte dicentes, quod potentia materiae est passiva, et potentia divina activa. Hoc enim excluditur multis modis. Nam potentia passiva dicitur respectu transmutationis suscipiendae ab agente. Sed potentia

[1] *Nat. Auscult.* iv. 8, § 13.

existendi in pluribus non hoc respicit, et ita differunt. Item potentia existendi in pluribus simul quae non est arctata nec limitata attestatur summae nobilitati, et ideo datur Deo et animae quae est in pluribus partibus corporis tota. Sed Deo et animae non datur propter aliquam rationem patiendi, ergo nec materiae. Item quaecunque sit haec potentia, sive activa sive passiva, oportet quod substantia materiae sit infinita si haec potentia est infinita; et si hoc, tunc materia est Deus vel aequalis Deo, quia nulla res habet substantiam et potentiam infinitam nisi Deus.

Mathematical disproof.

Sit ergo potentia infinita *A* linea, et linea *B* sit ejus finita substantia, et *C* linea sit potentia aliqua finita, quae sit pars potentiae infinitae. De omni enim quanto dato contingit partem per intellectum abscindere, ut vult Aristoteles primo Coeli et Mundi, et *D* linea sit substantia ejus quod est *C* potentia; multiplicetur ergo *D* quousque aggregetur substantia quae sit aequalis ipsi *B*, hoc enim est possibile, quoniam *D* et *B* sunt finita, et haec substantia sic aggregata sit *E* linea. Similiterque multiplicetur potentia *C* in tantum ut excrescat potentia proportionalis ipsi *E*, et sit *F* linea, hoc enim est possibile quia *C* et *D* sunt finita, et ideo potest ad utrumque addi in tantum, ut aggregata sint proportionalia, sicut *C* et *D*. Sed *F* potentia est finita, quia ex aggregatione finita nascitur, et habet substantiam *E* aequalem ipsi *B* substantiae, cui correspondet *A* quod est potentia infinita; ergo potentia *F* finita et *A* potentia infinita habebunt aequales substantias. Quare finitum et infinitum aequabuntur, quia aequalibus substantiis debentur aequales potentiae, et pars toti aequabitur, quoniam *F* est pars *A* sicut *C*, ut positum fuit a principio de *C*, et *F* crescit ex aggregatione finita *C*, quare *F* erit pars *A*, et ita pars aequabitur toti.

A	Potentia infinita.
B	Substantia ejus finita.
C	Potentia finita pars ipsius *A*.
D	Substantia potentiae *C*.
E	Substantia aequalis ipsi *B* substantiae
F	Potentia proportionalis ipsi *E*.
G	Potentia major quam *F*.

Et ex hoc ulterius sequetur, quod pars habebit majorem

essentiam, et plus de essentia quam totum. Quoniam accipiatur aliqua potentia finita major ipsa *F*, et resecetur a potentia infinita quae est *A*, et sit *G* linea. Oportet ergo quod *G* potentia, cum sit major quam *F*, habeat plus de substantia, seu fundetur in majori et nobiliori essentia, quam fundatur *F* potentia. Sed *F* habuit aequalem essentiam ipsi *A*, ut probatum est. Ergo *G* habebit majorem quam *A*; et ita parti respondet plus de essentia et nobilius quam toti, quod est impossibile. Et etiam sequetur, quod finitum habebit plus de essentia sibi respondente quam infinitum; et omnia haec sunt impossibilia.

Item nullum infinitum potest habere potentiam finitam. Ergo per oppositum nullum finitum potest habere potentiam infinitam. Antecedens probatur per consimilem demonstrationem priori, quia sit *a* linea illud infinitum, et *b* linea sit illa potentia finita. Accipiatur aliquod finitum vel pars ipsius *a* et sit *c* linea. Ergo habebit minorem potentiam; et sit *d* linea. Multiplicetur ergo *d* quousque aequetur ipsi *b*, quod possibile est, quia utrumque est finitum; et ideo tantum potest addi ad *d*, ut aggregatum sit aequale potentiae quae est *b*; et sit illud aggregatum *e* linea. Similiter ergo multiplicetur *c*, et addatur ei quousque poterit habere potentiam illam quae est *e*, et illud aggregatum sit *f* linea; hoc enim bene possibile est, quoniam *c* substantia et *d* potentia ejus sunt finitae, et ideo si ad *d* tantum potest addi, ut fiat *e* potentia major quam *d* ad *c*, tantum potest addi de substantia ut aggregatum excrescat, cui *e* potentia erit proportionalis, et illud aggregatum habebit illam potentiam. Sed *e* potentia est aequalis potentiae *b*, ut probatum est. Ergo *f* quod est finitum, quia ex aggregatione finiti et infiniti nascitur, habebit potentiam aequalem ipsi *a*, quod est infinitum. Quod patet esse omnino impossibile, et sequetur ex hoc, quod pars aequetur toti, scilicet *F* ipsi *A*, et etiam

a	Substantia infinita.
b	Potentia finita.
c	Substantia finita pars ipsius *a*.
d	Potentia ipsius *c*.
e	Potentia aequalis ipsi *b* potentiae.
f	Substantia potentiae *e*.

quod pars est major toto, ut patuit in priori ratione; et quod finitum habebit majorem potentiam quam infinitum, ut patet in terminis, et haec omnia sunt impossibilia.

Item si potentia materiae est infinita, sit haec potentia per *A* lineam designata infinitam, quae si dividatur in puncto *a*, tunc partes divisae sunt aequales, quia utraque ab ipso puncto *a* vadit in infinitum. Signetur ergo punctus *b*, et puncta *c*

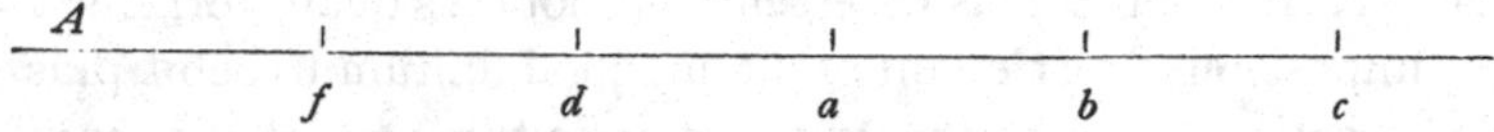

et *d*. Partes ergo divisae sunt aequales, sed *b a f* in infinitum decurrens est major quam *a f* per *a b*. Ergo *b c* linea in infinitum extensa, est major quam *a f*. Ergo est major aequali *a f*, quod est *a b c*. Ergo pars est major suo toto. Quapropter impossibile est potentiam materiae esse infinitam. Nec est haec demonstratio applicanda divinae potentiae, eo quod non est divisibilis, sicut potentia materiae corporalis. Sicut vero radices a parte efficientis traxi ad exempla rerum naturalium, sic similiter possunt aliqua notari quae consequuntur materiam in rebus mundi, in quibus miranda geometriae potestas elucescit. Nam si materia mundi una esset numero, oporteret necessario, quod forma esset una numero, sicut posuerunt Parmenides et Melissus, contra quos Aristoteles primo libro Physicae et libro de generatione se opponit. Certo enim ex unitate materiae numerali sequitur illa potentia falsissima, quam prius eliminavi per leges fractionum.

Capitulum IX.

An corpora se tangant in puncto.

Are the celestial spheres continuous? A line traversing their boundaries

Quoniam autem pro hac positione sunt geometricae falsigraphiae, quae omnem hominem possunt perturbare, ideo volo aliquas inferre. Si enim corpora sint diversa in hoc mundo, ut coelum et elementum, et imaginemur lineam transire per medium eorum, non secabunt illam lineam in diversis punctis, quia tunc longitudo linearis esset in medio, et ita superficialis latitudo et profunditas corporalis, et sic corpus tertium a coelo et elemento esset inter ea, quod falsum est. Ergo in eodem puncto intersecabunt lineam illam. Sed continua sunt quorum terminus est idem. Ergo sunt continua. Multorum et maximorum virorum fuerunt hic responsiones variae, et omnes adhaerebant isti parti, quod in diversis punctis intersecarent vel naturaliter vel mathematice, vel utroque modo, distinguentes inter continuitatem naturalem et mathematicam secundum Averroem in quinto Physicae. Sed istud nihil est; necesse enim est quod in eodem puncto lineae intersecent se, ut sit virga ferrea imaginata; sed ille punctus non continuat corpora secantia, et ideo non sunt continua, sed partes virgae vel lineae transeuntis, de quarum continuatione non potest concludi corpora esse continua per quae transit virga; sed si in diversis punctis virgae daremus sectionem, tunc necessario pars virgae, quae est corpus, esset media; et hoc est impossibile. Item argui potest subtilius, quod quibuscunque corporibus in hoc mundo positis, si sint diversa et contigua, non continua, supraponatur vel supponatur tertium corpus, ad quod aequidistanter ducantur lineae duae a mediis punctis datorum corporum. Si ergo illae lineae terminentur ad puncta diversa in tertio corpore, tunc pars illius corporis est media inter eas. et aequaliter semper distant et appropinquant, et ex altera parte in medio duorum corporum distabunt per corpus interpositum, quod omnino est falsum. Ergo continuabuntur ad unum punctum in tertio corpore, et eadem ratione ad extremitates suas in conjunctione corporum positorum, quapropter unum corpus erunt quae posita sunt, quia unus punctus non potest esse in diversis corporibus. Et hic turbati sunt multi. nec est tempus renovandi opiniones singulorum, praecipue cum veritas pateat. Si dicatur quod in eodem puncto corporis terminabuntur illae lineae nec sint aequidistantes, quia non distant in aliquo, eo quod corpora simul posita sunt sine medio, et quia omnis distantia tollitur, oportet quod ad unum punctum in tertio corpore terminentur, et quoniam duo puncta non habent aliquid longitudinis, quoniam non sunt quanta, ideo non faciunt distantiam aliquam sicut nec unus punctus. Unde tantam indivisionem important duo puncta simul juncta, sicut

is cut in two adjacent points.

But two such points occupy no

more space than one.

unum per se positum, quia puncto addito puncto non crescit quantitas, et ideo possunt remanere diversa puncta in extremitatibus duarum linearum simul junctarum in corporibus duobus sine medio, quamvis ad unum punctum in tertio corpore terminentur. Sed si in illis corporibus simul junctis ducantur duae lineae ab illis punctis infra corpora et descendat una linea super extremitates earum ad angulos rectos, necesse est per xiv primi Elementorum Euclidis quod lineae extensae in corporibus sint una linea continua. Ergo et corpora similiter, nam talis est sententia illius propositionis.

A mathematical line and a physical line differ only in the point of view.

Et ex hac propositione deceptus fuit Averroes quinto Physicae, et omnes sequentes ipsum, volentes quod una sit linea mathematica, et per consequens unum corpus mathematicum, quamvis diversa sint naturaliter. Nam, ut dicit ibi, contiguitas naturalis transit in continuitatem mathematicam ; sed hoc est impossibile, nam quantitas mathematica et naturalis idem sunt secundum esse, et secundum rem, sed differunt solum secundum considerationem, quia geometer considerat lineam naturalem, non prout est in naturali materia, et ideo dicitur mathematica. Et hanc eandem lineam considerat philosophus naturalis, ut est in materia naturali, sicut in ferro, vel lapide, vel alia re naturali. Et quia eadem res est secundum esse, et secundum veritatem existendi, naturalis et mathematica, ideo si hic esset linea una vel corpus unum mathematice, tunc eodem modo esset naturaliter. Dico ergo quod hic sunt lineae secundum veritatem diversae et corpora similiter, nec hoc repugnat intentioni Euclidis. Nam non vult quod sit una continua, sed quod quantum ad angulos rectos faciendos duae tantum faciunt quantum una, postquam illae duae jacent in eadem directione longitudinis ac si ibi essent duae partes ejusdem lineae continuatae ; et sic intendit in pluribus aliis locis ubi accipit unam pro duabus, cum idem faciat ad suum propositum una sicut duae. Et hoc est quando loquitur de communi differentia superficierum vel communi sectione, eam enim vocat unam lineam, cum tamen in veritate sint duae : sed una obtinet earum potestatem, et ideo facilius loquimur de una, quam de duabus.

Si tamen objiceretur, quod si quis ponat diversitatem in

corporibus, accipiantur igitur duae tabulae planae superficiei et circularis figurae, et elevetur una ab altera sub eadem figuratione, tunc aer citius repleret partes exteriores inter eas quam centrales. Quapropter esset vacuum in partibus interioribus ad tempus. Et hic multae stultitiae solebant dici. Nam aliqui dixerunt aerem in instanti moveri usque ad centrum ex lege naturae universalis, ne fieret vacuum; alii quod omnia corpora tangentia se in aere vel aqua habent superficies humidas, ut Aristoteles dicit secundo de Anima, et hanc humiditatem aeris vel aquae interceptam posuerunt rarefieri subito per totum, ne fieret vacuum. Sed hoc quod dicitur, ne fiat vacuum, est pura negatio[1]. Nulla vero negatio pura est causa affirmationis; et ideo responsiones hujusmodi sunt falsae. Propter quod dicendum est quod sub eadem figuratione non potest una ab alia elevari, sed oportet quod declinet una cum elevatur ab alia, et sic aer ingreditur paulatim. Hoc de plano potest quilibet experiri in scypho vitreo demisso in vas plenum aqua; nam pro mundo non potest elevari in eadem figuratione suarum partium. Cujus causa est, ut aqua ingrediatur ejus locum paulatim. Et haec est affirmativa ad quam ex consequenti excluditur vacuum.

Refutation of atomic theory of Democritus.

Si ergo oportet quod sint plures materiae corporales et plura corpora in hoc mundo, cum quodlibet corpus est divisibile in infinita, non tamen propter hoc mundus erit compositus ex partibus materialibus infinitis, quae vocantur atomi, ut posuerunt Democritus et Leucippus, quorum positione fuit Aristoteles magis impeditus et omnes naturales, quam per aliquid aliud erroneum. Quod tamen per geometricam potestatem eliminatur omnino; nam nihil fortius potest argui contra hoc, quam quod tunc diameter quadrati et latus essent commensurabilia, id est, haberent communem mensuram, scilicet aliquam partem aliquotam pro communi mensura, cujus contrarium Aristoteles semper docet. Et patet per demonstrationem ex ultima parte septimae propositionis decimi libri Elementorum, per quam demonstratur, quod si aliqua mensura

[1] A striking instance of Bacon's positivity. No natural action takes place in order that some other action may not follow. The remark is repeated in the next chapter.

ut pes vel palmus mensuret costam, non mensurabit diametrum, nec e contra; ut si diameter sit ex decem pedibus non erit costa ex aliquot pedibus. Et non solum sequitur ex hac positione, quod essent commensurabilia, sed aequalia. Quod patet evidenter in hac figura. Nam si latus habeat decem atomos, vel duodecim, vel plures, tunc trahantur tot lineae ab illis atomis ad totidem in opposito latere, et latera quadrati aequalia sunt; quapropter tot lineae occupabunt totam superficiem quadrati[1]; et ideo cum diameter transit per illas lineas, et non possunt plures assignari in quadrato, singulas capiet atomos a singulis lineis, et ideo non plures erunt in diametro, quam in costa, et sic habent partem aliquotam pro communi mensura, et tot partes habet costa sicut diameter, quorum utrumque est impossibile.

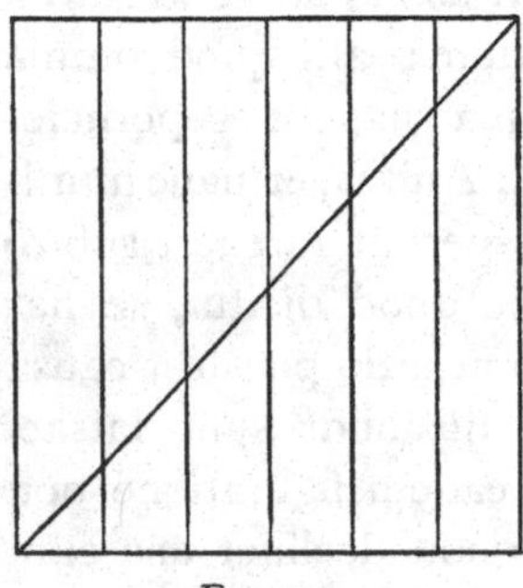

FIG. 14.

CAPITULUM X[2].

De figura mundi.

Proof that the world is spherical.

Quoniam vero necesse est corpora mundi esse plura, et divisibilia, et quanta, oportet quod figurationem habeant debitam, ad hoc ut mundus consistat. Figuratio vero est passio materiae, et invenitur in rebus ratione materiae, sicut et quantitas. Nam figura uno modo est quantitas clausa lineis; alio modo dicitur ipsa clausio quantitatis. Necesse est vero mundum extra habere figuram sphaericam. Nam quaecunque alia detur, accidet vacuum vel possibilitas ad

Revolution of any other figure would create a vacuum.

[1] The fallacy lies in supposing that lines could occupy superficial space; as Bacon would have seen if he had tried to occupy the surface of the square with small circles or squares.

[2] Cf. *Opus Tertium*, cap. 40.

ipsum. Sed natura nec sustinet vacuum, nec possibilitatem respectu ejus. Si enim esset alicujus figurae angularis, tunc de necessitate accideret vacuum in moto suo: nam ubi modo esset unus angulus nihil esset donec alius angulus ibidem veniret[1]. Figurae autem aliae quae maxime competerent essent vel ovalis figurae vel consimiles ei vel lenticularis et ei conformes secundum Aristotelem, libro Coeli et Mundi. Sed negat coelum habere hujusmodi figuram, sed causam non exprimit. Figura vero lenticularis est figura illius leguminis quod vocatur lens. Habet enim laterales superficies gibbosas, deficiens a vera sphaericitate propter breviorem diametrum quae transit per illa latera. Si vero esset figurae ovalis, aut pyramidalis rotundae, aut columnaris, aut alicujus hujusmodi, et moveretur super breviorem diametrum, adhuc accideret vacuum in actu, si vero super longiorem, non accideret vacuum in actu, sed possibile esset vacuum accidere, quoniam tanta esset possibilitas in mundo, quantum ad figuram ut moveretur super breviorem diametrum, sicut super longiorem. Si vero esset lenticularis figurae vel casealis, vel hujusmodi, et moveretur super longiorem diametrum, accideret vacuum in actu, et si super breviorem, non accideret in actu sed possibilitas relinqueretur. Nam aeque possibile esset mundo, quantum est de se, ut moveretur super unam diametrum sicut super aliam. Nec potest dici quod si esset ovalis figurae moveretur semper super longiorem, et si lenticularis moveretur super breviorem, ne vacuum accideret. Refellendum est istud sicut superius, scilicet quod impossibile est quod pura negatio sit causa alicujus affirmationis; sed, ne sit vacuum, pura negatio est. Et ideo oportet quod mundus sit sphaericae figurae, in qua sola corporali sunt omnes diametri aequales, ut possit libere volvi secundum omnem diametrum, et sic nullum sequitur inconveniens. Similiter intra necesse est, ut sit sphaericae figurae et concavae: nam planae figurae non potest. Quoniam

Equality of radii of sphere.

[1] This is from Aristotle. *De Coelo*, ii. 4 εἰ γὰρ ἔσται εὐθύγραμμος, συμβήσεται καὶ τόπον ἔξω εἶναι καὶ σῶμα καὶ κένον. Κύκλῳ γὰρ στρεφόμενον τὸ εὐθύγραμμον οὐδέποτε τὴν αὐτὴν ἐφέξει χώραν, ἀλλ' ὅπου πρότερον ἦν σῶμα, νῦν οὐκ ἔσται, καὶ οὗ νῦν οὐκ ἔστι, πάλιν ἔσται διὰ τὴν παράλλαξιν τῶν γωνιῶν. Ὁμοίως δὲ κἂν εἴ τι ἄλλο σχῆμα γένοιτο μὴ ἴσας ἔχον τὰς ἐκ τοῦ μέσου γραμμάς, οἷον φακοειδὲς ἢ ᾠοειδές.

ducantur a centro terrae tres lineae et una ducatur perpendiculariter ad superficiem coeli, ut *a b*, illa erit brevior aliis per xviii primi Elementorum et per xxxii ejusdem. Ergo coelum non aequaliter distabit a terra: sed oportet ut sit ejusdem naturae in omni parte. Ergo quaelibet nata est elongari aequaliter a terra. Nec potest esse convexa interius,

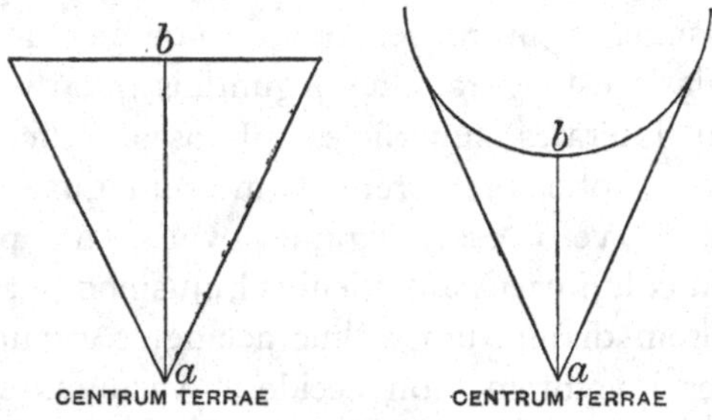

FIG. 15.

ut patet per viii tertii libri, quae dicit quod a puncto extra signato si ducantur ad circulum plures lineae, illa quae caderet in diametrum esset brevior aliis. Ergo relinquitur, quod si a centro terrae ducantur tres lineae ad convexitatem coeli, una erit brevior, scilicet illa, quae est perpendicularis super sphaeram, ut patet in figura, et ideo coelum non aequaliter distaret a terra, quod tamen oportet ut dictum est. Et iterum si esset mundus convexae figurae, intra non esset natus continere omnia: sed mundus natus est continere omnia. Quod si non potest esse planae figurae, nec convexae, oportet quod sit concavae, cum non sit alia. Sed concava potest esse multis modis, aut sphaerica, aut columnaris, aut pyramidalis, aut lenticularis, aut alia multiplex. Non est autem possibile, quod sit alicujus nisi sphaericae concavae, propter hoc quod in hac sola figura sunt omnes lineae aequales, quae ab uno puncto ducuntur ad superficiem. Non enim est in aliis possibile dare punctum, a quo omnes lineae ductae ad superficiem aequentur: nam diametri sunt inaequales. Sed oportet partes coeli aequaliter distare a terra, propter naturae aequalitatem. Ergo de necessitate erit sphaericae.

Item inter omnes figuras isoperimetras sphaera ipsa maxime

capit, sicut proponit viii propositio libri Isoperimetrorum[1]. Superficiales vero figurae dicuntur isoperimetrae, ut triangulus et quadrangulus, et circulus, quando latera trianguli extensa in continuum et directum tantum habent in longitudine, quantum latera quatuor quadranguli extensa et quantum circumferentia circuli si extenderetur, et sic de quibuscunque figuris superficialibus. Unde dicitur isoperimeter ab ἴσον quod est aequale et περί quod est circum, et μετρόν quod est mensura, quasi aequalis circummensurationis. Et inter omnes istas superficiales isoperimetras circulus maxime capit, sicut dicit vii propositio de isoperimetris. Corporales vero tunc dicuntur isoperimetrae, ut sphaera, cubus, et columna, et quaecunque, quando superficies sphaerae extensa in continuum et directum tantum habet in longum et latum quantum superficies sex ipsius cubi, et quantum superficies columnae rotundae, et sic de aliis. Sed inter omnes istas sphaera maxime capit, sicut demonstratur in libro supradicto. Cum ergo coelum debet omnia continere, oportuit quod esset sphaericae figurae. Item nobilitas mundi, et hujus figurae dignitas correspondent. Nam haec figura est prima figurarum corporalium, quia una superficie contenta, omnes autem aliae habent plures. Ergo competit corpori primo ut coelo. Item haec est simplicissima, quoniam sine angulis et cono et lateribus et omni diversitate. Ergo debetur corpori simplicissimo, quod est coelum. Item aptissima est motui. Ergo debetur primo mobili. Item est elongata ab occasionibus et impedimentis, quia non habet angulum in quem aliquid offendat. Ergo maxime competit corpori quod impedimentum et occasionem offensionis capere non potest. Item est perfectissima, quia nihil addi potest ei; sed omnibus aliis potest aliquid addi. Ergo corpori debetur perfectissimo.

Of figures of equal surface the sphere has the greatest content.

Spherical form adapted for motion.

Quod autem corpora contenta in coelo habeant figuram sphaericam, hoc demonstratur de aqua, quae jacet in medio,

The elements within the

[1] The author of this book is Zenodorus, who probably belongs to the generation succeeding that of Archimedes. Its principal propositions, fourteen in number, are to be found in the fifth book of Pappus, and also in Theon's *Commentary on the Almagest of Ptolemy*. See Cantor, vol. i. pp. 308–9, and 379–80. Cantor shows, vol. i. pp. 605 and 635, that Zenodorus was known to the Arabs.

heavens of similar form.

Water.

ut per consequens pateat de aliis. Ducantur lineae undique ad superficiem aquae a centro terrae, planum est quod aqua semper currit ad inferiorem locum propter suam gravitatem ut videmus. Ergo si una illarum esset brevior altera, aqua curreret ad extremitatem illius donec aequaretur. Ergo omnes lineas ductas undique a centro mundi ad superficiem aquae aequari necesse est. Sed ad planum aequari non possunt, per xxviii et xxxii primi Elementorum, ut superius dictum est, nec ad convexam per viii tertii. Ergo oportet, quod superficies aquae continens terram sit concava, et non cujuscunque concavitatis, sed sphaericae, quoniam in sola illa figura omnes diametri sunt aequales. Et haec demonstratio non solum tenet de aqua interius, sed exterius. Nam exterius fluit ad inferiorem locum semper sicut interius. Et ideo oportet, quod sit convexa exterius, nam neque ad planam neque concavam exterius possent omnes lineae ductae a centro esse aequales, secundum formam demonstrationis prioris, et hoc patet per experimentum. Nam sit navis *g d*, et portus *a*, et *c* sit superficies navis ubi figitur malus, et *b* sit extremitas mali, et ducatur *c a* linea perpendiculariter a portu ad extremitatem mali. Planum ergo est per xix et xviii primi Elementorum quod *a b* linea est longior quam *c a*. Ergo si mare esset planae figurae, tunc oculus existens in *c* videret portum melius, quam existens in *b*, quoniam *b* plus distat ab *a* quam *c*. Sed per experientiam scitur, quod ille qui est in summitate mali potest videre portum citius quam ille qui est in superficie navis. Ergo relinquitur quod aliquid impedit visum illius qui est in navi. Sed nihil potest esse, nisi tumor sphaericus aquae. Ergo est sphaericae figurae. Sed si hoc, tunc terra est sphaericae figurae convexae, nam aliter non elongaretur aequaliter a coelo, neque appropinquaret centro mundi aequaliter; sed hoc oportet fieri. Item esset vacuum ubi se non contingerent: quoniam si esset planae vel concavae non contingeret concavitatem aeque, ut patet, et ideo vacuum esset inter eas.

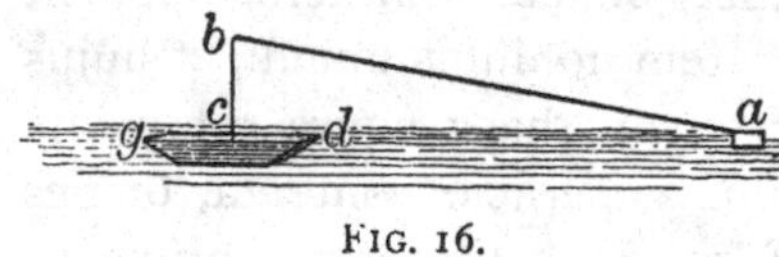

FIG. 16.

Similiter de sphaericitate aeris concava intra et extra, omnino negotiandum est sicut in prima demonstratione de aqua; quia aer est gravis[1] in sphaera sua sicut aqua, ut prius habitum est, et ideo currit ad inferiorem locum. Et praeterea vacuum esset inter aerem et aquam, si aer esset alterius figurae. Nam, si esset planae, tunc non tangeret aerem nisi in puncto, per tertiam primi libri Theodosii; quia sphaera non tangit planum nisi in puncto, ut ibi dicitur. Si vero convexae, iterum non tangit nisi in puncto per duodecimam tertii Euclidis. Nam signentur duo circuli in sphaera aquae convexa, et aere convexo, non se tangent illi circuli nisi in uno puncto, sicut probatur in illa duodecima. Ergo nec corpora in quibus signantur; nam si corpora se tangerent in pluribus locis, et illi circuli, ut patet ad sensum. Et ita oportet, quod utroque modo sit vacuum. Et si esset alterius concavitatis quam sphaerae, non omnes lineae ducerentur ab uno puncto ad superficiem corporis aequales. Igitur oportet superficiem aeris esse sphaericam intra et extra. Air.

Deinde patet per has demonstrationes ultimas de igne, sicut de aere. Sed illa quae vadit per descensum ad inferiorem locum non tenet hic, quia ignis non est gravis in sphaera sua. Oportet tamen omnes lineas ductas a centro mundi ad superficiem ignis esse aequales, quia ignis est contrarius terrae et summe leve, et ejusdem naturae in toto et partibus singulis. Quapropter aequaliter elongatur a loco gravium, et a centro mundi, et hoc exterius et interius. Ergo oportet quod sit concavus intra, et convexus extra, et hoc sphaerice, sicut priora concludunt. Fire.

CAPITULUM XI.

Quod plus aquae contineat vas inferiori quam superiori loco positum.

Sed nunc per figuram aquae magnum naturae miraculum A vessel filled with

[1] The view as to the gravity of air deserves notice. It was maintained however by Aristotle. Cf. *De Coelo*, iv. 4, § 5. Ether, or fire, on the contrary, was regarded as imponderable, as the next paragraph shows.

water varies in content according to distance from earth's centre.

potest suscitari; quoniam si scyphus continens aquam ponatur in loco inferiori, poterit plus capere de aqua, quam in loco superiori, ut in cellario et solario. Nam propter inclinationem naturalem aquae ad centrum mundi ubicunque sit, sive in loco superiori, sive inferiori, partes ejus semper currunt ad inferiorem locum; et ideo per lineas aequales semper distant a centro, et ideo oportet semper quod pars superior aquae sit portio unius sphaerae describendae circa centrum mundi, licet in fundo scyphi retinet figuram vasis, quia ibi solum tangit vas et non superius. Quare pars suprema figurabitur secundum legem gravitatis aquae, et hoc est respectu centri mundi, et ideo superior pars erit portio sphaerae imaginandae circa centrum mundi. Sed constat quod in loco inferiori erit portio minoris sphaerae et in superiori portio majoris, quia magis tunc distabit a centro; nam sphaera superior continebit inferiorem, ut patet in circulis circa idem centrum. Diameter autem scyphi erit chorda utriusque portionis, si scyphus impleatur utrobique quantum potest capere. Ergo illa diameter resecabit de majori sphaera minorem portionem, et de minori majorem. Et per xxxviii propositionem triangulorum Jordani[1], in circulis inaequalibus eadem chorda resecat de majori circulo minorem portionem, et de minori majorem, et ita erit de sphaeris.

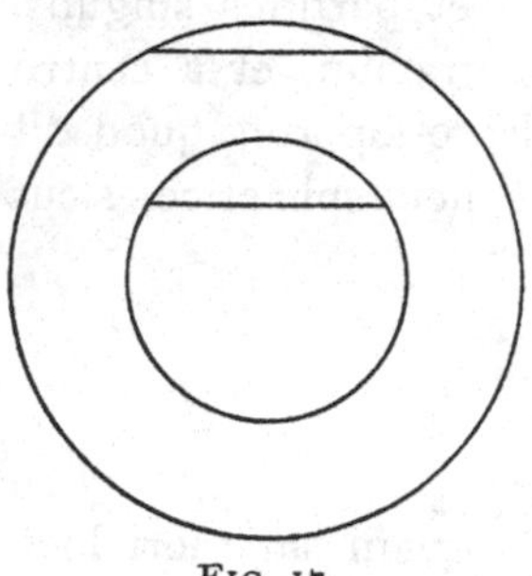
Fig. 17.

Nam signentur illi circuli in sphaeris illis, et portiones circulorum in portionibus sphaerarum, et patet quod idem est. Ergo portio aquae super diametrum scyphi erit major, quando vas est in inferiori loco, quam quando est in superiori, et ideo gibbositas major et tumor altior; quare oportet quod plus de aqua ibi sit, si scyphus sit omnino plenus, quam quando est in superiori loco. Quapropter ad eandem aquam potest plus infundi de aqua in scyphum, quando est inferius, quam quando est superius. Nam inferius aqua quae est super

[1] See note on this writer in chap. 16, p. 169.

diametrum scyphi contrahet se a lateribus vasis, et coangustabit se in portionem minoris sphaerae, quia propinquior est centro mundi, et ideo chorda ejusdem aquae fiet minor, quam quando fuit superius. Et erit pars diametri scyphi ejus chorda, non tota diameter, sed ex utroque latere abscindetur aliquid de diametro, ut residuum fiat chorda portionis aquae. Quapropter in lateribus aquae a loco abscissionis diametri usque ad vas erunt duo spatia parva, ubi poterit plus de aqua infundi quam quando est in superiori loco. Et hoc totum facit inclinatio aquae secundum legem suae gravitatis figurantis se secundum sphaeram minorem et majorem respectu centri mundi; cum tamen vulgo studentium videatur hoc esse omnino impossibile, imo magnis viris qui geometriae nesciunt potestatem.

CAPITULUM XII.

An figurae quinque corporum regularium mundo conveniunt, ut voluerunt Platonici.

Plato's view of correspondence of heaven and four elements to the five regular solids.

Ex istis vero figurationibus sphaericis corporum mundi aperitur magnum fundamentum in certificatione rerum naturalium, et evacuantur violentae falsitates. Nam Platonici [1], in quorum tempore viguit geometria, ut Averroes dicit tertio Coeli et Mundi, aestimaverunt quod corpora mundi principalia, scilicet coelum et quatuor elementa, figurarentur figuris quinque corporalibus quae regulares vocantur, et sunt aequiangulae et aequilaterae et inscriptibiles spherae et circumscriptibiles eidem, et nullae aliae. Et propter has causas sunt nobilissimae figurarum praeter sphaeram, quas lator praesentium de facili potest praesentare, et sunt tres ex superficiebus triangularibus, et quarta ex quadratis, et quinta ex pentagonis; et non possunt esse plures, quod est mirabile. Prima habet quatuor superficies triangulares, et vocatur tetraedrum a τετράς quod

[1] See *Timaeus*, 54. It does not appear that Bacon had read Plato.

est quatuor, seu pyramis quatuor basium triangularium. Secunda habet superficies octo triangulares, et ideo vocatur octaedrum, nam octo purum est Graecum, non Latinum. Tertia habet viginti superficies triangulares, et vocatur icosaedrum, ab εἴκοσι quod est viginti. Et non possunt esse plures figurae regulares ex basibus triangularibus. Nam nullus angulus corporalis potest valere quatuor rectos superficiales, ut docet xxi undecimi libri Elementorum. Sed sex anguli triangulorum aequiangulorum valent quatuor rectos: nam tres valent duos rectos, ut patet ex xxxii propositione primi libri quae vulgata est. Et ideo sex anguli triangulorum non possunt componere angulum corporalem; et ideo nec figura corporalis potest fieri ex superficiebus triangularibus, quarum sex anguli semper concurrant ad angulum unum corporalis figurae. Sed bene potest fieri, quod quinque anguli vel tres triangulorum constituant minus quam quatuor recti. Et si ex tribus angulis triangulorum constituatur angulus corporalis, tunc oportet quod sint quatuor superficies triangulares in corpore illo. Et si ex angulis quatuor triangulorum fiat angulus corporalis, tunc oportet quod sint octo trianguli in figura corporali. Si vero quinque anguli triangulorum faciunt angulum corporalem, tunc oportet quod in figura corporali sint viginti superficies triangulares undique, ut patet ad sensum in figuris corporalibus. Ex duobus autem angulis superficialibus non potest fieri angulus corporalis, quia omnis angulus talis est ad minus ex tribus superficialibus, ut dicit Euclides in principio xi; et quilibet geometer scit hoc. Ergo tantum tria corpora regularia erunt ex triangulis. Ex quadratis vero non potest esse nisi unum; nam angulus quadrati rectus est, et ideo tres tantum tales congregati possunt facere angulum corporalem, quia si quartus addatur jam non potest esse angulus corporalis, quia omnis angulus solidus est minor quatuor rectis. Sed si tres anguli quadratorum concurrant ad angulum solidum, tunc in corpore constituto erunt sex superficies quadratae, ut est in taxillo; et vocatur haec figura cubus et hexaedrum ab ἕξ Graece, quod est sex Latine. Si vero accipiantur anguli pentagonorum regularium, tunc tres faciunt angulum solidum, et non plures, quia si quatuor acciperentur, jam esset plus quam quatuor recti, eo quod angulus pentagoni est et rectus et quinta recti, ut patet ex xxxii propositione primi Elementorum, et sic essent ibi quatuor recti et quatuor quintae. Sed omnis angulus corporalis est minor quatuor rectis. Ergo tantum erit una figura corporalis ex pentagonis superficiebus, et oportet quod habeat duodecim superficies pentagonas, sicut patet in fabricatione illius corporis, et hoc vocatur dodecaedrum, id est, figura corporalis duodecim basium pentagonarum aequilaterarum.

Et omnes istae figurae corporales regulares fiunt ex superficiebus regularibus quae sunt aequiangulae et aequilaterae. Non est autem possibile quod ex superficiebus hexagonis fiat aliqua figura corporalis regularis, quia nullus angulus corporalis potest fieri ex angulis talium hexagonorum, propter hoc quod angulus talis hexagoni valet rectum et tertiam recti, ut patet ex xxxii primi Elementorum. Ergo tres tales faciunt tres rectos et tres tertias unius recti, sed tres tertiae valent integrum. In nulla enim re possunt esse nisi tres tertiae. Ergo tres anguli hexagonorum regularium valent quatuor rectos, sed nullus corporalis valet quatuor rectos. Ergo nulla figura corporalis potest fieri ex superficiebus hexagonis, et longe minus ex heptagonis et octogonis et supra, quia majores angulos obtinent quam hexagoni.

Quoniam vero dodecaedrum patitur in se inscriptionem omnium aliarum, ut patet ex quinto decimo libro Elementorum, ideo dederunt Platonici hanc figuram partibus coeli, propter hoc, quod coelum noscitur omnia continere. Unde dixerunt partes coeli concurrere in punctum unum secundum hanc figuram, ut constituatur corpus coeli. Et quia ignis ascendit in figura pyramidali, ideo dederunt partibus ignis talem figuram. Quoniam autem octaedrum maxime assimilatur pyramidi, et aer praecipue assimilatur igni, dederunt ei figuram illam. Sed quoniam partes aquae revolvuntur anfractibus et fluxibus multifariis, ideo dederunt ei figuram icosaedrum, quae multiformitate laterum et angulorum circumvolvitur in sphaera. Partibus vero terrae cubum dederunt, quia illa figura stabilis est et fixa inter omnes, sicut terra inter mundi corpora stabilitatem obtinuit et fixionem.

Aristotle's objection that vacua would be left.

Sed Aristoteles venit contra istos in tertio Coeli et Mundi[1], et probat, quod ex hac figuratione vacuum erit in sphaera aquae, et in sphaera aeris, et in sphaera coelesti. Nam solus cubus et pyramis possunt locum replere, quando congregantur circa punctum unum. Nam replere locum est duobus modis; uno, sicut dicimus vulgariter, quod omne corpus replet locum, nisi sit instantia in ultimo coelo, et sic non accipitur repletio loci in proposito. Alio modo dicitur esse repletio loci non solum corporaliter, sed superficialiter, et magis quia paucitas superficierum implentium sua loca est causa paucitatis corporum replentium sua loca, ut dicit Averroes super tertium Coeli et Mundi. Replere vero spatium superficiale est implere quatuor angulos rectos, quia non possunt plures esse circa punctum unum in superficie, ut patet ex intersectione duarum linearum ad angulos rectos hoc modo. Et sic quadrata propriissime possunt locum replere superficialem, scilicet quatuor quadrata, quia angulus quadrati rectus est, et sex trianguli, quia sex anguli tales valent quatuor rectos, et tres hexagoni, quia eorum anguli tres valent quatuor rectos, ut prius tactum est. Sed pentagoni non possunt locum replere, quia tres anguli eorum valent tres rectos et tres quintas, et hoc totum minus est quatuor rectis, et quatuor anguli pentagonorum valent quatuor rectas et quatuor quintas, quod est plus quatuor rectis, et ideo haec figura regularis non potest locum replere. Similiter nec heptagonus, nec aliqua alia, ut patet. Et ideo tantum tres superficiales locum replent. Et propter hoc paucae erunt corporales, quae locum replebunt corporaliter, et hujusmodi repletio corporalis non est per unum corpus, sed quando plura corpora congregantur circa unum punctum undique, ita ut impleant spatium corporale circa illud punctum. Et hoc spatium habet octo angulos corporales, et duodecim angulos superficiales rectos distinctos secundum rem, licet sint viginti quatuor secundum rationem, quoniam quilibet angulus corporalis est ex tribus superficialibus, et ideo quoad hoc computantur ter octo, qui sunt viginti quatuor. Sed saepe iterantur aliqui, quia hi anguli sunt conjuncti; si enim essent divisi,

[1] *De Coelo*, iii. 8, § 1 Ὅλως δὲ τὸ πειρᾶσθαι τὰ ἁπλᾶ σώματα σχηματίζειν ἄλογόν ἐστι, πρῶτον μὲν ὅτι συμβήσεται μὴ ἀναπληροῦσθαι τὸ ὅλον.

tunc oporteret esse viginti quatuor[1] secundum veram distinctionem, ut quilibet angulus corporalis ab alio distinctus haberet tres rectos sibi proprios. Omnia haec patent ex tribus lineis intersecantibus se ad angulos rectos, ut in tribus festucis vel aliis. Quoniam vero angulus cubi est ex tribus rectis, ideo octo tales possunt propriissime replere locum circa punctum unum. Et ideo in sphaera terrae secundum figurationem nunc dictam, non erit vacuum, quia octo cubicae partes terrae congregatae circa centrum mundi replent locum totum necessario circa illud centrum. Angulus vero pyramidis est ex tribus angulis triangulorum, quapropter valet duos rectos, et ideo anguli sex tales valent quatuor angulos cubicos, nam utrobique valent duodecim rectos, et alii sex valent alios quatuor angulos cubicos. Quapropter concludit Averroes in tertio Coeli et Mundi, quod duodecim anguli pyramidum congregati circa punctum unum implebunt totum locum corporaliter, sicut octo cubici anguli, et ideo in sphaera ignis non est vacuum. Sed aliae figurae congregatae circa punctum unum non possunt secundum Aristotelem et Averroem replere locum. Quotquot enim congregarentur habebunt majus vel minus octo angulis cubicis, et ideo locum non replebunt. Et ideo in sphaera aquae et aeris et coeli accidit vacuum necessario secundum figurationem Platonicorum. Sicut vero cubus in corporali repletione respondet quadrato in superficiali, quia cubus sit ex quadratis superficiebus, propter quod utraque figura propriissime replet locum, sic pyramis respondet triangulo regulari, quia fit ex triangulis, et utraque figura locum replet. Sed tertiae figurae, scilicet hexagono superficiali, non respondet figura corporalis replens locum, quia ex superficialibus hexagonis non potest figura hexagonalis regularis constitui, ut demonstratum est prius.

Hexagon in bee-hives and in crystal.

Et tamen apis facit domus hexagonas ne vacuum intercipiatur; et natura in ventre terrae generat crystallos omnes hexagonas in unum congregatas. Et sic lapides, qui vocantur irides et in insulis Hiberniae et India dicuntur ab autoribus inveniri, congregantur in figura hexagona. Et dicuntur lapides iridis, quia repraesentant colores iridis et arcus coelestis, quando

[1] J. has XV.

ponuntur ad radios solares. Et sic est de omnibus generatis in hoc mundo, quae per superficies suas congregantur, ut retineant figuras hexagonas, ut vacuum excludatur, et hoc est mirabile. Sed tamen non est vera loci repletio secundum quod Aristoteles accipit in hoc loco: nam talis est secundum omnem situm corporum et superficierum, ut taxilli quatuor superficialiter secundum omnem situm replent locum, et octo corporaliter, qualitercunque mutentur anguli vel latera, nam aequalitas plena est in illis angulis et lateribus. Et sic est de pyramidalibus corporibus, et de triangulis superficialibus et quadratis et hexagonis. De aliis non contingit hoc secundum omnem situm sed secundum aliquem, et ideo hic non computantur. Nam si aliquibus domibus apum erectis aliae ponantur secundum alium situm, non est loci repletio, sed vacuum spatium relinquitur, et ideo non sunt de replentibus locum, ut absolute et simpliciter dicatur repletio. Magna est ac profunda consideratio de his figuris replentibus locum propter rerum naturalium figurationem. Sed quantum sufficit ad praesens in universali de hac figuratione in corporibus mundi principalibus declaravi.

CAPITULUM XIII[1].

An possint esse plures mundi, et an materia mundi sit extensa in infinitum.

If there were two universes a vacuum would be left.

Et transeo ulterius ad duo exempla breviter annotanda in corporibus mundi, quae fundatur super geometricam potestatem, et sunt adhuc annexa materiae corporali eorum. Nam Aristoteles dicit primo Coeli et Mundi[2], quod mundus occupat totam suam materiam in uno individuo unius speciei, et sic de quolibet corpore mundi principali, quoniam unus mundus est numero, nec possunt plures mundi esse in hac specie, sicut nec plures soles nec plures lunae, licet multi posuerunt contrarium. Nam si esset alius mundus, esset sphaericae

[1] Cf. *Opus Tertium*, cap. 41.

[2] *De Coelo*, i. 9 ἐξ ἁπάσης γάρ ἐστι τῆς οἰκείας ὕλης ὁ πᾶς κόσμος.

figurae, sicut iste, et non potest esse distantia inter eos, quia tunc spatium vacuum sine corpore esset signabile inter illos, quod falsum est. Quapropter oportet ut se tangerent, sed non possunt tangere se nisi in puncto uno per xii tertii Elementorum, ut prius declaratum est per circulos. Ergo alibi quam in illo puncto erit spatium vacuum inter eos. Aliud est, quod corporalis materia mundi non est extensa in infinitum, ut multi posuerunt. Nam geometrica potestas hoc excludit. Quoniam conjungantur[1] duae lineae *a* et *b* angulariter in centro mundi, et a concursu earum extendantur in infinitum, et tertia uni illarum ducatur aequidistanter et terminetur ad aliam, et sit *c a*. Ergo *a* et *b* sunt aequales, et *o*, quae est pars *a*, et *c* sunt aequales, quia ab eodem puncto scilicet *o* vadunt in infinitum. Sed *c* linea aequatur *b* lineae. Ergo *o* linea aequabitur ipsi *a* totali, scilicet pars suo toti, quod est impossibile.

CAPITULUM XIV[2].

De unitate temporis.

Time concerned not with matter, but with motion represented by space of one dimension.

Multae autem aliae demonstrationes geometricae possent ad hoc adduci, atque veritates aliae in rebus mundi possent notari quasi infinitae, in quibus geometrica virtus elucescit. Sed haec sufficiunt persuasioni, et solum evacuabo duas falsitates, quae ad unitatem materiae numeralem consequuntur secundum opiniones vulgatas. Nam ponunt quod tempus sequitur ad materiam rerum, et aevum ad formam, et ideo sicut materia est una numero et non plures, sic tempus est unum numero simul et semel, et sicut forma variatur in rebus, sic aevum multiplicatur in aeviternis. Unde plura dicuntur esse aeva, et unum tempus, ut secundum numerum angelorum sit numerus aevorum. Sed cum probatum est quod materia non potest esse una, tunc falsum est tempus habere unitatem ab ea. Deinde tempus non potest sequi, nisi ad subjectum suum.

[1] The diagram in O. has been omitted as unintelligible.

[2] Cf. *Opus Tertium*, cap. 41, which contains an application of the view here stated to the doctrine of Transubstantiation.

Sed motus est subjectum suum, non materia; et subjectum motus non est materia, sed corpus compositum ex materia et forma.

The present moment, being a point, may correspond to any number of motions.

Postremo res geometricae nobis ostendunt causam unitatis in tempore, et demonstrationem addunt super hoc. Nam corpus quia habet undique dimensionem, ideo non compatitur secum aliud corpus: ubique enim habet corpus, unde aliud excludat secundum longum et latum et profundum. Ergo superficies secundum longum et latum excludet aliam superficiem, sed non secundum profundum, quia sic est indivisibilis et caret dimensione. Et linea secundum longum excludit aliam, sed non secundum latum et profundum quia sic non habet dimensionem. Ergo punctus cum omni careat dimensione, non habet unde excludat aliud a suo loco indivisibili; sed imaginato primo puncto in suo loco secundus adveniens habebit eundem locum in mente, quia non est distantia media, et sic de tertio puncto, et de infinitis. Motus vero non habet nisi linearum dimensionem a priori in posterius secundum longitudinem spatii, et hoc est a praeterito in futurum. Ergo solum secundum hunc decursum, scilicet a priori in posterius, seu a praeterito in futurum, unus motus excludet alium, scilicet prior posteriorem, et praeteritum excludet futurum. Sed comparatio motus ad praesens est alia quam secundum decursum a praeterito in futurum. Ergo respectu praesentis nullus motus habet dimensionem nec divisibilitatem, et ideo non habebit unde excludat alium a praesenti. Et ideo infinitos secum potest pati praesentes; et ideo unum tempus praesens sufficit omnibus motibus praesentibus, et propter hoc habetur hic vera causa unitatis temporis, et non propter materiam. Deinde ex istis elici potest unitas vera aevi, sicut temporis. Nam aevum vel solum habet dimensionem linearum, si ponamus aevum esse divisibile et habere partes, ut multi aestimant contra totam philosophiae potestatem, atque contra Augustinum et Dionysium, quanquam Anselmus velit contrarium. Et si hoc sit verum, tunc[1] sic est de aevo sicut de tempore, propter quod erit unum et non plura. Aut aevum erit indivisibile, et tunc erit ad aeviterna, sicut locus indivisibilis ad

[1] nec, J.

puncta et atomos, et idem numero est locus unius puncti et plurium, ut prius habitum est. Ergo unum erit aevum omnium aeviternorum, et hoc est necessarium, et nulli perito in philosophia dubium. Nec est contra sanctos et doctores principales, sed conveniens sententiae eorum.

CAPITULUM XV.

An motus gravium et levium excludat omnem violentiam. Et quomodo motus gignat calorem. Itemque de duplici modo sciendi.

Quoniam vero motus est subjectum temporis, et tempus est mensura motus, possumus adhuc videre magnam geometriae potestatem in motibus corporum istius mundi. Aestimant vero naturales, quod motus gravium deorsum sit naturalis omnino, et motus levium sursum est similiter omnino naturalis, ita ut non habeant de violentia. Sed figuratio geometrica ostendit nobis contrarium. Nam sit *d b c* lignum vel lapis in aere, et *a* centrum mundi, *g h* diameter mundi. Cum ergo *d b c* sint semper in suo toto aequaliter distantes descendant ad centrum per lineas aequidistantes. Ergo *d* descendet per lineam *d e*, et *b* per lineam *b a*, et *c* per lineam *c o*, quapropter *d* cadet extra centrum mundi in diametro *h g* versus coelum, scilicet in *e* puncto, et *c* in *o*, quare in hoc descensu *d* declinabit a centro *a* versus centrum per altitudinem *a e*, et *c* per altitudinem *a o*. Sed omnis declinatio gravis a centro versus centrum est violenta. Ergo *d* et *c* moventur violenter, et sic de omnibus partibus *d b c* praeter *b* quae sola vadit in centrum. Quapropter multum erit hic de violentia. Caeterum incessus rectus et naturalis ipsius *d* est per lineam *d a*, unde si separetur *d* a suo toto caderet in *a* per rectum incessum, quia omne grave tendit in centrum. Omnis autem declinatio

In a falling object the central point alone follows a free path.

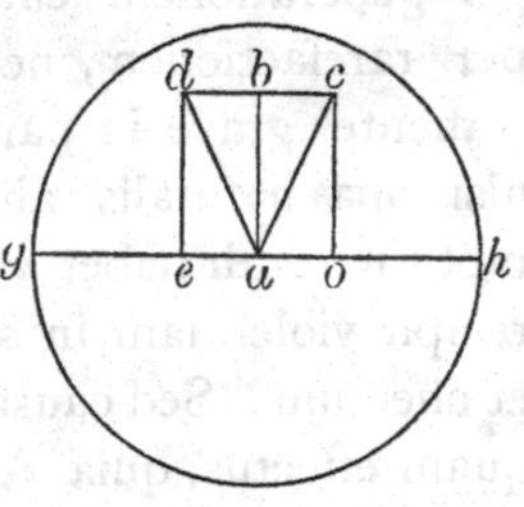

FIG. 18.

gravis ab incessu recto est violenta, sed quanto magis *d* movetur super lineam *d e*, magis recedit ab incessu recto, ut patet ad sensum, quia *d a* et *d e* lineae magis separantur inferius quam superius. Ergo *d* quanto magis descendit deorsum, tanto magis movetur per violentiam, et similiter *c*, et ideo quaelibet pars ipsius totius *d b c* gravis, praeter *b* quae sola semper descendit secundum incessum rectum. Manifestum est ergo quod magna et multiplex violentia est in motu naturali ipsius gravis. Et ex hoc sequitur quaedam veritas in rebus naturalibus, scilicet quod motus naturalis generat calorem; nam quum demonstrata est violentia, et constat grave naturaliter inclinari deorsum, planum est quod duae virtutes sunt in gravi moto deorsum inclinantes ipsum in partes contrarias. Ergo una distrahit partes gravis in unam partem, et alia in aliam, et ad has distractiones necesse est rarefieri partes gravis. Sed rarefactio est dispositio immediata ad calorem, unde per experientiam scimus quod grave descendens deorsum calescit. Potest ergo hoc hic adverti, sicut in prioribus, quod causae rerum naturalium debent assignari per mathematicae potestatem. Et potest homo videre quod in rebus naturalibus sunt duo modi arguendi, unus per demonstrationem quae procedit per causas, et alius per demonstrationem ad effectum, ut cum prioribus demonstrationibus probatur per causam, quod violentia accidit gravi in suo motu naturali, postea demonstratur hoc idem per effectum, scilicet per generationem caloris. Nam non generaretur calor nisi per rarefactionem, nec rarefactio ista nisi per virtutes distrahentes grave in partes contrarias, et hae non possunt esse nisi una naturalis, altera violenta, quapropter grave in suo motu naturali habet violentiam. Et sic haec conclusio, grave recipit violentiam in suo motu naturali, probatur per causam et effectum. Sed causa sola facit scientiam aut longe majorem quam effectus, quia Aristoteles dicit primo Posteriorum quod scire opinamur cum causas cognoscimus. Ergo cum demonstratio, ut ibidem docet, est syllogismus faciens scire, necesse est quod demonstratio per causam sit longe potentior, quam per effectum; et hoc vult Aristoteles libro Posteriorum[1].

[1] Et . . . posteriorum, om. in O. Cf. Anal. Post. i. 2, § 1.

Quapropter cum in rebus naturalibus demonstratio habetur per causam per vias mathematicae, et demonstratio per effectum habetur per vias naturales, plus potest mathematicus in rebus naturalibus sciendis, quam ipse philosophus naturalis. Et maxime hoc planum est, quod motus simpliciter violentus, ut motus gravis sursum, generabit calorem, et longe magis quam naturalis motus; quia in violenter moto sunt duae virtutes motrices omnino contrariae, et secundum totum, et in contrarias partes omnino, ut virtus naturalis gravis tendit deorsum, et virtus violenta tendit omnino sursum. Et ideo magna est distinctio partium rei motae violenter, et major quam in naturali.

Capitulum XVI.

De motu Librae.

Researches of Jordan on gravity.

Et cum jam dictis expedit altius aperire geometricam potestatem in motibus; et hoc propter intellectum universalem scientiae de ponderibus, quae est pulchra et difficilis nimis hominibus non habentibus experientiam causarum in motibus gravium et levium. Dicit ergo Jordanus in libro de ponderibus[1], quod si aequilibris fuerit positio aequalis,

[1] This was Jordanus Nemorarius, the most original, if we except Leonardo Fibonacci, of the mathematicians of the thirteenth century. There is good ground for believing that he is identical with Jordanus Saxo who, on the death of St. Dominic, 1221, succeeded to the generalship of the order. His principal works are, (1) *De Ponderibus*, of which an edition was published in 1533, by Peter Apianus; (2) a treatise on Arithmetic, edited in Paris, 1496 and 1514, by Faber Stapulensis; (3) *Algorithmus Demonstratus*, edited by Schöner, of Nuremberg, in 1534; (4) *De Numeris Datis*, printed in Zeitschrift, Math. Phys. xxxvi histor. literar. Abtheilung, 1891; (5) *De Triangulis*, printed 1887, by the Copernicus-verein für Wissenschaft und Kunst of Thorn. Most of these works were evidently known to Bacon. The treatise, *De Ponderibus*, which consists of a short preface followed by thirteen propositions, is interesting as one of the earliest studies, by a mathematician of great originality, of the mechanics of a particle forming part of a rigid system. With regard to gravity we find, of course, the doctrine, still awaiting Galileo's refutation, that heavy bodies fall more rapidly than light. The arithmetical treatise, containing many algebraic problems, had also been studied by Bacon, as his fragment on the Principles of Mathematics clearly shows. [See Cantor, *Gesch. der Mathem.* vol. ii. pp. 49–54.]

aequis ponderibus appensis, ab aequalitate non discedet, et si ab aequidistantia separatur ad aequalitatis situm revertetur. Et istud videmus ad sensum in lance utraque, quarum virga sit ex parte utraque aequalis in longitudine et in pondere, et omnino appendantur pondera aequalia, et libra aequaliter teneatur per appendiculum, ut stet appendiculum ad angulos rectos super regulam librae in centro revolutionis, nam hic punctus vocatur centrum revolutionis a quo appendiculum exit ad angulos aequales. Et dicitur centrum revolutionis, quia quando per violentiam manus deprimentis alterum ponderum aequalium, aut propter inaequalitatem appensorum unum eorum facit nutum, aliud elevabitur, et hic motus descensus et elevationis describet circulum unum, cujus ille punctus a quo exit appendiculum est centrum, et ideo dicitur centrum revolutionis. Quod ut planius sit fiat figura. Nam sit regula seu baculus librae *a b*, et *c d* sit appendiculum, tunc centrum revolutionis a quo exit appendiculum erit *d*, et in circumferentia istius circuli appensa movebuntur, nam illud quod descendet describet circulum inferiorem, et illud quod ascendet describet circulum superiorem. His suppositis, arguitur[1] sic. Cum alterum brachiorum librae aequalibus appensis nutum faciat per manum deprimentis, fit, secundum Aristotelem quarto Coeli et Mundi, gravius, quia grave quanto adquiret magis de loco gravis, tanto magis adquirit de forma gravitatis, ut ipse dicit. Ergo quod descendit fit gravius, quantumcunque parum descendat a situ aequalitatis, et ideo quanto magis descendit, tanto erit gravius. Ergo fiet inaequale reliquo appenso et ponderosius eo. Ergo licet fuerint in situ aequalitatis aequalia, tamen cum recedunt ab illo situ fient inaequalia in pondere; quare semper descendet illud quod nutum facit, et aliud semper ascendet, et ideo nunquam ad situm aequalitatis revertentur.

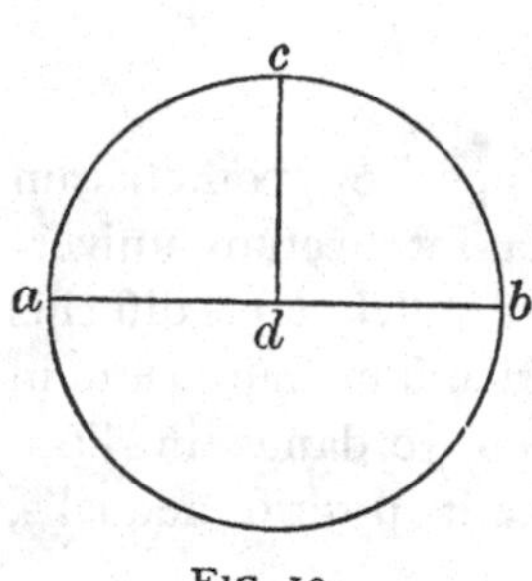

FIG. 19.

Gravity of particles in the arm of a balance varies with their position.

[1] J.'s reading, argumentor, for arguitur, suggests that what follows is Bacon's opinion: which it is not.

Sicut quando duo pondera inaequalia ponuntur in brachiis, statim recedunt a situ aequalitatis, et nunquam ad eundum situm revertentur, sed semper descendit quod est ponderosius. Ergo similiter hic, quod est contra Jordanum et contra sensum. Item Jordanus dicit, quod inter quaelibet gravia est velocitatis in descendendo et ponderis eodem ordine sumpta proportio, sed istud grave quanto magis descendit, tanto fit ponderosius. Ergo tanto velocius descendit. Ergo nunquam revertetur per naturam ad situm aequalitatis. Item Jordanus dicit, quod minus grave secundum situm est, quod descensum alterius sequitur motu e contrario, id est, quod ascendit quando descendit, et e contra. Sed appensum nutum faciens est minus grave secundum situm, ut probabo. Quare sequetur descensum alterius appensi motu contrario, et ascensum similiter. Quapropter secundum quod unum descendit, reliquum ascendit, et e contra : quare nunquam in situ aequalitatis quiescent.

Quod autem appensum faciens nutum sit minus grave secundum situm, manifestum est per hoc, quod minus capit de directo descensu in diametro transeunte per centrum revolutionis versus centrum mundi : quapropter secundum Jordanum erit minus grave secundum situm. Et hoc exigit ipsa veritas per figuram declaranda. Et hujusmodi figuratio solvet objecta, nec potest habere remedium intellectus nisi per figuram. Describatur ergo circulus super centrum revolutionis, quod est *o*, in cujus circumferentia appensa revolventur, et trahatur diameter *a b* aequidistans horizonti, et lineetur alia diameter intersecans hanc quae tendat in centrum mundi, et sit *d c*, et signentur arcus aequales in utroque semicirculo ab utraque parte diametri aequidistantis horizonti, et hoc a parte utriusque termini ejus, et a terminis arcuum ducantur in utroque semicirculo lineae aequidistantes sibi invicem, et diametro aequidistanti horizonti, quae sunt *f h*, *g p*, *t q*, *s r*, quae omnes secant diametrum cadentem in centrum mundi. Oportet ergo secundum Jordanum et commentatorem ejus, quod illae lineae aequidistantes secent de diametro quae vadit in centrum mundi, partes inaequales, ita ut illa aequidistans, quae propinquior est diametro aequidistanti horizonti, secet majorem partem diametri alterius, quam remotior aequidistans, ut *t q* separabit majorem partem diametri *d c* quam *s r*, ita ut pars diametri *d c*, quae est inter *a b* et *t q* sit major quam pars ejusdem diametri quae est inter *t q* et *s r*, et eodem modo pars diametri *d c*, quae est inter *a b* et *g p*, erit major quam illa quae est inter *g p* et *f h*. Et secundum hoc oportet quod sumpta una aequidistante in semicirculo uno, et alia in alio, quae aequaliter distant a diametro eis aequidistante, illae secabunt partes aequales de diametro descendente in centrum mundi ut *t q* et *g p* secabunt partes aequales de *d c*, et similiter *s r* et *f h*, sicut dicit vicesima sexta propositio de triangulis Jordani. Si ergo partes diametri cadentes in centrum mundi divisae per aequidistantes sunt inaequales, ita ut illae partes diametri quae dividuntur per aequidistantes propinquiores diametro aequidistanti horizonti sint majores; tunc ergo intelligamus regulam librae jacere in diametro aequidistante horizonti, et appendiculum sit erectum in diametro cadente per centrum, ut libra sit in situ aequalitatis et brachia ejus, deinde postea moveatur libra, et elevetur pars una librae usque ad primam aequidistantem in semicirculo superiori, et alia deprimatur usque ad terminum primae aequidistantis in semicirculo inferiori, ut regula sit in situ *t p* lineae, et pars librae altior sit in *t*, et reliqua in *p*. Si ergo *p* descendat usque ad terminum alterius aequidistantis *h*, transibit de diametro cadente in centrum, partem ejus quae est inter aequidistantes *g p* et *f h*, quod minus est quam illa pars diametri, quae est inter *t q* et *a b*, ut patet ex praedictis. Ergo si descenderet usque *a* caperet plus de descensu recto in diametro cadente in centrum mundi quam *p*, dum descendit in *h*; quare *t* est gravius secundum situm quam *p*. Et iterum *t* descendit versus centrum mundi. Sed *p* propter declinationem circuli recurvatur a centro, et saltem minus tendit in centrum, ut patet ad sensum. Ergo relinquitur, quod ex hac causa adhuc erit minus grave. Et quia sic est, ideo solvitur

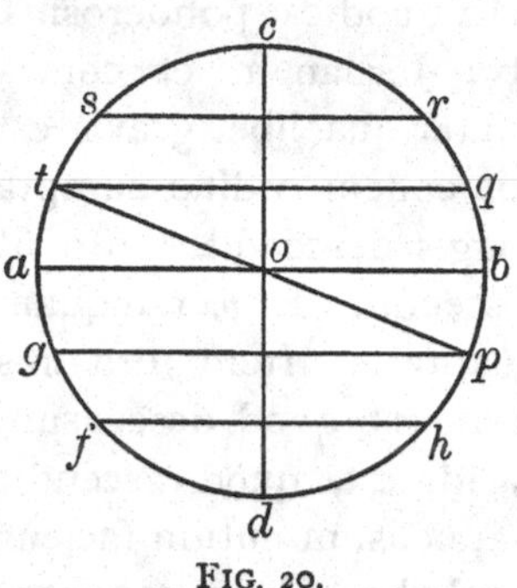

FIG. 20.

prima objectio; nam licet pars regulae[1] *p* descendens est propinquior centro mundi, ut patet, si linea recta trahatur ab eo in centrum *o*, quia illa linea est brevior quam linea quae trahitur a *t*, ut patet ad sensum, et ita gravior sit in quantum plus habet de loco deorsum, tamen quia *p* secundum illam rectam lineam, quae est *p o* non movetur versus centrum, sed secundum circulationem circumferentiae circuli, et illa circulatio facit eam minus capere de incessu recto in diametro cadente in centrum quam capiat *t*, atque obliquat et incurvat ipsum *p* a centro, ut non ita recte tendat in centrum sicut *t*, quando descendit, oportet quod *t* sit gravior quam *p* dum sunt in tali situ. Et ideo si dimittantur sibi ipsis *t* descendet, et ejus descensum sequetur *p* motu contrario, scilicet ascendendo usque ad situm aequalitatis. Et ideo *p* non semper descendet, sed praevalent duae causae gravitatis hic assignatae contra illam de qua objectio fecit mentionem. Iterum illa gravitas non cogit, nam modica est et insensibilis, et ideo non operatur hic, sicut nec si pluma una apponeretur ad alterum appensorum aequalium, cum sunt in situ aequalitatis, faceret nutum, et tamen secundum veritatem illud brachium ubi fuerit pluma est gravis, quia pluma habet aliquid gravitatis. Et ideo similiter est hic; quia enim brachium descendens, quando est in ultimo descensu parum distat a situ aequalitatis, ideo valde modicum et insensibile est quod adquirit de gravitate, et ideo gravitas acquisita non est computanda.

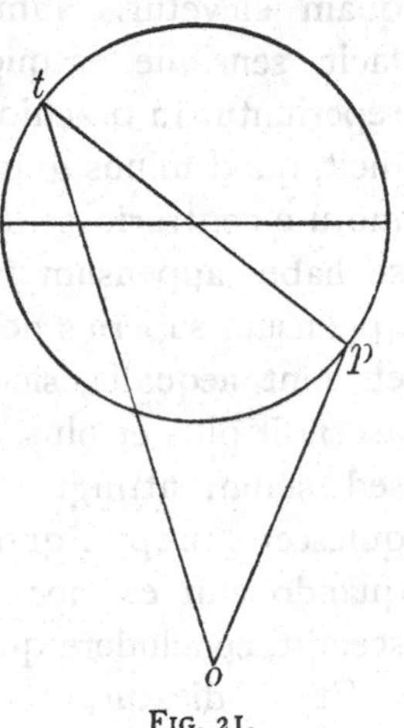

FIG. 21.

[1] Jebb's figure (Tab. i. 26) is wrong, as he puts *p* in the upper instead of the lower end of the balance. Further, he draws *p o* and *o t* to the centre of the circle, in which case both are, of course, equal. The lines should be drawn to the centre of the earth, as in fig. 21. What is meant is that *p* is nearer to the earth's centre than *t*, and therefore moves with greater force: but that this is counterbalanced by the fact that the path of *p* tends to become more horizontal than that of *t*; 'minus capit de incessu recto in diametro cadente in centrum quam capiat *t*.'

Et tunc patet aliud, quod hujusmodi ponderositatem majorem concludit quod semper velocius descendet ut nunquam elevetur. Jam enim patet, quod haec gravitas nihil facit sensibile, atque duae causae gravitatis praetactae reperiuntur in brachio altiore. Cum vero in tertio argumento dicit, quod minus grave est quod descensum alterius sequitur motu e contrario, scilicet ascendendo, bene concedo. Nam sic se habet appensum inferius, quia movetur sursum, quando appensum superius descendit, ut quando sibi ipsis dimittuntur, et sunt aequalia, sicut positum est. Sed non propter hoc ascendit plus et plus, nec reliquum descendit quantum potest, sed solum attingunt situm aequalitatis, et ibi nata sunt quiescere, propter quod objectio concludit plus quam deberet, quando vult ex hoc, quod unum ascendit et reliquum descendit, concludere quod transeant situm aequalitatis.

Oscillation of balance due to atmosphere.

Et si dicatur, quod ibi est motus titubationis et ideo brachium superius descendit ultra situm aequalitatis et qua ratione transit parum et multum, quia hic transitus est unius naturae, et similiter de reliquo brachio ut ascendat semper, postquam transit situm aequalitatis. Sed dicendum est, quod hic descensus brachii altioris ultra situm aequalitatis non est propter naturam ipsius appensi, sed propter reinclinationes partium aeris impetuosas. Cum enim aer receperit motum, retinet ipsum bene, et ideo diu titubant partes ejus huc atque illuc, et non permittunt statim appensum quiescere in loco aequalitatis[1].

[1] The remaining portion of Part IV is not divided into chapters, with the exception of the final section on astrology, which in the Bodleian MS. is divided as though it were a distinct treatise. What follows consists first of a disquisition on Chronology, secondly, of a review of geographical knowledge, and thirdly, of the treatise referred to.

In the *Opus Tertium*, before those subjects are entered upon, there occur some discussions of matters not treated of in the *Opus Majus* (chapters 42–52). These deal (*a*) with the question of Vacuum : (*b*) following on this, with the question of growth and nutrition : (*c*) of place and motion with regard to immaterial beings : (*d*) of *aevum*, or created eternity.

MATHEMATICAE IN DIVINIS UTILITAS.

Connexion of Mathematics with theology.

Postquam[1] manifesta est necessitas mathematicae in rebus hujus mundi et in scientiis humanis nunc potest istud idem ostendi in divinis. Et hoc est magis considerandum, quia humana nihil valent nisi applicentur ad divina. Cum igitur ostensum sit quod philosophia non potest sciri nisi sciatur mathematica, et omnes sciant quod theologia non potest sciri nisi sciatur philosophia, necesse est ut theologus sciat mathematicam. Caeterum Deus posuit res creatas in scriptura sua, qui solus novit potestatem creaturarum quas condidit, nec potest falsum sentire, nec decet suam veritatem. Ergo cum omnes res a Deo et angelis et summis coelorum usque ad terminos eorum ponantur in scriptura, vel in se vel in suis similibus vel in suis contrariis, et contrariorum est eadem scientia, ut dicit Aristoteles, et verum est vel in universali vel in particulari, necesse est theologum scire res hujus mundi, si textum sacrum debet scire.

Praeterea nos videmus, quod sensus literalis stat in cognitione naturarum et proprietatum creaturarum, ut per convenientes aptationes et similitudines eliciantur sensus spirituales. Nam sic exponunt sancti et omnes sapientes antiqui, et haec est vera et sincera expositio, quam Spiritus Sanctus docuit. Quapropter oportet theologum scire optime creaturas. Sed ostensum est, quod sine mathematica sciri non possunt. Ergo mathematica omnino est necessaria sacrae scientiae.

Mathematical knowledge of patriarchs.

Et hoc tertio per propria potest ostendi. Et cum multis modis probabitur quod intendo, primo tamen per occupationes sanctorum persuadere conabor, cum exclusione infamiae mathematicae quam multi imprudenter allegant, quia sanctorum testimonia non intelligunt. Patriarchae enim et Prophetae ante diluvium et post invenerunt ipsam et docuerunt caeteros homines, 1. Chaldaeos, 2. Aegyptios; et ab Aegyptiis ad Graecos descendit; et non ita evidenter scribitur quod

[1] Cf. *Opus Tertium*, chap. 54.

sic laboraverunt in aliis scientiis. Sed cum per istos viros data est nobis lex divina, et fuerunt sanctissimi, non occupaverunt se nisi in scientia quae maxime est utilis legi divinae; ergo mathematica est maxime consona legi divinae. Minor propositio habet secum suam probationem. Major probatur per auctoritatem triplicem. Primo per historiographos, et praecipue per maximum eorum Josephum. Primo enim Antiquitatum libro in tribus locis de his scientiis faciens mentionem, expresse narrat totum quod propositio major proponit. Nam dicit, 'quod filii Adae invenerunt geometriam, astronomiam, arithmeticam, et musicam; et Noe et filii ejus docuerunt Chaldaeos: deinde Abraham docuit Aegyptios.'

Appreciation of it by the fathers. [1] Secundo, verificatur hoc per beatum Hieronymum et Cassiodorum et alios doctores sacros, ut vulgus etiam theologorum non ignorat; et ratificant sancti, quod Josephus asseverat. Tertio, per philosophorum assertionem. Quoniam Albumazar quinto libro majoris introductorii in astronomiam undecima doctrina sive undecimo capitulo refert Sem filium Noe hujus scientiam alios docuisse, atque in prologo compositionis astrolabii Ptolemaei dicitur, quod a filio Sem, qui divina memoria commonitus erat, aut fortasse divino nutu commotus, hujus scientiae studium in orbem derivatum affirmatur. Hoc idem indicat sanctorum nostrorum post adventum Domini occupatio, ut Augustini, Cassiodori, Isidori, Hieronymi, Orosii, Bedae, Origenis, Eusebii Caesariensis. De his enim scripserunt, et in his se et alios exercitaverunt solum, aut magis quam in aliis. Cum ergo isti fuerunt doctores sacrae scripturae et viri sancti, manifestum est, quod hujusmodi scientiae maxime valent sacrae scientiae professoribus. Quod autem ipsi de his scripserunt, manifestum est per Cassiodorum et Isidorum, qui

[1] This passage is much abbreviated in O. Thus: 'Secundo verificatur per doctores sanctos quos recitant sancti: quod Josephus asseverat. Quod autem, de his scripserunt manifestum est per Cassiodorum et Isidorum quum tractatus suos de omnibus istis quatuor composuerunt. Augustinus etiam de numeris et musica libros diversos conscripsit. Sed non solum isti sed alii multi. Et Cassiodorus dicit, Mathematica dicere valeamus quaecunque docentur, hoc tamen signum commune propter sui excellentiam proprie vindicavit. Et in tractatu,' &c. O. has many other omissions in this section of the work, which are supplied by the two Cottonian MSS., Tib. C v, and Jul. D. v. I have indicated some of them, but not all.

tractatus suos de omnibus istis quatuor composuerunt. Augustinus etiam de numeris et musica libros diversos conscripsit; Hieronymus in libris diversis, et Orosius ad Augustinum, et Eusebius de mundi locis, quorum certificatio eruitur ex fontibus astronomiae, sicut certum est, et sequentia declarabunt. Beda de cursu solis et lunae, et de tota temporis diversitate, atque Origenes et Eusebius, quae astronomiae subjacere noscuntur. Non solum vero de his scripserunt, sed et idoneos docuerunt haec, ut essent acuti contra haereticos, sicut patet per Cassiodorum in mathematica sexto libro Ecclesiasticae Historiae; etiam partes geometriae et arithmeticae disciplinae, ut ibidem dicitur, quae magis a divinis elongantur quam astronomica et musicalia. Sed non tantum occupaverunt se scribendo et alios docendo, imo veritates theologicas per virtutem harum scientiarum exposuerunt; sicut per omnia originalia sanctorum sine contradictione manifestum est, ut evidenter patet in numeris et locis mundi et rebus coelestibus et aliis quae ad dictas scientias pertinere noscuntur, sicut patebit inferius.

Sed non solum sic exposuerunt sacra de facto, sed verbo asserunt has scientias per omnem modum valere rebus divinis, et super alias scientias extollunt in hac parte. Cassiodorus quidem in praefatione de artibus et disciplinis secularium studiorum sic ait, 'Mathematicam Latino sermone Doctrinalem possumus appellare, quo nomine licet omnia doctrinalia dicere valeamus quaecunque docentur, haec tamen sibi commune vocabulum propter sui excellentiam proprie vindicavit.' Et in tractatu mathematicae sic ait, 'Disciplinales sunt, quae nunquam opinionibus deceptae fallunt, et ideo tali nomine nuncupantur. Has dum frequenti meditatione revolvimus, sensum nostrum acuunt, limumque ignorantiae detegunt, et ad illam speculativam contemplationem, Domino largiente, perducunt; quas merito sancti patres legendas persuadent, quoniam ex magna parte per eas a carnalibus rebus appetitus abstrahitur, et faciunt desiderare, quae solo Domino largiente corde possumus respicere. Duplex quidem pars mathematicae tam a sanctis quam a philosophis ipsis numeris attribuitur vulgato sermone, videlicet arithmetica de numeris absolute, et

Cassiodorus.

musica de proportionibus et rationibus caeteris quae in numeris reperiuntur, prout ad sonum et gestum referuntur; quia sonus et gestus sunt praecisum musicae subjectum, discussi tamen per proportiones numerorum in musicae scientia.'

Augustine's view of arithmetic, Augustinus autem in secundo de Ordine Disciplinae dicit, 'ad divinarum scripturarum neminem aspirare debere sine scientia potentiae numerorum.' Atque in secundo de Doctrina Christiana dicit, 'numerorum imperitia multa facit non intelligi translate et mystice posita in scripturis'; et ponit exempla plurima, adjungens, 'quod ita in aliis multis numerorum formis secreta in sanctis libris ponuntur, quae propter numerorum imperitiam legentibus clausa sunt.' Et ideo propter utilitatem numerorum maximam Isidorus [1] in tractatu arithmeticae dicit, 'Adime seculo computum, et cuncta ignorantia caetera complectitur; nec differre possunt a caeteris animalibus, quae calculi nesciunt rationem.'

of music, Quamvis autem jam communiter cum arithmetica patet laus musicae, propter hoc quod utraque proportiones quae in numeris sunt considerat, tamen numeros, ut sunt in sonis, maxime laudat Augustinus ad Omerium dicens, 'In omnibus rerum motibus quid valeant numeri facilius consideratur in vocibus, eaque consideratio quibusdam gradatis itineribus nititur ad superna itinera veritatis, in quibus viis se ostendit sapientia hilariter.' Et in libro Retractationum dicit [2], 'Sextus maxime innotuit, quoniam in eo res digna cognitione

[1] Born in Carthagena, 570. In 601 he became bishop of Seville, succeeding his brother Leander, the 'apostle of Spain,' and held the bishopric till his death in 636. His great powers were principally devoted to organizing the work of the Church in Spain. But he was a man of wide learning, and is placed by Dante (*Paradiso*, x) in the Sphere of the Sun with Albert, Thomas Aquinas, and other great theologians. His principal work is a Cyclopaedia, founded on the works of Martianus Capella and Cassiodorus, entitled *Origines*, in twenty books, from which much is to be gathered as to the very limited scientific knowledge of Christendom in the seventh century. The passage here quoted is in lib. iii. cap. 4. (Cf. Cantor, vol. i. pp. 705–7.) This work continued to be of great authority throughout the middle ages.

[2] O. has, 'Maxime pervenitur a mutabilibus rebus ad immutabiles ac si Dei invisibilia per ea quae facta sunt conspiciantur. Et in libro de Doctrina Christiana posuit exempla de psalterio decachordo, et admittit quod musicam in pluribus locis et in sanctis scripturis honorabiliter positam invenimus. Cassiodorus in tractatu musicae sic ait, Musica disciplina,' &c.

versatur, quomodo a mutabilibus numeris perveniatur ad immutabiles, ac si invisibilia Dei per ea quae facta sunt conspiciantur.' Item in libro de Doctrina Christiana dicit, 'Non pauca claudit atque obtegit in sanctis libris rerum musicarum ignorantia,' et ponit exempla in psalterio decachordo et cythara et hujusmodi, et adjungit, 'quod musicam in plerisque locis in sacris scripturis honorabiliter positam invenimus.' Cassiodorus quidem et in mandatis Dei et moribus et scriptura sacra et in omnibus rebus creatis eam valere dijudicat. Unde in tractatu musicae sic ait, 'Musica disciplina per omnes actus vitae nostrae diffunditur; primum si creatoris mandata faciamus, et puris mentibus statutis ab eo regulis serviamus, musica quippe est scientia bene modulandi. Quod si nos vitam bona conversatione tractamus tali disciplinae probamur semper esse sociati. Quando vero iniquitatem gerimus, musicam non habemus. In ipsa quoque religione valde permixta est; unde decalogi decachordus, tinnitus cytharae, tympana, organi melodia, cymbalarum sonus; ipsum quoque Psalterium ad instar instrumenti musici nominatum non dubium est, eo quod in ipso contineatur virtutum coelestium suavis nimis et grata modulatio; et ut breviter cuncta complectar, quicquid in supernis sive in terrenis rebus convenienter secundum actoris sui dispositionem geritur, ab hac disciplina non refertur exceptum. Gratissima vero nimis utilisque cognitio, quae sensum nostrum ad superna erigit et aures modulatione permulcet.'

De utilitate et scientia astronomicorum dicit, 'Astronomiam si casta et moderata mente perquirimus, sensus nostros, ut veteres dicunt, magna claritate perfundit. Quale enim est ad coelos animam subjicere[1] et totam illam machinam supernam indagabili ratione discutere, et inspective mentis subtilitate ex aliqua parte colligere, quod tantae magnitudinis arcana velaverunt.' Et subjungit, 'Ex quibus, ut mihi videtur, climata nosse, horarum spatia comprehendere, lunae cursum pro inquisitione Paschali, ne simplices aliqua confusione turbentur, qua ratione fiat advertere, non videtur absurdum. Est et alia quoque de talibus non despicienda commoditas, si

of astronomy;

[1] Caelos animo subire, J.

opportunitatem navigationis, si tempus arandi, si aestatis caniculam, si autumni suspectas imbres inde discamus.' Et Augustinus in libro secundo de Doctrina Christiana de utilitatibus istius scientiae loquens, vult triplicem utilitatem ejus esse, videlicet, 'demonstrationem praesentium, et cognitionem praeteritorum, et rationabiles conjecturas futurorum. Demonstratio praesentium in assignatione proprietatum rerum coelestium consistit. Et praeter demonstrationem praesentium habet praeteritorum narrationi simile aliquid, eo quod a praesenti cognitione motuque siderum in praeterita eorum vestigia regulariter licet recurrere. Habet etiam regulares conjecturas futurorum non suspiciosas et ominosas, sed ratas et certas,' ut ait. Et cum sic sentiant sancti de tribus partibus posterioribus mathematicae, necesse est eos de parte prima laudabiliter sentire, quae scilicet est geometria. Ex hujus enim notitia caeterae dependent, cum prima sit omnium et radix caeterarum. De cujus laude scribens Cassiodorus sic dicit, Etenim si fas est dicere, sancta divinitas, quando creaturae suae diversas species formulasque dederit, quando cursus stellarum potentia veneranda distribuit et statutis lineis facit currere quae moventur, certaque sede quae sunt fixa constituit, quicquid bene disponitur ac completur, potest disciplinae hujus qualitatibus applicari.'

of geometry.

Seven respects in which mathematics are of use to theology. (1) Knowledge of the heavens.

Et si velimus descendere ad propria studii theologiae inveniemus mathematicam omnino necessariam propter septem magnas causas. Una est notitia coelestium rerum; nihil enim est ita conveniens theologiae et ejus professoribus. Nam theologia est coelestis divinitus; et ideo nulla scientia humana speculativa conveniet ei in tantum sicut coelestis. Et per totam scripturam a terrenis revocamur et ad coelestia excitamur. Et conversatio nostra secundum Apostolum in coelis est si sumus vere Christiani, atque aspiramus et credimus nos fore mansuros corporaliter in coelo et perpetue. Quapropter nihil deberet tantum sciri a nobis sicut coelum, nec aliquid in humanis tantum desiderari. Et si nos gaudemus exponere scripturam, justum est exponi per rerum inferiorum proprietates quae in scriptura ponuntur et aliter sciri non possunt. Quare similiter cum in scriptura sint multa de coelestibus et difficilia, necesse est theologum scire coelestia. Praeterea cum rerum magnitudo excitet nos ad reverentiam creatoris, et non est comparatio rerum inferiorum ad coelestium magnitudinem, tunc cognitio inferiorum non habebit comparationem superiorum respectu finis, qui est laus et reverentia creatoris. Dicit enim Avicenna in nono Metaphysicae, 'quod ea quae sunt sub circulo lunae sunt pene nihil comparatione eorum, quae sunt supra.' Et Ptolemaeus ostendit in Almagesto et omnes sciunt astronomi, quod tota terra cum omnibus inferioribus se habet respectu coeli sicut centrum ad circumferentiam. Sed centrum non habet aliquam quantitatem. Quapropter consimiliter concludunt de terra respectu coeli, licet in se sit magnae quantitatis. Et minima stellarum visu notabilium, ut dicit Alfraganus in principio sui libri, est major terra; sed minima stellarum respectu coeli non habet quantitatem, de qua sit vis. Et cum ex octavo Almagesti et ex Alfragano pateant sex stellarum fixarum magnitudines, quaelibet illarum quae sunt in prima magnitudine, est aequalis terrae circiter centies et septies. Et illarum quaelibet quae sunt in sexta magnitudine, est aequalis terrae decies octies. Et sol est centies septuagies fere major tota terra, sicut probat Ptolemaeus quinto Almagesti. Et secundum eum una stella non complet motum suum semel in circuitu coeli nisi in triginta sex millibus annorum propter coeli magnitudinem, cum tamen stella moveatur incredibili velocitate. Sed terra tota posset perambulari infra tres annos. Quapropter magnitudo inferiorum non habet comparationem ad coelestia. Similiter nec utilitas; quia tota utilitas inferiorum causatur ex superioribus. Duplex enim allatio solis sub obliquo circulo cum aspectibus planetarum est causa omnium quae fiunt hic inferius.

Si igitur descendamus ad coelestia secundum considerationes theologiae, patet quod theologi quaerunt in sententiis et tractatibus super sententias, an orbes coelestes sint continui vel discontinui, et de numero coelorum, maxime propter nonum et decimum, et de figura eorum et circulis epicyclis et eccentricis, et motibus in illis, et de differentiis positionum in coelis, ut de dextro et sinistro, ante et retro, et sursum et deorsum, et de proprietatibus coelorum, ut de luce et perspicuo et hujusmodi, atque de influentia coelorum in haec inferiora, et de differentia coeli et naturae elementaris praecipue propter ignem. Nam Augustinus et alii aliquando secundum opinionem Platonis nominant coelum esse igneae naturae. Quaerunt etiam de locis mundi propter paradisum, an sit sub aequinoctiali circulo vel non; et de inferno ubi sit; et utrum coelestia habent posse super generabilia et corruptibilia, et utrum super animam rationalem; et de fato et hujusmodi quae ad judicia astronomiae noscuntur pertinere; et alia innumerabilia multiplicantur quotidie in quaestionibus theologiae. Sed non solum tractatus sententiarum, sed ipse textus sacer cum expositionibus sanctorum hoc requirit. Nam capitulum primum Genesis multas habet difficultates propter coelestia, ut ex ipso textu patet, et per expositiones sanctorum, praecipue Basilii, Ambrosii, Bedae, in libris suis qui Exemeron vocantur. Et in Josue, propter diei longitudinem, sole stante, praecipue per operationem ad longitudinem diei, qua retrocessit sol decem lineis ad verbum Isaiae prophetae. Nam videtur esse contradictio in his locis. Atque cum Salomon dicat in Ecclesiaste quod sol omni die secundum expositionem Hieronymi in originali flectitur ad aquilonem, vix aliquis mathematicorum potest hoc intelligere. Quoniam sciunt quod a solstitio hyemali usque ad solstitium aestivale flectatur omni die ad aquilonem per gradum unum fere. Sed in alia medietate anni e converso. Et de altitudine firmamenti in Ecclesiastico; et quomodo sol in meridiano exurit terram, ut ibidem dicitur, indigent astronomica capacitate. Atque Hyades et Pleiades et Arcturus atque Orion, et interiora Austri de quibus beatus Job loquitur, habent magnam difficultatem, praecipue cum dictat beatus Hieronymus super Isaiam, quod Orion habet xxii stellas, quarum novem primae sunt in tertia magnitudine, et novem aliae in quarta magnitudine, et quatuor residuae in quinta magnitudine, nec plus dicit. Sed haec sciri non possunt nisi ex octavo Almagesti, ubi sex gradus magnitudinis stellarum assignantur, et determinantur quae stellae sunt in qualibet illarum. Et quia infinita sunt alia in scriptura et in expositionibus sanctorum, quae scientiam coelorum et astronomiae judicia tangunt:

quapropter necesse est theologum bene scire coelestia, tum propter occupationem quaestionum in sententiis et summis, tum propter ipsum textum.

(2) Astronomy verifies Biblical geography.

Secunda radix astronomiae respectu theologiae et proprie respectu textus consistit in locorum mundi consideratione. Nam totus textus est plenus his locis, et ideo nihil magnificum sciri potest nisi sciantur haec loca. Tota enim series scripturae decurrit penes regiones, civitates, deserta, montes, maria, et caetera loca mundi, quorum certitudo non potest haberi nisi per scientias praedictas, quia harum proprium est distinguere partes habitabiles a non habitabilibus, et habitabile dividere in tres partes magnas, Europam, Africam, et Asiam, et istas tres in septem climata nota, praeter alia climata irregularia quamplura. Et haec climata certitudinaliter nisi per virtutem illarum scientiarum nescit quisquam separare in provincias et regiones et caetera loca, ut inveniantur civitates notae et famosae, sicut Jerusalem, Babylon, Meroe, Alexandria, Antiochia, Ephesus, Athenae, Tarsus, Roma, et caeterae prae aliis notatae ab astrologis secundum debitam distantiam ab invicem, et ab oriente et ab occidente, septentrione et meridie, quibus inventis poterunt et regiones famosae ab eis denominatae inveniri, et maria et deserta et montes et omnia quae in sacris literis continentur. Hic enim est magna utilitas istarum scientiarum in sacra scriptura. Et forte nihil utilius de philosophia poterit inveniri; quoniam qui ignorat loca mundi, ei multoties non sapit cortex historiae per loca infinita, et maxime propter falsitatem multiplicem bibliarum novarum; atque per consequens ad intellectus spirituales impedietur ascendere et non nisi imperfecte poterit eos explicare. Qui vero imaginationem bonam locorum habuerit, et situm eorum et distantiam et altitudinem et longitudinem latitudinem et profundum cognoverit, necnon et diversitatem eorum in caliditate et siccitate, frigiditate et humiditate, colore, sapore, odore, et pulchritudine, turpitudine, amoenitate, fertilitate, sterilitate, et aliis conditionibus expertus fuerit, et optime placebit ei historia literalis, et de facili atque magnifice poterit ingredi ad intelligentiam sensuum spiritualium. Non enim est dubium quin viae corporales significent vias spirituales, et loca corporalia significent terminos viarum spiritualium et convenientiam locorum spiritualium, quoniam locus habet proprietatem terminandi motum localem et rationem continentiae ; et ideo istorum locorum cognitio et literam facit, ut dictum est, intelligi, et vias parat ad intelligentias spirituales : quod multipliciter confirmatur per dicta et facta et scripta sanctorum.

Primo, quia dicit Hieronymus in prologo secundi Paralipomenon, 'Sanctam scripturam lucidius intuebitur, qui Judaeam oculis contemplatus est, et antiquarum urbium memorias, locorumque vel eadem vocabula vel mutata cognoverit.' Secundo, quia laboraverunt sancti in videndo loca illa et circumeundo. Propter quod beatus Hieronymus dicit praedicto prologo, 'Nobis curae fuit cum eruditissimis Hebraeorum hunc laborem subire, ut circumiremus provinciam, quam universae Christi Ecclesiae sonant.' Hoc autem non fecisset, nisi propter sacrarum literarum intelligentiam. Tertio, quia libros multos scripsit de locis mundi, distantiam et situm eorum et caeteras conditiones magna certitudine determinans. Orosius etiam ad Augustinum haec loca mirabili utilitate ac sincera veritate explicat evidenter. Isidorus autem in pluribus locis regiones et civitates utilius, si potest dici, quam priores determinat. Cassiodorus autem in climata distinguere non omittit. Eusebius etiam Caesariensis, ut narrat Hieronymus in libro de locis, post chirographiam terrae Judaeae et distinctas tribuum sortes ipsius quoque Jerusalem templique in ea cum brevissima expositione picturam ad extremum laboravit, ut congregaret nobis de sacra scriptura omnium pene urbium, montium, fluminum, viculorum, et diversorum locorum vocabula, quae vel eadem manent vel immutata vel aliqua ex parte corrupta. Origenes quidem Adamantius super Josuam in originali, sicut et in glossa super xviii Josuae recitatur, loquens de multitudine locorum in scriptura positorum, et inter caeteras laudes locorum istorum, sic nos alloquitur, dicens, 'Ne cum fastidio haec legatis, et putetis vilem scripturam ex multis nominibus contextam ; sed scitote in his contineri majora mysteria quam potest humanus sermo proferre, vel auditus mortalis audire.' Cum ergo nostri sancti expositores et doctores sacri in istis locis tantum laboraverunt, et tanta mysteria contineri fatentur, non est dubium quin eorum cognitio per omnem modum sacrae scripturae est necessaria. Sed proprium est astrologiae et astronomiae dare rationes locorum mundi et plenam certitudinem. Quapropter istae scientiae sunt valde necessariae in hac parte. Possunt autem pulchre verificari in exemplis. Qui enim audit historias versantes circa Jordanem, Jericho cum planitie sua, montem Oliveti, vallem Josaphat, Jerusalem, et non habet imaginationem locorum istorum et proprietatem eorum, sensum literalem ignorabit, et ei non immerito non multum sapiet cursus historiae, atque per consequens sensus spirituales obteguntur. Qui vero scit longitudines, latitudines, profunditates, altitudines, varietatem qualitatum, ut calidi, frigidi, sicci et humidi, necnon et eorum quae consequuntur ad haec quatuor, ut mollis, duri, grossi, subtilis, asperi, laevis, aridi, liquidi, lubrici, et aliorum innumerabilium, quae quarto Meteorologicorum determinantur; necnon et colores, et sapores, odores, pulchritudines, turpitudines, amoenitates, sterilitatem, fertilitatem, naturam infectivam, corruptivam, et eis contrarias, et caetera quae in locis considerari habent, potest sensum literalem ad purum concipere et in eo delectari, atque ad sensus spirituales gloriose et placide transire.

The spiritual meaning of Biblical geography.

Paucas enim conditiones locorum praedictorum considerando, possumus magnificos sensus exprimere moraliter et allegorice et anagogice. Jordanis quidem decurrit ab aquilone ad austrum, et jacet in oriente respectu Jerusalem, quae est in occidente non longe a mari magno posita, et inter illa duo primo a parte Jordanis est Jericho civitas cum sua planitie, deinde mons Oliveti, tertio vallis Josaphat, et tunc sequitur Jerusalem. Mundus autem, dicunt sancti, significatur per Jordanem secundum rationem interpretationis ejus et propter ipsius proprietates, quia currit in mare mortuum, quod est instar inferni, et propter multa alia. Et Jericho carnem significat, ut volunt sancti. Mons Oliveti significat excellentiam vitae spiritualis propter montis excellentiam, et dulcedinem devotionis propter Oliveti rationem. Vallis Josaphat significat humilitatem propter rationem vallis, et viam coram oculis majestatis, propter hoc quod interpretatio hujus nominis Josaphat est, in conspectu Domini. Et Jerusalem significat visionem pacis, et moraliter est anima sancta quae habet pacem cordis; allegorice significat Ecclesiam militantem; anagogice Ecclesiam triumphantem.

Qui igitur a principio vitae suae seu ab ortu nativitatis suae ab oriente rationis deliberantis seu usu rationis vult devenire saltem in occidente vitae suae et in senectute ad pacem cordis, et hoc moraliter; et ut sit membrum fidele et perfectum Ecclesiae, sub cujus umbra jaceat in pace contra insultus hostis maligni, et hoc allegorice; et ut sit ejus conversatio in coelesti Jerusalem in hac vita, et transferatur in morte ad illam civitatem sanctam, ubi sedebit in pulchritudine pacis in tabernaculis fiduciae vel requie opulenta; debet primo Jordanem, id est, mundum, aut sibi subjiciendo relinquere, ut sancti seculares, aut omnino renunciando recedere, ut religiosi. Ibi enim est primus gradus vitae spiritualis et aliis facilior. Quo facto, oportet aggredi carnem, quia ipsam non est ita facile vincere sicut mundum. Est enim pestis familiaris et non relinquens subjectum. Debet autem ipsam non destruere et cum impetu frangere, sed paulatim et discrete ejus superbiam domare. Propter quod consideratur Jericho cum sua planitie; et ideo debet poenitens plana via procedere, ut rationabile sit ejus obsequium, ne si carnem stulte obruat, non possit spiritus ad altiora pertingere. Hoc enim est contra multos ad poenitentiam conversos, qui primo anno vel secundo corpora sua destruunt, et postea fiunt inutiles, ut nec se nec alios valeant juvare. Postquam vero homo mundum subjugaverit, et carnem ut oportet domuerit, tunc et non ante est aptus ut ascendat ad excellentiam vitae spiritualis et dulcedinem devotionis. Ex tunc enim potest ascendere ad montem Oliveti et ad cacumen perfectionis attingere, atque in suavitatem orationis et contemplationis se immergere. Cum vero fuerit sufficienter exercitatus in ascensu et circuitu istius celsitudinis, adhuc [1] oportet vallem Josaphat transire, hoc est, totam vitam suam in perfecta humilitate debet concludere, ut sit pauper et humilis spiritu in conspectu Dei, non in oculis suis vel hominum. Multi enim apparent humiles sibi et aliis, et sunt coram

[1] J. has, 'tunc esset in Jerusalem nisi quod oportet.'

Deo et angelis superbissimi. Quando vero totam compleverit vitam perfecta humilitate, tunc est in Jerusalem, secundum ejus sensum triplicem. Habebit enim pacem cordis, quia pax talis sequitur vitae spiritualis perfectionem. 'Non enim est pax impiis,' dicit Dominus. Sed sanctis est pax Dei, quae exsuperat omnem sensum, et in pace Ecclesiae militantis quiescit securus, qua pace carent infideles et peccatores in statu damnationis existentes, quos agitat diabolus et turbat de peccato in peccatum, et de poena peccati ad novam poenam. Et[1] ut dictum est, participabit etiam in hac vita in spe certa et revelationibus illa beata visione pacis supernae Jerusalem, quam per gratiam Dei in morte consequetur.

Non solum autem haec loca inter Jordanem et Jerusalem cognita et historiam declarant et sensus spirituales explicant, sed alia loca innumerabilia quae inter duos hos terminos reperiuntur in scriptura. Si quis etiam velit ulterius considerare alias conditiones enumeratas, multo magis et quasi incomparabiliter poterit sensus divinos elicere, ut patet intuenti. Sed modo sufficit innuere quomodo ex paucis multa ex parvis magna ex planioribus obscura contingat elicere. Sed non possunt loca mundi sciri, nisi per astronomiam; quoniam oportet nos primo scire longitudines et latitudines locorum. Latitudo est ab aequinoctiali, et longitudo ab oriente, quatenus sciamus sub quibus stellis quae loca coaptentur, et quantum sunt a via solis. Nam secundum haec videmus sensibiliter res hujus mundi variari, et non solum in naturalibus, sed in moralibus. Oportet etiam per astronomiam scire, qui planetae dominentur quibus regionibus; nam secundum hoc potenter immutantur regiones. Et multa hujusmodi consideranda sunt per astronomiam, ut sciamus naturas locorum in scriptura; et non solum propter loca, sed propter res in eis locatas. Et rerum omnium cognitio est necessaria, tam propter sensum spiritualem, quam literalem, ut patet ex dictis.

(3) Astronomy verifies chronology.

Tertia radix est de temporibus. Totus enim cursus scripturae currit per tempora et secula et aetates a principio mundi usque ad Christum Dominum, et omnia sunt ordinata propter

[1] This and the following sentence are omitted in O.

ipsum, ut alius legislator non expectetur, sed quod ille solus sit salvator mundi per suam legem; quatenus error Judaeorum evacuetur de expectatione Messiae; et error Saracenorum de Mahometo qui secutus est Christum; et error eorum qui adhaerebunt adhuc legis latori nefariae qui venturus est, ut Albumazar docet libro Conjunctionum, qui in veritate erit Antichristus[1]; quatenus etiam omnes sectae paganorum, idololatrarum, Tartarorum, haereticorum, et caeterorum infidelium, qui per mundum dispersi sunt per sectas pene innumerabiles, deleantur per certificationem temporis salvatoris, ut nec ante eum nec post aliquis alius teneatur, per quem salus humani generis habeatur. Sed nullus potest certificare de temporibus, nisi astronomus, nec aliqua scientia habet de his certificare nisi astronomia. Omnes enim mendicant reliquias ejus in hac parte, ut manifestum est. Et si consideremus inveniemus multis modis quomodo astronomia est hic necessaria. Nam tempora ista, quaedam sunt lunaria, quaedam solaria et lunaria, quaedam habent principium determinatum, ut apud Judaeos astronomos. Incipiunt enim a lunatione Octobris annum, quia sunt usi ab antiquo tabulis et canonibus ad occasum solis civitatis Jerusalem. Sed adhuc gaudent uti eis propter terram quae data fuit eis a Deo. Quaedam vero habent principium indeterminatum, ut tempora solemnitatum apud Hebraeos, et tempora Arabum, et hoc differenter multum. Quaedam vero tempora sunt solaria, et horum quaedam semper habuerunt quartum ultra dies integros, ut tempus Graecorum et Latinorum; quaedam nunquam, ut tempora Persarum; quaedam aliquando sic aliquando non, ut tempora Aegyptiorum. Et principia annorum variantur apud ipsos, sicut canones astronomiae docent et Almagesti, et alia multa. Cum igitur in scriptura contineantur anni lunares et solares et Graecorum et Latinorum et hujusmodi, et volumus omnia tempora reducere ad annos solares, et ad annos Latinorum, qui sunt anni Christi, necesse est nobis in sacra historia scire horum temporum diversitatem, et ut sciamus quid est proprium cuilibet et quomodo aequantur ad invicem, et quomodo possumus extra-

Reduction of all Calendars to Christian standard.

[1] What follows to the end of the paragraph is omitted in O.

here majus de minori, et contrario, et quodlibet de quolibet. Sed impossibile est hoc fieri, ut exigit scriptura, nisi per canones et tabulas et caeteras considerationes astronomiae [1].

Caeterum computantur anni a principio mundi secundum Hebraicam veritatem, et computantur secundum LXX interpretes; sed discordant penitus, et omnes historiae et chronicae auctores et sancti hic contradicunt sibi invicem, non solum in toto tempore a principio usque ad Christum, sed in aetatibus particularibus. Sed diversitas haec non potest certificari, nisi per aliquam radicem certam. Nulla vero scientia potest hic invenire nec habet unde cogitet de tanta certitudine nisi astronomia, cujus est considerare revolutiones certas et ratas eclipsium et conjunctionum planetarum et caeterarum revolutionum coelestium, stante ordine naturae. Quapropter oportet quod astronomia apponat hic diligentiam. Et hoc possumus videre per libros astronomorum. Nam Ptolemaeus in Almagesti certificat nos de tempore a Nabugodonosor usque ad Alexandrum, et ab eo usque ad Octavianum Augustum, et ab eo usque ad Adrianum principem. Et Ecclesia tenet quod 41mo [2] anno Augusti fuit Dominus natus. Quapropter per hanc viam certificatur multum tempus a Nabugodonosor usque ad Christum [3]. Et si consideremus sententias Albumazar in libro Conjunctionum, videbimus quod ipse ponit principium mundi et primum hominem, scilicet Adam, et ab eo numerat annos usque ad diluvium, et ponit diem et horam quibus incepit diluvium, et per revolutiones planetarum et per eorum conjunctiones determinat sequentia secula, scilicet quando fuit Nabugodonosor, et quando Alexander, et quando Dominus Christus, et quando Mahometus, et sic de multis. Et ideo per artem Albumazar et Ptolemaei et caeterorum mathematicorum oportet resolvere tempora usque ad principium mundi, ut [4] sciamus quot anni sint a principio usque ad

[1] The whole of the foregoing paragraph, except the first two sentences, omitted in O.

[2] Sic in O. J. has xlii.

[3] This is given somewhat more fully in *Op. Tert.* ch. 54 (Brewer, p. 208). The time from the creation of Adam to the Deluge is stated to be 2,226 years, 1 month, 23 days, and 4 hours.

[4] The following page and half, down to the word dicentes, is omitted in O., but is supplied by the two Cottonian MSS.

Christum non solum in summa, sed per singulas aetates ac secula.

Date of creation.

Secundo, possumus persuadere de utilitate astronomiae propter temporis principium determinatum, scilicet an a lunatione Octobris seu aequinoctio autumnali, aut ab aequinoctio vernali. Nam multi voluerunt secundum sententiam vulgi, quod mundus fuerit creatus circiter aequinoctium vernale; sed alii apud aequinoctium autumnale; quia in veritate secundum Hebraicam veritatem, annus, quantum ad seriem temporis naturalem, incipit circiter aequinoctium autumnale. Et hoc manifeste potest probari per textum Exodi, ubi dicitur, quod scenopegia celebretur in exitu anni; id est in mense post exitum anni. Nam 23 habetur sic, 'Sollemnitatem quoque in exitu anni, quando congregaveris omnes fruges de agro.' Ergo post principium novi anni tunc incipit. Et 34 de eodem dicitur, 'Facies sollemnitatem, quando redeunte anni tempore cuncta conduntur.' Et in glossa Hieronymi Ezechielis primo habetur, quod October est primus mensis anni et Januarius est quartus. Et Nehemiae primo, 'Et factum est in mense Casleu, id est Decembri anno vicesimo'; et secundo, Factum est in 'mense Nisan anno vicesimo.' Si enim Nisan, id est Aprilis, esset caput anni non diceretur anno vicesimo sed anno primo et vicesimo, ut patet. Item, per praecepta de seminandis et metendis agris propter quietem septimi anni. Nam si Nisan, id est Aprilis, sit caput anni, tunc cum seminatur in Septembri vel Octobri sexti anni, non poterit meti in septimo anno, quia ferialis est secundum legem, et sic peribit seges. Item cum totus annus septimus ferialis est, tunc si incipiatur annus ab Aprili usque ad Septembrem non haberent quod comederent, quia fruges sexti anni non abundabunt nisi pro septimo anno et pro octavo et pro semine octavi, ut dicit Josephus, et non pro nono anno. Item per Hieronymum in epistola de solemnitatibus illud idem patet. Nam dicit sic, 'In fine anni solaris apud Hebraeos in septimo mense, quando congregantur fructus in horrea sive in cellaria, tunc sollemnia celebrare lege praeceptum est, prima die tubarum, decima die expiationum celebrari debere sabbata. Et a quinto decimo die usquedum finiantur octavae tabernaculorum, feriae esse praecipiuntur.'

Caeterum per Bedam libro temporum, et praecipue per Josephum et omnes Hebraeos a principio usque nunc patet, quantum ad initium sollemnitatum, quod Moyses constituit annum ab Aprili propter Pascha, quod est prima sollemnitas, et propter mysterium novi temporis, scilicet Christiani, cujus anni ab incarnatione computantur, quae fuit circiter aequinoctium, et in quo tempore nunc gloriosum Pascha celebratur. Sed in venditione et emptione et caetera gubernatione, quantum ad anni principium, prioris seculi Moyses decreta servavit, ut dicit Josephus primo antiquitatum. Prius autem seculum fuit a Moyse redeundo ad principium mundi. Ergo cum principium primi anni mundi et principium mundi fuerunt idem, concludunt isti quod mundus incepit circiter aequinoctium autumnale, ut post spoliationem fructuum veteris anni incipiat cultura novi. Haec autem alteratio est gravis valde, et ideo recurrunt homines periti ad scientiam cujus est certificare tempora, scilicet ad astronomiam. Et illi qui tenent primam opinionem volunt se tueri per astronomiam, dicentes quod mundus fuit creatus in meliori dispositione propter hominis et rerum generationem, et ideo debuerunt planetae esse in meliori situ suo respectu gubernationis mundi. Quapropter ponunt quod sol fuit in medio mundi creatus, ut in aequinoctiali circulo, ut aequaliter se haberet ad totum mundum. Et in Ariete eum posuerunt non in Libra, quia dicunt astronomi quod sol ibi habeat suam exaltationem, quae est major ejus dignitas vel secunda post majorem. Habet enim quinque dignitates et fortitudines, scilicet, exaltationem, domum, triplicitatem, terminum, et faciem. Et iterum ponunt, quod cum oporteat planetas habere eccentricos, necesse fuit mundo ut lux solis fuisset in Ariete, quia locus augis est longe nobilior quam alia pars eccentrici. Nam quum sol et luna et caeteri planetae sunt in augibus suis, tunc sunt fortioris et melioris operationis in hoc mundo, sicut determinant astronomi, et sicut experientia docet. Et ex his omnibus sequitur secundum hos quod mundus incipiebat ab aequinoctio vernali. Et quod objicitur eis, quod omnia terrae nascentia fuerunt creata in maturitate fructuum, quod non accidit per naturam, dicunt quod hoc fuit ex vi creatoris non ex vi naturae, cum tamen adjungant quod in multis regionibus australibus est calor veris magis conveniens pro fructibus maturandis quam aestatis. Et non solum in vere, sed iterum in autumno habent fructus propter temperamentum aequinoctii utriusque. Alii vero, qui ad plura respiciunt, scilicet ad rationes praedictas de principio anni et mundi, efficaces et insolubiles eis qui praedictam opinionem tenent, istorum sententias nituntur reprobare per vias astronomiae. Nam allegant, quod antiqui astronomi ponunt principium anni circiter principium Octobris, sicut patet in expositione tabularum, quae Almanac vocantur. Et dicunt quod Ptolemaeus invenit certitudinaliter augem solis in Geminis tempore suo. Sed si in principio fuisset in Ariete, sicut in exaltatione solis, ut in decimo quinto gradu vel decimo nono, in quibus ponitur solis exaltatio, moveretur aux secundum motum planetarum, scilicet secundum signa contra motum primi coeli. Sed hoc non potest esse, quia tunc regiones habitabiles super quas est aux fierent inhabitabiles per successum temporis, quando scilicet oppositum augis eveniret super eas; et e contrario inhabitabiles fierent habitabiles quod est absurdum: propter quod ponunt quod aux moveatur secundum motum coeli stellati, non quemcunque imaginatum, sed per motum Indorum et Thebit, scilicet per descensum et ascensum polorum ejus, vel per motum capitum Arietis et Librae coeli stellati in parvo circulo circa capita Arietis et Librae fixa, quae sunt in coelo nono. Quo quidem motu capita Cancri et Capricorni moventur progrediendo et regrediendo in superficie eclipticae zodiaci immobilis nunc ad orientem, nunc ad occidentem, sicut apparet ex imaginatione Thebit, qui super opera Ptolemaei addidit in hac parte juxta sententias Indorum. Sic enim assignatur motus octavae sphaerae. Et in hoc Arzachel in tabulis et canonibus concordat, et Albumazar in libro Conjunctionum, et omnes astronomi sic modo utuntur. Sed et sic ponunt augem solis moveri progrediendo et regrediendo, et ita non recedit a Geminorum signo, propter quod non vadit in circuitu terrae, ut oppositum augis aliquando contingat terras habitabiles.

Caeterum illi, qui ponunt modo praedicto, scilicet quod sol fuit creatus in Ariete, dicunt quod paradisus est sub aequinoctiali propter temperamentum maximum, quia uterque locus est nobilior mundi, ut ponunt. Sed si aux fuerit ibi in principio mundi, tunc fuit oppositum augis in loco Librae, quapropter in aequinoctio autumnali ejusdem anni fuit sol super eadem loca. Quare tunc fuit solutum temperamentum, et bonitas complexionis paradisi et regionis totius aequinoctialis violata singulis annis, quod est contra proprietatem illorum locorum secundum positionem istorum. Nam sic factae sunt inhabitabiles cito propter oppositum augis, quia tunc sol est prope terram et comburit eam. Sed ipsi volunt quod temperatior locus mundi sit sub aequinoctiali, et maxime in paradiso, quod stare non potest cum praedictis. Et ideo hi, qui sic objiciunt, volunt solem fuisse in Libra, quatenus statim peccatum Adae sequeretur poena, ut separatus a paradiso non posset pertingere ad ipsum propter combustionem regionis ultra aequinoctialem. Et ad hanc combustionem faciendam, statim post ejectionem Adae a paradiso usque citra aequinoctialem, necesse fuit ut sol foret in Libra, quia ibi jam constitutus fuerit versus oppositum augis ultra medietatem eccentrici, quatenus statim incipiet combustio, quod non accidisset in Ariete. Et multae considerationes astronomiae sunt hic. Et ideo oportet theologum in hac parte scire bene radices astronomiae. Position of Paradise.

Et tertio indiget theologus considerationibus astronomiae nunc tactis propter longaevitatem Adae et filiorum ejus usque post dies Noae, de quibus loquitur scriptura. Nam mirum est quomodo declinavit aetas humana, cum in principio fuit tantae longitudinis et tot tempora transivit. Et ideo multi imponunt bonitati constellationis coelestis in principio mundi, et circiter hoc. Et investigant illam bonitatem per dignitates planetarum, ponentes, ut dixi, solem esse in aequinoctiali et in Ariete et in auge, et planetas alios esse in bonis locis et convenientibus aspectibus, ut fortificaretur natura humana per bonitatem constellationis a principio, quatenus vivere posset ita diu. Sed paulatim deficiente bonitate dicta per mutationes situs planetarum et stellarum Longevity of patriarchs.

abbreviata est vita hominis ut ad aliquem terminum veniret ultra quem non contingat transire; quia omnium natura constantium est ratio et terminus, ut Aristoteles vult, quem terminum ponit scriptura circiter octoginta annos in vigore, sed amplius est labor et dolor. Sed jam tactum est, quomodo contra haec objiciunt alii mathematici. Et magna disputatio est hic; et forsan invenietur quod declinatio aetatis non est per recessum a bona mundi dispositione in principio, sed ob alias causas determinatas, ut suo loco inferius exponetur.

Date of deluge.

Et quarto necesse est principium anni certificari propter diluvium. Nam, ut dicit Josephus, diluvium fuit in Novembri. Dicit enim haec verba, 'Contigit autem haec passio sexcentesimo anno nativitatis Noae, mense secundo, qui a Macedonibus Dios nuncupatur, ab Hebraeis autem Maresvan.' Sed Dios, ut dicit Beda libro temporum, est November, non Maius, ut vulgus aestimat theologorum, et ideo Maresvan est November, sicut idem Beda dicit. Et est secundus mensis, quia October est primus in naturali ordine temporis, ut prius probatum est. Unde Beda dicit, Tisseri enim, qui est October, qui Maresvan praecedit, propter collectionem frugum et celeberrimas festivitates, et hunc Tisseri novum annum appellant Hebraei. Sic enim Aegypto in annum constituerunt, ut Tisseri, id est October, esset principium anni. Et ideo magister in historiis[1] et quidam glossatores non intellexerunt Josephum, cum credebant quod Dios et Maresvan essent Maius. Et ideo omnes sequentes eos decepti sunt propter Graeca et Hebraica vocabula mensium Graecorum et Hebraeorum, quae non intellexerunt, ut manifestum est inquirenti eorum opinionem, qui probant principium anni et mundi fuisse circiter aequinoctium anni autumnale. Et ideo hic cadit eadem perplexitas, quae prius. Unde necesse est, ut diluvium fuerit in Novembri, secundum quod probatum est superius tempus naturaliter incipere ab Octobri. Et Josephus hoc

[1] This is Peter Comestor (the devourer of books), author of the *Historia Scholastica*, and commonly called *Magister Historiarum*. The *Historia Scholastica*, compiled in the latter half of the twelfth century, became the classical text-book for Biblical students.

evidenter ostendit omni homini, qui scit vocabula Graeca et Hebraica de mensibus anni.

Non solum vero de principio mundi et anni naturaliter accidit dubitatio apud theologos; sed de principio diei naturalis, an scilicet nox praecesserit diem artificialem vel e contrario. Et hoc est quintum hic inducendum circa substantiam temporis. Et multi dicunt diem praecessisse noctem, et exponunt scripturam ut possunt. Sed secundum Hieronymum super Jonam et super Matthaeum, nox praecessit diem. Nam, ut ait Alfraganus[1] in astronomia sua, 'Omnes nationes, quae utuntur mensibus lunaribus, incipiunt diem ab occasu solis.' Sed Hebraei et scriptura utuntur mensibus lunaribus et annis, sicut potest probari modis multis. Ergo Hebraei et scriptura utuntur die naturali cujus nox praecedit diem. Et ideo tabulae Hebraeorum astronomicae, quibus Hebraei usi sunt in certificatione temporum, factae sunt ad occasum solis civitatis Jerusalem, sicut tabulae astronomorum Latinorum factae sunt ad meridiem civitatis Toleti vel alterius. Propter quod in lege determinatur, ut a vespera dies incipiat. Nam Levitici xxiii dicitur 'a vespere ad vesperum celebrabitis sabbata vestra.'

(4) Definition of chronological periods.

Quarta vero radix mathematicae respectu theologiae est penes accidentia et passiones temporum, cujusmodi sunt primationes et caeterae aetates lunae et embolismi et hujusmodi. Textus enim et expositiones doctorum requirunt magnam istorum cognitionem, et maxime secundum Hebraeorum considerationes tam astronomicas quam usuales. Differt autem haec radix a praedicta, quod illa consistit penes substantiam temporum, haec vero penes proprietates et accidentia. Considerari ergo oportet radicales veritates circa hujusmodi passiones, antequam convertatur sermo ad scripturam, quia aliter persuasio esset omnino inintelligibilis. Dico ergo quod primatio lunae secundum astronomos non dicitur a visione novae lunae apud Hebraeos, ut aliqui de theologis dixerunt, quoniam hoc tempus non est aequale, sicut ostendit Alfraganus. Sed lunatio una aequatur alteri. Aliquando enim in mane videtur novacula lunae veteris in suo decremento,

[1] See note on Alfraganus in the geographical section

et in eodem die in vespere videtur novacula lunae novae, et aliquando per spatium trium dierum distant, ut experimentum docet et Alfraganus declarat. Et ideo Hebraei antiquitus per astronomiam certificaverunt primationem lunae, et cum non fuerat in visione novae lunae, nec potuit per visum cognosci, accenderunt faces in Jerusalem in monte alto, ut sciretur quod tunc fuit tempus primationis, quatenus homines essent parati facere solemnitates et festa quae habebant expedire. Et neque consideratur lunatio penes conjunctionem solis et lunae veram, quoniam hoc tempus non est aequale, sicut certum est astronomis. Sed considerabitur luna penes conjunctionem solis et lunae mediam, sicut dicit Alfraganus. Nam hoc tempus est aequale semper. Nec tamen dicitur prima luna in conjunctione, sed post conjunctionem, quando luna in tantum separatur a sole ut sit de se visibilis, licet videri non possit. Tunc enim est prima accensio lunae, quamvis non videatur in illa hora. Et haec diversitas accidit propter latitudinem lunae diversam ab orbe signorum, et secundum quod est in signis obliqui descensus vel recti, et secundum diversitatem regionum septentrionalium et meridionalium, ut docet Alfraganus in vicesimo quinto capitulo sui libri. Currit autem aetas lunae a prima in viginti novem et parum plus. Ex quibus aetatibus aggregatur tempus lunationis mediae, seu aequalis, quem vocant astronomi Hebraeorum et Arabum mensem lunarem, licet aliis multis modis dicatur mensis lunaris. Et quamvis peritissimi astronomi in tabulis et canonibus ponant tempus aequalis lunationis esse viginti novem dies et triginta et unum minuta unius diei, et quinquaginta secunda, ut patet per Arzachelem in tabulis Toletanis; tamen Hebraei astronomi consideraverunt subtilius et melius. Tempus enim dictum continet viginti novem dies, et duodecim horas, et quadraginta quatuor minuta unius horae, sicut opus algoristicum expediet. Sed Hebraei dividunt unam horam in mille octoginta partes, et quodlibet minutum horae continet octodecim partes horae, ut patet ex reductione fractionum unius generis ad fractiones alterius. Et ideo tempus lunationis aequalis[1] apud Hebraeos,

[1] The mean synodic period of the moon, i. e. the mean period of its return to the position in which it is in the same direction with the sun, as determined

secundum quod respondet praecise lunationi Arabum, non potest esse plus quam viginti novem dies et duodecim horae et septingenti nonaginta duae partes unius horae. Sed Arabes in tabulis et canonibus computant diminute, et deficiunt in omni lunatione per tria secunda, et quindecim tertia, et quadraginta quatuor quarta, quod patet per examinationem legitimam. Et ideo Hebraei astronomi, volentes complere lunationem, apposuerunt unam partem, quia minus non potuerunt ponere secundum hanc divisionem qua usi sunt. Et ideo computant usque nunc in una lunatione viginti novem dies et duodecim horas et septingentas nonaginta tres partes horae. Et longe certior est eorum consideratio quam astronomorum utentium tabulis et canonibus apud alias nationes, quanquam et plus aliquantulum computant quam praecise exigat lunatio. Nam excedunt in quatuor tertiis et sexdecim quartis unius horae. Sed hoc longe minus est quam defectus Arabum praedictus. Quapropter satis melius computant Hebraei. Nec est curandum de excessu Hebraeorum praedicto; quoniam in maximo tempore minimus error contingit, et de quo non est curandum. Menses autem Hebraeorum, et scripturae usuales diversificantur. Nam unus est triginta dierum et alius viginti novem, quia vulgus non potest computare nisi per integros dies.

Lunar cycle of Jews.

His igitur consideratis propter scripturae et sanctorum intellectum, occurrit consideratio ulterior, ut sciamus quod Hebraei utuntur cyclo lunari, cujus consideratione indiget scriptura exponenda. Nam sicut nos utimur cyclo decemnovennali, sic ipsi cyclo lunari, cujus primus annus incipit in quarto anno nostro. Et ideo falsum dicunt qui non posuerunt Hebraeos uti cyclo; habent enim embolismos in suo cyclo, sicut nos in nostro; immo nos habuimus ab eis; et per embolismos aequant annos lunares solaribus colligendo undecim dies ter, ut faciant in tertio anno mensem embolismalem, id est superexcrescentem. Et nos hunc modum embolismorum traximus ab eis. Et ideo male dicunt famosi

by calculation from an eclipse observed by the Chaldaeans 720 B.C., and from one observed in Paris in 1771, is 29^{d}, 12^{h}, 44^{m}, $2^{s}.8$, a result agreeing almost exactly with the value given by the Jewish astronomers.

viri, qui negant Hebraeos usos fuisse cyclo. Et colligunt Hebraei tredecim cyclos lunares et faciunt tabulam et canonem ad hoc; qui tredecim cycli continent ducentos quadraginta septem annos, quia in tanto tempore redeunt omnes observationes festorum legalium ad idem temporis principium. Currit igitur observantia legalis penes hoc multipliciter, necnon alia quamplura. Nam neomeniae et calendae, in quibus est festum sacrificiorum, et epularum solemnium, de quibus dicitur primo Regum xx 'Cras calendae erunt, et requiretur sessio tua,' exigunt ut sciamus quod mensis lunaris vulgaris incipiat ab occasu solis. Sed lunatio ipsa non habet principium determinatum. Quare si contingat luna prima in occasu vel ante in aliqua hora diei naturalis praecedentis computabitur in vespera sequente novilunium, et neomenia et calendae et novus mensis, quia jam est luna prima. Si vero post occasum solis venerit, ut in secunda hora diei et ultra, non dicetur illa die naturali novilunium nec neomenia nec calendae, quantum ad initium calendae. Considerandum tamen quod mensis primus durat ab occasu solis primae diei usque ad occasum solis tricesimae diei, et tamen lunatio[1] non durat nisi a principio noctis usque ad mane tricesimae diei quantum ad dies integros, licet aliquae fractiones sint ultra. Non igitur incipit secundus mensis ante occasum solis tricesimae diei, sed lunatio ejus incipit in mane tricesimae diei, et ideo duae calendae attribuuntur secundo mensi, in quibus fiebant epulae et sacrificia, scilicet in die artificiali tricesimae diei mensis primi et in die naturali prima et tricesima, quia isti duo dies sunt de lunatione secundi mensis, licet secundus eorum tantum sit pars mensis secundi. Propter quod primo Regum xx dicitur, quod sedes David die secunda post calendas vacua apparuit. Unde accidit quod menses pares habent semper duos dies epularum, sed menses impares habent unum tantum. Et ex his patet, quod Ecclesiasticus dicit, 'A luna signum diei festi mensis secundum nomen ejus est.' Et ex his posset videri an luna fuit prima in principio mundi, vel plena, ut multi dixerunt. Nam Judaei et scriptura utuntur mensibus lunaribus. Ergo principium

[1] luna, J., an erroneous reading.

primi mensis et principium mundi fuit idem. Sed mensis lunaris a primatione lunae incipit. Quapropter in principio mundi luna prima fuit[1].

Noah's exit from the ark.

Et jam dicta necessaria sunt consideranti egressum Noae de arca. Nam glossae involvunt nos in dubitatione gravi, propter quod magister in historiis deceptus fuit, cum voluit Noae fuisse egressum xxviii luna eodem die quantum ad calendas quo intravit. Si enim hoc esset verum, tunc non tantum per annum, sed per annum et diem ibi fuit, eo quod annus solaris constat ex diebus ccclxv, qui complentur a prima die Januarii usque ad ultimam diem Decembris, qui est finis anni. Et hoc est quod dicit Beda de Temporibus, 'Noe cum ejus familia decima septima die secundi mensis arcam ingressus et xxvii ejusdem mensis die post diluvium egressus asseritur. Claret igitur annum solis integrum, id est ccclxv dierum, esse descriptum, quia videlicet luna, quae praesentis anni verbi gratia per nonas Maias decima septima existit, anno sequenti vigesima septima pridie nonas occurret.' Haec Beda. Quia quota est luna, si undecim addas tota erit eadem die, anno revoluto. Ut si hodie est prima, eadem dies revoluto anno erit duodecima. Hoc autem verum est, ut in pluribus. Tamen aliquando revolutio ejus est tantum undecima, aliquando tredecima. De xi, verbi gratia, si hodie pridie nonas Aprilis fuerit prima, eadem die revoluto anno erit xi. De xiii, verbi gratia, si hodie quinto nonas Maii fuerit prima, eadem die revoluto anno erit xiii. Numeret quis, et ita per se inveniet. Ista tota intricatio magistri ex glossa Strabi exordium sumit. Ut ergo concordemus glossam Strabi cum Beda dicimus, quod istud Strabi eadem die debet intelligi eadem feria, ut si dominica intravit, dominica exivit. Et quod sequitur, si praesenti diei addantur undecim, praesenti die computato cum xi, bene dicit, et hoc probatur per quod subdit, ideo post annum xi additis, fuit xxvii dies, vel xxvii luna. Nam si ipsa dies xxvii excluderetur, non xxvii, sicut ipse in glossa ponit, sed xxviii fuisse probaretur. Et hoc sic apparet. Nam cum xvii luna secundi mensis in arcam ingressus est, patet quod sequenti anno xvi luna

[1] Cf. *Op. Tert.* ch. 54 (Brewer, pp. 214-15).

secundi mensis annus lunaris cccliv dierum expletus est; a xvii autem die usque ad xxvii sunt xi dies, qui si praedictis cccliv addantur, fiet annus solaris ccclxv dierum, sicut Beda superius computavit. Sic ergo egressus est eadem qua ingressus fuerat feria, sed non eadem calenda. Et sic uterque verum dixit. Sed magister tamen dictum Strabi non bene glossavit. Quare vero xvii luna intra xi debeat concludi, patet per aliam glossam, quae xvii exclusive decem additis computat xxvii. Ergo si Strabus computat xi additis tantum xxvii et iste, x additis, similiter computat xxvii, patet quod ille inclusive, iste exclusive, intellexerit. Cum autem magister vult hic excusare se et dare rationem quare dixit egressum xxviii, cum Beda dicat xxvii, dicens, 'potuit enim esse quod vespera xxvii lunae egressus sit, jam imminente xxviii, media vero tempora quolibet nomine extremorum saepe nominantur'; illud nihil est, quia licet conjunctio solis et lunae vera vel media omni hora tam diei quam noctis evenire possit tamen quia Hebraei et Arabes computant secundum menses lunares, et luna appropriatur nocti, sicut sol diei, ideo dies et menses a vespera incipiunt semper. Et haec dicit Beda de temporibus; 'Quacunque hora accendatur luna, priusquam vespera veniat, non dicetur prima.' Si vero post occasum solis accendatur, non prima in praecedenti vespera, sed tricesima aestimabitur.

Date of law-giving on Sinai.

Quod etiam super Exodum de adventu filiorum Israel in solitudine Sinai pro lege Dei recipienda magister dicit, in exponendo scripturam xxx dies pro mense ponimus; hoc est falsum. Nam si quilibet numeratur xxx dierum, cum sint in anno xii menses, erunt in anno lunari ccclx. Sed patet quod superflue numerantur vi dies. Constat enim quod in anno lunari non sint nisi cccliv, cum annus solaris lunarem xi diebus superet, habens ccclxv dies. Ergo in duobus mensibus non sunt nisi lix dies. Et hoc est quod Beda dicit libro de Temporibus, 'Nonnullo moveor scrupulo quomodo majores nostri diem qua data est lex, quae est iii mensis tertii, quinquagesimam ab agni occisione computent, ponentes primi mensis residuos dies numero xvii, quia xiii priores fuerunt ante pascha transacti, secundi xxx, tertii iii, qui fiunt simul dies quinquaginta. Et constet duos menses lunares non lx sed lix diebus terminari; at per hoc in summa temporis memorati non plus quam xlix inveniri.' Haec Beda. Igitur secundum Bedam prima dies Junii non fuit xlvii, imo xlvi, et sic si lex data fuit quinquagesima die a pascha, non fuit data iii die mensis Junii, sed iv. Judaei vero dicunt quod die sexta fuit lex data, et quarta die adventus eorum in solitudine Sinai. Dixit enim Dominus, 'Vade ad populum, et sanctifica eos hodie et cras, et sint parati in diem tertium.' Et ita iste dies tertius fuit sextus eorum adventus; et secundum hoc quinquagesima die fuit lex data, non incipiendo a pascha, nec a prima die azymorum, sed ab altera die azymorum. Et hic computus consonat illi de quo habetur Levitici xxiii 'Numerabitis ab altera die sabbati, id est solemnitatis azymorum, in quo obtulistis manipulos primitiarum septem hebdomadas plenas usque ad alteram diem expletionis hebdomadae septimae, id est, quinquaginta dies.' Hic igitur magna involutio invenitur apud lectores propter errorem magistri in quantitate mensis et in sua computatione, sicut patet consideranti subtilius, et quia nescitur consideratio Hebraeorum his temporibus. Et tabula de cyclis lunaribus multum rectificat observantias legis. Quoniam secundum illam tabulam necesse est, quod licet principium anni sit lunatio Octobris secundum veritatem, tamen propter solemnitates legis implendas oportet quod triplicem annum habeant; unum communem, qui est secundum veritatem annus lunaris, scilicet habens cccliv dies, et unum diminutum, scilicet qui habeat cccliii dies; et unum superfluum, qui habeat ccclv dies; et sic variatur accidentaliter principium anni. Et hoc necessarium est, quia in die dominica non potest annus incipere. Quoniam si in dominica, tunc in xv die mensis septimi esset dies dominica, et in vigilia illius festi colliguntur rami de arboribus, quod non licet facere in sabbato. Similiter nec in die Mercurii nec Veneris potest esse anni principium. Nam si in die Mercurii, tunc decima dies mensis esset in die Veneris, in quo non licet aliquid facere, quoniam nec cibum parare, cum sit par sabbato. Quapropter tunc oportet eos facere cibaria.

Majora vero accidunt hic dubitabilia, nisi juvemur magnifice

Dates of Nativity and Passion.

per astronomiae potestatem. Nam incidunt quaestiones contra sententias omnium theologorum Latinorum. Sed propter rerum magnitudinem procedam opponendo ad utramque partem, et qui potest solvere gaudeat de solutione. Et tamen nihil contra opiniones communes Latinorum arguam, nisi quod nescio dissolvere. Utinam inveniam qui solvat, si conclusio falsa est. Sin autem vera possit esse, tunc non esset solutio necessaria. Sed nullus tantae auctoritatis est in ecclesia praeter summum pontificem, qui ausus esset dare sententiam contra sententias vulgatas in hac parte quamvis essent falsae. Tenet ergo vulgus Latinorum quod Dominus fuit natus secundo anno cycli decemnovennalis, et decimo anno cycli solaris, et in his non est dubitatio, et quod passus fuit viii calendarum Aprilis, et quod luna fuit xv in die passionis, de quibus est quaestio magna, atque Latini redarguunt Graecos, qui posuerunt passum Dominum xiv luna. Hoc magistri dicunt omnes, et Augustinus, Hieronymus, Beda, dant auctoritates ad hoc. Contra hoc potenter arguitur. Nam si viii calendarum Aprilis fuit passus et luna fuit xv, non potest hoc esse, ut Beda scribit libro temporum, nisi fuisset xiii annus cycli decemnovennalis; et hoc est verum. Quia secundum hoc oportuit quod aureus numerus fuerit xiii, ut luna diceretur prima in calendario, ubi xiii scribuntur, quatenus ab illo loco computetur aetas lunae, ut inveniatur xv in viii calendarum Aprilis; sicut quilibet potest experiri in calendario. Sed cum Dominus fuit natus secundo anno cycli, tunc in fine illius primi cycli habuit xviii annos secundum cyclum, quibus si xiii qui fluxerunt usque ad passionem addamus de cyclo secundo, erunt xxxi anni secundum cyclos. Sed hi anni secundum cyclum non sunt nisi xxix anni aetatis suae, et de xxx quantum est a nativitate usque ad passionem; quoniam prope finem secundi anni ipsius cycli fuit natus, scilicet ante principium tertii anni per vii dies tantum, quia annus incipit in circumcisione Domini. Quapropter in fine primi cycli non habuit Christus annos aetatis suae nisi xvii et dies vii, quibus si addantur xiii alterius cycli, erunt triginta; ita quod in xxx fuit passio. Ergo non habuit Dominus nisi xxix annos, et tantum de trigesimo quantum est usque ad passionem ; quod est contra evangelium Lucae, qui dicit 'Johannem Baptistam baptizasse Jesum incipientem quasi xxx annorum.' Et iterum secundum fidem evangeliorum certum est, quod praedicavit pluribus annis.

Theory of Bede.

Ita Beda nititur probare quod tribus annis et dimidio ultra xxx, ut in xxxiv anno fuerit secundum eum passus, id est, xviii anno Tiberii Caesaris. Et alia objectio fortior sequitur; nam si passus fuerit viii calendarum Aprilis et luna fuerit xv, tunc, ut prius dictum est, aureus numerus fuit xiii; sed vulgus Latinorum sequitur beatum Dionysium [1] abbatem Romanum, qui primus in chronicis suis instituit annorum computationem fieri a Christo, cum prius computabatur a tempore Dioclesiani sacrilegi, ut Beda scribit, et certum est; ita, quod transacto dxxxi anno ab incarnatione vel nativitate incipiebat suum cyclum magnum componere, qui tenet dxxxii annos, ex multiplicatione cycli decemnovennalis in solarem procreatus. Et incipit suum cyclum a dxxxii anno post incarnationem, et non a dxxxiii, quia Dominus fuit natus secundo anno cycli, et ideo oportuit novum cyclum incipere a dxxxii, scilicet transacto quingentesimo tricesimo primo anno a nativitate. Et quod tantum fuit, Beda docet. Nam dicit, 'Dionysius paschales scribit circulos, incipiens ab anno dominicae incarnationis quingentesimo vicesimo septimo.' Si igitur inveniamus literam tabularum, et revolvamus ab ea cyclum Dionysii bis et quantum ultra hoc est usque ad annum praesentem a nativitate Domini, inveniemus viii calendarum in passione esse in dominica, sicut quilibet potest experiri. Sed constat ipsum non fuisse passum die dominica, sed die

[1] Dionysius, commonly styled Exiguus, was the first to substitute the era A.D. for A.U.C. He placed the date of the Nativity four years later than that accepted by previous writers, namely 754 from the foundation of Rome, instead of 750. His Paschal Cycle of 532 years, resulted from the multiplication of the lunar cycle of nineteen years, and the solar cycle of twenty-eight years. Bede speaking of the era introduced by him, says (*De temporis ratione*, cap. xlvii), 'Primo decennovenalis circuli versu temporum ordo praefigitur quem Graeci calculatores a Diocletiani principis annis observatore. Sed Dionysius venerabilis abbas Romanae urbis et utriusque linguae, Graecae videlicet et Latinae non ignobili praeditus scientia, paschales scribens circulos noluit eis, sicut ipse testatur, memoriam impii et persecutoris innectere, sed magis elegit ab Incarnatione Domini nostri Jesu Christi annorum tempora praenotare.'

Veneris. Quapropter multi diligentes in chronicis, sed praecipue Beda, et Marianus Scotus, et Gerlandus[1] famosus apud omnes, dubitationem nobis relinquunt maximam penes computationem secundum cyclos Dionysi. Nam Beda dicit computanti, 'Gratias age Deo, si annum passionis Domini sic inveneris. Quare si non invenias, incuriae chronographorum aut tuae tarditati ascribas.' Et in duobus obviant ei, principaliter in hoc, quod posuit Dominum esse natum secundo anno cycli; et in hoc, quod posuit praecise dxxxi annum transivisse ab incarnatione vel nativitate Domini.

Theory of Marianus.

Marianus ergo in suis chronicis concedit, quod contingere potest, et verum est, quod sit luna xv in viii calendarum Aprilis et xiii in annis cycli. Sed tunc vel erit xii annus aetatis Christi, vel in cclix; sed neque fuit passus xii anno aetatis suae, neque in cclix. Quapropter non videtur salvari posse Dionysius. Et ideo Marianus diligenter attendens defectus Dionysii, volens adhuc salvare opinionem Augustini et Hieronymi dicit, quia octavo calendarum Aprilis fuerit passus, et similiter quod luna fuerit xv, sed non secundo anno cycli decemnovennalis, sed xviii, ita quod de illo cyclo sint accipiendi duo anni Christi, ut xviii cycli sit primus annus Christi, et xix sit secundus, et alter cyclus integer, qui simul juncti faciunt xxi annum. Et tunc xiii de tertio cyclo addantur et fient xxxiv secundum cyclum. Sed anni aetatis Christi sunt pauciores, scilicet xxxii anni et iii menses, et bene stant cum hoc numero. Et ideo hic Marianus addidit xxii annos ad tempus Dionysii. Et ideo in computatione

[1] Bede wrote two works on chronology. The first, *De temporibus*, was complained of by his disciples as being too short. At their instance, he wrote several years afterwards the much more elaborate treatise, *De temporum ratione*.

Marianus Scotus was an Irish monk of the eleventh century who entered an Irish monastery at Cologne, and lived subsequently in Fulda and in Maintz. He is the author of an universal chronicle beginning with the Christian era, and continued to the year of his death in 1082. He considered the Dionysian date of the Nativity to be twenty-two years too late. Marianus is not to be confounded with the abbot of St. Peter's, Ratisbon, a contemporary of the same name.

Gerlandus was a student in the Benedictine monastery of Besançon, of which he became prior in 1131. He wrote a *Computus* (i.e. a work on the rectification of the Calendar), and also an arithmetical treatise, mentioned by Cantor, vol. i. p. 769.

nostra a Christo, si volumus sequi Marianum, addemus semper xxii annos plures quam computet ecclesia secundum Dionysium. Sed secundum Bedam et secundum computationem certam annorum imperatorum, manifeste patet Marianum esse superfluum. Nam secundum eum a nativitate Christi usque ad cyclum Dionysii sunt dliii; quia ad dxxxi jam addit xxii. Sed superius patuit quod ab incarnatione Christi usque ad Dionysium non fuerunt tot, sed dxxxi. Quantumcunque enim largius computentur anni imperatorum, inveniemus quod ipse superfluit in xxii annis, vel in xxi, vel ad minus in xx, et ideo non est imitandus. Patet etiam per beatum Augustinum Marianum esse superfluum. Dicit enim Augustinus decimo octavo libro de Civitate Dei, sic computantur ccclxv anni a passione Christi usque in consulatum Honorii et Euticiani. Ergo iste annus, scilicet consulatus Honorii et Euticiani, a nativitate est cccxcvii, addendo annos a nativitate usque ad passionem, scilicet xxxii annos secundum Marianum et iii menses secundum superiorem computationem ; sed secundum computationem Mariani, sicut ipse in chronicis suis posuit, ccccxix fuit. Constat ergo eum esse superfluum xxii annis. Nam si ex ccccxix annis xxii dempseris, remanent cccxcvii.

Theory of Gerlandus.

Gerlandus autem famosus, quem omnes computistae sequuntur et astronomi in tractando ea quae ad computum pertinent, videns Dionysium non posse salvari nec sequentes ejus, nolens etiam Augustino et Hieronymo adhaerere de calenda passionis, sed beato Theophilo, qui fuit vicinus apostolis, ponenti quod decimo calendarum fuit Dominus passus, dempsit de computatione Dionysii vii annos, et posuit Dominum fuisse passum nono anno cycli. Nam cum Dionysius posuit Dominum natum secundo anno et dempsit unum annum cycli a nativitate demit Gerlandus illum i et ii usque ad viii inclusive. Et ideo Dominus fuit secundum eum xi annorum, completo illo cyclo, quibus si addatur unus cyclus integer et v anni, quia v est aureus numerus ut xv luna sit ibi, erunt xxxv anni secundum cyclos, et isti bene stant cum annis nativitatis Domini. Sed sicut Marianus fuit superfluus, sic Gerlandus est diminutus. Nam per computationem superius suppositam secundum Bedam, et secundum annos imperatorum, a nativitate Domini sunt tantum dxxxi vel circiter, ita quod ad minus defecit Gerlandus in v annis, quantumcunque largius annos imperatorum computemus.

Magister vero in historiis dubitationibus aliorum apponit novum dubitabile. Nam cum dicit in fine capituli de coena Domini, 'si revolvamus tabulam computi, inveniemus lunam xxii in calendis Aprilis et diem Veneris in tempore passionis: ergo viii calendas Aprilis fuit xv luna et dies Veneris,' multipliciter oberrat. Nam si sua sententia esset vera, tunc numerus aureus esset xiii, et tunc, sicut prius tactum est, fuisset passus ante xxx annos completos, quod est falsum. Item si tabulam revolvamus, ut magister dicit, inveniemus G esse literam dominicalem in tempore passionis, et G esse in viii calendarum. Quare fuit passus in dominica, quod est falsum et contra evangelium. Sed multi, ut Beda scribit, et maxime Victorius, ut patet in epistola sua ad Papam Hilarium de paschali observatione, dicunt Christum fuisse passum vii calendarum Aprilis et resurrexisse v. Sed tunc, ut Beda dicit, oportet quod fuerit annus secundus cycli decemnovennalis, quia binarius erit numerus aureus, supponendo quod Dominus fuerit passus xv luna, sicut in prioribus semper suppositum est. Sed quia xv anno Tiberii posuit Dominum passum, ut ex ejus opusculis patet, ideo merito redarguitur a Beda et aliis.

The Passion was on the fourteenth day of the moon.

Haec igitur secundum considerationem chronographorum discussa sunt, quae omnia sequuntur ex hoc, quod supponitur lunam fuisse xv in die passionis. Sed astronomi solliciti in hac parte non possunt invenire xv lunam nec viii calendarum in tempore passionis, nec x nec vii, ita ut ibidem inveniatur dies Veneris, quod oportet secundum Evangelia, nec etiam a xxx anno Domini usque ad finem vitae suae possunt invenire in aliqua calendarum. Quapropter summa dubitatio accidit in hac parte. Nam diligenter haec discussi tam per me quam secundum consensum peritorum in astronomia. Sed de xiv luna bene invenitur. Quapropter multa secundum scripturam sunt mihi et multis insolubilia in hac parte in contrarium, per quae ostenditur quod fuit passus xiv luna, sicut Graeci ponunt et Hebraei concordant. Nam dicitur in Matthaeo 'Non in die festo.' Sed dies festus est dies azymorum, qui instabat, et dies azymorum est xv. Ergo ante fuit occisus. Item Joannes xviii, 'Non introierunt in praetorium ut non contaminarentur, sed manducarent Pascha.' Ergo in vespera proximo ventura manducarunt Pascha; si igitur accipitur ibi Pascha pro agno paschali, in vespera illa incipiebat xv dies, et fuit luna xv, et tunc computabatur. Ergo ante illam vesperam fuit xiv luna. Quare occisus fuit Dominus xiv luna, sicut agnus paschalis in lege. Cum autem dicitur, quod Pascha ibi non sumitur dicto modo, sed aliter, hoc non potest habere auctoritatem ex scriptura, et ideo eadem facilitate contemnitur secundum Hieronymum, qua probatur. Caeterum cum dicunt Pascha hic accipi pro azymis, hoc esse non potest. Nam immundi, licet prohibeantur edere Pascha, id est agnum paschalem, non tamen prohibebantur edere azyma, si post comestionem agni fierent immundi. Immo si aliquis fermentum comederet, dicit lex, Exodi xii, quod periret de coetu Israel. Praeterea, nec inveniebatur fermentum in domibus eorum in illis diebus, quare tunc non comederent panem per vii dies, quod est omnino absurdum. Et ideo non habet haec responsio locum. Item Joannes xix, 'Erat autem parasceve Paschae.' Ergo eadem die ad vesperam paraverunt Pascha. Sed quando paraverunt Pascha incipiebat xv luna. Ergo cum fuit passus ante, fuit xiv. Item Joannes eodem, 'Ibi ergo, propter parasceven Judaeorum, quia juxta erat monumentum, posuerunt Jesum.' Ideo enim acceleraverunt sepelire, ne in die xv sepelirent, nam in die azymorum non sepelissent eum, quia nullum sepeliebant in praecipuis festis, sicut sunt Pascha, Pentecoste, Scenopegia, et hujusmodi. Item Lucae xxiii, 'Et revertentes mulieres,' scilicet in die crucifixionis, 'paraverunt aromata, et sabbato quidem siluerunt secundum mandatum.' Ergo illo die non fuit dies azymorum, sed xiv; non enim licuit eis parare aromata in die azymorum. Nam Exodi xii de prima et ultima die azymorum dicitur, 'Nihil operis facietis in eis exceptis his, quae ad vescendum pertinent.' Qua ratione enim siluissent die sabbato propter mandatum, eadem ratione siluissent in die Veneris, si fuisset dies azymorum. Nam praeceptum cadit super utrumque, licet sabbatum sit sanctius. Et Augustinus libro de quaestionibus novi et veteris testamenti dicit, quod fuerit passus xiv luna. Haec autem et multa alia hic adduci possunt cum exclusione falsarum responsionum.

From this with proper astronomical tables we should be able to fix the date.

Sed haec nunc sufficiant ut excitemur ad duo, scilicet ut sciamus quota luna Dominus fuit passus, an xiv vel xv, et si una illarum stabiliatur, tunc inveniatur dies passionis per tabulas astronomiae, ut tactum est. Sed haec omnia habent summam difficultatem, magis tamen propter hoc, quod theologi ignorant astronomiam et computum, et hujusmodi, quam propter difficultatem rei in se. Si enim essent periti in his, pro certo bene invenirent aetatem lunae et diem passionis, et mutarent multas sententias quas solemnizant. Nam peritissimi in istorum consideratione tenent quod Dominus fuit passus xiv luna. Et hoc verificato, facile est invenire calendas per tabulas speciales ad hoc compositas. Quatenus vero excitatio mentis fiat ad hoc, ponam hic unam tabulam, in qua secundum tabulas primationum inventa est oppositio solis et lunae per omnes annos Domini usque ad xxxviii, ut scilicet videatur in quo anno accidit in Martio circiter passionem Domini, et in quo in Aprili. Non tamen propter certificationem hujus rei pono hanc tabulam, sed pro exemplo, ut videatur modus persuadendi in hac parte ; nam certificatio est valde difficilis propter hoc, quod motus coelorum non sunt omnino certificati, nec tabulae qualescunque sufficiunt in hoc casu. Multi enim in astronomia periti laboraverunt hic, ut has oppositiones solis et lunae invenirent, et non potuerunt invenire annum passionis a xxx in xxxv, ubi in Martio esset oppositio in die Veneris, nec dies ante oppositionem nec proximus post, ita quod cum passione concordaret. Nec ego potui invenire adhuc. Ubi tamen poterit oppositio cum die Veneris inveniri secundum tabulam praesentem [1], patebit per ejus expositionem.

[1] This is evidently not the table referred to in *Opus Tertium*, chap. 54 (Brewer, p. 215). He says in that passage, '[Hebraei] posuerunt unam tabulam ex tredecim cyclis talibus (i. e. cyclis lunaribus), qua revoluta complentur omnes, et omnia redeunt ad idem temporis principium. Et hic cyclus cum canonibus suis et expositionibus est apud eos loci computi et kalendarii apud nos quantum ad

	Anni Domini.	Literae sextarum feriarum.	Dies proximus ante oppositionem.	Dies oppositionis [et ideo passionis signum, et similiter in alia tabula].	Dies proximus post oppositionem.	Dies sequens proximum.	Post quot dies de Martio et horas et minuta fuit media oppositio, quantum ad meridiem Novariae. Dies.	Horae.	Minuta.	Secunda.	Tertia.	Quarta.
	1	g	b	c	d	e	27	14	31	2	0	0
	2	f	e	f	g	a	16	23	19	41	8	48
	3	e	b	c	d	e	6	8	8	20	17	36
	4	c	f	g	a	b	24	5	41	2	42	8
	5	b	b	c	d	e	13	14	29	41	50	56
	6	a	e	f	g	a	2	23	18	20	59	44
	7	g	c	d	e	f	21	20	51	3	24	16
	8	e	f	g	a	b	10	5	39	42	33	4
Principium cycli alia via, sc. rationis.	9	d	c	d	e	f	28	3	12	24	57	36
	10	c	g	a	b	c	18	12	1	4	6	4
	11	b	c	d	e	f	7	20	49	43	15	12
	12	g	g	a	b	c	25	18	22	25	39	44
	13	f	c	d	e	f	14	27	11	4	48	32
	14	e	g	a	b	c	4	11	59	43	57	20
	15	d	e	f	g	a	23	9	32	26	21	52
	16	b	g	a	b	c	11	18	21	5	30	40
	17	a	d	e	f	g	1	3	9	44	39	28
	18	g	b	c	d	e	20	0	42	27	4	0
	19	f	e	f	g	a	9	9	31	6	12	48
Principium cycli secundum computistas.	20	d	b	c	d	e	27	7	3	48	37	20
	21	c	e	f	g	a	16	15	52	27	46	8
	22	b	b	c	d	e	6	0	41	6	54	56
	23	a	f	g	a	b	24	22	13	49	19	28
	24	f	a	b	c	d	12	7	2	28	28	16
	25	e	e	f	g	a	2	15	51	7	37	4
	26	d	b	c	d	e	20	13	23	50	1	36
	27	c	f	g	a	b	20	22	12	29	10	24
	28	a	c	d	e	f	10	19	45	11	34	56
	29	g	g	a	b	c	28	4	33	50	43	44
	30	f	c	d	e	f	18	13	22	29	52	32
	31	e	a	b	c	d	7	10	55	12	17	4
	32	c	c	d	e	f	26	19	43	51	25	52
Hic.	33	b	g	a	b	c	14	4	32	30	34	40
	34	a	e	f	g	a	4	20	5	12	59	12
	35	g	a	b	c	d	23	10	53	52	8	0
	36	e		d	e	f	0	19	42	31	16	48
	37	d	a	b	c	d	19	17	15	13	41	20
	38	c	e	f	g	a	9	2	3	52	54	8

Dies proximi ante oppositionem [mediam].	Dies oppositionis mediae.	Dies proximi post diem oppositionis mediae.	Secunda dies post diem oppositionis.	Post quot dies de Aprili et quot horas et minuta et tertia fuit media oppositio, quantum ad meridiem Novariae; tamen in duobus annis, 17. sc. et 36. ponitur post quot de Martio; et secundum hanc computationem incipit dies in medio diei praecedentis, et terminatur in medio sui.						
d	e	f	g	26	2	55	5	15	44	April.
g	a	b	c	15	12	3	44	24	32	April.
c	d	e	f	4	20	52	23	33	20	April.
g	a	b	c	22	18	25	5	57	52	April.
d	e	f	g	12	3	13	45	5	40	April.
g	a	b	c	1	12	2	24	15	28	April.
e	f	g	a	20	9	35	6	40	0	April.
g	a	b	c	8	18	23	45	48	48	
d	e	f	g	26	15	56	28	13	20	
b	c	d	e	17	0	45	7	21	48	
e	f	g	a	6	9	33	46	30	56	
b	c	d	e	24	7	6	28	55	28	
e	f	g	a	13	15	55	8	4	16	
b	c	d	e	3	0	43	47	13	4	
f	g	a	b	21	22	16	29	37	36	
b	c	d	e	10	7	5	8	46	24	April.
e	f	g	a	30	15	53	47	55	12	Mart.
c	d	e	f	18	13	26	30	19	44	April.
f	g	a	b	7	22	15	9	28	32	
c	d	e	f	25	19	47	51	53	4	
g	a	b	c	15	4	36	31	1	52	
c	d	e	f	4	13	25	10	10	40	
a	b	c	d	23	10	57	52	35	12	
b	c	d		10	19	46	31	44	0	
g	a	b	c	1	4	35	10	52	48	
d	e	f	g	19	2	7	53	17	20	
a	b	c	d	9	10	56	32	26	8	
e	f	g	a	27	8	29	14	50	40	
a	b	c	d	16	17	17	53	59	28	
e	f	g	a	6	2	6	33	8	16	
b	c	d	e	34	23	39	15	32	48	
e	f	g	a	13	8	27	54	41	36	
a	b	c	d	2	17	16	33	50	24	
f	g	a	b	21	14	49	16	14	56	
b	c	d	e	10	23	37	55	23	44	April.
e	f	g	a	30	8	26	34	32	32	Mart.
c	d	e	f	18	5	59	16	57	4	April.
f	g	a	b	7	14	47	56	5	52	

Tabulae explicatio.

Prima igitur linea in prima tabula inferius occurrens tenet annos Domini usque ad xxxviii; quia certum est infra hos ipsum subiisse passionem. Secunda linea habet omnes literas feriarum sextarum, quae contigerunt in illis xxxviii annis. Tertia linea tenet dies proximos ante oppositionem solis et lunae. Quarta, diem oppositionis. Quinta, diem post oppositionem immediate. Sexta, diem sequentem, et sumitur hic oppositio solis et lunae media. Septima linea cum sibi annexis tenet tempus transactum de Martio ante diem passionis. Reliqua vero tabula negotiatur circa Aprilem, ut inveniatur oppositio in Aprili circiter diem passionis, excepto quod xvii anno et xxxvi accidit in Martio. Sciendum autem quod hae tabulae factae sunt ad meridiem civitatis Novariae, licet fuerit facta Parisius; sed causa fuit, magis enim secreta est, et meridies ibi praecedit meridiem Parisius per xxv minuta unius horae. Si igitur de tempore habito subtraxerimus xxv minuta, relinquetur tempus oppositionis post medium diem civitatis Parisius. Secundum ergo hanc tabulam passus fuit Dominus iii nonas Aprilis die Veneris in oppositione solis et lunae mediae xv anno cycli decemnovennalis et xiv cycli solaris, anno xxxiii ab incarnatione secundum cyclum Dionysii, et hoc est, xxxii secundum veram aetatem Domini. Et istud in tabula secunda accidit super b literam in directo xxxiii anni post duos dies de Aprili et xvii horas, et xvi minuta, et xxxiii secunda, l tertia, xxiv quarta, et sic sapientissimi in his considerationibus aestimaverunt, qui

multa. Et hanc tabulam literis Hebraicis misi in *Opere Majori* cum ejus expositione et canonibus suis, secundum quod pertinet ad computum eorum.' He refers to this table again in the following chapter, p. 220. The table given in the *Opus Majus* is the one spoken of in the *Opus Tertium*, p. 223.

What seems probable is that Bacon included with the *Opus Majus* the whole or part of a work written by him in 1263, entitled *Computus*, of which a good description will be found in Charles's work on Bacon, pp. 336–8. The *Computus* contains a full discussion of the solar and lunar year and month, of the lunar cycle of nineteen years and its insufficiency; and a series of tables comparing the Mahommedan and Christian calendars.

The table here given contained in the first edition several errors which have been corrected by collation of O. with Tib.

multum laboraverunt ad hoc probandum. Unde secundum eos, quod viii calendas Aprilis passus fuerit, non est opinio ecclesiae certificata, sed vulgata, sicut multa alia vulgantur, quae indigent majori certificatione. Si igitur oppositio media et xiv luna ab accensione concurrant in unum in passione, res manifesta est secundum hanc tabulam et secundum aestimationem sapientum. Si vero oppositio praecessit in passione xiv lunam per diem integrum, oporteret recurrere ad tabulam accensionis novae lunae factam consimiliter huic, et tunc magis excluderetur dubitatio. Sed hujus rei certificatio, sicut aliorum quae scribo, desiderat Vestrae Celsitudinis assensum, quatenus periti in hujusmodi veritatem firmiter stabilirent. Occasionem vero dedi in hoc capitulo qualiter ad hujus rei certificationem poterimus pervenire, et nego me in praesenti tractatu de tanta difficultate certificasse.

(5) Geometry throws light on many theological questions.

Dictum est, quod septem sunt radices mathematicae, quibus necesse est uti mathematica. Una est de coelestibus. Alia de locis mundi. Tertia de temporibus quantum ad substantiam. Quarta de passionibus et accidentibus temporum, de quibus dictum est. Nunc volo afferre quintam radicem, et est de figurationibus geometricis penes lineas, angulos, et figuras tam corporales quam superficiales. Impossibile enim est quod sensus spiritualis sciatur, nisi fuerit literalis scitus. Sed sensus literalis sciri non potest, nisi homo sciat significata terminorum et rerum significatarum proprietates. Nam in illis consistit profunditas sensus literalis, et ex eis extrahitur sensuum spiritualium altitudo per convenientes adaptationes et similitudines, sicut sancti docent, et patet ex proprietate scripturae, et omnes sapientes antiqui sic tractaverunt scripturam.

The Ark.

Cum igitur opera artificialia, ut arca Noae, et tabernaculum cum vasis suis et omnibus, atque templum Salomonis et Ezechielis et Esdrae et hujusmodi alia pene innumerabilia ponantur in scriptura, non est possibile ut literalis sensus sciatur, nisi homo ad sensum habeat haec opera depicta, sed magis figurata corporaliter; et sic sancti et sapientes antiqui usi sunt picturis et figurationibus variis, ut veritas literalis ad oculum pateret, et per consequens spiritualis. Nam in veste Aaron, ut dicit scriptura, erat descriptus orbis terrarum et parentum magnalia. Et ego vidi Aaron sic figuratum cum veste sua. Sed nullus posset de hujusmodi corporum figuratione cogitare nec ordinare, nisi optime sciret libros Elementorum Euclidis et Theodosii et Millei[1] et aliorum geometrarum. Nam propter horum ignorantiam apud theologos non solum decipiuntur in maximis, sed in minimis. Dicunt enim cum magistro in historiis quod sphaerulae candelabri fuerunt corpora circularia, non habentes plenam sphaericitatem. Sed non dicuntur sphaerulae diminutive propter hoc, sed quia fuerunt parvae sphaerae, habentes tamen completam definitionem sphaerae Euclidis et Theodosii. Nam Hebraei in Gallico suo illas sphaerulas vocant pomeas a rotunditate pomali et sphaerica, et sic de aliis infinitis. O quam ineffabilis luceret pulchritudo sapientiae divinae et abundaret utilitas infinita, si haec geometricalia, quae continentur in scriptura, figurationibus corporalibus ante nostros oculos ponerentur. Nam sic mundi malitia diluvio gratiae deleta, attolleremur in sublimi cum Noe et filiis et omnibus animantibus suis locis et ordinibus collocatis. Et cum exercitu Domini in deserto excubaremus circa tabernaculum Dei, et mensam propositionis, et altare, et sancta sanctorum, ac cherubim obumbrantia propitiatorium, et caetera illius antiqui populi insignia tanquam praesentia videremus. Deinde tabernaculi vacillantis instabilitate vacuata fixum Domini templum Salomonica sapientia fabricatum intraremus. Et cum Ezechiele in spiritu exultationis ad sensum intueremur, quod ipse tantum spiritualiter intellexit, ut tandem reparata nova Jerusalem cum Esdra et Nehemia intraremus majorem domum pleniori gloria decorandam. Certe ipsa visio sensibilis esset pulchra, sed pulchrior quando figuram nostrae veritatis videremus praesentialiter, pulcherrima vero quando

Aaron's vestments.

The tabernacle and temple.

[1] Milleius is the Arabic transformation of Menelaus of Alexandria, a mathematician and astronomer of the first century A.D. Two of his astronomical observations, made in the first year of Trajan, are mentioned in the *Almagest* (vii. 3). His work on Spherical Trigonometry (for his discussion of spherical triangles may be so called) is not extant in the Greek original, but has been preserved in Hebrew and Arabic translations. It was edited by Halley in 1758.

scripturae intellectum spiritualem et literalem contemplantes gauderemus visibilibus instrumentis excitati, quod scimus omnia nunc in ecclesia Dei esse completa, quae ipsa corpora sensibilia nostris oculis exhiberent. Et ideo nihil reputo dignius studioso in sapientia Dei, quam hujusmodi figurationes geometricas ante ejus oculos exhiberi. Utinam jubeat dominus quod haec fiant. Et sunt quatuor homines vel tres, qui ad haec sufficerent, sed sunt peritissimi Latinorum; et merito debent esse periti, quoniam ineffabilis difficultas hic latet propter obscuritatem textus sacri, et propter contrarietates sanctorum et diversitates caeterorum expositorum.

The rain-bow.

Sed aliter patet geometrica utilitas respectu sapientiae divinae, tam in textu quam in questionibus; et non solum in his, sed in comparationibus pulchris respectu gratiae et gloriae et poenae futurae, et cautela vitiorum. De quolibet ponam aliquod exemplum. Et pro omnibus in generali revocemus ad memoriam, quod nihil sciri potest de rebus hujus mundi sine geometrica potestate, ut prius probatum est. Et rerum cognitio necessaria est in scriptura propter sensum literalem et spiritualem, ut superius expositum est. Nam proculdubio tota rerum mundi veritas jacet in sensu literali, ut dictum est, et maxime rerum geometricarum, quia nihil est nobis ad plenum intelligibile, nisi figuraliter ante oculos nostros disponatur; et ideo in scriptura Dei tota rerum sapientia figurationibus geometricis certificanda continetur et longe melius quam ipsa philosophia possit exprimere. Nec mirum, cum ipse Deus auctor totius sapientiae suam ordinavit scripturam. Quare pro infinitis exemplis volo unum ad praesens afferre. Nam Aristoteles magis omnibus philosophantibus nos involvit suis obscuritatibus in tractando de iride, ut nihil per eum quod dignum sit intelligamus, immo multa falsa in translatione Latinorum continentur, sicut ex interpretum varietate contendimus. Nam quod in codicibus Latinorum habetur, quod iris non accidit ad radios lunae nisi in quinquaginta annis bis, manifeste falsum est, sicut quilibet potest in plenilunio quando pluit experiri, dummodo claritas lunae non impediatur per nubium densitatem. Et Avicenna dux ac princeps philosophiae post Aristotelem, ut clamant omnes, seipsum ignorasse iridis naturam humiliter confessus est. Et sic de omnibus philosophis certum est, quod nullus potuit scientiam iridis obtinere. Nec mirum, cum sacram scripturam non ita diligenter perscrutati sunt ; sicut fuerat eis necesse. Nam omnes philosophi ignoraverunt causam finalem iridis. Sed finis imponit necessitatem eis quae sunt ad finem, ut Aristoteles dicit secundo Physicorum, et certum est in omnibus. Finem autem, propter quem est iris, solus textus Dei explicat evidenter, scilicet, cum dicitur, 'Arcum meum ponam in nubibus coeli,' &c. Ex quo habetur, quod contra diluvium et abundantiam aquarum est arcus Dei ordinatus. Ergo oportet, quod quandocunque apparet hic arcus in coelo, sit valida consumptio humiditatum aquearum ; et hoc est verum. Nam nubes resolvuntur abundanter, et rorationes fiunt infinitae, sicut philosophi dicunt, et nos videmus in magna parte. Sed consumptio humiditatis aqueae non est nisi propter aliquid quod habeat virtutem consumendi. Nihil autem in generatione iridis invenimus nisi radios solis et nubes. Nubium congregatio est causa materialis ; ergo radiorum projectio est causa efficiens. Sed radii incidentes non possunt magnas et mirabiles operationes perficere, quia non concurrunt ad invicem ; concursus autem virtutum exigitur ad hoc, quod valida operatio educatur. Sed concursus non potest fieri nisi per reflexionem et fractionem. Quapropter oportet, quod iris generetur per infinitas reflexiones vel fractiones in stillicidiis infinitis sine intervallo cadentibus, ut sic tam colorum quam figurae veritas per hujusmodi multiplicationes penes figuras, angulos, ac lineas inveniatur, et non per diversitatem materiae nubis ut in textu Latinorum continetur et omnes credunt, sicut certis experimentis explicabo cum de scientiis experimentalibus faciam mentionem. Sicut ergo philosophi propter ignorantiam scripturae sacrae non potuerunt scire veritatem iridis, sic eodem modo impossibile est philosopho infideli attingere ad plenam certitudinem creaturae alicujus

[1] Lunar Rainbow. Aristotle's words are (*Meteorologicorum*, iii. cap. 2) : *ἐν τῇ πανσελήνῳ γὰρ γενέσθαι ἀνάγκη τὸ μέλλον ἔσεσθαι, καὶ τότε ἀνατελλούσης ἢ δυνούσης· διόπερ ἐν ἔτεσιν ὑπὲρ τὰ πεντήκοντα δὶς ἐνετύχομεν μόνον.* Bacon had good cause to complain of Aristotelian translation.

propter scripturae ignorantiam. Nam in veritate quaelibet creatura ibi accipitur secundum ultimam sui dignitatem, scilicet secundum veram definitionem et descriptionem ejus, quia Deus fecit creaturas quas posuit in scriptura, et solus ipse scit eas sicut sunt. Cum igitur geometrica potestas requiratur ad cognitionem omnis creaturae corporalis, non est dubium quin ineffabili modo valeat sapientiae divinae propter rerum cognitionem.

Explanation of Ecclesiasticus, cap. 43. v. 4.

Sed rediens spiritualiter ad propositum pono exemplum de scriptura, quae dicit, 'sol tripliciter exurit montes[1],' &c. Nam et incidunt radii infiniti ad omne punctum montis, propter quod infiniti exeunt a quolibet puncto solis, et lux est causa caloris maxime cum congregatur, ut sensibiliter scimus, et infiniti reflectuntur a superficie montis, quia a denso fit reflexio, et congregantur in omni puncto aeris, et distrahunt omne punctum aeris, rarefacientes aerem prope montes, et sic secundo calefaciunt montes. Et per medium nubium franguntur radii duplici fractione, primo in superficie nubium versus solem, dein in aere inter nubes et montes. Et per hanc duplicem fractionem est necesse omnes radios solis venientes ab uno puncto concurrere ad invicem in punctum unum in monte vel aere prope montes, praecipue in bene altis montibus et Alpibus[2]. Et sic fiet calor in montibus, licet non in altissimis; quoniam montes qui ascendunt usque ad nubes vel prope sicut sunt montes Italiae et Hispaniae, et mons Caucasi et Caspiae et alii innumerabiles, habent frigus magnum et nives quasi perpetuas; quia nimis accedunt ad medium interstitium aeris, qui est locus frigidissimus in hoc mundo. Et sic a singulis punctis solis radii venientes ac dupliciter fracti incedunt et augmentant calefactionem montium. Et haec omnia patent per figurationes mathematicas, sicut accidit de beryllo rotundo vel crystallo, vel urinali, vel quocunque perspicuo grossiori aere. Quando enim haec corpora

[1] This passage in the Revised Version is given thus: 'A man blowing a furnace is in works of heat; but the sun three times more, burning up the mountains:' a rendering which would dispense with Bacon's laborious attempt to explain *tripliciter*.

[2] The following sentence seems to imply that Alps are spoken of in the sense usual in Switzerland, of the lower mountain pastures.

occurrunt radiis solaribus, accidit per duplicem fractionem quod possit ad oculum et ad tactum eligi punctus aeris post hujusmodi corpus, in quo stuppa sicca vel aliquid bene combustibile accendatur, ut prius habitum est. Et haec triplex calefactio potest vocari triplex combustio, de qua loquitur scriptura ; vel aliter possumus dicere quod facit unam montium combustionem de tribus de quibus scriptura intendit, cum dicit, 'sol tripliciter exurit montes.' Nam alio modo consideratur combustio penes varietatem angulorum, quia multiplicatio lucis vel est ad angulos rectos, vel obliquos, vel nullos. Si ad obliquos, est fortis actio ; si ad rectos, fortior est per xviii primi Elementorum Euclidis et per alias vias ; si ad nullos, fortissima propter omnimodam aequalitatem, ut prius habitum est. Sed ab omni puncto solis ad quodlibet punctum montis objecti soli veniunt radii, et infiniti cadunt ad angulos rectos, infiniti ad obliquos, ut manifestum est. Quamvis autem superficies montium non est ubique polita et lenis, nec sphaerica, tamen in multis partibus ejus est hujusmodi, et infinities per portiunculas diversas, et ideo ratione sphaericitatis hujusmodi accidit casus ad nullos angulos, et ideo est triplex incidentia. Quapropter accidit triplex radiorum casus penes angulum et privationem anguli. Et in omnibus modis est calefactio per casum lucis et congregationem. Quapropter secundus modus combustionis, de quo loquitur scriptura, potest hac via nunc dicta assignari. Tertio vero casus lucis variatur penes figuras, nam cadit in sphaerica figura et pyramidali cujus conus est in sole et basis in monte, atque in pyramidali cujus conus est in puncto montis et basis in sole. Et omnibus his tribus modis figuratur multiplicatio lucis a sole in monte, ut certum est ex supradictis. Sed multiplicatio lucis sphaerica licet possit esse fortis, tamen pyramidalis, cujus conus est in sole, est fortior ; et pyramidalis, cujus conus est in puncto montis, est fortissima, quia, ut dictum est per illam pyramidem a tota superficie solis objecta monti venit lux ad quodlibet punctum montis, quoniam basis illius pyramidis est superficies solis, et infinitae tales pyramides veniunt ab eadem solis superficie, sicut manifestum est figuranti. Et natura non requirit plures figuras in multiplicatione lucis, nec in alicujus virtutis vel speciei multiplicatione, ut a quocunque agente fiat species multiplicanda, ut certum est scienti naturae vias per geometricam potestatem. Et per has tres multiplicationes figuratas oritur combustio tertia, de qua loquitur scriptura. Sed haec sciri non possunt, nisi homo optime sciat potestatem geometricam.

Propter vero quaestiones pono unum exemplum pro mille, ubi geometrica requiritur potestas, licet vulgus non advertat. Nam de luce et multiplicatione ejus multa quaerunt theologi, scilicet utrum sit substantia vel accidens, utrum sit corpus in medio, utrum subito vel successive feratur in medio et caetera. Sed haec nullo modo sciri possunt sine eis quae dicta sunt de multiplicatione secundum lineas angulos et figuras. Nam multiplicatio lucis est sicut multiplicatio omnis alterius speciei cujuscunque agentis. Lux enim in medio est species lucis quae est in corpore luminoso. Et ideo regulae, quae dictae sunt de speciei multiplicatione sumptae generaliter, intelliguntur in luce et in qualibet specie determinata. Cum ergo multiplicatio speciei absolute consideratae requirat maximam geometriae potestatem, ut prius ostensum est abundanter, manifestum est quod ad lucis multiplicationem intelligendam eadem geometriae potestas necessaria est, quamvis theologi non utantur.

Symbolic meaning of direct, refracted, and reflected rays.

Deinde de spiritualibus rebus exprimendis per res geometricas pono exemplum in gratia et gloria et salvandis ac damnandis, ut videamus quomodo lineae rectae, fractae, et reflexae, valeant hujusmodi spiritualibus adaptari. Et cum gratiae infusio maxime manifestatur per lucis multiplicationem, expedit per omnem modum ut per multiplicationes lucis corporales manifestentur nobis proprietates gratiae in bonis, et repulsa ejus in malis. Nam in bonis perfectis infusio gratiae comparatur luci directe incidenti et perpendiculari, quoniam non reflectunt a se gratiam, nec frangunt per declinationem ab incessu recto, qui attenditur secundum viam perfectionis vitae. Sed infusio gratiae in imperfectos, licet bonos, comparatur luci fractae; nam propter imperfectiones eorum non tenet gratia in eis incessum omnino rectum. Peccatores autem, qui sunt in peccato mortali, reflectunt et repellunt a se gratiam Dei, et ideo gratia apud eos comparatur luci repulsae seu reflexae. Sed sicut corporum, a quibus reflectitur lux, quaedam sunt aspera, a quibus lux reflexa dissipatur nec apparet; et quaedam sunt polita et aequalis superficiei et laevis, a quibus fit reflexio sensibilis, ut patet in speculis; sic peccatores existentes in mortali peccato sunt dupliciter. Quidam enim sunt, qui sic repellunt gratiam, quod nihil boni apparet in eis, sed apud se et apud proximos manifestum est quod dissipatur totus gratiae effectus in eis, et isti sunt aperte mali qui peccata sua non abscondunt. Alii vero sunt, qui licet boni non sint, tamen occultant peccata sua, et apparent boni in conspectu hominum, ut hypocritae. Et potest aliter fieri comparatio ista. Nam peccatores in hac vita comparantur corporibus reflectentibus, ut dictum est, et homines in gratia in hac vita comparantur illis quibus est fractio lucis propter imperfectiones humanae fragilitatis, quia quantumcunque sit homo perfectus in hac vita, tamen habet multas obliquitates, et imperfecta est caritas in eo et notitia Dei. Sed existentes in patria comparantur illis rebus quae recipiunt lucem penitus secundum rectum incessum omni exclusa declinatione, quantum possibile est creaturae, et quantum exigit ordo divinae justitiae, qui reddit unicuique secundum quod meruit. Quoniam vero damnati omnino carent gratia et gloria Dei, et glorificati ratione imperfectionis meriti in multis atque ex lege creationis (cum omnis creatura habeat imperfectionem respectu gloriae conditoris) sunt quodammodo imperfecti, potest fieri tertia comparatio secundum statum damnatorum et glorificatorum et Dei, quatenus apud damnatos teneat omnino comparatio lucis reflexae, et apud glorificatos valeat similitudo lucis fractae, et divinae gloriae summa rectitudo et perfectio notentur, quomodo melius probatur trinitas personarum et unitas, per incessum lucis directum, secundum quod possumus et secundum quod licet nos exemplificare per creaturas ea quae sunt apud creatorem.

Geometrical symbol of Trinity.

Et quod valde notandum est adjungam, scilicet quod impossibile est beatam trinitatem et essentiae unitatem aptius a nobis repraesentari in exemplo creaturae sensibilis quam per res geometricas. Nam in solo triangulo inter omnes res factas invenitur unitas essentiae cum distinctione trium occupantium eandem essentiam. Quoniam idem spatium numero et totum capit quilibet de angulis trianguli, ut patet ad sensum, et tamen veraciter sunt anguli distincti, quod est mirabile in creatura, nec alibi reperitur nisi in summa trinitate. Et cum super datam lineam necesse est triangulum aequilaterum collocare, ut prima propositio Euclidis denunciat, quid magis proprie potest assumi ut intelligamus quod data persona Dei patris necesse est trinitatem personarum aequalium exhiberi?

Symbolic sense of rational and irrational numbers.

Haec et hujusmodi multa possunt ex geometricis ad sapientiam salutarem pertrahi, et maxime de decimo libro, ubi rationalitas et communicatio quantitatum declaretur. Nam haec et ad divina et ad virtutum figurationes cum exclusione vitiorum utiliter et praeclare possunt adaptari. Nam quicquid de rationabilibus quantitatibus continetur ibi potest pulchre et proprie ad vitam rationabilem, quae in virtute consistit, facile pertractari. Et quod de irrationabilibus quantitatibus exponitur potest moribus irrationabilibus et bestialibus peccatorum conformari. Quicquid etiam de communicatione quantitatum dicitur potest convenienter trahi ad actus caritatis. Et quod de incommunicatione proponitur, ad odium et divisionem animorum noscitur pertinere. Sed longiorem sermonem haec requirunt. De cautela vero peccatorum magnum adjutorium est per considerationem hujusmodi geometricarum multiplicationum. Nam cum res delectabiles hujus mundi sint nobis sicut muscipula, quibus capiamur ad peccatum, ut in libro Sapientiae scribitur; et sicut est in hamo diaboli, qua strangulemur, nisi quantum Deus dignatur nos custodire, et sensus nostri super res delectabiles delati nuntiant intellectui occasionem peccati; primum et principale remedium quod in homine sit est ut in quinque sensus suos species rerum delectabilium, ut mulierum, et ciborum, et divitiarum, in quibus humana cupiditas aestuat et ambitio superba gloriatur, non recipiat secundum multiplicationem principalem; nam accidentalis sufficit homini ad damnationem, quando recipitur avide et abundanter. Sed magis adhuc cavenda est multiplicatio recta sensibilis quam fracta vel reflexa, propter fortiorem actionem prius expositam. Maxime vero, ne species delecta-

And of propagation of rays.

bilis recta cadat ad angulos aequales in sensum, quia tunc fortissima est, praecipue si conus pyramidis brevis occurrat. Sic Eva recepit speciem soni serpentis et pomi visibilis et suavis odoris. Et Adam allectus est ut se et totum genus humanum specierum sensibilium multiplicatione damnaret. Sic David sanctus propheta per speciem Betsabeae deceptus de adulterio cecidit in homicidium. Sic senes presbyteri quos judicavit Daniel specie mulieris decepti sunt. Et certum est omnem hominem decipi speciebus rerum hujus mundi; quia secundum Johannem, 'omne quod est in mundo, vel est concupiscentia carnis, vel concupiscentia oculorum, vel superbia vitae,' a quorum amore nullus se potest abstinere, qui gaudet in sensu specierum quae veniunt a rebus. Nam capitur sicut bestia. Et ideo homines sanctitati vacantes avertunt sensus ab omnibus rerum delectabilium speciebus quantum possunt, et maxime de illis quae pertinent ad sensum tactus et gustus, quae sunt gulae et luxuriae, ut non solum non tangant, sed nec videant nec audiant de his fieri mentionem, ne species multiplicata in sensus spiritum cogat servire carnis illecebris; et praecipue cavent ne prope sint, ut vitent pyramides breviores, atque multiplicationes principales et rectas et ad angulos aequales, ut praedixi. Et haec maxime cavenda sunt quando oportet vel in confessione vel ob aliam aliquam causam loqui cum mulieribus. Nam omnes homines quantumcunque sanctos species fortes in hac parte turbarent, etsi non vincerent quos Deus custodire dignatur.

(6) Sixfold utility of arithmetic in theology.

Jam dictum est de quinque radicibus mathematicae respectu scripturae; nunc remanent nobis duae aliae, scilicet numeri et musicalia. Quoniam vero sancti abundanter ostendunt utilitatem istorum respectu theologiae, et magis exprimunt horum comparationes et laudes respectu divinorum quam de prioribus, ideo plura de aliis tetigi, et de his solum innuam quasdam vias, quarum explicatio in libris sanctorum copiose invenitur. Numeri vero ad sensum scripturae valent quatuor modis. Nam non est numerus in scriptura positus, quin accipiatur in sensu literali secundum omnes ejus proprietates arithmeticas, quatenus per congruas similitudines eliciantur sensus spirituales, sicut sancti infinitis quasi modis manifestant. Secundo, propter historias sciendas et certificandas necesse est theologum abundare in potestate numerandi, ut sciat omnes modos algoristicos, non solum in integris sed in fractionibus, quatenus sciat numerare, addere, minuere, mediare, multiplicare, dividere, et radices extrahere, tam fractiones quam integra; et iterum quod non solum sciat vulgares fractiones penes medietates, tertias, quartas, quintas, et sic in infinitum; sed quod sciat astronomicas fractiones per minuta, secunda, tertia, quarta, quinta, et sic in infinitum; quia in his historialibus oportet recurrere ad motum solis et lunae, in quibus tales fractiones considerantur principaliter, de quibus satis exemplificatum est superius. Et non solum oportet fractiones Latinorum scire et Arabum, sed Hebraeorum, qui frangunt unam horam in mlxxx partes de quibus tactum est prius. Caeterum illum, qui uti debet his, scire necesse est reducere diversa genera fractionum ad invicem; quia si inter integra accidant septem quintae, et decem septimae, et viginti octavae, et hujusmodi infinitis modis, non potest homo se expedire, ni istas diversitates fractionum reducat ad unum genus fractionis, ut sic reducantur ad integra. Sed in his fractionibus et integris utendis magna est subtilitas, et pulchra sapientiae lumina relucescunt, praecipue cum per haec tota sapientia divina et humana dirigitur et rectificatur, secundum quod auctoritates Cassiodori et Isidori prius manifeste expresserunt, qui crudeliter nos redarguunt ex eo, quod omnes ignorantes numerorum potestatem asserunt non differre a brutis animalibus, quae nec divina nec humana cognoscunt. Tertio, valet consideratio numerorum in scriptura, scilicet in divisione et distributione sortium praedarum et portionum facienda sacerdotibus, Levitis, principibus, et singulis, secundum quod Moyses peritissimus in numerationibus algoristicis distribuit unicuique quod suum est, infinitam quasi multiplicationem rerum distribuens in partes certissimas operibus algoristicis propriissimis tam in fractionibus quam in integris. Et nullus literalem sensum, nec per consequens spiritualem potest elicere, ut exigit dignitas scripturae, nisi operum horum sciat rationem et noscat operationes numerandi per quas Moyses et alii in scriptura procedunt.

Quarto[1], necesse est optime scire rationes numerandi propter corruptionem numerorum in scriptura, quia quasi infinitis modis corrumpuntur; fere enim omnes numeri sunt corrupti. Quae corruptio ad veritatem reduci non potest, nisi per omnimodam numerandi potestatem tam in fractionibus quam in integris. Fere enim aut pro majori parte omnes numeri in sacro textu et in libris sanctorum sunt scriptorum et correctorum vitio depravati, de quibus pauca exempla subjungam pro infinitis. Nam undecimo capitulo Genesis habetur, 'Vixit Arphaxat, postquam genuit Sale, ccciii annis.' Sed in Hebraeo habentur cccciii anni. De Graeco vero non potest haberi veritas, quia inter Arphaxat et Sale interponit Cainan, quem nos non habemus, sicut nec Hebraeus. Et in eodem capitulo, 'Vixit Reu xxxii annos' secundum Hebraeum et antiquos codices Latinorum. Sed in exemplari vulgato habentur xxxv, ut historiae veritas continuari non possit. Et xviii capitulo in penultima parte ejus, 'Quid si minus quinquaginta justis quinque fuerint, delebis propter quinque.' Sic Hebraeus, sic antiqui libri habent. Moderni autem posuerunt xlv propter sensum planiorem. Sed non licet sine sede apostolica mutare textum quem recepit et jussit per omnes ecclesias derivari. Et xxxvii capitulo, 'Joseph cum esset xvii annorum' secundum Hebraeum et Graecum. Sed moderni habent xvi, et quod in eodem capitulo non sit venditus xxx argenteis, sed xx, probatum est superius multiplici testimonio. Et in Exodi xiv capitulo, 'tulitque sexcentos currus,' secundum Hebraeum et Graecum et Josephum et antiquos codices. Sed pars glossae truncata fere in omnibus libris facit quod exemplar vulgatum habet ccc currus. Nam communiter habetur hoc tantum in glossa 'contra fidem trinitatis pugnaturus.' Unde propter nomen trinitatis abraserunt sexcentos et posuerunt ccc. Sed in pluribus antiquis glossatis invenitur sic, 'contra fidem trinitatis numero senario se armavit pugnaturus.' Et ideo secundum

Arithmetical errors in received version of Bible.

[1] This paragraph supplies further illustrations of the corruption of the Paris text of the Bible, spoken of on p. 77, and also in the *Opus Minus* Brewer, pp. 330-349). It is also one of many proofs of the care with which Bacon had collated the Septuagint and the Hebrew text.

glossam possunt ibi esse sexcenti. Et Exodi xxxii habetur in Hebraeo et Graeco 'quasi tria millia.' Sed exemplar vulgatum habet viginti tria millia. Et dicit magister in historiis quod hoc est de antiqua translatione. Sed antiqua translatio est LXX interpretum, et est de Graeco, Graecus autem non habet hoc, sed sicut Hebraeus. Et hic error habet apparentiam ex prima ad Cor. x. Sed non est intelligendum pro hoc loco, ut patet in textu et in glossa. Quoniam cum dicit, 'Neque idololatrae efficiamini,' &c. quod ad hunc locum pertinent, non ponit numerum interfectorum. Sed statim cum dicit, 'Neque fornicemur' &c. sequitur, 'et ceciderunt viginti tria millia,' quod intelligitur secundum glossam de his, qui fornicati sunt cum Madianitis, ut legitur xxv Numerorum. Nolo plura exempla aggregare ad praesens, cum non sit ex principali intentione corruptionem textus probare. Sed haec ideo intuli, ut videatur necessitas computandi in scriptura, quatenus sciamus verificare corruptiones numerorum.

Properties of number three.

Et adhuc potest quinta ratio de numerorum utilitate signari. Nam unitas in se ducta cubice, id est, ter, ut semel unum semel non multiplicat essentiam, sed eadem manet licet aequalitate trium laterum deducatur. Et sic familiari exemplo theologi notant beatam Trinitatem. Quamvis autem multiplex numeri perfectio inveniatur secundum quod denarius dicitur esse perfectus, et septenarius, et senarius, tamen maxime ternarius sibi vindicat perfectionem; quoniam perfectio praecipua quae numeris aliis attribuitur est quod omnes partes aliquotae simul sumptae reddant suum totum, ut patet in senario. Sed in ternario solo accidit, quod pars aliquota et non aliquota simul sumptae reddunt ipsum, scilicet unitas et binarius, quod non potest in alio inveniri. Atque cum in omni numero sint duo, scilicet discretio partium et unitas congregationis earum, qua numerus dicitur esse unus; nec aliquid aliud facit numerum nisi haec duo; ista praecise reperiuntur in ternario. Per binarium enim est discretio, et per unitatem ipsa unitas habetur. Et ideo tota perfectio numeri in ternario et non in alio reperitur. Et propter hoc hic numerus convenit creatori magis quam alius. Et sicut ab unitate omnis multitudo procedit, sic ab ipso Deo. Et arith-

Three kinds of proportion.

metica docet in fine suo omnes medietates investigari in proportionibus numerorum. Nam nec geometra nec musicus nec astrologus de his tractat, sed supponit quod arithmetica, quae est scientia numerorum, edocet. Theologi autem nituntur has medietates ad proprietates divinarum personarum exaltare, et praecipue Ricardus de Sancto Victore in libro de trinitate, qui liber magnam habet difficultatem ob istarum medietatum ignorantiam apud vulgus. Et multi propter eandem abutuntur comparationibus istarum medietatum ad divina. Triplex vero est medietas, scilicet arithmetica, geometrica, musica. Arithmetica vero consistit in identitate differentiarum primi ad secundum, et secundi ad tertium, ut quatuor, tria, duo. Nam unitas sola est differentia hinc inde, et huic simile reperitur in divinis personis. Nam differentia Patris ad Filium est quod Pater dat Filio et non accipit ab eo; Filius autem e converso. Similiter inter Filium et Spiritum Sanctum est consimilis differentia. Nam Filius dat Spiritui Sancto et non accipit ab eo; Spiritus autem Sanctus e converso accipit ab eo et non dat ei. Et cum geometrica medietas sit, quando est eadem proportio primi ad secundum, et secundi ad tertium, ut inter viii, iv, ii, ratio hujus medietatis per omnimodam aequalitatem et unitatem essentiae in tribus personis reperitur, et in omnibus essentialibus, quae sunt potentia, sapientia, bonitas. Nam credimus et firmiter tenemus quod qualis est Pater talis est Filius, et talis Spiritus Sanctus. Harmonica seu musica medietas consistit in identitate proportionis primi ad tertium, et differentiae primi et secundi ad differentiam secundi et tertii, ut vi, iv, iii. Nam dupla proportio est inter primum et ultimum; similiter inter binarium, qui est differentia primi et secundi, et unitatem, quae est differentia secundi et tertii. Et sic per similitudinem in personis divinis reperitur. Nam quae est proportio Patris ad Spiritum Sanctum quantum aequalitas; sic est inter differentiam Patris et Filii, et inter differentiam Filii et Spiritus Sancti. Nam inter Patrem et Filium est differentia, quod Pater dat et non accipit, Filius accipit et non dat; et haec eadem est inter Filium et Spiritum Sanctum, ut patet. Sic igitur breviter assignantur hae medietates, quamquam et aliter a multis assignentur, quos non possum intelligere; quia videtur mihi quod non possunt secundum arithmeticae judicium salvari. Sed non est temporis praesentis opiniones singulorum explicare.

Dimensions of heavenly bodies.

Et sexta ratio de utilitate arithmeticae potest sumi in scriptura penes res hujus mundi, quibus ipsa utitur. Nam altitudinem et magnitudinem et spissitudinem et numerum coelorum et stellarum tangit et requirit certificationem istorum Et sancti in expositionibus suis multa loquuntur. Nam Hieronymus dicit super Isaiam, quod Orion habet xxii stellas, quarum ix primae sunt in tertia magnitudine, et ix residuae in quarta, et iv ultimae in quinta magnitudine, et non exponit se. Oportet ergo theologum haec non ignorare. Sed specialiter arithmetica rectificat in his, et ideo volo hic has veritates magnificas aperire per numeros, quatenus pulchre et utiliter apparet comparatio numerorum. Ad hoc autem considerandum necesse est ponere aliquam radicem notam. Haec autem est quantitas arcus terrae, quae respondet uni gradui in coelo, secundum quod docet Alfraganus capitulo viii.

Terrestrial arc corresponding to a degree.

Et Averroes consentit in fine secundi Coeli et Mundi. Isti vero dant modum certificandi hoc in numero milliariorum et partium ejus. Ptolemaeus autem quinta Almagesti dictione procedit per viam demonstrationis diffuse in quantitate altitudinum solis et lunae et in eorum magnitudine. Sed non dat quantitatem certam per numerum milliariorum; nec de aliis corporibus coelestibus determinat quantitatem. Oportet igitur supponere, quod cubitus aequalis et geometricus contineat pedem et dimidium, et milliare continet 4,000 cubitorum, et sic accipit Alfraganus in sua consideratione. Omnis autem circulus sphaerae potest dividi in 360 partes, quae gradus vocantur. Intelligamus igitur maximum circulum in sphaera coelesti, qui transeat per centrum et dividat sphaeram in duas partes aequales. Dico ergo quod uni gradui[1] istius

[1] The first attempt to compare angular astronomical magnitudes with terrestrial distances was that of Eratosthenes (B.C. 276–196), who found that when the sun was vertical at Syene, it was one-fiftieth part of a great circle (i.e. 7° 12′) from the zenith at Alexandria. The distance between the two places was

circuli in coelo respondebunt multa milliaria in terra, cujus experientiam innuit Alfraganus in hoc, quod accepta in aliquo loco elevatione poli super horizonta, si perambuletur ad septentrionem vel meridiem, usquequo elevatior polus appareat vel depressior per gradum unum, invenitur quantitas milliariorum terrae, quae respondeat uni gradui in coelo. Nam si in nocte clara stellata quis per foramina quadrantis vel astrolabii vel alterius instrumenti perspexerit stellam nauticam et ipsum polum, et notaverit gradus quos attingit extremitas virgae in dorso astrolabii vel filum in quadrante, processeritque in terra ad septentrionem donec in altera nocte stellata viderit eundem polum elevari plus super horizonta per unum gradum, ille arcus terrae, quem perambulaverit, respondebit uni gradui, et erit ei similis, ut sumuntur arcus similes in sphaeris diversis secundum Theodosium, sed non erunt propter hoc aequales.

Cum autem per semidiametrum terrae, quae continet 3,250 milliaria, doceat 21° capitulo mensurari diametros orbium coelestium et distantias augium et oppositorum earum, oportet quod veram quantitatem semidiametri accipiat. Nam aliter magnus accideret error in distantiis augium, eo quod quantitas semidiametri terrae multotiens replicata faceret magnum errorem in illis distantiis, nisi praecise sumeretur. Cum igitur fatendum est Alfraganum percepisse hunc errorem, manifestum est quod ipse accipit veram semidiametri quantitatem, scilicet 3,250, et veram diametrum, scilicet 6,500. Ergo oportet quod supponat radicem veram et completam, quae est in quantitate arcus terrae respectu gradus coeli, licet non exprimat eam perfecte. Quapropter ipse supponit quod sit 56 milliaria, et duo tertiae milliarii, et 27 nonagesimae, et una sexcentesima tricesima, vel 56 milliaria et 2,984 cubiti et

Dimensions of the Earth.

regarded as 5,000 stadia. But the uncertainty as to the Greek measures of length, and the coarseness of their astronomical instruments (independently of the fact that Alexandria and Syene are not on the same meridian), make it impossible to deduce any precise result from this observation. The Arabian instruments were better, but were obviously insufficient as a basis for solving the problem here discussed by Bacon, of the Earth's magnitude.

It may be noted here that in the oldest MSS. of the *Opus Majus*, as well as in those of a later date, Arabic numerals are very commonly employed in dealing with large numbers.

quinque septimae unius cubiti. Et ideo, si quis bene consideret, ipse respectu diametri et semidiametri, quibus utitur, omittit octavo capitulo in hac radice, quae est quantitas arcus terrae respectu gradus in coelo, 50[1] sexcentesimas tricesimas unius milliarii, sive quod idem est, 317 cubitos, et tertiam cubiti, 28 sexagesimas tertias unius cubiti, quoniam non exprimit ibi nisi quod arcus iste terrae est 56 milliaria et duae tertiae unius milliaris. Sed licet radicem plenam supponat, tamen fractiones aliquas omittit propter taedium numerorum. Mos enim ejus est in libris suis multotiens omittere fractiones et similiter faciunt alii autores. Si igitur volumus huic quantitati diametri, scilicet 6,500 qua utitur, adaptare radicem, dicemus quod arcus terrae respondens uni gradui in coelo continet 56 milliaria, et duas tertias unius milliaris, 27 nonagesimas milliarii, et unam sexcentesimam tricesimam. Et si volumus computare per cubitos, erit arcus terrae respondens uni gradui in coelo 56 milliaria, et duo millia cubitorum, nongenti octoginta quatuor cubiti, et 8 sexagesimae tertiae unius cubiti, quae 8 sexagesimae tertiae sunt plus quam una octava cubiti per unam quingentesimam quartam unius cubiti, quod de facili patet. Nam octo sexagesimae tertiae valent 64 quingentesimas quartas, et 63 quingentesimae quartae sunt octava totius; ergo 64 quingentesimae quartae excedunt octavam in una quingentesima quarta; et ita octo sexagesimae tertiae unius cubiti excedunt octavam ejus in una quingentesima quarta. Et si triplicaverimus diametrum hanc 6,500 et ejus septimam addiderimus, habebitur circumferentia totius terrae, et erit praecise viginti millia milliariorum quadringenti et viginti octo milliaria, et duo millia ducenti octoginta quinque cubiti, et quinque septimae unius cubiti; vel penes alias fractiones, erit numerus hic vigesies mille quadringenti viginti octo milliaria, et quatuor septimae unius milliarii. Et secundum hoc tota terrae superficies erit centies trigesies bis mille millia milliaria sexcenties mille milliaria. Et secundum hanc radicem perfecte computatam erit quarta terrae habens 33,150,000 milliaria in sua superficie. Et octava terrae

[1] The omission is greater than Bacon states; i. e. it is $\frac{190}{630}$ of a mile.

habebit 4,143,750 milliaria. His enim duabus quantitatibus indigemus, sicut et caeteris praedictis.

Dimensions of the celestial spheres.

His visis, consideranda est altitudo coelestium, et similiter magnitudo et spissitudo. Nam Alfraganus dicit xxi capitulo, quod Ptolemaeus et alii sapientes posuerunt medietatem semidiametri terrae esse quantitatem qua metiti sunt longitudines a centro terrae, et posuerunt corpus terrae quantitatem qua mensuraverunt corpora stellarum. Et hoc satis patet ex demonstrationibus Ptolemaei quinta dictione Almagesti. Sententiat ergo Alfraganus ex comparatione semidiametri terrae ad semidiametrum orbis stellati, quod distantia orbis stellati a centro terrae est vigesies millies centies decies aequalis medietati diametri terrae, quod est sexaginta quinque mille millia trecenta quinquaginta septem millia quingenta milliaria, quod si duplicetur, erit diameter totius orbis stellati scilicet 130,715,000 milliaria. Et cum multiplicatur hoc in tria et septimam diametri erit rotunditas maximi circuli in coelo stellato, scilicet 410,818,571 milliaria, et tres septimae milliarii, hoc est, 1,714 cubiti, et 2 septimae cubiti. Et si hunc numerum diviserimus in 360 partes, una pars, scilicet quae erit quantitatis unius gradus coeli stellati, habebit milliaria 1,141,162 et 251 trecentesimas sexagesimas milliarii, hoc est 2,788 cubiti, et octo nonae unius cubiti. Et si duxerimus diametrum in rotunditatem, erit superficies totius coeli stellati, 53,700,149,508,265,000, videlicet, quinquagesies ter millesies millesies millesies millesies mille millaria, septingentesies millesies millesies millesies mille milliaria, centies quadragesies novies millesies millesies mille milliaria, quingentesies octies millesies mille milliaria, ducenties sexagesies quinque millia milliaria[1]. Semidiameter autem coeli stellati est longitudo longior orbis Saturni, quia junguntur sine medio. Sed ejus longitudo propinquior terrae est 46,816,250 milliaria, quod est longitudo longior orbis Jovis, cujus longi-

[1] The manipulation of these large numbers is interesting; and they may, perhaps, to some extent account for the contentment of the ancient and mediaeval mind with a limited universe. Archimedes, in his *Psammites*, had assigned a wider limit for the diameter of the universe; viz. ten thousand million stadia, or one thousand million miles, as against one hundred and thirty million in the computation of Alfraganus and Bacon.

tudo propinquior est 28,847,000 milliaria, quod est longitudo longior orbis Martis, cujus longitudo propinquior est 3,965,000 milliaria, quod est longitudo longior orbis solis, cujus longitudo propinquior est 3,640,000, quod est longitudo longior Veneris, cujus longitudo propinquior est 542,570, quod est longitudo longior Mercurii, cujus longitudo propinquior est 208,541 et duae tertiae milliarii, et hoc est 2,666 cubiti et duae tertiae cubiti, et haec est longitudo longior Lunae, et haec, ut dicit Alfraganus, est 64 vicibus et sexta vicis unius aequalis medietati diametri terrae, et longitudo propinquior Lunae est 109,037 et medietas milliarii, hoc est, 2,000 cubiti, et haec est 33 vicibus et semis et medietate decimae, id est, una vicesima unius vicis aequalis medietati diametri terrae. Diametri quidem singulorum orbium habentur per duplum semidiametri; rotunditas cujuslibet habetur per triplicationem diametri cum additione septimae partis, et tota superficies cujuslibet orbis habetur per ductum diametri suae in rotunditatem suam, ut exemplificatum est in terra et in orbe stellato. Et quilibet potest haec experiri per computationem, et ideo omitto haec propter prolixitatem. Quoniam vero subtracta longitudine propiore a longiore remanet spissitudo orbis, ideo patet, quod spissitudo orbis Lunae est 99,504 milliaria, et spissitudo Mercurii 334,209, et Veneris 3,097,250, et Solis 325,000, et Martis 24,882,000, et Jovis 17,969,250, et Saturni 18,541,250. Haec ex radicibus Alfragani et Ptolemaei in Almagesti sumuntur.

Sciendum vero quod in omnibus istis altitudinibus sumitur distantia a centro terrae. Unde licet sapientes aliquando dicant longitudines has fieri a terra, intelligunt tamen quod a centro terrae, quia medietas terrae nihil facit sensibile. Cum ergo longitudo propinquior orbis Lunae sit 109,037 milliaria et medietas milliaris, subtracta medietate diametri terrae, quod est 3,250, a numero praedicto integrorum milliariorum, scilicet 109,037, relicta medietate, residuum distantiae orbis Lunae a terra erit 105,787. Et ponantur 20 milliaria pro dieta. Ergo si iste numerus dividatur per 20 exibit numerus dietarum 5,289, et restant 7 milliaria. Et si istae dietae dividantur per 365 dies unius anni, exibunt anni 14, et restant dietae 179, quae si dividantur per 30 quae sunt dies mensis perfecti, exibunt menses perfecti 7, et restat unus mensis imperfectus, id est, 29 dierum. Patet igitur quod sumendo dietam in 20 milliaribus posset homo pertransire spatium usque ad orbem Lunae in 14 annis et quinque mensibus perfectis et uno mense imperfecto, et adhuc restarent 7 milliaria et medietas milliarii, quae a principio relicta fuit.

De nono vero coelo et decimo nihil potest per instrumenta sensibilia sciri in altitudine et spissitudine et magnitudine, sicut de aliis coelis, ut de spissitudine octavi coeli quia haec omnia latent sensum, et ideo super haec cessat certificatio quantitatum et altitudinum, et spissitudinum. Et per longitudinem propinquiorem Lunae est manifesta longitudo longior orbis ignis, sed longitudo longior aeris non est nota. Nam philosophi haec neglexerunt, quia non est utilitatis notabilis. Quantum tamen est usque ad locum nubium verificaverunt, quoniam demonstratur in libro de Crepusculis[1] quod altitudo illa est per 51 milliaria et duas tertias. Plinius[2] autem secundo Naturalium recitat nubes elevari stadiis 400 et 900 secundum diversa philosophorum testimonia. Ergo ad minus extenditur aer in tantum et plus; quamvis Albumazar et quidam alii aestimant vapores non elevari nisi per duo milliaria et decimam et tertiam milliarii. Et Ptolemaeus in libro de Dispositione Sphaerae dicit, quod decem stadiis est ultima elevatio vaporum. Et Martianus in Astronomia in hoc concordat. Nam montem altissimum ponunt decem stadiis elevari, ut Olympum, in

Height of clouds.

[1] This book is generally attributed to Alhazen, and, like the treatise on Optics of the same author, was translated into Latin from Arabic by Gerard of Cremona. It is an attempt to determine with geometrical precision (*a*) the position of the Sun with reference to the horizon when evening twilight ceases or morning twilight begins: (*b*) assuming twilight to be caused by solar rays passing through terrestrial vapours, the height above the Earth to which these vapours rise. He found that at the extreme limit of twilight the Sun was between nineteen and twenty degrees below the horizon. From this datum, and from the three data of the Earth's radius, the Sun's radius, and the distance of the Earth from the Sun (as given in the current astronomy), he concluded that the extreme altitude of terrestrial vapours was somewhat less than fifty-two miles.

[2] Pliny (*Hist. Nat.* ii. 23) makes the limit of cloud vary from 40 to 900 stadia; the stadium being estimated at 125 paces or 625 feet. Bacon, in the following sentences, shows himself aware of the extreme uncertainty of these measures of length. But the estimate of the stadium hardly varied to the extent stated by him.

cujus cacumine non invenitur ventus nec vapor, quod probant homines per experientiam. Sed quoniam Aristoteles[1] libro Meteorologicorum vult vapores elevari usque ad confinium aeris et ignis, ideo tenenda est sententia saltem libri de Crepusculis, et aliae glossandae sunt per diversam quantitatem stadii et milliaris. Nam aliquando invenitur, quod stadium est octava pars milliaris, et alias quod est longe plus, et similiter de quantitate milliaris est magna diversitas; scitur enim quod montes Italiae et alii, ut Caucasus, in cujus cacumine apparent radii solis usque ad tertiam partem noctis, habent plus quam octies 125 pedes, et ideo habent plus quam octo stadia, secundum quod stadium dicitur esse 125 pedum. Et ideo alibi inveni in expositione libri de Ormesta Mundi, quod stadium continet duas leucas, et duas partes unius leucae, et sic unum stadium continet quinque milliaria et tertiam milliarii. Per hujusmodi ergo diversitates solvuntur praedicta. Quod autem certius est de altitudine maxima montium est, quod est octo milliaria, secundum quod docetur libro de Crepusculis, et ideo apud librum Ptolemaei potest esse falsitas translationis vel scripturae, quando altitudo maxima ponitur in octo stadiis.

Postquam jam habemus coelorum altitudines, planum est nos habere et stellarum secundum considerationes Ptolemaei et omnium mathematicorum. Nam longitudo propinquior Lunae est longitudo orbis ejus, et longior longitudo similiter est una, et sic de aliis. Sed de magnitudinibus et spissitudinibus stellarum non est notum per hanc viam. Qui igitur posset invenire diametros, de facili inveniret rotunditatem et totam superficiem per doctrinam quam prius tetigi in orbibus. Antiqui vero quidam, ut Martianus in sua Astrologia, nisi sunt invenire diametrum Solis et Lunae per fluxum aquae per foramen vasis, dum oritur stella, et dum volvitur totus orbis coeli; et ita per illas duas aquas inveniebant quanto arcui diameter corporis stellae subtenditur.

[1] Aristotle observes, *Meteorolog.* i. 3, § 10, that clouds form in a region not too near the heavenly bodies to be dissipated by their heat, nor too near the Earth to be affected by the rays of heat reflected from its surface. *Γίνονται γὰρ αἱ τῶν νεφῶν ἀθροίσεις οὗ λήγουσιν ἤδη διὰ τὸ σχίζεσθαι εἰς ἀχανὲς αἱ ἀκτῖνες.*

Measurement of apparent diameters of Sun and planets.

Instrumenta autem aquae erant, sicut nunc fiunt quaedam horologia. Sumunt enim caldarium in fundo habens foramen unum, et illud implent aqua, ita tamen quod in exitu primae guttae per foramen incipiat Sol oriri, et in occasu Solis considerabant quantum aquae exierat, quam dividunt in duodecim partes secundum numerum duodecim horarum, et ideo lapsus cujuslibet partis per dictum foramen est revolutio quindecim graduum, quae facit unam horam diei. Volentes igitur per hoc quantitatem Solis vel Lunae accipere, prima parte Solis oriente primam guttam aquae diurnae exire permittunt, et ultima parte Solis orta, foramen obturant. Quae igitur est proportio aquae nunc elapsae ad totam aquam diurnam, eadem est diametri Solis sive arcus coeli huic correspondentis ad totum arcum diei. Alii vero per astrolabium vel quadrantem consideraverunt quantum movetur regula, dum extremitates Solis et Lunae aspiciuntur, et considerant tunc, quod quae est proportio illius arcus in instrumento ad totam circumferentiam ejus, eadem est proportio stellae ad circumferentiam coeli. Sed isti modi non habent sufficientem certitudinem. Et propter hoc Averroes dicit in fine secundi Coeli et Mundi, quod investigatur quantitas diametrorum stellarum per instrumentum, quod dicitur, habens duas regulas. Illud enim instrumentum terminat angulum, quem respicit diameter stellae, et hoc instrumentum continetur ex duabus regulis ad modum circini ad quantitatem quatuor cubitorum, et sunt in extremitatibus earum primae perforatae, ut videantur duae extremitates Solis vel Lunae. Per angulum enim inter duas regulas, quando notus est, scitur diameter stellae quoniam est basis illius anguli. Et hoc instrumentum invenit Abrachis[1], ut vult Ptolemaeus quinta dictione Almagesti xiv capitulo. Ptolemaeus vero abhorruit primos modos propter insufficientiam, et hunc modum Abrachis invenit habere defectum. Nam secundum hoc instrumentum Abrachis aestimavit diametrum

[1] Abrachis, the Arabian transformation of Hipparchus. It may be remarked, as an illustration of the imperfection of the astronomical instruments used by the ancients, that Archimedes, who constructed an instrument for measuring the apparent diameter of the Sun, had to content himself with the conclusion that it was not less than $\frac{90^\circ}{200}$ and not more than $\frac{90^\circ}{164}$; in other words, he left a margin of error of nearly six minutes.

Lunae in omni longitudine sua esse aequalem secundum aspectum diametro Solis propter aequalitatem angulorum. Sed Ptolemaeus consideravit quod diameter Lunae non aequatur secundum aspectum diametro Solis, nisi quando Luna est in longitudine sua maxima. Et hoc est quando Luna est in auge epicycli, et epicyclus in auge eccentrici, et hoc est iterum quando est plena. Tunc enim angulis aequalibus diametri Solis et Lunae subtenduntur secundum aspectum. Sed quando Luna est in longitudine alia non est sic, et ideo quantitas anguli non certificatur apud existentiam Lunae in quocunque loco; quod ideo fit, quoniam tabula elevatur a regula, et oculus est super superficiem regulae, et ideo non recte protenditur oculus ad foramen tabulae, sed obliquatur aliquantulum, et deberet tunc accipi quod est inter duo foramina, et accipitur quod est inter duas regulas, et ideo est dubitatio de quantitate anguli. Et propter hoc Ptolemaeus voluit certificare quantitates angulorum per eclipses Lunares, et descendit ad diversas eclipses, in quibus manifestat propositum. Certificatur igitur per instrumentum et per eclipses quantitas anguli cui subtenditur diameter Lunae quando est in longitudine sua maxima et plena. Et ideo per quantitatem anguli certificatur diameter Lunae[1], et per triplicationem diametri cum additione partis septimae certificatur quantitas rotunditatis Lunae, et per ductum diametri in rotunditatem certificatur tota superficies Lunae, et totum ejus corpus.

The true diameter of the Moon.

Et quoniam medietas diametri terrae est radix ad hoc, ut Ptolemaeus utitur xv capitulo dictionis praedictae, ideo percipitur quantitas diametri Lunae per comparationem ejus ad diametrum terrae, et quantitas corporis Lunae per comparationem ad quantitatem corporis terrae. Et ideo Ptolemaeus in xvi capitulo concludit per demonstrationem suam et figurationem quod secundum quantitatem qua erit diameter Lunae pars una, erit diameter terrae tres partes et duae quintae fere. Diameter igitur terrae erit in longitudine triplum diametri Lunae, et duae quintae ejus. Et ideo si a diametro

[1] Bacon passes from the apparent to the real diameter of the Moon without explaining how the latter is found, i.e. from the Moon's mean horizontal parallax.

terrae, quae est 6,500 milliaria, separaverimus diametrum Lunae, per computationem erit diameter Lunae praecise 1,911 milliaria, et tres quartae unius milliaris, et una decima septima unius quartae; et triplicata diametro cum additione septimae partis habebimus ejus rotunditatem, quae erit 6,006 milliaria, et sex decimae septimae quartae partis unius milliaris; et ducendo diametrum in circumferentiam habebimus totum corpus Lunae quod erit 45,927,882 milliaria, et hoc erit quasi una de xxxix partibus totius terrae, sicut docet Ptolemaeus. Nam secundum quantitatem qua erit corpus Lunae pars una, erit magnitudo corporis terrae trigintuplum nocuplum et quarta ejus fere; quoniam cum quantitas diametri Lunae fuit pars una, quantitas diametri Terrae fuit tres partes et duae quintae. Si igitur istae quantitates ducantur in se cubice et corporaliter, patet quod unum ductum in se cubice non est nisi unum, sed tria et duae quintae ducta in se cubice faciunt trigintuplum nocuplum et quartam fere. Quapropter corpus terrae erit trigintuplum et quarta fere respectu corporis Lunae. Et hoc patet, quia quantitas praedicta de corpore Terrae investigata superius est trigintupla nocupla ad quantitatem corporis Lunae et fere quarta, ut patet consideranti.

The diameter and other dimensions of the Sun.

Quantitas autem diametri Solis et totius corporis ejus dependet secundum Ptolemaeum a duobus, scilicet a quantitate diametri Lunae et a quantitate semidiametri Terrae. Nam longitudinem Solis a Terra ostendit per longitudinem Lunae jam notam, et per angulos quibus diametri Lunae et Solis subtenduntur apud aspectum, cum fuerit Luna in longitudine sua maxima in plenilunio. Tunc enim videtur corpus Lunae cooperire totum Solem nec plus, et ideo tunc dicitur Sol eidem angulo subtendi apud aspectum cui subtenditur Luna. Nec considerat longitudinem Solis diversam, quia propter superfluam distantiam non accidit diversitas in aspectu, sive Sol sit in longitudine propinquiore, sive longiore. Et invenit per has considerationes, quod secundum illam quantitatem qua diameter Lunae est pars una, et diameter Terrae tres partes et duae quintae fere, erit diameter Solis 18 partes et 4 quintae unius partis. Diameter ergo Solis erit decuplum octuplum ad diametrum Lunae, et quatuor quintae ejus;

eritque diameter Solis ad diametrum Terrae quintuplum et medietas ejus fere. Si ergo per diametrum Terrae notam, quae est 6,500, sumpserimus diametrum Solis erit 35,941 milliaria, et 12 decimae septimae unius quartae. Et si haec quantitas triplicetur cum additione septimae partis, erit quantitas rotunditatis Solis 119,803 milliaria et una tertia, et 12 decimae septimae unius quartae. Et si duxerimus diametrum in rotunditatem erit tota superficies Solis 292,783,785,375 milliaria. Et si quantitas diametri Solis respectu diametri Lunae sumpta multiplicetur in se cubice, erit hic cubus sexies mille et sexcentuplum et quadragintuplum et quadruplum respectu corporis Lunae, et medietas ejus fere; et ideo secundum quantitatem, quae est corpus Lunae pars una, erit magnitudo corporis Solis respectu Lunae sexies mille sexcentuplum quadragintuplum et quadruplum et medietas ejus fere. Quod si diameter Solis comparata ad diametrum Terrae ducatur in se cubice, fiet cubus qui erit centuplum et septuagesimum fere. Et ideo Sol est centies septuagesies fere major corpore Terrae. Et istud quod dicit fere exponitur a Thebit in libro suo, qui intitulatur, De iis quae indigent expositione antequam legatur Almagestum. Sed nec omnino praecise. Nam dicit quod Sol est centies sexagesies sexies aequalis Terrae. Et Alfraganus xxii capitulo sui libri addit super hoc dictum Thebit quartam et octavam Terrae, quas prius determinavi propter hunc locum certificandum. Est igitur Sol major tota Terra centies sexagesies cum quarta et octava Terrae superadditis. Et istud est praecisa quantitas Solis secundum considerationes auctorum.

Dimensions of the planets.

Ptolemaeus vero non determinavit nisi quantitatem Solis et Lunae. Sed Alfraganus dicit xxii capitulo consimilem modum esse in aliis, quem Ptolemaeus accepit in his. Nam secundum quod invenit quod diameter Lunae, quando est in longitudine sua maxima et plena, est in aspectu aequalis diametro Solis, similiter est cum alii planetae sunt in medio longitudinum suarum, diametri eorum habent comparationem ad diametrum Solis in quantitate certa. Diameter enim Mercurii in aspectu est xv pars diametri Solis, et Veneris est decima pars diametri Solis, Martis est xx, Jovis xii, Saturni xviii pars diametri Solis; quoniam diameter corporis Mercurii est vicesima octava pars diametri Terrae, et diameter Veneris est[1] tertia pars diametri Terrae et tertia partis, et diameter Martis est similis diametro Terrae semel et sexta, diameter Jovis est aequalis diametro Terrae quater et semis et xvi vicis unius, et diameter Saturni est aequalis diametro Terrae quater et semis, quod est 29,250 milliaria. Ex quibus sequitur quod corpus Mercurii est una pars de 22 millibus partibus Terrae, et Venus est 39 pars Terrae secundum Alfraganum, sed secundum Thebit libro dicto est una de 32 partibus Terrae, et Mars est similis Terrae semel et semis et medietas octavae unius vicis, et Jupiter est aequalis Terrae nonagesies quinquies et Saturnus est similis Terrae nonagesies et semel. Et si quis per diametrum Terrae quantitatem diametri cujuslibet in milliaribus et per quantitatem diametri velit circumferentiam invenire, et per utramque totam superficiem in milliaribus, facile est per praedicta.

De stellis autem fixis 1,022 dat Alfraganus eandem artem. Nam tot sunt notae sapientibus de his quae sunt a septentrione usque ad ultimam, quae apparuit eis in climate tertio. Nam per instrumenta possunt quantitates istarum certificari. Sed harum 1,022 sunt sex gradus, et differentiae magnitudinum, secundum quod Ptolemaeus viii Almagesti, et Thebit in libro memorato, et Alfraganus et alii sapientes docent. Sunt igitur sex magnitudines, et prima est major inter alias, et sic per

The fixed stars.

[1] The omission in J.'s edition of the diameter of Venus in terms of the Earth's diameter has made the subsequent estimates of the size of the planets unintelligible. I have restored the reading as given in Tib. In Jul. the whole passage relating to this subject is omitted.

It will be observed that the expression *in aspectu* is used for the apparent diameter of the planets in terms of the Sun's apparent diameter. Then follows the statement of their real diameter in terms of that of the Earth. Throughout this estimate Bacon follows very closely that of Alfraganus, as given in his *Elementa Astronomica*, cap. 22 (ed. Golius, Amsterdam, 1669). Alfraganus however gives the size of Venus as $\frac{1}{37}$ of the Earth. Supposing her diameter as 1 to $3\frac{1}{3}$, this would make her cubical dimension $\frac{1}{37}$ of the Earth, nearly, and this is the estimate of Alfraganus; 'moles Veneris est pars una de Terrae partibus 37.'

It may be interesting to compare these results with those of modern astronomy, in which, taking the Earth's diameter as unity, that of the Moon is ·27, of Mercury ·4, of Venus ·97, of Mars ·5, of Jupiter 11·5, and of Saturn 4·25. The Sun is 110, which contrasts strongly with Bacon's estimate of 5·5. But Bacon's view of the lunar diameter is near the truth.

consequens ordinate. In prima vero magnitudine sunt xv stellae fixae, ut sunt Canis et Vultur cadens, et cor Leonis et hujusmodi, quarum cujusque diameter in aspectu est xx pars diametri Solis, et cujuslibet istarum diameter est aequalis diametro Terrae quater et semis. Et ideo quaelibet illarum est aequalis Terrae centies et septies secundum Alfraganum. Et ex mensura harum stellarum patet quantitas aliarum. Sunt ergo in magnitudine secunda xlv stellae, ut sex stellae Ursae majoris et duae stellae Ursae minoris, et unaquaeque istarum est aequalis Terrae nonagesies. Et in magnitudine tertia sunt 208, ut aliquae stellae Ursae minoris, secundum Ptolemaeum, et multae aliae, et harum quaelibet est aequalis Terrae septuagesies et bis. Et in quarta magnitudine sunt 474, ut 4 stellae Ursae minoris et multae aliae, et quaelibet istarum est aequalis Terrae quinquagesies quater. Et in magnitudine quinta sunt 217 quarum quaelibet est aequalis Terrae trigesies sexies. Et in magnitudine sexta sunt 62, quarum quaelibet est decies octies major tota Terra. Deinde sunt stellae aliae infinitae[1], quarum quantitas non potest sciri per instrumenta, et tamen sunt visu notabiles, et ideo habent quantitatem sensibilem respectu coeli, sicut pars respectu totius. Sed Terra non habet aliquam quantitatem sensibilem respectu coeli, sicut probat Ptolemaeus in principio Almagesti, et certum est hoc, quoniam se habet sicut centrum ad circumferentiam circuli. Quapropter quaelibet istarum stellarum visu notabilium est major tota terra, et hoc dicit Alfraganus capitulo iv. Ex his igitur omnibus, quae dicta sunt de magnitudinibus corporum, patet quod majus omnibus, exceptis orbibus omnibus praeter orbem Terrae, est Sol; deinde secundo stellae primae magnitudinis, tertio Jupiter, quarto Saturnus, quinto reliquae stellae fixae omnes secundum gradus et ordines suos, sexto Mars, septimo stellae fixae visu notabiles, octava Terra, nono Venus, decimo Luna, undecimo Mercurius.

(7) Importance of music to theology.

Musicalia[2] vero secundum quod sancti determinant neces-

[1] This recognition of numerous stars other than those included in the catalogue of Hipparchus is noteworthy.

[2] This subject is more fully treated in *Opus Tertium*, chapters 59–64. Besides music properly so called, the discussion includes gesture, accentuation, aspiration, punctuation, and prosody.

saria sunt theologiae in multis. Nam licet non oportet propter scientiam scripturae quod habeat theologus usum cantus et instrumentorum et aliarum rerum musicalium, tamen debet scire rationem omnium istorum, ut sciat naturas et proprietates harum rerum et operum, secundum quod musica speculativa et practica docent. Scriptura enim plena est vocabulis musicalibus, sicut jubilare, exultare, cantare, psallere, cythara, cymbala, et hujusmodi diversi generis. Praeterea continet in se multa genera canticorum tam in novo quam in veteri testamento. Item multa genera metrorum continentur in textu sacro Hebraico, quae sancti notant in expositionibus suis multipliciter. Sed musicae est dare causas et rationes horum, licet grammaticus doceat, quia est de his. Praeterea tota pronuntiatio scripturae consistit penes accentus, longitudines, et brevitates, et penes cola, commata, periodos; et haec omnia pertinent causaliter ad musicam, quia horum omnium musicus dicit propter quid, grammaticus vero quia est tantum. Sic enim auctores philosophiae determinant, et Augustinus in musicalibus de his determinat. Nam musica alia vertitur circa audibile, alia circa visibile, ut auctores determinant; quae vero est circa audibilia habet partes duas; una est circa vocem humanam, alia circa instrumenta. Quae vero est circa vocem humanam est quadruplex. Nam alia est melica, ut in cantu; alia metrica, quaè considerat naturam et proprietates omnium carminum et metrorum et pedum; tertia est rhythmica, quae omnem varietatem proportionum in rhythmis considerat; quarta est prosaica, quae accentus et alia praedicta considerat in sermone prosaico. Nam accentus est quidam cantus; unde dicitur accentus ab *accino*, *accinis*, quia omnes syllabae habeat suum sonum proprium elevatum aut depressum aut compositum, et omnes syllabae unius dictionis adaptantur sive accantantur uni syllabae supra quam residet sonus principalis. Et sic longitudo et brevitas et caetera omnia quae in recta pronuntiatione requiruntur ad musicam reducuntur, et de eis determinant auctores musicae, sicut patet apud Censorinum de accentibus et apud Martianum et apud multos. Et haec patent per Isidorum et Cassiodorum in musicalibus. Et

Augustinus reducit metra et pedes et hujusmodi ad musicalia. Horum igitur omnium potestatem deberet theologus perfectus habere, quia in sensu literali et spirituali istorum naturae et proprietates requiruntur, ut sancti multipliciter determinant. Musica vero in instrumentis determinat compositionem instrumentorum et usum. Et ideo cum scriptura sit plena hujusmodi instrumentis, oportet theologum perfectum scire saltem in universali compositiones instrumentorum, et scire quomodo usus eorum habeat fieri propter sensus mysticos infinitos praeter literales. Musica vero consistens circa visibile necessaria est; et quod sit talis, patet ex libro de ortu scientiarum. Nam quicquid potest conformari sono motibus consimilibus et figurationibus competentibus, ut fiat delectatio plena non solum audiendo sed videndo, pertinet ad musicam. Et ideo exultationes et omnes flexus corporum ad gestum reducuntur, qui est una radix musicae, quoniam haec conformatur sono motibus consimilibus et configurationibus competentibus, ut vult auctor libri praefati. Et ideo Aristoteles dicit septimo Metaphysicae quod ars saltandi non vadit in finem suum sine alia arte, et hoc est sine altera specie musicae cui ars saltandi configuratur. Exultationes igitur et caeteri flexus corporum, quos et quas exercuerunt Maria soror Moysi, et Delbora et aliae cantatrices, et David et caeteri cantores de quibus scriptura multipliciter tractat, debent sciri a theologis, ut omnes proprietates eorum sciant exprimere, quatenus sensus spirituales angelicae devotionis eructent. Et quamvis sic appareat satis ut nunc oportet quod istae scientiae sint necessariae respectu theologiae et philosophiae, tamen longe major est earum utilitas consideranda per alias vias magnificas quae pertinent ad sapientiam non solum absolute, sed ut ipsa debet ecclesiam Dei regere, et caetera tria aliquotiens tacta procurare, ut inferius tangetur.

Distinction of mathematics from magic.

Manifestato quomodo mathematica necessaria est sapientiae tam divinae quam humanae, adhuc necesse est ad certificationem praecedentium, ut evacuentur quaedam cavillationes in contrarium et exponantur quaedam dicta sanctorum, ut omnis dubitatio tollatur circa mathematicae utilitatem. Et illud in quo maxime percutitur mathematica est propter judicia astronomiae. Quoniam igitur contentio multorum ignorantium philosophiae potestatem et maximam utilitatem theologiae, tam relate quam absolute reprobantium mathematicorum considerationes, impedit studium sapientiae et laedit gravissime in hac parte; ideo volo in praesenti verificare intentionem eorum, et evacuare infamiam verae mathematicae. Theologi igitur multa invenerunt a sanctis effusa contra mathematicos, et aliqui eorum propter ignorantiam mathematicae verae et mathematicae falsae nesciunt distinguere veram a falsa, et ideo tanquam auctoritate sanctorum culpant veram cum falsa. Vocabulum enim verae mathematicae scribi per *t* aspiratum, et ab hoc nomine *mathesis* media correpta [1], quod scientiam designat, derivari a multis refertur auctoribus, et certum est ex Graeco; quia *matheo* verbum idem est quod disco, et *mathetes* est discipulus, et *mathesis* disciplina. Unde *mathematica* est disciplinalis scientia et doctrinalis, sicut Cassiodorus dixit superius. Sed vocabulum falsae mathematicae sine aspiratione scribi asseritur ab eisdem auctoribus et a *mathesi* media producta, quod divinationem notat, descendere, vel quod certius est, a *mantos* vel a *mantia*, quae sunt idem quod divinatio, sicut Hieronymus dicit in originali ix Isaiae. Quicquid vero sit

[1] Bacon corrected this error in his later writings. Of the true mathematica he says, 'Haec scribitur cum aspiratione et dicitur a mathesis, quae est disciplina, quae habet mediam productam, ut in mea grammatica Graeca exposui; licet totum vulgus Latinorum credat quod media sit correpta, secundum quod fiunt versus falsi, Scire facit Mathesis, sed divinare Mathesis.'

This passage is from the unpublished MSS. on Mathematics (Brit. Mus. Sloane, MSS. 2156, fol. 72, c), a section of the encyclopaedic work projected by Bacon, but in all probability never completed. A similar passage will be found in his commentaries on the apocryphal Secretum Secretorum (Bodleian, Tanner MSS., 116). 'Mantia vero Graece est divinatio Latine qua magia ac mathematici utuntur, secundum quod haec nomina, mathematica et mathematicus, derivantur a matesi media correpta, quae matesis est magica imponens necessitatem rebus contingentibus et libero arbitrio. Contra quam et contra matematicos ab ea dictos egregie beati Augustinus et Gregorius in Omeliis de Epiphania disputant et alibi . . . ut Aristoteles, Avicenna, Ptolemaeus et omnes alii veraciter philosophantes. Sed tamen hi veri judicatores dicuntur mathematici a mathesi media producta quae et per aspirationem scribitur.'

de hac scriptura et derivatione, tamen falsa mathematica est ars magica. Nam numerantur quinque species artis magicae, scilicet mantice, mathematica, maleficium, praestigium, sortilegium.

False mathematic excludes free-will.

Mathematica ergo est secunda pars artis magicae. Haec sibi usurpat considerationem coelestium characteribus, carminibus, conjurationibus, sacrificiis superstitionis, et fraudibus variis deformatam. Et ponit per virtutem constellationum omnia de necessitate contingere, nihil ad utrumlibet, nihil a casu nec fortuna, nihil a consilio, de bonitate tamen essentiae, et in adjutorium constellationum efficacius ordinavit singulis constellationibus propria figmenta characterum et aliorum praedictorum. Et haec expresse asseruntur in libris magicis. Unde haec scientia ista omnia ponit per coelum de necessitate contingere, et praesumit per hanc necessitatem infallibiliter de omnibus judicare futuris. Sed ista mathematica damnata est non solum a sanctis, sed a philosophis, ut dicit Isidorus in tractatu astrologiae, asserens unam partem astronomiae esse superstitiosam, scilicet quae est magica, et dicitur mathematica falsidica. Unde Aristoteles et Plato, testante Isidoro, eam damnaverunt; et Plinius[1] per diversa loca Naturalis Historiae eam saepius percutiens propter errores quos haec fantasia scripsit in naturalibus et medicinalibus, tandem eam nimis abhorrens xxx libro originem illius aperit, et quomodo totum mundum defoedavit evidenter ostendit. Tullius etiam in libro Divinationum magis in particulari ad ejus malitiam descendens ostendit quod cultum divinum destruxit, rempublicam violavit, et medicinam infecit et naturalem philosophiam et omnes bonas artes subvertit. Ptolemaeus etiam et Aristoteles et Avicenna et Messehalac et Haly et

[1] The introduction to the thirtieth book of Pliny's *Natural History* discusses the origin of magic. It arose, he says, at a very remote period of history, from the threefold root of medicine, religion, and mathematic. Every country was infested by it, but specially the East. Zoroaster may be looked upon as its founder. Osthanes introduced it into Greece at the time of the Persian war, when it spread like wild-fire; although Greek philosophers in the course of their travels had already looked into it. Pliny proceeds to trace its progress in Italy, Gaul, and Britain. For Cicero's account and criticism of Chaldaean astrology see *De Divinatione*, lib. ii. cap. 42–47.

Albumazar, qui prae aliis majori auctoritate de his locuti sunt, necessitatem absolutam in rebus inferioribus non ponunt per virtutem coeli, quia liberum arbitrium non subjacet rebus naturalibus, nec aestimant judicium debere fieri infallibile, immo nec aliquam libero arbitrio ponunt necessitatem, cum nec rebus naturalibus eam ascribant, ut patebit. Et ideo philosophi universaliter damnant istorum falsorum mathematicorum insanias.

Incantations and invocations of spirits.

Nec solum damnant quantum ad principale, scilicet propter errorem quem habent de coelestibus, sed quia mathematici isti daemones advocant in adjutorium coelestium dispositionum per conjurationes et sacrificia, quod est omnino nefandum, atque nihilominus maculant suas considerationes in coelestibus per circulos et figuras et characteres vanissimos et carmina stultissima, et orationes irrationabiles in quibus confidunt. Praeterea fraudes operum adjungunt, scilicet per consensum, per tenebras, per instrumenta sophistica, per subtilitatem motionis manualis, in quibus sciunt illusionem esse, et multa stultis miranda faciunt per haec in quibus virtus coeli nihil operatur, et ideo sibimetipsis contradicentes, quod coelo attribuunt coram aliis, apud seipsos sciunt non habere veritatem. Similiter licet alias in quibusdam conjurationibus et sacrificiis et carminibus et characteribus et figuris variis confidunt tanquam cooperantibus constellationi secundum eorum judicium, tamen pluries fraudulenter ista componunt secundum species fraudis predictas, et coram sibi credentibus ascribunt plurima constellationi, respectu quorum ipsa nullam habet potestatem. Et hoc eos non latet. Propter igitur has stultitias mathematicae sophisticae philosophi eam damnaverunt, et sancti ac viri catholici haec percipientes eam simul cum philosophis reprobaverunt.

Christian miracles attributed to magic.

Sed praecipua ratio sanctorum fuit ad hoc, quod tales mathematici impediverunt a principio ingressum fidei in hunc mundum, quia non solum imbuti hac fantasia erraverunt in fide et finxerunt mores suos in coelestibus, tanquam per coelestia et caetera dicta de necessitate fierent iracundi vel mansueti, casti vel luxuriosi, et sic de aliis; sed opera miraculosa fidem Christi probantia ascripserunt isti mathematicae, dicentes, Christianos esse mathematicos et magos populum seducentes. Sicut enim ipsi per hanc daemoniacam calliditatem potuerunt multa coram populo rudi facere per quae detinebant eos ad errores et dominabantur eis, sic imposuerunt apostolis et martyribus et caeteris praedicatoribus fidei quod non ex parte Dei fecerunt vera miracula, sed per artem magicam qua ipsimet usi sunt. Et ideo hanc mathematicam et ejus professores sancti et ecclesia penitus damnaverunt.

True mathematicians have always condemned these errors.

Sed veri mathematici, quos in hac parte vocamus astronomos et astrologos [1], quia indifferenter a Ptolemaeo et Avicenna et aliis pluribus sic vocantur, non redarguuntur de sacrificiis conjurationibus carminibus characteribus, ut etiam plebs studentium non ignorat, sed solum super judicio infallibili et rerum necessario eventu notantur. Non autem possumus eorum sententiam deprehendere nisi ex libris eorum testimonia propria eruamus, ut sic vel errantes damnemus ex propriis verbis, vel excusatos a dentibus vulgi imperiti liberos extrahamus. Quoniam igitur maxime imponitur eis error judicii, tamquam infallibiliter velint cum divina certitudine contendere, ut plebs eis imponit, adducantur philosophi nobiliores, scilicet Aristoteles, Avicenna, Ptolemaeus, Hali commentator ejus et Messehalac, Albumazar, quibus magis incumbit hoc onus. In universali igitur considerando quid secundum philosophos tenendum sit hic, Ptolemaeus in principio Centilogii pronuntiat dicens, 'Astronomus non debet dicere rem specialiter sed universaliter, ut qui eminus videt aliquam rem,' et addit, 'judicia quae trado tibi sunt inter necessarium et impossibile.'

Proofs from Hali,

Et dicit Hali super hunc locum, 'Hoc ideo dixit, quia haec scientia non est nisi per probationem et opinionem; quia materia ad quam pertinet opus totum stellarum convertibilis

[1] In his mathematical treatise, Bacon speaks of astrology as the speculative side, astronomy as the practical side of the science. 'Astrologia componitur ex hoc nomine Astron, quod est stella, et hoc nomine Logos, quod est verbum vel ratio vel sermo, quia est sermo de stellis. Astronomia vero dicitur lex stellarum, et Nomos est lex. Unde quia lex universaliter sonat in practicam, ut in morali philosophia lex est ipsa practica, ita similiter Astronomia est practica Astrologiae.' (Sloane MS. 2156, fol. 82, c.)

est ad unum et ad aliud.' Propter quod capitulo lix Centilogii secundum unam translationem dicit, 'cum interrogatus fueris de absente, non judices de eo mortem, donec removeas ab eo ebrietatem ; nec vulnus, donec tollas ab eo sanguinis minutionem ; nec substantiam adquisitam, donec pecuniam sibi commissam removeas; est enim in omnibus idem judicium.' Et dicit Hali, 'qui tractat de judiciis, judicat per formas quae sunt prope veritatem, quia non invenitur in fortitudine judiciorum partitio vel differentia inter sopitum et mortuum, et inter vulneratum et minutum, nec inter illum cui commissa est pecunia et illum qui adquisivit' : et addit exemplum, 'quod patre suo se abscondente cum aliis a facie imperatoris, astrologus visitans eum quotidie praecepit poni concham maximam aeneam plenam aqua, supra quam poneretur scabellum, consuluitque ei ut desuper sederet in majori parte diei ; et hoc ideo praecepit, ut faceret errare astrologos imperatoris in existentia absconditorum. Dixerunt enim quod esset in medio maris propter similitudinem aquae conchae, cui superinsederat. Distinguere enim inter hoc et illud non poterant propter continentis et contenti similitudinem. Aqua enim undique continebatur, et vasis concavitas alvei marini concavitatem expressit.'

Similiter Ptolemaeus in libro de Dispositione Sphaerae dicit, 'quod judicationes astronomi de rebus inferioribus non sunt inventae ex scientia artificiali, quare sunt fines earum necessariae, sed sunt ex eis quae sunt secundum plurimum.' Et iterum dicit, 'astronomi affirmaverunt illud non secundum viam artificialem, nec secundum ingenium determinatum, sed per convenientiam cadentem super rem propinquiorem.' Et tertio dicit, 'quod manifestum est quod illud quod affirmant non est affirmatum nisi secundum rem plurimam, non secundum viam artificialem et non est res faciens illud necessarium.' Propter illud ergo cadet multotiens error in eis, ut affirmat. Et Ptolemaeus in primo capitulo[1] quadri-

From Ptolemy.

[1] The works entitled *Centilogium* and *Quadripartitum*, to which frequent reference is made in the *Opus Majus*, are both of them astrological works. The Greek text of both was printed in Nuremberg in 1535 ; the title of the first being Κλαυδίου Πτολεμαίου Βιβλίον ὁ καρπός, of the second, Κλαυδίου Πτολεμαίου

partiti dicit, 'hujus autem tam excellentis artis occulta profunditas ejusque tam augusta et incomprehensibilis exercitatio, ut ab humano ingenio eam impossibile sit penitus amplecti, ex ineffabili subtilitate, ejusque admiranda quasi quadam divinitate videtur contingere. Longe enim ab humano sensu remota est, et corpoream transcendens naturam quasi ultra hominem eminus transcenditur.' Et tertio capitulo dicit 'quod ex nimia ipsius artis profunditate quicunque error incidit.' Quapropter secundum Ptolemaeum, cui nunc temporis maxime ascribitur ratio judicandi, manifestum est quod astrologus non potest dare plenam certitudinem suorum judiciorum, praecipue in singulis. Quod etiam non solum ex his liquidum est, sed ex aliis ejus sermonibus, quibus licet innuat possibilitatem judicandi de multis cum rationabili certitudine, tamen tantam difficultatem simpliciter arti inesse pronuntiat, ut facile pateat ipsum penitus definire, quod non debet astronomus de sufficienti certitudine in singulis gloriari. Unde quadripartiti primo capitulo dicit, 'quamvis enim res haec in tantum sit ardua, ut nullatenus ad finem omnino perduci valeat, non tamen videtur conveniens desistendum, sed potius ea quae possibilitati nostrae non repugnant exponere dignum judicavimus.' Et post pauca intendit, 'quod talis artis est difficilis et laboriosa assecutio, ut absque diuturna exercitatione penitusque secularibus libera

μαθηματικῆς τετράβιβλον συντάξεως. The Latin translation used by Bacon was probably from the Arabic; but, so far as I have read the Greek original, its meaning is preserved in the translation with fair accuracy. The passage here referred to begins, *οὐ γάρ ἐστι δυνατὸν τὸν ἐπιστήμονα τὰς μερικὰς ἰδέας τῶν πραγμάτων ἀναγγεῖλαι, ὥσπερ οὐδὲ ἡ αἴσθησις δέχεται τὴν μερικὴν ἰδέαν τοῦ αἰσθητοῦ, . . . μόνοι γὰρ οἱ ἐνθυσιῶντες προλέγουσι καὶ τὰ μερικά.*

Ptolemy (supposing him to be the author of those treatises) takes it for granted that there is a force diffused from the eternal Aether upon the elements, and through them on the earth and the living things contained in it. *Ὅτι μὲν* (he says in the opening of the *Tetrabiblon*) *διαδίδοται καὶ διικνεῖταί τις δύναμις ἀπὸ τῆς αἰθερώδους καὶ αἰδίου φύσεως . . . πᾶσιν ἂν ἐναργέστατον καὶ δι' ὀλίγων φανείη.* We should not reject astrology, he observes, on account of the numerous errors made in particular judgements, any more than we should reject the art of navigation because sailors often blunder. Some of the mistakes are due to the inherent difficulty of the subject, some to the imperfect knowledge and skill of those who teach it. In any case the study of astronomical conjunctions rests upon certain facts, even though the inferences from the facts be erroneous.

et infructuosis expedita curis ad hanc nequaquam valeat quis aspirare.' Et quia liberis a curis aliis et studiosis est bona spes proficiendi in hac scientia, ideo in secundo capitulo dicit, 'nobis quidem minime videtur consentaneum, si aliunde error inciderit, artem a nobis penitus excludere.' Et adducit exemplum, dicens, 'Nautae enim maris scopulos minus providentes, si quo decepti errore naufragium patiantur, restitutis navium armamentis de pristina non deficientes salute alacri mente spe confirmatione naves conscendunt. Ad hunc etiam modum nos hanc astrorum scientiam firmam et certam et in sua veritate immobiliter persistentem, quantum possibilitas humani permittit ingenii, assequi studeamus, ut in benevolentiam tam studentium quam aliorum reducere conemur, quatenus artis ipsius observata dignitate, ea ducente, quae futura sunt praevideant, quae autem sua impossibilitate ars ipsa prohibet relinquantur.' Et tertio vult capitulo, 'quamvis quicunque hic error incidit negligentia praetermitti aut oblivione deleri inconveniens penitus reputandum est.' Et post pauca, 'cum autem evidentissime constat hanc futurorum praenotionem non cum errore teneri, nec omnes illius species erroribus implicari, quid demum ex hac efficacia, quibus etiam modis quae innovantur repelli valeant, diligentius restat aperire. Sin autem singula tradere non valeat, quantum tamen possibilitas contulit exhibere non fugiat. Quod si minimum id reperietur, minoris profectus ut in brevius redigatur non censemus.' Et ponit exemplum, 'licet enim physica non omni homini sit salutis occasio, cum sit tamen quorundam, appetenda sua utilitate videtur.'

Ex quibus verbis et consimilibus manifestum est, quod de intentione Ptolemaei non est quod astrologus det in particulari certum judicium et sufficiens in singulis; sed in universali, et medium inter necessarium et impossibile, et non in omnibus potest determinatum dare judicium. Propter quod Avicenna, qui opera Ptolemaei complevit, ut ipsemet recitat in prologo libri Sufficientiae, manifestat in decimo Metaphysicae, quod astrologus non potest certificare in omnibus nec debet, propter instabilitatem materiae generabilis et corruptibilis, quae non in omnibus semper obedit virtuti

From Avicenna and others.

coelesti, ut dicit Messehalac ponens exemplum de magnete, quia virtus ejus non habet potestatem super ferrum nisi in debita distantia et in aliis conditionibus quae exiguntur ad attractionem fuerit adaptatum.

Et iterum sciunt, quod ordinatio divina potest omnia mutare secundum sui voluntatem, et propter hoc adjungunt semper in suis sententiis in fine hoc verbum, quod 'sic erit, si Deus voluerit.' Caeterum ipsimet sciunt et testantur, quod anima rationalis potest multum mutare et impedire de effectibus stellarum, ut de infirmitatibus, et de pestilentiis frigoris et caloris, et fame, et multis, secundum quod Ptolemaeus in Centilogio dicit et docet. Et Hali exponit ipsum; et Isaac in libro febrium evidenter ostendit; et Aristoteles in libro Secretorum confirmat et declarat hoc idem. Nam quando praeviderint haec mala, possunt praeparare remedia. Unde dicit Isaac, quod 'non accidit malum homini, nisi sit detentus ignorantia scientiae coelestis,' et ponit exemplum in pestilentiis et in infirmitatibus et in voluntariis, in quibus haec scientia potest dare remedia, si complete sciretur. Ex his igitur manifestum est, quod philosophi non ponunt eventum rerum inevitabilem per coelestia in omnibus, nec infallibile judicium in singulis, sed secundum possibilitatem hujus scientiae; praecipue cum etiam addunt quod alia scientia, quae experimentalis vocatur, certius adhuc judicat quam astronomia vulgata. Ita docet Ptolemaeus in libro de Dispositione Sphaerae, et patet de hac in sequentibus.

His igitur et hujusmodi diligenter consideratis, manifestum est quod veri mathematici et astronomi seu astrologi, qui philosophi sunt, non ponunt necessitatem et infallibile judicium in rebus contingentibus de futuro. Et ideo quicunque eis haec attribuunt, manifeste de ignorantia philosophiae convincuntur, et veritatem reprobant quam ignorant. Unde in duobus peccant, videlicet quod ea quae ignorant tractant, et nihilominus in veritatem blasphemant. Sed illi qui sic vera tuentur ut falsa reprobant condemnant mathematicos magicos, qui non philosophantur sed tam philosophiae quam fidei contradicunt, sicut a principio tactum est. Et ideo sancti contra hos locuti sunt, et non contra veros mathematicos.

What the fathers condemn is the compulsion attributed to stellar influence.

Quod manifestum est per obviationes sanctorum. Nihil enim reprobant nisi quod per stellas necessitas rerum contingentium et maxime morum et actuum humanorum fiat, ac infallibile in omnibus judicium. Sed soli mathematici magici haec ponunt, et veri mathematici non sic dicunt. Quapropter quicquid Basilius in Hexaemeron, et Ambrosius in Hexaemeron, et Augustinus et Gregorius in homilia Epiphaniae et Augustinus super Psalmos lxi et xci, et super Joannem et multis locis ipse et alii doctores dicunt contra mathematicos, totum est directe contra eos qui fingunt necessitatem imponi eis quae sunt ad utrumlibet, et praecipue eis quae a libero arbitrio proficiscuntur, qui soli sunt mathematici falsidici vel fraudulenti, superstitiosi non philosophi, ut manifestum est. Et ideo nimis errant qui suas insanias in veros mathematicos nituntur sanctorum auctoritate tueri, et falso abutentes dictis sanctorum damnare veritatem praesumunt. Unde suas opiniones licet erroneas statuere moliuntur, quas fingunt ex ignorantia mathematicae verae et falsae. Et cum nullam rationem afferre possunt eo quod contra omnem rationem et philosophiae virtutem est quod contra mathematicos veros dicunt, rapiunt et furantur auctoritates sanctorum, quae nihil faciunt ad propositum, ut manifestum est inquirenti.

They do not deny that astrology throws light on future events.

Et adhuc considerandum est, quod si consideremus dicta sanctorum, nos inveniemus manifeste quod non solum non reprobant mathematicos veros, sed approbant in futurorum cognitione. Quoniam Augustinus dicit libro secundo de Doctrina Christiana, quod mathematica habet futurorum regulares conjecturas, non suspiciosas et ominosas, sed rectas et certas tam de futuris quam praesentibus et praeteritis. Et Cassiodorus concordat, ut prius dictum est. Et ad Januarium ostendit multipliciter, quomodo futura possunt cognosci per coelestia. Et Basilius in Hexaemeron similiter declarat hoc multis modis; unde certissimum judicium in aliquibus esse declarat. Et Ambrosius in Hexaemeron consentit; et Isidorus in suo opere de astronomia. Et sancti confirmant quod dicunt per evangelium, 'de coelo rubicundo in sero et mane,' et iterum, 'erunt signa in sole et luna,' &c. Et in Joele, 'sol convertetur in tenebras,' &c. Non igitur est mathematica infamis ex futurorum cognitione quam fideliter assignat; licet a principio sancti minus usi sunt et minus commendaverunt eam, propter errorem falsae mathematicae, cui nomine et quadam consideratione coelestium respectu futurorum vera mathematica familiaris est et conveniens, sicut determinat Augustinus in libro de Doctrina Christiana.

But subsequent theologians have confounded true mathematics with false.

Sed postquam in ecclesia fuit evacuata falsitas magicae mathematicae, venit in usum catholicorum doctorum consideratio mathematicae verae, et duravit usque ad adventum quorundam theologorum, qui philosophiae potestatem et artis magicae fallacias ignoraverunt, et ob hoc eam in lectionibus praedicationibus et collationibus publicis et privatis damnantes, primo in maximam philosophiae destructionem, deinde in laesionem theologicae majestatis, et in damna ecclesiae et totius reipublicae fidelium et infidelium conversionis, ut prius pro magna parte ostensum est, et residuum postea ostendetur; quia ignorata mathematica, tota philosophiae potestas ignorabitur. Et si veritas philosophiae laedatur, damnum infertur theologiae, cujus est uti potestate philosophiae, non solum absolute, sed prout ordinat ecclesiam, dirigit rempublicam fidelium, et juvat ad infidelium conversionem praedestinatorum et reprobationem praescitorum. Sed non solum errant in hoc quod futurorum cognitionem per mathematicam fiendam imprudenter damnant, sed quia pro parte, quam licet ex sua abhorrent ignorantia, totum reprobant. Quamvis enim in partibus philosophiae sint plurima inutilia et minus bene tacta et aliqua erronea, nihilominus tamen tota philosophia a nullo damnatur, nec debet nec potest. Quapropter cum in mathematicae partibus omnibus aliis a judiciaria omnia utilia proponantur respectu theologiae et ecclesiae ac reipublicae et conversionis infidelium aut reprobationis et totius philosophiae, et nihil reprobatum determinatur etiam in illa parte cui judicium ascribitur, multa etiam praeter artem judicandi praeclara de proprietatibus coelestium rerum et secundis stellarum et caeteris rebus hujus mundi tanguntur, manifestum est eos omnino errare, qui pro una parte, quam falso damnant, totum residuum quod maximas habet utilitates et in nullo habet calumniam ausi sunt reprobare. Et iterum licet pars judiciaria de humanis rebus esset culpanda, tamen alia pars de naturalibus et coelestibus calumniam fidei non importat.

Sed in rebus humanis veri mathematici non praesumunt certificare, sed considerant quomodo per coelum alteratur corpus, et alterato corpore excitatur anima nunc ad actus privatos nunc publicos, salva tamen in omnibus arbitrii libertate. Quamvis enim anima rationalis non cogitur ad actus suos, tamen fortiter induci potest et excitari ut gratis velit ea ad quae virtus coelestis inclinat, sicut nos videmus homines per societatem consilia timorem et amorem et hujusmodi multum mutare de proposito, et gratis velle ea quae prius non volebant licet non cogantur, sicut ille qui spe salutis projicit merces in mare carissimas. Caeterum nos videmus quod species seu virtutes rerum inferiorum immutantes sensus nostros, etiam species visibilium et audibilium quae debiliter immutant corpus, ita fortiter excitant homines ad volendum quae prius non curabant, quod aliquando nec mortem nec infamiam nec timorem aestimant dummodo suas compleant voluntates, sicut sunt illi qui vident et audiunt suos inimicos eis occurrere, et feruntur omni casu contingente ut se vindicent. Similiter voluptuosi, opportunitate accepta complendi suas delicias circa res visas et auditas, quasi bruta animalia moventur contra judicium rationis, gratis eligentes ea ad quae excitantur.

Astrology does not predict action: it only indicates a motive.

Sed longe magis possunt virtutes coelorum et species eorum et stellarum fortes imprimere in corpus et organa, quibus vehementer alteratis excitabitur homo fortiter ad actus de quibus non curavit prius, salvo suae libertatis arbitrio; quoniam virtutes coelorum fortiores sunt quam istorum inferiorum visibilium et audibilium et multorum aliorum sensibilium, et possunt mutare substantiam non solum accidentia, et corrumpere et destruere res omnes inferiores; quoniam secundum allationem solis sub obliquo circulo accidunt generatio et corruptio in rebus, ut Aristoteles dicit. Et non solum haec allatio solis absolute considerata, sed cum virtutibus aliorum planetarum et stellarum; et ideo valida est alteratio corporum nostrorum per virtutes coelestes, et per consequens animus fortiter excitatur ad actus suos licet non cogatur, et secundum

Temperament is affected by stellar influence.

hoc currit judicium astronomi, et non per infallibilitatem et necessitatem. Et astronomus in hoc multum juvatur, quod videt homines in actibus suis sequi multum suas complexiones quas habent, ut cholericus movetur de facili ad iram, nec potest refraenare primos motus omnes, et sic de aliis, secundum quod homines diversificantur in complexionibus. Et ideo astronomus, cum videt homines sequi suas complexiones, quae oriuntur a coelesti operatione, sicut et tota generatio, non est mirum si se extendat ad considerationem actuum humanorum.

National character moulded by climate.

Caeterum ipse videt manifeste, quod secundum diversitatem habitantium sub diversis coeli parallelis variantur mores, sicut habitantes versus polum, ut Scythae, alios habent mores quam habitantes versus meridiem, sicut Aethiopes ; et alios ab his habent illi, qui in quarto climate ; immo secundum diversitatem cujuslibet climatis et etiam partium climatis variantur mores habitantium, ut in climate scilicet septimo Picardos, Gallicos, et Normannos, et Flamingos, et Anglicos, videmus manifeste in moribus discrepare, cum tamen sint continui ad invicem et propinqui. Hoc autem non est ex ipsis hominibus a parte diversitatis animae rationalis, sed propter complexiones corporum innatas a natura coeli, sub cujus parallelis diversis et stellis situantur, et secundum diversitatem situs eorum respectu planetarum. Et non solum est haec diversitas secundum latitudinem regionum ab aequinoctiali circulo versus polum, sicut nunc tactum est, sed secundum longitudinem, licet causae sint occultiores. Per experientiam enim videmus, quod sub eodem parallelo variantur regiones, secundum quod magis ab occidente vel oriente distant, et non solum ipsae regiones sed partes regionum. Et non potest in terra nec in hominibus inveniri causa principalis hujus rei, sed in coelo requiritur secundum omnes sapientes ; unde, sicut prius dictum est, ad omne punctum terrae incidit conus unius pyramidis virtuosae a toto coelo. Et coni isti sunt diversae in natura, et pyramides similiter, quia diversas habent bases propter diversitates horizontum, quoniam quilibet punctus terrae est centrum proprii horizontis. Et ideo oportet omnium rerum diversitatem magnam ex hac causa oriri, etiam quantumcunque propinqui sunt, ut gemelli in eodem utero ; et sic de omnibus, prout videmus quod a duobus punctis terrae proximis oriuntur herbae diversae secundum speciem. Et hic sumit astronomus fundamenta sui judicii, et merito, quia diversitas plena rerum per coelum sic invenitur.

Predictions as to communities more easy than as to individuals.

Quapropter potest astronomus peritus non solum in naturalibus sed in humanis rebus multa considerare de praesenti et futuro et praeterito, et ideo saltem super regna et civitates potest judicare per coelestia et secunda coelestium quae per virtutes speciales coelorum renovantur, ut sunt cometae et hujusmodi, quia facilius judicium est super communitate quam super singulari persona. Nam judicium communitatis est judicium universale, et astronomus potest bene in judicia universalia. Et quia ad mores et consuetudines cujuslibet civitatis famosae tota provincia circumjacens inclinatur, tum propter refugium ad eam et communicationem negotiorum vitae, tum propter potentiam civitatum super loca vicina; (et similiter est de regno potentiori respectu regnorum circumjacentium,) tum propter communicationem, tum propter violentiam; prudens astrologus potest multa considerare utiliter in hac parte super moribus et legibus et sectis et guerris et pace et hujusmodi, quae pertinent ad rempublicam civitatum, provinciarum, et regnorum, quamvis difficultatem recipiat majorem in actibus singularium personarum judicandis.

But even with individuals bodily changes may be forecast.

Et si velit considerare diligenter et sine errore horas conceptionum et nativitatem singularium personarum, ut sciatur dominium coelestis virtutis ad horas illas, et diligenter consideret quando ad eas dispositiones venient coelestia secundum singulas partes aetatis cujuslibet, potest de omnibus naturalibus, sicut de infirmitatibus et sanitate et hujusmodi judicare sufficienter, quandocunque debent occidere et qualiter terminari, secundum quod auctores non solum astronomiae sed medicinae, ut Hippocrates, Galenus, Hali, Isaac, et omnes auctores determinant. Sed medici hujus temporis pauci sciunt astronomiam et ideo nec auctores suos multi intelligunt nec possunt intelligere, et ideo negligunt meliorem partem medicinae. Sed de his longus sermo fieri potest et utilis, cum fuerit opportunum. Et cum secundum complexiones infirmitates et sanitates varientur voluntates hominum et desideria et considerationes, licet non cogantur sed fortiter inducantur, ut manifestum est, tunc prudens astronomus potest de actibus moralibus singularis personae prudenter judicare, salva tamen omnibus arbitrii libertate, et poterit in multis habere judicium certum secundum possibilitatem materiae quam tractat. Nam quia ista est contingens et non necessaria, non potest dicere quod de necessitate contingent haec vel illa, potest tamen dicere in multis quod contingent, et quod veritas est de futuro, licet non necessaria. Aliud enim est esse verum, et aliud necessario verificari. Et hoc est judicium medium inter necessarium et impossibile. Et in quibus non potest habere hujusmodi judicium, habebit de facili judicium universale, aut medium inter universale et particulare; per judicium tamen universale et secundum quod possibile est de persona publica, ut principe et consiliario principis in civitate vel regione potest saepius habere judicium particulare de factis reipublicae; quia, ut dictum est, facilius est, judicare de communitate quam de singulari persona, et secundum judicium principis regulantur civitates et regna; quod enim principi placet, legis habet vigorem.

The horoscope of the ruler throws light on the future of his realm.

Et ideo super negotiis civitatum famosarum et provinciarum et regnorum potest multum astronomus, dummodo sciat bene dominium coeli in conceptione et nativitate principis, et bene sciat quod complexio variatur per hoc, et qualiter ad mores inclinatur juxta proprietatem complexionis. Et tunc notet quando constellatio veniet quae stimulet complexionem et fortificet ut stimuletur et excitetur animus ad actus consimiles; et similiter quando in contrarium mutatur complexio per contrarium coeli dispositionem, ut ad contraria excitetur animus principis. Secundum enim hujusmodi variationes constellationum et complexionum et voluntatem principum et praelatorum oriuntur in populo auctoritate superiori innovationes consuetudinum et mutationes legum et morum. Deinde de facili oriuntur aliquando discordiae et dissensiones ad quas sequuntur bella, vel aliquando propter legum honestatem et utilitatem oritur concordia civium et aliorum, et fit pax. Et ideo astronomus peritus potest de facili judicare de hujusmodi negotiis communibus civitatum et regionum, quia non solum per vias proprias eis habeat unde procedat, sed per conditiones personarum, quae principantur. Et similiter per proprietates earum personarum quae principibus et praelatis assistunt, et quorum consiliis innituntur, quia principes mali ducuntur ad bona reipublicae per bonos consiliarios, et principes boni pervertuntur per malos consiliarios. Si igitur ex constellatione nativitatis et conceptionis complexio alicujus principis, vel alterius cui innititur, inveniatur disposita ad perversitatem morum et discordias et guerras, et astronomus simul cum hoc videat quod trahunt hujusmodi in consuetudinem et fortius excitantur quando constellatio consimilis accidit, potest rationabiliter judicare de infortuniis civitatis et regni quibus praesunt, quando dispositio coeli et eorum quae per coelestia renovantur, ut cometae et hujusmodi, conformiter accidunt. Et optima judicia fieri possunt, secundum oppositas conditiones principum et eorum quibus per omnia credunt, quando similiter coelestis dispositio conformis invenitur. Et quoniam tales personae in uno regno paucae sunt, et sunt publicae et notae omnibus, quorum mores relucent toti regno, multum potest astronomus certificare de eis, et utiliter judicare de factis publicis per proprietates talium personarum. Haec igitur sunt quae volui recitare pro infamia mathematicae tollenda in hujusmodi judiciis, ex quibus patet omni sapienti quod non sit vera mathematica in hac parte vituperanda, sed omnino amplectenda et diligenda propter gloriosas utilitates quae possunt evenire ex judiciis mathematicae verae, quae in nullo veritati contradicit.

Astrology strengthens our faith in the stability of the Church.

Postquam potestas mathematicae respectu scientiarum philosophiae et rerum istius mundi et theologiae, et sic respectu totius sapientiae ostensa est, prout ipsa sapientia secundum se consideratur absolute, nunc volo ostendere illud idem secundum quod refertur haec sapientia ad ecclesiam Dei et rempublicam fidelium et conversionem infidelium et repressionem eorum qui converti non possunt. Et quia in multis quae nunc numerari non possunt necessaria est ecclesiae, nunc volo tres casus proponere, qui sunt quasi infiniti miraculi et utilitatis ineffabilis. Primus consistit in certificatione fidei quam tenet ecclesia. Magnum enim solatium fidei nostrae possumus habere, postquam philosophi qui ducti sunt solo motu rationis nobis consentiunt, et sectam seu professionem fidei Christianae confirmant et nobiscum concordant in stabilitate hujus sectae; non quia quaeramus rationem ante fidem, sed post fidem, ut duplici confirmatione certificati laudemus Deum de nostra salute quam indubitanter tenemus. Et per hanc viam mathematicae non solum certificamur de professione nostra, sed praemunimur contra sectam Antichristi, de qua simul cum secta Christi fit consideratio in mathematica. Et hujusmodi nobilissima perscrutatio fit per revolutionem omnium sectarum[1] principalium a principio mundi, nec possunt esse plures, et sunt sectae Hebraeorum, et Chaldaeorum, et Aegyptiorum, et Agarenarum seu Saracenorum, qui fuerunt de Agar et Ismaele, secta Christi, ac secta Antichristi. Nec mirum si locuti sunt philosophi de his, quoniam fuerunt post patriarchas et prophetas et instructi per filios et libros eorum, ut prius ostensum est.

The moral influence peculiar to each planet.

Quanto igitur planius et plenius possum ad praesens, recitabo sententias mathematicorum in quibus auctores concordant. Dicunt igitur Jovem et Venerem esse planetas benivolos et fortunatos, Saturnum et Martem malivolos et infortunatos. Mercurium dicunt medio modo se habere, quia cum bonis est bonus, cum malis malus, quia convertibilis naturae est. De benivolis vero fortunatis dicunt Jovem meliorem esse, et majorem fortunam ei deberi, minoremque Veneri. Et ideo cum duae vitae sunt, praesens et futura, et plus valet futura quam praesens, sicut aeternum quam temporale, dicunt Venerem significare super fortunas hujus vitae, quantum ad ludos et gaudia atque laetitiam et hujusmodi, et Jupiter respectum habet ad bona alterius vitae, quae majora sunt. Et significat super sapientiam et intellectum et solutionem somniorum et divinum cultum fidem et legis doctrinam, religionem et venerationem et Dei timorem et aptationem morum et multa talia ut astronomi narrant.

The heaven divided into *houses*.

Praeterea distinguunt totum coelum in xii partes, quae

[1] The list of religions given in the fourth part of Bacon's *Moralis Philosophia* is somewhat different; viz. Pagani, Tartari, Idololatrae, Saraceni, Judaei, Christiani.

vocantur domus, quae distinguuntur per meridianum circulum et horizontem, et alios quatuor circulos intersecantes se in eorum sectionibus, ita ut prima domus incipiat ab horizonte et sub eo sit constituta, deinde succedunt ei secunda domus et tertia usque ad angulum terrae, qui est punctus in coelo sub terra oppositus puncto meridei super terram. Et ab illo puncto sunt aliae tres domus usque ad lineam occidentis. Deinde supra horizonta incipit vii et sequitur viii et ix usque ad punctum meridiei. Deinde x et xi et . xii usque ad horizontem. Primam igitur domum dant Saturno, secundam Jovi, et sic ulterius secundum ordinem planetarum, ita quod octava iterum datur Saturno, et nona Jovi. Consideraveruntque omnes concorditer quod domus nona est domus religionis et fidei. Et Ptolemaeus et Albumazar et Altavicus et Messehalac et alii omnes assignaverunt istis domibus proprietates suas, quia prima est domus vitae, secunda domus substantiae, et sic de caeteris, juxta suas proprietates et dispositiones naturales. Unde nona domus, ut dicunt, est peregrinationum atque itinerum fidei et deitatis et religionis, ac domus culturae Dei sapientiae librorum epistolarum et legatorum narrationum ac rumorum et somniorum. Et ideo merito, ut dicunt, attribuitur Jovi, qui habet respectum ad bona alterius vitae, quia illis bonis debentur fides et religio et cultura Dei et consideratio sapientialis, et librorum et epistolarum multitudo, ut patet ex lege divina; et legatorum, ut prophetarum et apostolorum et praedicatorum copia narrantium rumores idoneos de nobilibus conditionibus illius vitae, et revelationes frequentes habentium in somnis et extasi et raptibus de hac vita.

Religious significance of Jupiter in conjunction with other planets.

Dicunt igitur planetas conjungi et complecti sibi invicem, et hoc est quando fuerunt in eodem signo et praecipue quando in eodem gradu et in xvi minuto illius gradus et infra. Volunt ergo philosophi Jovem ex sua conjunctione cum aliis planetis significare super sectam religionum et fidei. Et quia sunt sex planetae quibus complecti et conjungi potest, ideo asserunt sex fore debere in mundo sectas principales. Unde periti auctores praedicti et alii dicunt, si complectatur

Jupiter Saturno, significat libros divinos, et significat de sectis Judaicam, quia est antiquior aliis et prior, sicut Saturnus pater planetarum et remotior et prior in exitu planetarum et ordine in esse. Et ipsam omnes confitentur, et ipsa nullam aliam, sicut Saturno omnes planetae complectuntur et ipse nulli, propter tarditatem sui motus. Quia quando planeta est ante eum ad orientem, nunquam ipse Saturnus consequitur aliquem, sed in tantum invalescit alius planeta, quod consequitur aliquando Saturnum et conjungitur ei. Omnes quidem sectae appodiant se ad sectam Judaeorum, quia haec fuit prima et est radix aliarum, a qua omnes aliquod genus testimonii et constitutionis sectae habuerunt: unde philosophia accepit ab illa multa testimonia et multos modos constituendi sectam, sicut prius patuit. Si vero Jupiter complectatur Marti, tunc dicunt ipsum significare super legem Chaldaicam, quae docet adorare ignem, cujus naturae Mars est in naturali potentia et effectu. Si Soli, significatur lex Aegyptia, quae ponit coli militiam coeli, cujus princeps est Sol. Si Veneri, significare dicitur super legem Saracenorum, quae est tota voluptuosa et venerea, quam licet in scriptis Mahometus redegit, ipsa tamen per longa tempora in usu vitae habebatur a suis cultoribus; unde in libro qui ascribitur Ovidio[1] de vitae suae mutatione cum loqueretur de secta venerea, quam hominibus sui temporis legem dixit esse, dicit in metro suo,

'In qua, si libeat, quod cunque licere putatur,
Scripta licet super hoc nondum lex inveniatur;'

Quam postea per sexcentos annos et amplius scripsit Mahometus in libro qui dicitur Alcoran. Ovidius enim ante Christum et in temporibus Christi fuit, et secta Mahometi incepit per sexcentos annos et amplius post incarnationem Christi, sicut patet ex differentia annorum Christi et Arabum, quae est dcxxi annorum, et cxcv dies. Sed anni Arabum computantur a Mahometo, ut dicit Alfraganus et alii similiter.

Combination of

Si vero complectatur Mercurio, tunc est lex Mercurialis.

[1] The poem *De vetula seu de mutatione vitae suae*, attributed to Ovid, is believed to have been written in the twelfth century by Leo, protonotary of the palace of Byzantium.

Mercurius enim habet respectum, ut dicunt, ad Deitatem et oracula prophetarum et credulitatem et orationem, et maxime quando conjungitur ei Jupiter; quoniam tunc significat numerum psallendi et numerum librorum divinorum. Et dicunt, quod lex Mercurialis est difficilior ad credendum quam aliae, et habet multas difficultates supra humanum intellectum. Et hoc convenit propter motus Mercurii difficiles, cujus circuitus est in epicyclo et eccentrico et aequante, in quibus considerantur sui motus longitudinis et inflexus et reflexus in motibus latitudinis, per declinationem eccentrici ab orbe signorum ad septentrionem et meridiem, et epicycli ab eccentrico in partem septentrionis et meridiei, et sunt mirabiliores et difficiliores omnibus motibus planetarum, sicut patet ex dictis Ptolemaei, et planius ex sententiis Albategni, Thebit, et Archaselis, et probabiliter ex dictis Alfragani. Et propter hoc significat, ut dicunt, super legem quae habent difficiles articulos et occultas veritates, cujusmodi est lex Christiana. Sed quia Mercurius est significator scripturae et scriptorum, et profunditatis scientiae in libris profundis, atque facunditatis, sive dulcedinis locutionis et linguae, et rhetoricae et velocitatis ejus et explanationis sententiarum, significat quod tam authenticis scripturis et tot profundis scientiis et tanta potestate eloquentiae defendetur, quod stabit semper in robore suo, donec ultima lex Lunae perturbet eam ad tempus. Et dicunt, quod haec lex est prophetae nascituri de virgine, secundum quod omnes antiqui Indi, Chaldaei, Babylonii, docuerunt quod in prima facie Virginis ascendit virgo mundissima nutritura puerum in terra Hebraeorum, cui nomen Jesus Christus, ut dicit Albumazar in majori introductorio astronomiae. Et in moralibus recitabitur auctoritas ejus inter alias philosophiae auctoritates de iis quae pertinent ad philosophiam moralem, quia etiam ea quae dicuntur hic praeparat mathematica ad usum illius philosophiae, ut ibi declarabitur expressius. Et ortus prophetae de virgine multum convenit legi Mercuriali, quia Mercurius habet maximam potestatem in Virgine, secundum judicium astronomorum omnium. Creatus enim fuit in Virgine, et dignitates, seu potestates seu testimonia seu virtutes seu fortitudines quinque quae debentur

Jupiter with Mercury related to Christianity.

planetis ratione signorum habet Mercurius in Virgine, ut sunt scilicet domus, exaltatio, triplicitas, terminus, facies.

Double meaning of *House* in astrology.

Et domus nunc et prius dicitur aequivoce [1]; quoniam istae

[1] The two senses of the word Domus will be best illustrated by the accompanying figures taken from Bacon's commentary on the Secretum Secretorum

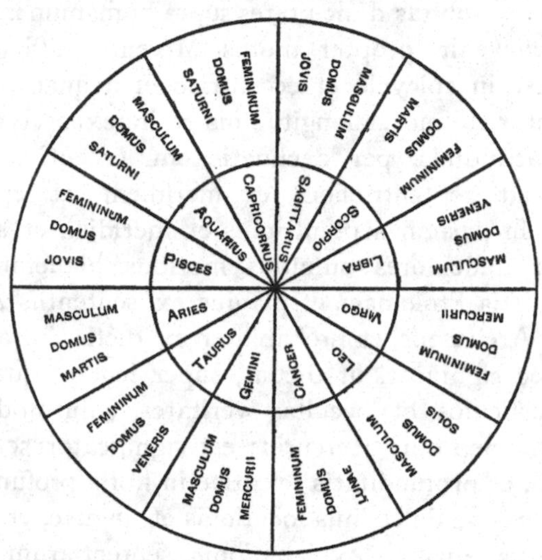

(Tanner MS. 116). In the lower figure which represents the *domus accidentales* or *situales* described on a previous page, the division into *houses* gives the state of the sky at any particular moment. The sign placed in the first *house* may be

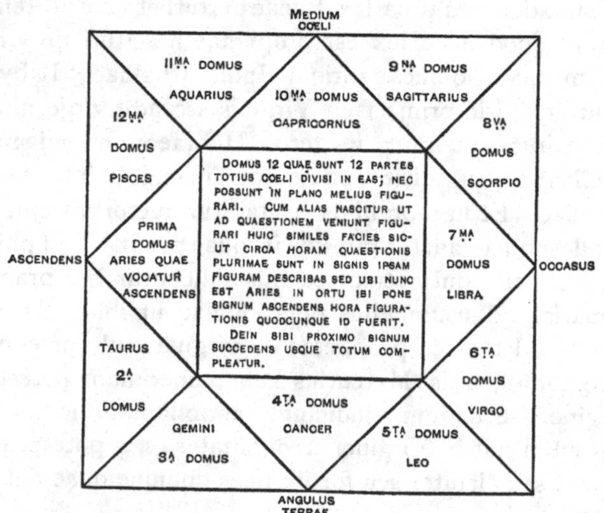

domus vocantur essentiales et naturales, priores vocantur accidentales et situales; quoniam hae domus sunt xii signa, quorum divisio naturalis est, quia sectiones zodiaci et coeli manent in suis locis de circulo coelesti, hoc est de firmamento, quam divisionem signorum faciunt sex circuli sese intersecantes in polis zodiaci, et dividunt totum coelum et mundum in xii partes aequales. Quae partes possunt considerari in zodiaco solum, et tunc sunt proprie signa, ut Aries et alia, aut possunt illi circuli imaginarie extendi ad polos, in quibus se intersecant, et tunc dividunt totum coelum in xii partes aequales habentes angustiam in extremitatibus circa polos et latitudinem in medio ad modum fundi naviculae, ita quod illa latitudo continet extremitates partium zodiaci, quas communi nomine vocamus signa, unde signum dicitur proprie, ut Aries, vel Taurus, &c. Et sumitur aliter pro tota coeli parte contenta inter duos circulos transeuntes, verbi gratia per fines Arietis, qui circuli concurrunt in polis mundi; et istud signum dicitur esse signum Arietis, quia ejus latitudo consistit in extensione Arietis, et sic stellae quae sunt extra corpus Arietis dicuntur esse in signo Arietis, quamvis sint juxta polos mundi. Sed aliae domus dicuntur accidentales, quia divisio earum est accidentalis, et non manent sectiones in eodem loco coeli, quia non sequuntur motum coeli, et ideo mutantur earum loca in circulo seu coelo in omni hora. Et sumuntur hae sectiones, ut dictum est, per circulum meridianum et horizontem cum aliis quatuor. Domus autem quae dicitur naturalis adhuc est duplex; quaedam est principalis, quaedam non principalis, unde dicitur accidentalis respectu principalis. Principalis autem domus planetae est in qua creatus fuit, ut Leo est domus Solis, Cancer Lunae, Virgo Mercurii, Libra Veneris, Aries Martis secundum quosdam, secundum alios Scorpius, Sagittarius Jovis, Capricornus

any of the twelve, according to the hour at which the horoscope is taken. The figure here given shows Aries rising, and Libra setting.

In the upper figure, representing the *domus essentiales et naturales*, special signs of the zodiac are allotted to each planet; one to the Sun, one to the Moon, and two to each of the other five. Each of these five has a masculine and a feminine sign; corresponding, I suppose, to the distinction in the text of *principales* and *minus principales*.

Saturni. Domus autem minus principales sunt, ut Aquarius Saturno datur, Pisces Jovi, Scorpius Marti secundum unam opinionem, secundum aliam Aries, Taurus Veneri, Gemini Mercurio ; ita quod quilibet quinque planetarum habet duas domos sed Sol et Luna non nisi singulas. Ita decrevit antiquitas sapientum.

Exaltation and Triplicity.

Exaltationes vero sunt hae. Sol exaltatur in Ariete, Luna in Tauro, Saturnus in Libra, Jupiter in Cancro, Mars in Capricorno, Venus in Piscibus, Mercurius in Virgine. Et sicut Sol exaltatur in Ariete, sic ejus descensio est in Libra, et sic de reliquis ; et similiter depressio Mercurii est in Piscibus, et ideo exaltatio Mercurii est in Virgine, sicut ejus domus, et est haec exaltatio in xv gradu Virginis. Triplicitas planetae dicitur, cum sit in signo in quo creatus est, vel in aliquo ejusdem naturae cum signo in quo creatus est. Unde sciendum est quod quatuor sunt triplicitates signorum. Una est calida et sicca quae continet tria signa calida et sicca cujusmodi sunt Aries, Leo, Sagittarius. Unde cum est Sol in aliquo istorum trium dicitur esse in sua triplicitate. Et alia est triplicitas secunda, ex Tauro, Virgine, et Capricorno, et haec est frigida et sicca ; et Mercurius, quando est in aliquo istorum, est in triplicitate sua. Quia licet domini istius triplicitatis in die sint primo Venus, deinde Luna, et in nocte primo Luna, postea Venus, et eorum particeps in nocte et die sit Mars, tamen Mercurius participatur eis in Virgine proprie, ut dicunt astronomi, et ideo triplicitatem habet in Virgine sicut exaltationem et domum. Tertia triplicitas est ex Geminis, Libra, Aquario, quae est calida et humida. Et quarta est ex Cancro, Scorpione, et Pisce, quae est frigida et humida.

Terminus and Facies.

Famosiores autem termini sunt Aegyptiorum. Jupiter habet sex primos gradus Arietis, Venus sex sequentes, Mercurius octo, Mars quinque, Saturnus quinque, Venus adhuc octo primos Tauri, Mercurius sex sequentes. Et sic mira diversitate variantur isti termini, ut patet in tabula terminorum, ita quod Mercurius habeat septem primos gradus Virginis pro termino, non solum secundum Aegyptios, sed secundum Ptolemaeum, et hoc est quod nunc quaerimus.

Facies autem signorum accipiuntur per divisionem cujuslibet signi in tres partes aequales; et unaquaeque constat ex decem gradibus, quae vocantur facies, et alio modo decani; quarum facierum initium est a primo gradu Arietis, et terminatur in decimo gradu ejusdem, et dicitur Martis. Secunda usque in vicesimum, et dicitur facies Solis, quia Sol succedit ei in ordine circulorum. Tertia est in finem Arietis et dicitur facies Veneris, et sic de caeteris secundum ordinem, ut patet in tabula facierum; ita quod Mercurius habet decem gradus Virginis ultimos pro facie. Et sic patet, quod Mercurius habet omnes istas potestates in Virgine. Et vocantur istae potestates per similitudinem. Unde planeta in domo sua comparatur regi in domo sua regia et in dominatione sua; et cum fuerit in exaltatione sua, est sicut vir in regno suo et gloria; et cum fuerit in triplicitate sua, est sicut vir in honore suo et inter auxiliatores atque ministros; et cum fuerit in termino suo, est sicut vir inter parentes suos et cognatos et gentem suam; et cum est in facie sua, est sicut vir in magisterio suo. Et domus dicitur habere quinque fortitudines, exaltatio quatuor, triplicitas tres, terminus duas, facies unam. Unde domus habet in se fortitudines quinque facierum, et exaltatio habet fortitudinem quatuor facierum, et sic ulterius.

Connexion of Mercury with constellation Virgo.

Ex his igitur patet quod hae potestates Mercurii essentiales et principales sunt omnes in Virgine. Atque addendum est, quod unusquisque planeta habet adhuc potestatem accidentalem quandam in signo sibi debito, quae vocatur gaudium. Unde Saturnus, cum intrat Aquarium, gaudere dicitur, ut Jupiter in Sagittario, Mars in Scorpione, Venus in Tauro, et Mercurius in Virgine. Et ideo nusquam dominatur Mercurius tantum, sicut in Virgine. Nec aliquis planeta habet tot in ea dominia, propter quod appropriatur Virgini Mercurius. Et ideo ex hac causa dicunt legem Mercurialem debere esse sectam prophetae nascituri de virgine: et ideo haec secta Mercurialis ponitur ab eis esse lex Christiana. Si vero complectatur Lunae, dicunt domini astronomiae, quod erit lex Lunae et ultima, quia circulus Lunae est ultimus, et haec erit lex corruptionis et foeda quae violabit omnes alias leges et suspendet eas, etiam Mercurialem ad tempus. Luna enim, ut dicunt, significat super nigromantiam et mendacium, et ideo lex Lunae erit nigromantica et magica et mendosa. Et propter corruptionem Lunaris motus et figurationum Lunarium significat super corruptionem istius legis, quae in se erit corrupta et alias corrumpens. Non tamen multum durabit, ut dicunt, quia Luna velociter mutatur a figuratione et luce sua et motu propter brevitatem sui circuli. Et hoc, ut dicunt, statuetur ab aliquo magno et potente qui praevalebit aliis, et aestimant astronomi fideles tam moderni quam antiqui quod haec est lex Antichristi, quia ille ultimo in fine mundi adveniet, et inducet legem corruptionis, et infatuabit mundum per artem magicam et mendacia sua.

These things point to supremacy of Christian Faith.

Sic igitur astronomi discutiunt sectas et praecipue Albumazar [1] in libro conjunctionum, et maxime primo et secundo libro, ut inveniantur sex sectae principales in quibus homines occupantur in hoc mundo. Et per hoc habetur quod secta Christi sit una de principalibus. Et si comparemus eam ad alias, manifestum est per nobiles conditiones legislatoris et ipsius sectae quod nulla alia digna est, sed sunt hominum figmenta. Et hoc de ultima lege statim patet, quia non est ibi veritas. Et in lege Venerea, quae est Saracenorum, delectatio peccati abundat secundum tenorem ejus praedictum. Sed philosophia excludit peccatum a lege. Similiter lex Aegyptiorum nulla est nec Chaldaeorum, quia docent colere creaturam, et hoc negat philosophia. Nam soli Deo cultus debetur, sicut exponetur in moralibus. Secta vero Judaeorum minus elongatur a veritate. Sed lator non fuit filius virginis,

[1] This astrological writer is frequently cited by Bacon. He is entered in the Museum Catalogue as Jafar ibn Muhammad (Al Balkhî). His principal works appear to be (a) the *Liber Conjunctionum*, otherwise called *Albumazar de magnis conjunctionibus, annorum revolutionibus, ac eorum profectionibus, octo continens tractatus*; and (b) *Introductorium in Astronomiam Albumazaris Abalachi, octo continens libros partiales*. Both were printed at Vienna, 1489. The latter treatise deals with objectors to astrology, and is specially severe on the more ignorant medical practitioners (plebaei medicinae professores) who neglect the science, as contrasted with the skilled and experienced physicians who acknowledge its value. See *Introduct. in Astron.* lib. i. cap. 5, entitled 'De Utilitate Astrologiae.'

sicut in lege Christiana; atque non habet confirmationem per tot scripturas authenticas, nec habet tam nobiles articulos secundum quod prius expositum est secundum philosophos. Et ideo oportet quod lex Christi obtineat principatum. Sed in morali philosophia hoc ex propriis erit planum. Nam non solum sic in universali investigant leges, sed determinant tempora inceptionum earum et finem aliquarum. Et hoc investigant per conjunctiones planetarum, et per revolutiones motuum eorum. Albumazar igitur in libro conjunctionum et caeteri astronomi determinant tres esse Saturni et Jovis conjunctiones, magnam scilicet, majorem, et maximam. Magna est, qua conjungitur in omnibus viginti annis in quocunque signo hoc sit. Jupiter enim perficit suum cursum in duodecim annis, et Saturnus quasi in triginta annis, et ideo fit ut post viginti annos junguntur in nono signo ab eo, in quo prius juncti fuerant; et post alios viginti in quinto a primo; et post tertios viginti iterum in illo primo. Et haec est conjunctio magna, quae fit in hac triplicitate duodecies, vel aliquando terdecies. Primum enim signum, quintum, et nonum faciunt triplicitatem. Et haec conjunctio dicitur significare pluries super sublimationem regum et potentum, et super gravitatem annonae, et super ortus prophetarum. Et postquam totiens in ista triplicitate conjuncti fuerint, ut ad aliam mutentur, tunc vocatur conjunctio major. Et hoc fit in omnibus ducentis quadraginta annis vel circiter, et significat super sectam et mutationem ejus in quibusdam regionibus. Et quando mutata fuerit conjunctio ab ista triplicitate in aliam, ut a fine Cancri ad initium Arietis, tunc dicitur maxima, per revolutionem Saturni triginta duabus vicibus, et fit omnibus nongentis sexaginta annis, et significat super mutationes imperiorum et regnorum, et super impressiones ignitas in aere, et super diluvium, et super terrae motum et gravitatem annonae. Et una major vel fere maxima fuit xxiv anno Augusti Caesaris, quam dixerunt sapientes astronomi significare super legem Mercurialem futuram. Et in libro, qui dicitur de mutatione vitae Ovidii, qui inscribitur de Vetula, propter quam mutatio fuerat facta, refertur Ovidius Naso locutus fuisse de hac conjunctione, et ex ejus dispositione prorupisse in admirationem sectae

Three kinds of conjunction of Jupiter and Saturn.

One of the greatest occurred at Christian era.

Mercurialis producendae in mundum per prophetam nasciturum de virgine absque maris commixtione, quam futuram esse praedixit post illam conjunctionem per annos sex, ita quod secundum ipsum nasceretur xxx anno Octaviani Augusti. Nam xii anni fluxerunt a morte Julii usque ad Actium bellum, in quo plene obtinuit imperium Octavianus Augustus. Nam ante magis laboravit ut acquireret regnum, quam possedit. Et tunc Christus fuit natus xxx anno Octaviani. Si vero computentur illi xii anni de regno ejus, tunc Christus fuit natus xlii anno Augusti, secundum quod alii computant Sed in idem redit. Gessit enim Augustus quinque bella civilia magna infra illos xii annos, ut narrant historiae, et maxime Orosius in libro de Ormesta[1] Mundi. Ultimum vero fuit ad Actium, in quo Antonium et Cleopatram devicit et quievit imperium in pace. Loquens igitur Ovidius de conjunctione majore et fere maxima dicit in metro suo hoc modo,

'Una quidem talis fatali tempore nuper
Caesaris Augusti fuit anno bis duodeno
A regni novitate sui, quae significavit
Post annum sextum nasci debere prophetam,
Absque maris coitu de virgine, cujus habetur
Typus, uti plus Mercurii vis multiplicatur.
Cujus erit concors complexio prima futurae
Sectae nam nusquam de signis sic dominatur
Mercurius sicut in signo Virginis.'

Et prima facies Virginis ascendebat in oriente, quando conjunctio illa facta fuit. Et fuit conjunctio illa prope caput Arietis. Si enim revolvamus motus Saturni et Jovis ad tempus illud, inveniemus eos fuisse conjunctos per medios cursos suos ante nativitatem Christi per sex annos, quinque dies, et tres horas; et erat medius cursus utriusque in Ariete decem gradus, lvi minuta, lii secunda. Medius vero

[1] The title, *Ormesta Mundi*, was often applied to the encyclopaedic work of Orosius. Ducange throws out the suggestion that Ormesta is a misreading for Orchestra, scene or spectacle. But this seems hardly a satisfactory explanation of the word.

motus octavae sphaerae erat ex signis circuli parvi decem gradus, quinque minuta, li secunda, xxvii tertia, et erat minuenda a locis omnium planetarum; unde remanserunt de Ariete ii gradus, xiv minuta, xlii secunda. Cum ergo differentia inter duas conjunctiones per cursus medios addat viii signa, ii gradus, xxv minuta, xvii secunda, sequitur quod praecedens conjunctio fuerat in Cancro xxix gradibus, li minutis, xxv secundis, et ita mutata fuit triplicitas a signo aquatico ad igneum. Si vero haec conjunctio fuisset propinquior capiti Arietis, fuisset maxima, et tunc erant anni Graecorum perfecti trecenti quinque et novem menses, et fere xviii dies, quod potest probari per tabulas annorum.

Influence of Saturnian revolutions on political and religious change.

Per revolutiones vero motuum planetarum considerant illud idem. Nam Albumazar octava differentia libri secundi de conjunctionibus dicit, quod mora sectae et regni et permutatio accidunt praecipue secundum quantitatem decem revolutionum Saturniarum, praecipue si Saturno conveniat mutatio ad signa mobilia, quae sunt Cancer, Libra, Capricornus, Aries, dummodo Jupiter fuerit cadens ab eo. Sed si Jupiter fuerit cum eo, aut aspiciat eum, minuet multum de malo propter ejus bonitatem. Quando enim fuerunt completae decem revolutiones Saturni in diebus Darii, fuit apparitio Alexandri magni, et destructio regni Persarum. Et circiter post decem alias revolutiones completas apparuit Jesus filius Mariae, super quem sunt orationes cum permutatione sectae. Et quando completae sunt decem aliae apparuit Meni [1], et venit cum lege quae est inter Paganos et Nazarenos. Et post decem alias venit Mahometus, et fortasse illud est ante complementum decem revolutionum, ut in revolutione nona, et forsan post, ut in undecima. Et illud est secundum quantitatem ejus, quod exigunt conjunctiones praemissae, quae sunt fortiores istis revolutionibus. Et similiter mutatur citius vel tardius secta secundum proprietates planetarum dominantium regnis diversis, ut Saturnus Indiae dominatur, Jupiter Babyloniae, Mars Thraciae, Sol Romanis et imperio eorum, Mercurius Aegypto, Luna Asiae.

[1] Meni, commonly written Manes, is the originator of the Manichaean heresy, which arose in Mesopotamia towards the end of the third century.

Destruction of Mahommedanism.

Et de destructione legis Mahometi pulchre et certitudinaliter loquuntur. Nam secundum quod Albumazar dicit viii capitulo secundi libri, non potest lex Mahometi durare ultra sexcentos nonaginta tres annos. Sed tantum valet durare et durabit nisi propter aliquam causam coincidentem abbrevietur tempus secundum quod prius tactum est, quod abbreviatio potest fieri major et minor ex causis diversis. Et nunc est annus Arabum sexcentesimus sexagesimus quintus [1] a tempore Mahometi, et ideo cito destruetur per gratiam Dei, quod debet esse magnum solatium Christianis. Propter quod laudandus est Deus, qui philosophis dedit lumen sapientiae, per quod lex veritatis confirmatur et roboratur, et per quod percipimus inimicos fidei destrui debere. Et huic sententiae concordat Apocalypsis xiii capitulo. Nam dicit quod numerus bestiae est 663, qui numerus est minor praedicto per xxx annos. Sed scriptura in multis locis subticet aliquid de numero completo, nam hic est mos scripturae, ut dicit Beda. Et hic forsan voluit Deus, quod non exprimeretur totaliter, sed aliquantulum occultaretur, sicut caetera quae in Apocalypsi scribuntur. Unde ante tempus ultimum quod isti sectae determinatur, secundum ejus causam principalem, prout determinat Albumazar, forsan continget quod Saraceni destruentur aut per Tartaros aut Christianos. Et jam major pars Saracenorum destructa est per Tartaros, et caput regni quod fuit Baldac, et Caliph qui fuit sicut papa eorum. Jam haec facta sunt xii annis elapsis [2].

Planetary changes not merely signs of the future, but co-operative influences, though not compulsory.

Et quamvis loquantur de sectis, et sectae dependent ex libertate rationis, tamen non imponunt aliquam necessitatem libero arbitrio, dicentes planetas esse signa innuentia nobis ea quae Deus disposit ab aeterno fieri sive per naturam, sive per voluntatem humanam, sive per rationem propriam secundum beneplacitum suae voluntatis. Ita dicitur in libro de cursibus

[1] 665 Arabic years of 354 days make very nearly 644 of our years, which added to 622 the date of the Hejira brings the date to 1267, the year in which Bacon was writing.

[2] The capture of Baghdad, called by Bacon and others Baldac, by Halagu, grandson of Chinghis Khan, was in the 656th year of the Hejira, which counting Arabic years, would bring the date to 1258 of our era. But Halagu's expedition had begun three years before.

planetarum. Et praeterea dicunt quod voluntas non cogitur, sed tamen corpus alteratur per virtutes coelorum, et tunc anima corpori unita excitatur fortiter et inducitur efficaciter, licet in nullo cogatur, ut velit gratis sequi inclinationes corporis ad actus privatos vel publicos, et ad bonos sicut ad malos, ut sic opiniones et sectae et mutationes consuetudinum inducantur per aliquem famosum in populo et potentem, secundum quod praevisum fuit et praecognitum a Deo; ita quod planetae sic non solum sint signa, sed aliquid faciant in excitando.

Et cum posuerunt Dominum Jesum Christum esse Deum et hominem, ut Ethicus astronomus manifeste dicit in Cosmographia, et Alchimus similiter, necnon et in illo libro qui inscribitur, *Ovidius de vetula*, Deum incarnari in Christo colligitur, attribuentes ei quod soli homini denegatur, volunt quod dispositio coelestis potuit esse in signum conceptionis Virginis, et nativitatis illius Hominis, in quantum homo, sicut stella praestitit signum in ejus nativitate, secundum quod dicitur in libro cursuum planetarum, quod planetae omnes et caeterae stellae aliter Deo homini facto, aliter puro homini, aliter creatori, aliter creaturis famulantur. Voluit ergo Deus res suas sic ordinare, ut quaedam quae futura praeviderit vel praedestinaverit rationabilibus per planetas ostenderentur, ideo scilicet ut mens humana Dei mirifica recognoscens in amorem sui conditoris succensa excresceret. Et ideo volunt quod innuere et significare possunt coelestia hoc opus deificum conceptionis et nativitatis, quatenus creatura attestetur suo creatori in carne venienti. Sicut contra naturam sol obscuratus fuit in passione; propter quod philosophi haec videntes dixerunt, quod aut aliquid Deus naturae patiebatur, aut tota mundi machina dissolveretur. Et proculdubio volunt quod impossibile sit Deum subjici creaturae, nec quod istud opus divinum in quantum fuit a virtute infinita et supra naturam, aliquo modo fuerit subjectum dispositioni coelesti: sed solum quod illud fuerat in signum. In quantum tamen Virgo mundissima fuit vera mater et naturalis Domini Jesu Christi, et operata fuerit ex virtute naturali in praeparando materiam et in fovendo post conceptionem et in hujusmodi, posuerunt virtutem coeli cooperatam fuisse virtuti naturali Virginis gloriosae, et excitasse eam in quantum naturaliter operabatur, quia homo generat hominem et sol. Si enim aliquid naturale fuerit in illa conceptione per materiae praeparationem et fomentum in utero et hujusmodi, in quantum fuerat naturalis et vera mater, non aestimant inconveniens ponere coelestem dispositionem esse plus quam signum, consideratis pure naturalibus. Sed quicquid dicunt in hac parte, hoc ad regulam fidei reducendum est, ut a catholica veritate non discordet. Et licet omnia et ad plenum non sufficiant ostendere secreta istius sectae, tamen an sit haec secta, et qualis sit in universali, pulchre attestantur, ut satis admirantes sapientiam eis datam facile excusemus eorum ignorantiam, quia defecerunt a plena certificatione ritus Christiani, cum in eo non fuerant instructi. Et laudare debemus, quod nobiscum concordant et confirmant nostram professionem. Sed in tertia parte moralis philosophiae de hac confirmatione abundantior fiet sermo.

Astrology may help us to foretell the time when Antichrist shall come.

Et quoniam post legem Mahometi non credimus quod aliqua secta veniet nisi lex Antichristi, et astronomi similiter concordant in hoc, quod erit aliquis potens qui legem foedam et magicam constituet post Mahometum, quae lex suspendet omnes alias, multum esset utile ecclesiae Dei considerare de tempore istius legis, an cito veniet post destructionem legis Mahometi, an multum longe. Et Ethicus philosophus in sua Cosmographia dicit expresse, quod gens quae fuit clausa infra portas Caspias irruet in mundum et obviabit Antichristo et eum vocabit Deum Deorum. Et proculdubio Tartari fuerunt infra portas illas et exiverunt. Jam enim fractae sunt portae, sicut certi sumus. Nam fratres minores[1], quos dominus rex Franciae Ludovicus qui nunc regnat misit, transiverunt cum Tartaris per medium portarum ultra longe inter montes, ubi fuerunt inclusi. Et notum est non solum omnibus nationibus orientis quod Tartari exiverunt a locis eis, sed et eis qui bene sciunt mundi dispositionem, et noverunt partes habitabiles et regionum diversitates per astronomiam, et per auctores alios ut Plinium et Martianum et caeteros qui mundi regiones describunt, et per historias. Nolo hic ponere os meum in

[1] Much will be said in the course of the work of this important mission, of which William Rubruquis was the leader.

coelum, sed scio quod si ecclesia vellet revolvere textum sacrum et prophetias sacras, atque prophetias Sibyllae, et Merlini et Aquilae, et Sestonis, Joachim et multorum aliorum, insuper historias et libros philosophorum, atque juberet considerari vias astronomiae, inveniretur sufficiens suspicio vel magis certitudo de tempore Antichristi[1].

Sed haec hactenus. Nunc vero inferam secundum[2] quod Reformation of the Julian Calendar.

[1] The subject of Astrology is again discussed at the end of the geographical section in what may almost be regarded as a distinct treatise.

It has often been remarked that Roger Bacon was supported by Albertus Magnus, by Aquinas, and indeed, all the best thinkers of his time in his conviction of the truth of astrology. To a believer in a limited and spherical universe with a terrestrial centre, nothing could seem more valid as a working hypothesis for explaining physical changes on the earth's surface than that alterations of the directions in which the planets were seen should be followed by corresponding alterations of terrestrial objects. The combinations of planetary bodies as seen in conjunction, in opposition, or in intermediate positions, offered a wide field of speculation, which became practically boundless when to the apparent relation of these bodies to one another were added their apparent relations (also ever varying) with the fixed stars. Human and terrestrial events complicated as they might be, were paralleled by equal complication in the play of celestial forces. It may be said on the whole that so far from belief in astrology being a reproach to Bacon and his contemporaries, to have disbelieved in it would have been in the thirteenth century a sign of intellectual weakness. It conformed to the first law of Comte's *philosophia prima* as being the best hypothesis of which the ascertained phenomena admitted.

When the universe was unlocked by the Copernican theory, it might have been supposed that astrology would speedily disappear. Yet it was not so. Francis Bacon, writing three centuries and a half after Roger Bacon, had not abandoned belief in it, and Campanella's work (*Astrologicorum libri sex*) published in 1629, is a treatise that Roger Bacon might himself have written. 'Sunt causae ipsae stellae,' says Campanella (lib. ii. 1), nedum signa, ut optime probat S. Thomas; corporearum mutationum per se, voluntariarum per accidens. . . . Omne judicium astrologicum aut est de mutationibus aeris et maris et telluris; unde de penuria et abundantia rerum nobis utilium et noxiarum ratiocinamur; aut de monarchia et rerum publicarum initiis et mutationibus, et de gentium et legum et morum translationibus et civitatum et provinciarum; aut de cujusque hominis ortu vita morte et eventibus, aut tandem de bonis malisque electionibus rerum agendarum.

[2] In *Opus Tertium*, chap. 68, p. 274, there is an almost exact repetition of what follows in the next sixteen pages. [In O. these are altogether omitted. The Cottonian MS. (Jul.) supplies them.] It is probable that the scribe copied the MS. of *Opus Tertium*; since in the passages of that work here referred to, Bacon explains that what had been said in the copy of *Opus Majus* sent to the Pope on the subject of the Calendar, was not in all respects correct, so that, he goes on to say, 'hic iterum feci transcribi et correxi.'

non solum expedit ecclesiae, sed quod maxime decet eam, et quod sine grandi periculo et confusione vitari non potest; quamvis tamen longis temporibus jam accidit multiplex abusus. Et quoniam totus hic error procedit ex pura ignorantia et negligentia considerationis, tanto est vilior coram Deo et hominibus sanctis, et apud omnes, non solum sapientes astronomos. Sed et computistae vulgati sciunt multiplicem errorem, et scribunt super hoc sicut et astronomi, quorum utrorumque scripta per Dei ecclesiam vulgata sunt, in quibus errores hi notantur, et de remediis datur consilium. Sed nullus propter concilium generale ausus est facere remedium. Quod autem intendo hic est de correctione calendarii quo utitur ecclesia. Julius quidem Caesar in astronomia edoctus complevit ordinem calendarii secundum quod potuit tempore suo; et sicut historiae narrant contra Achorium astronomum et Eudoxum ejus doctorem disputavit in Aegypto de quantitate anni solaris, super quam fundatum est calendarium nostrum. Unde, sicut Lucanus refert, ipse dixit,

'Non meus Eudoxi vincetur fastibus annus.'

That calendar makes the solar year too long by $\frac{1}{130}$ of a day.

Sed non pervenit Julius ad veram anni quantitatem, quam ponit esse in calendario nostro ccclxv dies et quartam diei integram; quae quarta colligitur per iv annos, ut in anno bisextili computetur unus dies plus in quarto anno, quam in aliis annis communibus. Manifestum autem est per omnes computistas antiquos et novos, sed et certificatum est per vias astronomiae, quod quantitas anni solaris non est tanta, immo minor. Et istud minus aestimatur a sapientibus esse quasi cxxx^{ma} pars unius diei[1]. Unde tamen in cxxx annis superflue computatur unus dies, qui si auferretur, esset calendarium correctum quoad hoc peccatum. Et ideo cum omnia quae sunt in calendario fun-

[1] That the length of the year was wrongly given in the Julian Calendar must have been known to the small group of Arabian men of science who studied Ptolemy's *Almagest*. But that the amount of the error was not matter of common knowledge half a century after the *Opus Majus* was written, is shown by the passage in Dante (*Paradiso*, xxvii. 142–3), where the error is spoken of as being the hundredth part of a day. The difference between $\frac{1}{100}$ and $\frac{1}{130}$ is considerable to those who know Dante's minute and precise way of dealing with such questions. The mean length of the equinoctial year is 365^d, 5^h, 48^m, $51^s \cdot 6$.

dentur super quantitatem anni solaris, necesse est ea vacillare, postquam est erroneum fundamentum.

Deinde accidit alius error major, scilicet de fixione aequinoctiorum et solstitiorum[1]. Nam hic error non solum ex quantitate anni exoritur, sed habet in se graves errores. Ponuntur autem aequinoctia et solstitia in diebus fixis, ac si semper fuissent ibi et debeant esse in aeternum. Sed certum est astronomis quod non figuntur, immo ascendunt in calendario, sicut per tabulas et instrumenta probatur sine dubitatione. Et in principio ecclesiae ponebatur solstitium hyemale viii calendas Januarii in die nativitatis Domini, et aequinoctium vernale viii calendas Aprilis in festo annuntiationis Virginis gloriosae, et solstitium aestivale viii calendas Julii, scilicet in die nativitatis beati Johannis Baptistae, et aequinoctium autumnale viii calendas Octobris. Et istius opinionis fuit Hippocrates medicus[2]; cui Christiani facile concordabant, propter hoc quod beatus Joannes dixit, Illum oportet crescere, me vero minui. Unde expositores aliqui sacrae scripturae posuerunt Dominum natum quando dies crescit, et hoc est in solstitio hyemali; et Johannem Baptistam nasci quando dies incipiunt minui, ut in solstitio aestivali. Et Isidorus fuit istius opinionis Hippocraticae, ut patet ex viii libro Etymologiarum, et sanctus Anatolius, qui in principio ecclesiae de hujusmodi disputavit. Sed postea mutaverunt haec viri ecclesiastici, et statuerunt aequinoctium vernale esse xii calendas Aprilis, sicut dicit Beda in libro temporum, quod usque nunc tenetur. Et hoc manifestum est omnibus qui aliquid sciunt de computo, et qui considerant usum ecclesiae. Nam sicut in lege veteri post aequinoctium vernale quando luna fuerat xiv celebrabatur Pascha, sic ecclesia ordinavit quod terminus Paschae sit ibi, et in dominica sequente celebraretur Pascha; quia oportet quod

It takes no account of changes in the time of equinoxes and of solstices.

[1] It should be noted that the time of the *solstitia* in the middle of the thirteenth century coincided almost exactly with that of the sun's apogee and perigee (aux et oppositum augis). In our own time the former occur about nine days before the latter. The perigee and apogee have a progressive motion on the ecliptic of 11″. 8 annually. The retrograde motion annually of the equator on the ecliptic (precession of the equinoxes) is 50″. 1.

[2] Hippocrates of Chios the geometer, not Hippocrates of Cos the physician, is presumably meant. They were contemporaries.

in die dominica fiat festum istud apud Christianos, et ideo non possumus celebrare Pascha xiv luna, sed in dominica sequente. Et posuerunt quod primum Pascha sit xi calendas Aprilis ut nullo modo sit ante, propter hoc quod aequinoctium dixerunt figi in xii calendas ubi luna potest aliquando esse xiv. Nam sicut dicit Beda in computo suo, ecclesia non utitur xiv luna quae praecedit aequinoctium, sed quae est in aequinoctio, vel post ipsum; et ipsum aequinoctium, ut dicit, est xii calendas Aprilis, et ibi est primus terminus Paschae, sicut omnes computistae sciunt; quod si sit in die sabbati, tunc in crastino potest esse Pascha quia dies dominicus est. Et ideo propter rationem aequinoctii fixam in xii calendas Aprilis, dicit ecclesia quod infimum Pascha est in xi calendas Aprilis.

These occur earlier than in the official tables at the rate of one day in 125 years.

Sed quamvis usus ecclesiae tenuit in principio aequinoctium esse viii calendas et postea mutavit, et teneat nunc aequinoctium fixum xii calendas Aprilis, tamen certum est quod aequinoctium istud non est in locis illis, sed jam ascendit in calendario longe ab his locis, et similiter solstitia et reliquum aequinoctium. Nam hoc anno fuit solstitium hyemale idibus Decembris per duodecim dies ante nativitatem Domini, et aequinoctium vernale iii idus Martii, et solstitium aestivale est xvii calendas Julii, et aequinoctium autumnale xvi calendas Octobris. Et hoc potest non solum astronomus certificare, sed quilibet laicus ad oculum perspicere per casum radii solaris, nunc altius, nunc inferius ad parietem vel aliud, secundum quod quilibet potest notare. Et mutabuntur ab his locis in temporibus succedentibus, quia post annos circiter xciv, scilicet anno Domino mccclxi, erit solstitium hyemale pridie idus Decembris et aequinoctium vernale iv idus Martii, et solstitium aestivale xviii calendas Julii, et aequinoctium autumnale xvii calendas Octobris, scilicet quodlibet eorum per unum diem antequam sit modo. Nam post annos circiter cxxv ascendit per unum diem. Et hoc accidit ex errore quantitatis anni, quia sol circiter tantum tempus deficit de quantitate anni, quo utitur calendarium, per unum diem. Et hoc est quod plus concordat cum annis quos computamus ab Incarnatione. Nam Ptolemaeus anno cxl ab Incarnatione invenit aequinoctium vernale xi calendas Aprilis et solstitium hyemale xi calendas Januarii, ut patet ex Almagesti. Sed ab hoc loco in calendario usque ad idus Decembris, ubi nunc est solstitium, sunt ix dies quibus ascendit hoc solstitium. Sed ab anno probationis Ptolemaei sunt nunc de annis Domini mcxxvii eo quod nunc sit annus Domini mcclxvii, a quibus si demantur cxl, qui fluxerunt ab Incarnatione usque ad probationem Ptolemaei, remanebunt mcxxvii anni. Sed in isto tempore nunc dicto cxxv anni reperiuntur novies et duo anni ultra. Quapropter hoc tempus, scilicet cxxv anni, satis convenit cum numero annorum Christi, ut semper unus dies in tanto tempore minuatur de quantitate anni, et denotet mutationem solstitii et aequinoctii.

Protestor tamen quod in tanta difficultate non loquor praecise, sed multum propinque veritati certificandae, usque quo currat finalis probatio super anni quantitate et mutatione solstitiorum. Per jam dicta tamen secundum probationem Ptolemaei non potuit in viii calendas esse solstitium hyemale in tempore nativitatis Domini, sed oportuit quod in decimo calendas fuerit; eo quod in clx annis a tempore nativitatis usque ad probationem Ptolemaei non potuit mutari solstitium per tres dies, nec per duos, sed per unum et parum de alio. Et ideo x calendas Januarii potuit esse secundum hoc quod Ptolemaeus invenit xi calendas. Et cum per eandem probationem Ptolemaei potuit aequinoctium primo anno nativitatis fuisse xi calendas Aprilis, non potuit esse viii sicut primo crediderunt in ecclesia primitiva, et longe magis non potuit esse xii calendas Aprilis, sicut nunc creditur secundum usum ecclesiae. Quia cum semper ascendit aequinoctium, et in tempore Ptolemaei fuit xi calendas Aprilis, tunc ante illud tempus fuit retro hunc locum magis versus Aprilem; et ideo x calendas secundum probationem Ptolemaei. Secundum haec igitur neque sunt aequinoctia et solstitia fixa, nec sunt fixa illis diebus quibus usus fuit in ecclesia. Nec fuit Hippocrates longe a veritate, quoniam ipse fuit ante Christum plus quam ccc annis, et ideo potuit

There are points which need further investigation.

esse aequinoctium temporibus suis viii calendas vel prope, scilicet vii.

Hence ensue grievous errors as to the time of keeping Easter.

Sed tertium inconveniens est longe majus. Nam, ut prius tactum est, veritas est, quod sine errore debet Pascha celebrari die dominica post xiv lunam quae invenitur vel in aequinoctio, vel post aequinoctium vernale, propter conformitatem legis Christianae ad legem antiquam propter Paschae solemnitatem quae primo fuit in lege veteri, et praecessit sicut figura novi Paschae. Cum igitur verum est quod aequinoctium sit iii idus Martii, et possibile est quod ibi sit luna xiv, scilicet in xiv anno cycli decemnovennalis, ut prima computetur pridie calendas Martii super C literam, oportet quod die dominica proxima post illam diem sit Pascha secundum veritatem. Sed non potest ab hac xiv luna dies dominica elongari plus quam usque ad xiii calendas Aprilis, ut patet in calendario. Sed hoc est ante xi calendas Aprilis ubi primum Pascha celebrat ecclesia. Quare in illo anno xiv cycli celebrabitur Pascha gloriosum tempore non suo. Et idem accidit in tertio anno cycli, ut patet per aureum numerum in calendis Martii. Nam xii calendas Aprilis vel citra erit Pascha. Et quoniam potest contingere dies dominica in his annis pridie idus Martii, et idibus, et sic ultra, usque ad xii calendas Aprilis, ideo multipliciter debet fieri Pascha in illis diebus, quod observari non potest si primum Pascha sit xi calendas Aprilis. Et quoniam aequinoctium verum ascendit plus et plus, ita quod circiter mccccIxxxi annum erit v idus Martii, et sic ascendendo ulterius versus principium Martii secundum computationem calendarii, et ultra Martium continue propter errorem de falsitate anni, necesse est quod Pascha fieret circa principium Martii vel in Februario, et sic semper antecedendo secundum antecessionem aequinoctii. Sed hoc est inconveniens maximum; quia sic non solum Pascha, sed Quadragesima et omnia festa mobilia recederent horribiliter a statu suo, et confunderetur totus ordo ecclesiastici officii. Praeterea cum secundum veritatem Pascha potest celebrari ante xi calendas Aprilis secundum aequinoctiorum veritatem, et hoc per multos dies et quot volumus secundum quod aequinoctium antecedit, oportet quod verum Quadragesimae principium similiter ascendat antequam incipiatur secundum usum ecclesiae, et ita in vera Quadragesima carnes per multos dies comedentur; et in tantum potest aequinoctium antecedere, quod tota Quadragesima vera erit in tempore quo Christiani comedent carnes, quod est absurdissimum. Et sic festum Paschale quo mundus salvatur, et festum Pentecostes quo gratia Dei diffusa est in ecclesiam, et caetera festa mobilia violantur: et per consequens alia festa immobilia, quia illa cedunt Paschati et festis aliis sui generis, ut notum est evidenter. Et cum haec sunt horribilia ex se, sunt magis stulta et derisione digna; quia propter ignorantiam et negligentiam diabolus procuravit quod sic accideret ecclesiae Dei. Nam non oporteret aliud, nisi quod certificarentur anni quantitas et aequinoctium. Et si peritissimis astronomis non denegatur certificatio quantitatis anni, tamen facilius est certificare aequinoctium, quia per centum et circiter viginti annos non mutatur nisi per unam diem secundum opinionem magis probatam nunc diebus. Et daretur ars de hoc, et etiam inveniretur aequinoctium per secula futura usque in finem mundi et ultra, quod facile est, et fierent tabulae et canones de hoc, et scriberentur per totam ecclesiam Dei, et ponerentur cum calendario, et tunc non accideret impedimentum in hac parte. Et similiter aliud aequinoctium inveniri posset, et solstitia de facili, et omnis error in his evacuaretur ad laudem et honorem Dei, et tollerentur infamia et scandalum, quae nunc multiplicantur apud omnes computistas et astronomos et sapientes propter errores qui in hac parte vulgari ab ecclesia permittuntur.

The error in the age of the moon as given in the calendar increases at the rate of nearly seven hours in 76 years.

Sed majus inconveniens accidit ex primatione designata per aureum numerum in calendario. Nam quilibet potest videre ad oculum, si aspiciat coelum, quod luna est prima secundum veritatem per tres dies vel quatuor antequam signetur in calendario, et in omnibus lxxvi annis recedit primatio a loco suo in calendario per xvi minuta unius diei et xl secunda, et hoc est plus[1] quarta unius diei, et prope tertiam unius diei,

[1] I have corrected this passage from *Opus Tertium*, cap. 70. *Minutum unius diei* means the sixtieth part of a day, i. e. twenty-four minutes. It does not

quia xvi minuta et xl secundae sunt sex horae et xl minuta unius horae. Et in omnibus ccciv annis recedit a loco primationis in calendario per unum diem et vi minuta unius diei, et xl secunda. Et post 4,256 annos dicetur luna secundum calendarium prima, quando est plena lumine. Et post 7,904 annos erit error unius lunationis integrae, excepto modico, scilicet exceptis xxxviii minutis, et xxx secundis. Et hic error potest devenire ad centum lunationes, et tunc iterum redibit error primus, ut sequantur sequentes per ordinem, et sic in infinitum hi errores revolvuntur. Et quod haec omnia sint vera non est dubium peritis astronomis. Etiam quilibet computista novit, quod fallit primatio per tres dies vel quatuor his temporibus, et quilibet rusticus potest in coelo hunc errorem contemplari.

The lunar cycle in use does not accurately correspond to nineteen solar years.

Quantumcunque vero brevius possum, aperiam grossam declarationem errorum praedictorum. Nostrum vero calendarium ponit cyclum decemnovennalem esse aequalem xix annis solaribus cum quarta integra sumptis, secundum quod calendarium utitur hujusmodi anno solari, ut prius habitum est. Et hic cyclus continet lunationes omnes, quae contingunt in xix annis lunaribus. Et computantur ccxxxv quia quolibet lunari anno, qui dicitur communis, habemus xii lunationes, et illis addimus vii lunationes in toto cyclo decemnovennali ad restaurationem defectuum, qui accidunt xii lunationibus respectu xii mensium solarium in anno solari, quia annus lunaris consistens ex xii lunationibus non habet nisi cccliv dies integros, et solaris habet xi plures, nam habet ccclxv. Et ideo annus lunaris citius finitur quam solaris per xi dies; et tunc colliguntur isti xi dies superflui usque ad tertium annum, et fit una lunatio seu mensis lunaris. Et sic in tertio anno cycli apponitur unus mensis lunaris, et vocatur embolismus et annus embolismalis, id est, superexcrescens, quia habet xiii lunationes, et sic ultra per totum cyclum semper colliguntur isti menses embolismales, ut fiant anni embolismales, quatenus cyclus

seem quite clear why Bacon chose the numbers 4,256 and 7,904, except that they were multiples of 76. The error in the first period would be 15^{d}, 12^{h}, 57^{m}, 36^{s}, and in the second, 28^{d}, 19^{h}, 32^{m}.

lunationum decemnovennalis aequetur xix annis solaribus. Sed istud est impossibile. Nam secundum Ptolemaeum in Almagesti et omnes astronomos mensis lunaris non est proprie a visione novae lunae, quia hoc tempus est inaequale; quia aliquando in mane ejusdem diei est novacula veteris lunae, et in vespere novacula novae lunae, et aliquando est spatium duorum dierum inter eas, et aliquando tres dies intercipiuntur, ut planum est sensui, et causae ab astronomis assignantur; sed menses debent aequari. Item nec a conjunctione solis et lunae secundum eorum cursum verum, quia hoc tempus est inaequale; et ideo considerabitur penes conjunctionem solis et lunae secundum utriusque eorum cursum medium et aequalem, quia hoc tempus est aequale et uniforme. Et secundum quod probat Ptolemaeus in Almagesti, hoc tempus est xxix dies, et xxxi minuta unius diei, et l secunda, et viii tertia, et ix quarta, et xx quinta. Et huic concordat Arzachel, qui fundavit tabulas suas super quantitatem anni lunaris, qui annus lunaris continet cccliv dies et quintam et sextam unius diei, id est, xxii minuta. Et secundum hoc erit tempus aequalis lunationis xxix dies, et xxxi minuta et l secunda, quae multiplicata per xii faciunt cccliv dies et xxii minuta. Sed Arzachel omisit tertia Ptolemaei et quarta et quinta, quia in maximo tempore parum quantitatis adjiciunt.

The Arabian cycle of thirty years of twelve lunar months is preferable.

Posito ergo quod tempus aequalis lunationis sit xxix dies et xxxi minuta et l secunda, accidit quod minimum tempus reducens integras lunationes ad idem temporis principium est xxx anni Arabum, qui continent ccclx lunationes integras; et continet dies 10,631 praecise. Quoniam cum una lunatio sit xxix dies, et xxxi minuta, et l secunda, xii lunationes quae faciunt unum annum Arabum et annum unum lunarem, continebunt cccliv dies et xxii minuta unius diei. Sed cccliv multiplicata per triginta faciunt 10,620, et xxii multiplicata per triginta faciunt xi dies integros, qui prioribus additi faciunt 10,631, reducentes primo lunationes aequales ad consimile temporis initium. Cum igitur xxx anni Arabum sint tempus minimum quod reducit integras lunationes ad consimile temporis initium, non est possibile ut aliud tempus hoc idem faciat, nisi sit ei aequale vel multiplex. Sed xix[1] anni calendarii nostri non aequantur xxx annis Arabum, nec sunt eis multiplices, quia xxx anni Arabum continent xxix annos solares et unum mensem et octo dies. Ergo relinquitur quod xix anni solares non possunt aequari vero cyclo primationum, nec per consequens xix anni cycli decemnovennalis. Et ideo cyclus decemnovennalis non est verus cyclus primationum.

Successive periods of nineteen solar years are not of the same length owing to leap-years.

Praeterea istud potest magis in particulari manifestari, ut appareant inconvenientia prius nominata. Nam singuli xix anni solares non sunt aequales ad invicem, eo quod in primo cyclo sunt tantum quatuor bisextiles anni, et in tribus aliis sunt quinque, quia quartus annus in primo cyclo bisextilis est, et viii et xii et xvi, eo quod quartus annus semper sit bisextilis. Et ideo in secundo cyclo primus est bisextilis et sic ulterius, ita quod in illo sunt quinque, ut patet consideranti, et similiter in aliis duobus; et tunc iterum redit cyclus qui habet quatuor bisextiles, et subsequuntur tres qui habent quinque, et sic semper currit ordo cyclorum. Primus autem cyclus annorum solarium xix praecise habet dies 6,939, et in illis assignantur ccxxxv lunationes. Sed si multiplicaverimus tempus aequalis lunationis, hoc est xxix dies, et xxxi minuta, et l secunda, in ccxxxv resultabunt 6,939 dies et xl minuta et l secunda, quae sunt plus quam duae tertiae unius diei. Quapropter completis xix annis habentibus tantum iv dies bissextiles, nondum completae sunt ccxxxv lunationes, sed desunt eis xl minuta et l secunda unius diei. Quilibet vero xix anni habentes quinque dies bisextiles habent dies 6,940. Unde cum ccxxxv lunationes aequales habeant 6,939 dies, et xl minuta, et l secunda, tunc xix anni habentes quinque annos bisextiles superant ccxxxv lunationes spatio xix minutorum, et x secundorum, quod est fere tertia unius diei. Quod patet si de uno die, quo superfluunt nunc dicti xix anni super ccxxxv lunationes, subtrahantur xl minuta, et l secunda, quibus illae lunationes superfluebant super xix annos habentes tantum quatuor dies bisextiles. Et ita patet, quod non quilibet xix anni solares aequantur sibi invicem. Sed

[1] xxix in J.

quater xix anni redeunt ad idem temporis principium, qui faciunt lxxvi annos, et semper aequantur aliis lxxvi annis. Si ergo aggregentur ter xix minuta et decem secunda, quae quilibet xix anni habentes quinque bisextos addunt super ccxxxv lunationes, provenient nobis lvii minuta et xxx secunda, quibus isti ter xix anni habentes quinque bisextos simul sumpti excedunt suas lunationes. Sed xix anni habentes tantum dies quatuor bisextiles sunt minores suis lunationibus in xl minutis et l secundis. Si igitur subtrahantur haec xl minuta et l secunda a praedictis lvii minutis et xxx secundis, relinquentur xvi minuta et xl secunda, quibus quater xix anni solares, qui faciunt lxxvi annos, excedunt suas lunationes.

Et hic est primus error notabilis, quem a principio notavi, ex quo sequitur secundus. Nam si accipiamus quater lxxvi annos, qui faciunt ccciv annos, superabunt suas lunationes in quater[1] xvi minutis et xl secundis. Sed haec sumpta quater faciunt unum diem et sex minuta, et xl secunda. Relinquitur ergo error secundus, quod ccciv anni superant suas lunationes in uno die et vi minutis unius diei et xl secundis. Et ideo accidit quod post ccciv annos dicamus secundum calendarium lunam primam, ipsa existente majoris aetatis quantitate unius diei et vi minutorum et xl secundorum. Et si haec accidunt, manifestum est inquirenti, quod multiplicato tempore multiplicabuntur inconvenientia. Et ideo post 4,256 annos dicetur luna prima secundum computationem calendarii, quando ipsa erit plena lumine. Et ulterius. quod una lunatio integra superabundabit; et tandem quod centum lunationes in maximo tempore superfluent, sicut patet per radices jam datas. Nec est necesse facere computationes majores pro praesenti persuasione, quia promptae sunt demonstrationes ad hoc, cum fuerit opportunum. Manifestissimus igitur est error primationum secundum calendarium, et non unus sed multiplex, nec parvus sed quasi infinitus.

Et cum ita sit error cycli decemnovennalis in singulis annis, sequitur quod similiter erit error in cyclis epactarum. Nam xi dies excrescentes in anno solari super annum lunarem

Error in the epacts.

[1] J. has quantitate.

dicuntur epacta sequentis anni solaris, quasi *epiaucta*, hoc est, superaugmentata; et per illos xi dies majoratur aetas lunae in primo die sequentis anni, et per eosdem dies majoratur aetas lunae in principio cujuslibet mensis futuri anni super aetatem suam in principio cujuslibet mensis prioris anni. Et quia hujusmodi errores contingunt, necesse est ut aetas lunae vera quaeratur aliter quam per hujusmodi cyclos.

Superiority of both Arabic and of Hebrew tables.

Et remedium horum omnium est, quod possumus cognoscere primationem lunae secundum veritatem astronomicam, si numeremus tempora secundum annos et menses Arabum; quia primus dies cujuslibet mensis anni Arabum est dies conjunctionis solis et lunae secundum eorum utriusque cursum medium. Unde si diem hunc volumus dicere primum diem aetatis lunae, tunc cognitis initiis mensium cognoscuntur initia primationum. Quod si velimus inchoare primationes nostras prima die visionis lunae vel secunda vel tertia, incipiemus primationes nostras a secundo die vel tertio cujuslibet mensis Arabum, et procedemus uniformiter in computatione lunationum et non proveniet nobis error. Et quia scientia annorum Arabum, et initia mensium eorum, dabit nobis veram cognitionem primationum, ideo non oportet nisi recurrere ad tabulas et canones annorum et mensium Arabum. Et hoc dico secundum doctrinam quae vulgata est apud astronomos Latinos et Arabes. Si tamen velimus alia via procedere, possumus eadem certitudine sed majori auctoritate probare quae volumus, scilicet per tabulas Graecorum, et maxime Hebraeorum. Nam a principio Hebraei fuerunt peritissimi in sapientia astronomiae; et omnes nationes habuerunt hanc scientiam sicut caeteras ab eis, sicut probavi in superioribus. Et ideo si quis consideret tabulas Hebraeorum ad occasum solis Jerusalem, inveniet plenam in hujusmodi veritatem.

In the present year, 1267, Lent begins and ends a week too late.

Sed quamvis errores quos enumeravi sint horribiles secundum se, tamen non est comparatio ad eos qui ex jam dictis sequuntur. Nam totus ordo ecclesiasticarum solemnitatum confunditur per hujusmodi primationes erroneas secundum calendarium, sicut per aequinoctiorum falsam fixionem. Et ne differam in alios annos ad evidentiam istius erroris, pono casum in hoc anno. Nam non solum media conjunctio solis et lunae hoc anno fuit vi calendas Aprilis, super B literam, sed prima accensio lunae et visio primae lunae. Ergo luna fuit xiv quinto idus Aprilis, super A literam, et xiv luna est terminus Paschae, ac dominica sequens est dies Paschae. Quare in B litera sequente, scilicet in crastino, videlicet iv idus Aprilis debet esse dies Paschae secundum veritatem. Sed modo transfertur usque ad octo dies ultra propter primationem sumptam juxta aureum numerum. Nam aureus numerus istius anni est xiv, qui ponitur iii calendas Aprilis super E literam, et ibi dicitur esse primatio secundum calendarium. Et ideo secundum hoc xiv luna, quae est terminus Paschae est pridie idus Aprilis, et in dominica sequente assignatur Pascha, scilicet xv calendas Maii, et sic per viii dies ultra veritatem. Quapropter solemnitas paschalis qua mundus salvatur non celebrabitur tempore suo, sed jejunatur hoc anno per totam septimanam Paschae veram. Nam jejunium extenditur per viii dies plus quam deberet. Et tunc sequitur aliud inconveniens, quod per octo dies tardius incipiebatur jejunium Quadragesimae; ergo Christiani comedebant carnes in vera Quadragesima per octo dies, quod est absurdum. Et iterum tunc nec Rogationes nec Ascensio, nec Pentecostes, celebrantur hoc anno suis temporibus. Et sicut hoc anno 1267 accidit, ita accidet anno sequenti. Nam secundum aureum numerum erit Pascha iv idus Aprilis, scilicet 1268 anno Domini. Sed esse debet tertio nonas Aprilis per octo dies ante, quia luna prima est per tres dies antequam assignatur per aureum numerum. Et ideo quarto calendas Aprilis, vel saltem tertio, erit decima quarta luna quae est terminus Paschae; quapropter in prima dominica sequente erit dies secundum veritatem. Et sicut hoc anno accidit, ita et saepius per antecessionem aequinoctii et primationis potest contingere quod Pascha non solum per octo dies, sed per multo plures celebretur antequam debeat, et caetera festa, sicut prius expositum est de aequinoctio. Nam per longitudinem temporis accidet lunaris diei primatio quando erit plena, et quando erit in quacunque distantia a sole, ut prius habitum est in erroribus primationum. Et ideo maximum inconveniens et intolerabile hic sequetur.

The remedy is of easy application.

Cum igitur omnes astronomi et computistae possunt videre hos errores, et omnis homo qui vult inclinare cervicem suam ad veritatis inquisitionem potest hos intueri, necessarium esset et debitum ac Deo beneplacitum et hominibus sapientibus desideratum ut remedium poneretur. Et remedium facile esset ; nam inveniretur verum aequinoctium per tabulas astronomiae et per instrumenta, et verificaretur primatio per easdem considerationes, ut evacuarentur omnes modi errorum praedictorum, et xiv luna ab aequinoctio sumeretur, sive esset in die aequinoctii sive post eum, et ibi fieret terminus Paschae ut in die dominica sequente fieret dies paschalis. Et ideo non esset sequendus aureus numerus, ut aliquid fixum super calendas. Et possent fieri tabulae de his primationibus et aequinoctiis, et secundum illas posset ordinari calendarium in anno et in mensibus secundum consimile artificium quo Hebraei utuntur.

Objection that the Council of Nice, and afterwards Pope Leo, have decided the matter.

Sed contra haec possit objici de synodo Nicaena, quae statuit primationes paschales inveniri juxta cyclum decemnovennalem. Et beatus papa Leo, discussione facta de hujusmodi contentione, tandem definivit adhaerere sententiae synodi Nicaenae. Atque Beda in libro temporum capitulo xliii nititur quantum potest roborare hunc cyclum et primationes per decursum hujus cycli accidentes. Nam auctoritatem Nicaenae synodi et Leonis papae introducit et miraculum interponit. Nam cum multi voluerunt quodam tempore Pascha debere celebrari viii calendas Aprilis, et alii x calendas Maias secundum quod ordo cycli decemnovennalis exigebat, elisa est haec contentio per quoddam baptisterium, in quo nocte Paschae singulis annis replebatur fons sacer aqua per se, et baptizatis hominibus sicut venerat recedebat ; quod accidit decimo calendas Maias secundum quod cycli ratio exigebat, et non octavo calendas Aprilis.

At that time the error had not become serious.

Sed haec si bene intelligantur non contradicunt veritati. Nam cum Eusebius Caesariensis episcopus primo ordinaverat hunc cyclum, ipse parum fuit ante synodum Nicaenam, ita quod cum ipse ordinavit veraciter hunc cyclum secundum quod cursus lunae tunc fuit, non potuit esse mutatio aliqua sensibilis in tempore quo celebrata fuit illa synodus sacrosancta. Et ideo sancti patres statuerunt hunc cyclum observari, quia non habuit tunc errorem, nec diu postea habuit falsitatem notabilem. Unde cum papa Leo[1] fuerit post synodum Nicaenam quasi cxx annis, sub Martiano principe, manifestum est non posse primationem ad plus recessisse a loco suo in calendario a tempore Eusebii nisi per unum diem. Et cum ipse papa discussit hunc errorem, saltem dedit occasionem posterioribus considerandi veritatem in hac parte. Caeterum novus error fuit et non multum notabilis, et synodus Nicaena magnae auctoritatis fuit, propter quod non expedivit tunc temporis definitioni hujus synodi contraire.

Scientific knowledge then was not sufficient, and yet some settlement was necessary.

Etiam non invenit astronomos in ecclesia sufficientes ad hoc, quia a principio fuit astronomia odiosa Christianis, propter causas superius annotatas, scilicet in parte prima hujus tractatus. Nec adhuc inventi sunt usque nunc qui darent remedia in hac parte. Nam bene fuerunt multi qui sciverunt probare hos errores per vias astronomiae, et quantum erratur; nec mirum, cum visus nobis multos ostendat; et in universali dixerunt quae essent remedia, scilicet quod certificaretur quantitas anni, quod aequinoctia et solstitia invenirentur veraciter, quod primationes similiter; et tangunt modos universales ad haec. Sed nullus adhuc expressit nobis veram quantitatem anni, cum plena demonstratione, in qua non sit dubitatio. Similiter nec de aliis. Et ideo non fuit mirum si non fuerunt haec certificata temporibus ecclesiae primitivae, quando mathematica reputabatur inter artes suspectas et inutiles Dei ecclesiae, sicut multae aliae nobiles scientiae

[1] Pope Leo was in doubt as to the day on which Easter should be celebrated in 455 A. D. According to the calculation of Theophilus of Alexandria, it should be the 8th before the Calends of May, that is on April 24. This seemed too late: for the view hitherto maintained was that Easter day should not be earlier than March 22, nor later than April 21. The Pope wrote on the subject to the emperor Martian who referred his letter to Proterius of Alexandria. Proterius confirmed the view of Theophilus, and the Pope to avoid schism in the matter accepted his authority, and wrote to the bishops of the Western church in accordance with it. To avoid similar difficulties in future, he urged Victorius, a Gaul of Aquitaine, to construct a new paschal canon. Victorius multiplying the lunar cycle of nineteen years into the solar cycle of twenty-eight, produced a cycle of 532 years, beginning with the consulate of the Gemini which he took for the date of the Passion, and terminating in 559 A. D. This canon was usually followed by the Latin church afterwards. (See Fleury, *Hist. Ecclesiast.* lib. xxviii. ch. 51.)

propter quinque rationes in prima parte signatas. Synodus igitur Nicaena decrevit cyclum observari dum haberet veritatem et quia illis temporibus non fefellit. Caeterum decrevit hoc teneri, ut vitaretur contentio juxta caput cujuslibet, donec accideret in ecclesia Dei potestas mathematicae, qua certificari possunt omnia de quibus est contentio. Minus enim malum est sine comparatione unum inconveniens ad tempus tolerari propter impossibilitatem remedii, quam quod quilibet opinionem suam promulget aeque falsam sicut est illa quae ab omnibus communiter sustinetur. Et sic loquitur Beda. Nam dicit, 'cum lunam sic secundum cycli rationem signabant, aliud majus periculum per hoc declinaverunt.' Et quod Beda adduxit miraculum, concedendum est quod illis temporibus non fuit error in cyclo, ut manifestum est, sed postea crevit et apparuit sensibiliter.

Quod etiam adducit exemplum de lapide selenite, cujus splendor crescit cum luna primi mensis et decrescit, per quae cognoverunt antiquitus primationes paschales, istud non est pro cyclo, nisi dum habuit veritatem, sed magis contrarium : quia ille lapis ostendit nobis verum aequinoctium, et decimam quartam lunam paschalem in eo vel post ipsum sumendam pro termino Paschae; quod non potest hic cyclus his temporibus facere, nec unquam faciet, nec fecit diu ecclesiae. Caeterum Beda fuit circa tempora exordii cycli. Nam ipse refert in computo suo se tunc attigisse annum Domini septingentesimum primum. Et ab Incarnatione usque ad tempora Constantini, sub quo fuit synodus Nicaena, fuerunt cccxxxii anni secundum Bedam. Et ab hoc tempore usque ad Martianum principem fuerunt centum viginti anni, sub quo Leo papa fuit, qui sunt in universo ccccliii. Ex quo patet, quod Beda non fuit per ccc annos postea, sed nec a synodo nisi per circiter ccclxix annos, et ideo non potuit lunatio multum recedere a loco suo. Nam ad plus per unum diem integrum et per aliquid de secundo, eo quod in ccc annis et iv fit mutatio unius diei. Et ideo propter causas prius tactas bene potuit cyclus observari in tempore Bedae, sicut in tempore Leonis papae. Sed tamen scrupulus dubitationis jam exortus fuit a tempore Leonis, et augmentabatur haec dubitatio in tempore beati papae Hilarii, qui praecepit Victorio novum cyclum componere, in quo quia non fuit certitudo, sicut nec in decemnovennali, adhaeserunt posteriores consuetudini antiquae et statuto concilii Nicaeni.

At present the imperfections of the Calendar bring discredit on the Church.

Patet igitur ex his, quod salva omni auctoritate Nicaenae synodi, potest cyclus hic immutari; quia tunc error non fuit, et sustinebatur ut vitaretur majus periculum ad tempus, donec ecclesia posset habere astronomicam potestatem, per quam solam potest remedium adhiberi. Nam primitiva ecclesia non habuit usum astronomiae. Et ideo postea usque nunc fuit omissa correctio hujus rei propter longam consuetudinem, et propter hoc quod usus astronomiae non fuit in usu praelatorum nec multitudinis studentium, nec adhuc est; licet aliqui sunt satis prompti et periti in hac parte. Debet autem nunc temporis remedium apponi propter istos errores manifestos et palpabiles, atque propter scandalum multiplex in ecclesia. Nam omnes literati in computo et astronomi sciunt haec, et derident ignorantiam praelatorum qui haec sustinent. Atque philosophi infideles, Arabes, Hebraei, et Graeci, qui habitant inter Christianos ut in Hispania, et Aegypto, et in partibus orientis, et in multis aliis mundi regionibus abhorrent stultitiam quam conspiciunt in ordinatione temporum quibus utuntur Christiani in suis solemnitatibus. Et jam Christiani habet peritiam astronomiae, per quam potest fieri certificatio. Potest igitur Vestra Reverentia jubere, et invenietis homines qui praeclara remedia apponent in hac parte; et non solum in praedictis, sed in totius calendarii defectibus. Nam tredecim sunt radicales, et habent ramos quasi infinitos. Si igitur istud opus gloriosum fieret temporibus Vestrae Sanctitatis, tunc una de majoribus rebus et melioribus et pulchrioribus consummaretur quae unquam in ecclesia Dei fuerunt attentatae[1].

[1] Paul of Middelburg, bishop of Fossombrone, in his work, *Paulina de recta Paschae celebratione* (A. D. 1513) discusses the question of the Calendar with great fullness. Speaking of the confusion resulting from the error as to the length of the year and from the imperfection of the lunar cycle, he says, 'Quod animadvertentes *majores nostri* exhortati sunt aliquando Romanos pontifices ut errori huic enormi succurrerent.' That he should not have mentioned Bacon's name in this connexion is the more strange that he does mention him in another part of his work, when he treats at some length of his conjecture as to the date

The value of mathematical science in the government of the world.

Postquam[1] declaratum est quomodo mathematica potenter requiritur ad philosophiam et theologiam et Dei ecclesiam, nunc manifestandum est qualiter est necessaria rei publicae fidelium dirigendae. Et duobus modis principalibus valet. Uno scilicet modo propter cognitionem futurorum, praesentium, et praeteritorum; alio modo in operibus utilibus. Cum vero humanum genus sit expositum infinitis periculis de futuro, summe necessarium est, ut habeat vias cognoscendi futura. Et cum Deus dedit homini majora, scilicet animam et corpus, et promittit vitam aeternam, non debuit denegare minora. Nam et sceleratis sol oritur et piratis patent maria; quapropter longe magis Deus debet bonis utilem cognitionem rerum, et praecipue multitudini, propter hoc quod in ea bonum publicum invenitur. Et quia semper aliqui boni et Deo placentes in mundo inveniuntur, ideo Deus mundo dedit multiplicem futurorum cognitionem, nec potest stare sine ea, ut docet Avicenna in primo de Anima, et decimo Metaphysicae. Prius vero tactae sunt radices de cognitione futurorum in illa distinctione in qua de excusatione mathematicae disputavi, et ostensum est quod possibile est sufficiens judicium in omnibus, quod est scilicet medium inter necessarium et impossibile, et inter universale et particulare. Nam per haec mens humana illustratur, ut possit prudenter in omnibus dissere et utilitates sibi et aliis providere. Postea vero descendi ut res postulabat ad judicia specialia in humanis, scilicet in distinctione de sectis. Et si in rebus humanis et praecipue in hujusmodi fiat verum et utile judicium, multo magis in rebus naturalibus potest hoc fieri, tam in particulari quam in propria disciplina.

Et licet, ubi actum est de excusatione mathematicae, atque superius de comparatione virtutum coelestium ad haec inferiora,

of the Passion. But the *majores* whom he speaks of here are Peter d'Ailly and Nicolas of Cusa. The first of these had, as will be seen presently, copied into his *Imago Mundi* an extract from Bacon's Geographical treatise: and with regard to the lunar cycle, he advocated Bacon's view that the Arab cycle of thirty lunar years should be adopted. It is abundantly clear that the impulse to this reform was given by Bacon. By no previous writer had any attempt been made to apply astronomical science to the rectification of the Calendar.

[1] Bacon now passes to the subject of Geography. A survey of the planet was needed in order to estimate the dangers to the Church from foreign foes; and generally for the exercise of the Pope's functions as universal bishop.

sit tactum de cognitione locorum mundi et rerum generabilium in eis per coelestia, tamen nunc uberius hoc exponam transeundo ad medicinam propter corpus humanum, cujus cognitio magis est homini necessaria quam alicujus rei alterius in hoc mundo. Et non solum manifesto quomodo res in diversis locis mundi cognoscuntur, sed quomodo in eisdem per temporum diversa curricula causantur. Effectus vero non cognoscitur nisi per suam causam, ut certum est apud omnes; sed coelestia sunt causae inferiorum, unde oportet quod sciant haec generabilia per ingenerabilia quae sunt coelestia. Quod vero coelestia non tantum sint causae universales, sed causae rerum inferiorum propriae et particulares, probatur per Aristotelem, qui dicit secundo de Generatione, quod elementa deterius agunt quam organa et instrumenta artificialia respectu artificis. Sed tota actio artificialis principaliter attribuitur artifici, ut aedificatori, non ejus organo, ut securi. Ergo manifestum est quod coelo attribuitur operatio principalis omnium inferiorum, cum non sint agentia nisi coelum et elementa quae sunt ejus instrumenta. Praeterea hoc patet inductive. Nam in omnibus inanimatis coelum est causa particularis sine contradictione; quia inanimata non generant aliquod, nec producunt individua suae speciei, quia lapis non generat lapidem, sicut homo hominem et asinus asinum. Manifestum est igitur quod virtus coeli incorporata in materiam elementorum producit omnia inanimata, et animata per putrefactionem similiter: nam circa eorum generationem non sunt nisi coelum et elementa. Praeterea dicit Averroes septimo Metaphysicae quod idem facit virtus solis in materia putrefacta quod virtus patrum in seminibus, quare necesse est coelum esse causam particularem usque ad generationem rerum ex propagatione. Sed quod similiter sit ibi probo. Nam in Vegetabilibus Aristoteles dicit primo de plantis, quod sol est pater plantarum et terra mater; et in Animalibus vult illud idem; quia de homine, de quo minus videtur, dicit secundo Physicorum, homo generat hominem ex materia et sol, et constat quod pater non continuat nec terminat generationem, sed incipit tantum per decisionem seminis; quapropter oportet quod continuans et perficiens generationem sit sol seu virtus coeli. Et non solum est coelum causa in

Effect of astronomic conditions on inorganic and organic nature.

recte generatis, sed in peccatis naturae et monstris. Nam Avicenna decimo octavo libro de Animalibus dicit; Si embryo non potest recipere humanitatem, recipiet animalitatem sicut in rebus monstruosis: ut quando filius hominis habuit caput arietis, et agnus habuit caput tauri: quoniam virtus in eo induxit formam secundum figuras coelestes, quae accidunt unicuique. Et si ulterius descendamus, possumus causas rerum inferiorum magis prope investigare per coelestia. Primo vero articulus hic est, quod quilibet punctus terrae est conus unius pyramidis virtuosae coeli.

General principles of geography.

Ut hoc autem certius planiusque videatur quod intendimus, necesse est considerare quae sit diversitas regionum mundi, et quomodo eadem regio in diversis temporibus variatur, et quomodo res diversae ejusdem regionis diversas recipiunt passiones in eodem tempore. Sed haec sciri non possunt, nisi quantitatem et figuram habitabilis terrae et climata ejus distinguamus. Quatenus vero ad haec deveniamus, oportet nos supponere mundum esse sphaericae figurae, sicut superius est demonstratum. Et imaginabimur tres lineas a mundi terminis ductas intersecantes se in centro mundi ad angulos rectos, ut una sit a dextro in sinistrum in coelo, et hoc est quod ab oriente in occidens per centrum mundi ducatur; alia a sursum in deorsum, id est a meridie in septentrionem, et hoc est a polo antarctico ad polum arcticum, et tertia ab ante et retro, id est a puncto medii coeli supra nos usque ad punctum oppositum in coelo sub terra. Et per quandam transumptionem vocabuli vocatur angulus terrae. Sic docet nos Aristoteles imaginari sex diversas positiones in coelo in secundo Coeli et Mundi.

Celestial and terrestrial equator and equinoctial colure.

Si igitur imaginabimur unum circulum transeuntem per oriens, et medium coeli, et occidens, et angulum terrae, hic dividit coelum in duas partes aequales, in medio ejus relinquens unam medietatem respectu unius poli, et aliam respectu alterius, et vocatur aequinoctialis, et quia habitantes sub eo habent perpetuum aequinoctium, et quia omnibus habitatoribus terrae est aequinoctium quando sol venit ad illum circulum et describit eam in die naturali: et hoc est in principio veris et in principio autumni, quando sol ingreditur capita Arietis et Librae. Si vero imaginemur alium circulum magnum, qui transeat per polos mundi et per fines orientis et occidentis, intersecantem priorem circulum ad angulos rectos, qui vocatur colurus transiens per aequinoctia, tunc coelum sic dividetur per quatuor quartas, quarum duae erunt supra terram in situ nostro, et aliae duae sub terra. Et erit una quarta septentrionalis supra nos, scilicet quae continebitur inter medietatem aequinoctialis circuli et duas quartas coluri dicti, terminatas ad polum arcticum ex una parte, et ex alia ad puncta orientis et occidentis apud aequinoctialem, ut patet in figura, et haec est quarta pars quae est supra. Consimili autem modo oportet nos imaginari terram sphaericam esse, et illae tres lineae praedictae transibunt per centrum terrae intersecantes se in eo ad angulos rectos; nam sunt perpendiculares super eam, et quia in earum intersectione est centrum coeli et totius mundi, necesse est quod idem punctus sit centrum mundi et terrae; quia intersectio ista non est nisi in puncto uno, per quintam primi libri Theodosii de Sphaeris. Nam si linea recta descendat a coelo perpendiculariter ad superficiem contingentem sphaeram terrae, in ea erit centrum terrae, per illam quintam, quae dicit, Si sphaeram plana superficies contingat, a puncto vero contactus recta linea ad contingentem superficiem perpendiculariter ducatur, in eadem centrum sphaerae esse necesse est: sed illa eadem linea erit perpendicularis ad superficiem contingentem sphaeram coeli, ergo in illa erit centrum coeli, et haec linea est una trium dictarum. Similiter et utraque aliarum per eandem rationem transibit per centrum coeli et centrum terrae: sed quaelibet istarum non intersecat aliam nisi in puncto uno, ergo in eodem erit centrum terrae cum centro coeli, et propter hoc terra jacet in medio mundi. Et ideo si imaginemur circulos duos in terra respondentes circulis in coelo praetactis, unum sub aequinoctiali circulo transeuntem per oriens et occidens, et punctum in terra sub puncto medii coeli; et alium, per occidens et oriens et puncta in terra respondentia polis; tunc per hos circulos dividetur terra in quatuor quartas, quarum duae erunt in superficie terrae in situ nostro; et alia ex altera parte terrae. Et una erit septentrionalis, scilicet a medio terrae sub aequinoctiali usque ad punctum terrae sub polo arctico contenta infra lineas quae ab oriente et occidente concurrunt in polum sive punctum terrae polo suppositum. Et haec est quarta quam quaerimus, in qua est habitatio nota, et est sub quarta coeli prius notata.

What proportion of the earth's surface is habitable.

Habitatio[1] vero dupliciter consideratur; uno modo respectu coeli, scilicet, quantum propter solem potest habitari, et quantum non. Et de hoc dictum est prius in universali, et tangetur posterius. Alio modo consideratur quantitas habitabilis respectu aquae, scilicet, quantum aqua impediat. Et hoc est modo considerandum. Ptolemaeus vero in libro de Dispositione Sphaerae vult quod fere sexta pars terrae est habitabilis propter aquam, et totum residuum est coopertum aqua. Et ideo in Almagesti secundo libro ponit quod habitatio nota non est nisi in quarta terrae, scilicet in qua habitamus; cujus longitudo est ab oriente in occidens, et est medietas aequinoctialis; et ejus latitudo est ab aequinoctiali in polum, et est quarta coluri. Sed Aristoteles vult in fine secundi Coeli et Mundi quod plus habitetur quam quarta. Et Averroes hoc confirmat. Dicit Aristoteles[2] quod mare parvum est inter finem Hispaniae a parte occidentis et inter principium Indiae a parte orientis. Et Seneca libro quinto Naturalium dicit quod mare hoc est navigabile in paucissimis diebus, si ventus sit conveniens. Et Plinius[3] docet in Naturalibus quod

[1] This paragraph, including half of that which follows, has a remarkable history. It is inserted without acknowledgement of its source in the *Imago Mundi* of Cardinal d'Ailly (Petrus Alliacus), who died 1425, and whose work was printed at Louvain, 1480. It forms the greater part of the eighth chapter, entitled, 'De quantitate terrae habitabilis.' From this work it was quoted by Columbus in a letter written in October 1498 to Ferdinand and Isabella from Hispaniola. See *Imago Mundi*, fol. 13, b; and Humboldt, *Examen Critique*, vol. i. pp. 61–70 and pp. 96–108: also *Cosmos*, vol. ii. p. 621 (Bohn's ed.). Humboldt remarks that the *Imago Mundi* 'exercised a greater influence on the discovery of America than did the correspondence with the learned Florentine Toscanelli.'

[2] Arist. *De Coelo*. ii. 14, § 15 *διὸ τοὺς ὑπολαμβάνοντας συνάπτειν τὸν περὶ τὰς Ἡρακλείους στήλας τόπον τῷ περὶ τὴν Ἰνδικήν, καὶ τοῦτον τὸν τρόπον εἶναι τὴν θάλατταν μίαν, μὴ λίαν ὑπολαμβάνειν ἄπιστα δοκεῖν*. Cf. Seneca, *Nat. Quaest.* i. Prolog., 'Quantum enim est quod ab ultimis litoribus Hispaniae usque ad Indos jacet? Paucissimorum dierum spatium, si navem suus ferat ventus, implebit.'

[3] Pliny, *Nat. Hist.* ii. 67.

navigatum est a sinu Arabico usque ad Gades: unde refert quendam fugisse a rege suo prae timore et intravit sinum maris Rubri qui vocatur Arabicus, qui circiter spatium navigationis annualis distat a mari Indico secundum Hieronymum in epistola, ut inferius exponetur. Et ideo latitudo terrae per quam decurrit mare Rubrum est magna valde ; ex quo patet principium Indiae in oriente multum a nobis distare et ab Hispania, postquam tantum distat a principio Arabiae versus Indiam. A fine Hispaniae sub terra tam parvum mare est quod non potest cooperire tres quartas terrae.

More than a fourth part of the earth is habitable.

Et hoc per auctoritatem alterius considerationis probatur. Nam Esdras[1] dicit quarto libro, quod sex partes terrae sunt habitatae et septima est cooperta aquis. Et ne aliquis impediat hanc auctoritatem, dicens quod liber ille est apocryphus et ignotae auctoritatis, dicendum est quod sancti habuerunt illum librum in usu et confirmant veritates sacras per illum librum. Et pluries in officio divino utuntur auctoritatibus illius libri. Et ideo, sive Esdras sive alius hunc fecerit, supponendus est pro auctoritate. Et propter hoc dico quod licet habitatio nota Ptolemaeo et ejus sequacibus sit coarctata infra quartam unam, plus tamen est habitabile. Et Aristoteles potuit plus nosse, quia auctoritate Alexandri misit duo millia hominum ad investigandum res hujus mundi, sicut Plinius dicit octavo Naturalium. Et ipsemet Alexander perambulavit usque ad finem orientis, et sicut patet ex historia Alexandri et ex epistolis quas Aristoteli conscripsit, semper mandavit ei de omnibus mirabilibus et insolitis quae inveniebat in oriente. Et ideo potuit Aristoteles plus certificare quam Ptolemaeus. Et Seneca similiter; quia Nero imperator discipulus ejus similiter misit ut exploraret dubia hujus mundi, sicut Seneca narrat in Naturalibus. Et ideo secundum haec quantitas habitabilis magna est et quod aqua cooperitur modicum debet esse. Versus enim polos mundi oportet quod aqua abundet, quia loca illa frigida sunt propter

[1] What Bacon calls the fourth book of *Esdras* answers to the second book in the *Apocrypha*. See ch. 6, v. 42, 'Upon the third day thou didst command that the waters should be gathered together in the seventh part of the earth: six parts didst thou dry up.'

elongationem a sole, sed frigus multiplicat humores; et ideo a polo in polum decurrit aqua in corpus maris et extenditur inter finem Hispaniae et inter principium Indiae non magnae latitudinis, et vocatur Oceanus; ut principium Indiae possit esse multum ultra medietatem aequinoctialis circuli sub terra accedens valde ad finem Hispaniae. Sed ne verum hic pro falso damnetur, oportet nos scire quod Hispania in hoc loco accipitur non pro citeriori sed pro ulteriori, de quo auctores certi loquuntur, ut Plinius [1] in Naturalibus et Merlinus in prophetia sua, et Orosius in libro de Ormesta Mundi, et Isidorus decimo quarto libro Etymologicorum. Quoniam docet quod inter Hispaniam quae nunc vocatur et Africam nunc dictam non fuit aqua decurrens, sed terra continua ab antiquis temporibus, sed postea oceanus irruit in profunda terrae et conjunxit se mari Tyrrheno, quod decurrit per littus Arragoniae provinciae et Italiae. Citerior igitur Hispania a Pirenaeis montibus usque ad Carthaginem porrigitur: sed ulterior transit Gaditanum fretum usque in provincias Africae. Unde extenditur ultra Gades Herculis et attingit montem Atlantem. Haec necessario recitavi secundum auctores dictos, ne Aristoteles et suus commentator per ignorantiam ulterioris Hispaniae deriderentur, cum dicunt ad probationem parvitatis maris inter Hispaniam et Indiam quod elephantes sunt tantum in illis duobus locis. Verum enim est quod circa montem Atlantem abundant elephantes, ut Plinius dicit, sicut et Aristoteles, et similiter in India, et ideo in ulteriori Hispania est frequentia elephantorum; sed Aristoteles dicit quod elephantes in illis locis esse non possunt nisi essent similis complexionis, et si essent multum distantes non haberent similem complexionem, et ideo nec elephantes essent in illis locis tantum. Quapropter concludit haec loca esse propinquiora, et ideo oportet quod mare sit parvum inter ea.

The sea occupies less than three-quarters of the earth's surface.

Non igitur mare cooperiet tres quartas terrae, ut aestimatur. Nam sit medietas terrae superior *a b c d*, in cujus una quarta, scilicet *a b c*, est habitatio nobis nota. Jam patet quod multum

[1] Pliny speaks of the tradition that Spain was originally continuous with the opposite African coast (lib. iii, proëmium). But he never speaks of any part of Africa as Hispania ulterior.

de quarta illa sub nostra erit habitatione, propter hoc quod principium orientis et occidentis sunt prope, quia mare parvum ea separat ex altera parte terrae. Et ideo habitatio inter orientem et occidentem non erit medietas aequinoctialis circuli, nec medietas rotunditatis terrae, nec duodecim horae, ut aestimant, sed longe plus medietate rotunditatis terrae et plus quam revolutio medietatis coeli. Quantum autem hoc sit, non est temporibus nostris mensuratum, nec invenimus in libris antiquorum ut oportet certificatum; nec mirum, quoniam plus medietatis quartae in qua sumus est nobis ignotum; nec sunt civitates a philosophis comprehensae, ut patebit ex sequentibus. Similiter si loquamur de aliis duabus quartis, et consideremus vias naturales secundum quod philosophia naturalis decurrit, non erunt illae coopertae aquis, ut vulgus mathematicorum aestimat. Nam cum poli et regiones versus eos sint ejusdem remotionis a sole et planetis secundum comparationem polorum ad vias planetarum in medio mundi inter duo tropica; necesse est quod secundum haec aequales dispositiones sint in quarta nostra, et in quarta ultra aequinoctialem versus alterum polum; et similiter in quarta sub pedibus nostris usque ad aequinoctialem; et in quarta ultra aequinoctialem erit consimilis complexio secundum tenorem praedictorum. Et ideo si quarta nostra non est cooperta aquis saltem usque ad latitudinem et distantiam ab aequinoctiali per 66 gradus, ut in fine insularum Scotiae et in regno Norguegiae, manifestum est quod similis causa naturalis erit in alia quarta ultra aequinoctialem in superiori parte terrae sicut in illa in qua sumus, quia elongatio a via solis inducit frigus, et frigus multiplicat humorem, et ideo circa polos erit congregatio naturalis aquarum, et in regionibus quae sunt prope. Et ideo in altera quarta ultra aequinoctialem secundum hoc debet esse multum habitabile; saltem usque ad regiones quarum latitudines sunt usque ad 66 gradus, sicut hic.

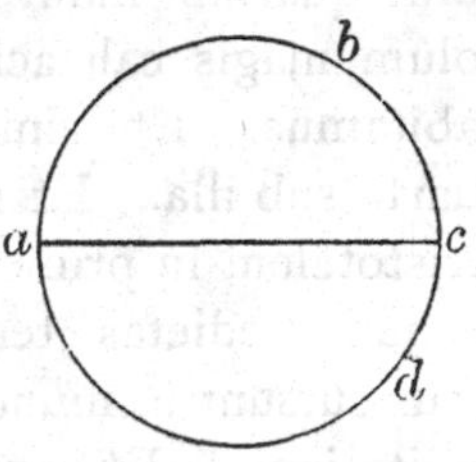

FIG. 22.

Caeterum potest argui secundum mathematicos major habitatio ex defectu aquae ibi quam in quarta nostra, The eccentricity of the solar orbit is likely to produce more dry land in the southern hemisphere.

quoniam in parte illa est oppositum augis solis, et sol descendit ad terram ibi multum. Unde oportet quod comburat quartam illam in aliqua parte sui, et residuas usque ad polum magis calefaciat quam partes nostrae quartae in qua habitamus. Et similiter contingit persuadere de reliqua quarta sub illa. Et iterum sumitur argumentum ad hoc per Aristotelem in primo Coeli et Mundi et per Averroem, quod reliqua medietas terrae ultra aequinoctialem circulum est locus sursum in mundo et nobilior et ideo maxime competit habitationi. Et propter hoc ex ordinatione naturae erit quod impedimenta habitationis magis excludantur, saltem in magna parte illius medietatis, scilicet longius ab opposito augis solis, si eccentricum ponimus, et ubique si non ponatur eccentricus, et hoc propter stellas nobiliores in illa parte, ut vult Averroes primo Coeli et Mundi. Et Ptolemaeus dicit in libro de Dispositione Sphaerae quod natura exigit ut sint duo genera Aethiopum, scilicet sub duobus tropicis. Ex quo arguunt aliqui quod habitatio est ultra aequinoctialem sicut citra: et secundum haec non erit figura habitabilis quartae sphaerae nec semicirculus descriptus in plano, nec aqua circuet in circuitu mundi per polos et oriens et occidens cooperiens tres quartas ejus, ut creditur: sed magis erit figura

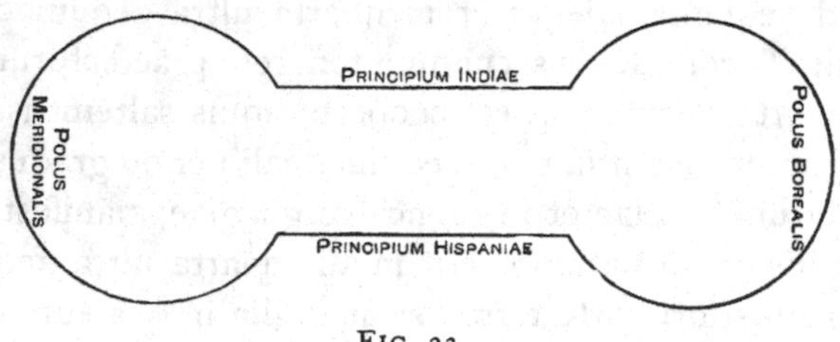

FIG. 23.

aquae hujus vel consimilis, ita quod hoc mare vocetur Oceanus, habens plurimum de aqua circa polos, cujus longitudo extenditur a polo in polum inter principium Indiae et finem Hispaniae, quae est mathematicis nota.

Description of the known world.

Et quoniam non est mathematicis nota habitatio nisi in quarta in qua sumus, et etiam non tota quae comprehenditur intra medietatem aequinoctialis et medietatem coluri transeuntis per polos et extremitates medietatis aequinoctialis[1] sub hac figura, ideo in executione sententiae mathematicorum oportet relinqui quod incertum est, et fieri sermonem de eo quod philosophis magnis innotescit. Ptolemaeus[2] inter omnes ab incarnatione Domini maxime certificavit in hac parte, et in secundo Almagesti quartam istam, in qua sumus, distinguit. Et ejus et Alfragani et aliorum divisio magis nota est per septem climata famosa. Clima vero apud eos est spatium terrae in quo dies prolixior excedit per dimidiam horam diem prolixiorem alterius partis terrae, vel exceditur ab eodem. Certum tamen est quod magis naturalis et verior distinctio esset penes quartam horae, secundum quod Ptolemaeus primo dividit: sed quia istae partes sunt parvae, ideo considerant philosophi majores et conjunxerunt duas illarum in uno climate. Et quoniam haec climata et civitates famosae in eis non possunt evidenter percipi sermone, oportet

The seven zones of Ptolemy and Alfraganus.

[1] The words, 'et medietatem coluri . . . medietatis aequinoctialis' are printed twice over in Jebb's edition.

[2] In the first printed edition of Ptolemy's *Cosmographia*, dedicated to Pope Sixtus IV, in 1478, there are twenty-seven maps: one of the World, ten of Europe, four of Africa, twelve of Asia. These correspond exactly to the description given in the eighth book of the *Cosmographia*. In all of them parallels of latitude and meridians are given, as in modern Atlases. In his map of the World, embracing as it did, on his system, 180 degrees of longitude and 80 degrees of latitude, a very elaborate and scientific attempt is made to project those curved lines accurately on a plane surface. It appears probable that these maps are the work of Agathodaemon of Alexandria a contemporary of Ptolemy. [Bunbury, vol. ii. p. 578.]

The parallels of latitude are twenty-five in number: four of South latitude, twenty-one of North. They are arranged according to the length of the day at the Summer solstice, by intervals of one quarter of an hour. Beginning with 16° 25′ S. where the longest day is thirteen hours, proceeding north to the Equator where it is twelve; thence northwards to lat. 45°. From this point the parallels are drawn at half-hourly intervals to lat. 58°, where the longest day is eighteen hours. The next parallel is drawn at 61°, the day being nineteen hours; the last at 63°, where it is twenty hours. This, says Ptolemy, is the latitude of Thule.

In Ptolemy's Atlas longitude is calculated in degrees and half degrees, taking as the starting-point the meridian of the Fortunate islands (Ferro in the Canary islands), which he wrongly estimated as being only 2° 30′ west of the Sacred Promontory (Cape St. Vincent). Further, owing to his acceptance of Posidonius' estimate of the Earth's magnitude, 180,000 stadia, his degrees of longitude as measured on the Equator, were only 500 stadia (i. e. fifty geographical miles instead of sixty).

quod figura sensui ministretur[1]. Primo igitur figuram hujus quartae cum climatibus suis ponam, et signabo civitates famosas in locis suis per distantiam earum ab aequinoctiali, quae vocatur latitudo civitatis vel regionis; et per distantiam ab occidente vel oriente, quae longitudo regionis vocatur. Et in divisionibus climatum atque in latitudinibus et longitudinibus civitatum utar auctoritate et experientia sapientum. Sed in signatione civitatis in loco suo per longitudinem et latitudinem suam inventas ab auctoribus, superaddam artificium, quo locus civitatis habeatur per distantiam ejus a meridie et septentrione et oriente et occidente. Et hoc artificium consistit in concursu lineae rectae aequidistantis aequinoctiali signatae in plano: secundum formam lineae rectae ductae a numero graduum latitudinis regionis signato in quarta coluri ducta ab aequinoctiali ad polum mundi in concursum, inquam, cum arcu circuli magni qui transit per polos mundi et per numerum longitudinis civitatis signatum in aequinoctiali circulo. Hic autem modus melior est et facilior, et sufficit considerationi locorum mundi in hujusmodi figuratione sensibili.

Equatorial regions south of the zones.

Ostendam etiam cum latitudine cujuslibet climatis quot milliaria quodlibet contineat in se, et quot gradus in coelo cuilibet respondeant, et quot horas habet dies prolixior. Elevatio autem poli supra horizonta in quolibet climate est latitudo ejus ab aequinoctiali, et distantia zenith capitis ab aequinoctiali, idem cum latitudine et cum elevatione poli: et pono numerum milliariorum totius spatii septem climatum. Sed licet non notata sunt a philosophis nisi septem climata, nihilominus tamen signant alia spatia terrae et ante climata[2] et post. Nam Ptolemaeus dicit in libro de Dispositione Sphaerae, quod perambulatum est auxilio regum Aegypti usque ad aequinoctialem circulum. A paucis tamen et raro pertransitum est hoc spatium ante climata propter distantiam ejus, sed magis propter negligentiam principum qui philosophos deberent juvare in hac parte. Signo ergo spatia tria ante

[1] It is evident from this and other passages that Bacon prepared a map of the world to illustrate this part of the *Opus Majus*. No trace of it, however, has been hitherto discovered.

[2] Ante climata, i.e. beyond 16° 25′, South latitude.

climata nota, quae plus de terra habent quam unum clima, et pono numerum milliariorum latitudinis istius spatii inter aequinoctialem et primum clima, et quot milliaria sunt ab aequinoctiali usque ad finem septimi climatis; deinde divido spatium quod est ultra climata. Et Ptolemaeus in secundo Almagesti distinguit istud spatium per excessum quartae unius horae adjectum super longitudinem diei in praecedenti regione usquequo veniat ad latitudinem regionis quae est lxi gradus. Et postea per medietatem unius horae usque ad latitudinem lxiv graduum; et ab illo loco dividit spatia per unam horam usque ad latitudinem lxvi graduum; ubi est nox continua in solstitio hyemali, nisi quod medietas solis fertur subito sub horizonte et est dies continuus in solstitio aestivali, et hoc est ultra Scotiam longe. Et deinde semper apparet sol in aquilone versus polum.

Sed tunc ulterius dividitur spatium notabiliter per quantitatem diei unius mensis, vel duorum, vel trium, vel quatuor, vel quinque, vel sex. Nam habitantes sub polo habent dimidium annum diem, hoc est, solem supra horizonta per sex menses et sub horizonte per alios sex; sed tamen crepusculum vespertinum durat per septem septimanas et unum diem, scilicet, a sexto decimo die Septembris inclusive, in quo sol ingreditur nunc temporis signum Librae usque ad sextum diem Novembris exclusive. Et in tanto spatio est claritas solis super terram, sicut apud nos in aestate accidit post occasum solis: quia in illo die sexto est declinatio solis sub horizonte per octodecim gradus et sex minuta; et crepusculum durat usque ad finem octodecim graduum et non plus. Et ab eodem die sexto Novembris inclusive usque ad 21 diem Januarii inclusive est nox obscura per decem septimanas et per quinque dies. Nam in illo 21 die est sol sub horizonte octodecim gradibus et sex minutis; et ideo non potest incipere crepusculum matutinum in illo, sed oportet quod sol prius transeat illa sex minuta. Sed ab illo die, scilicet 21 Januarii exclusive incipit aurora, et durat usque ad ingressum solis in signum Arietis, et hoc est 13 die Martii nunc temporis, unde durat per septem septimanas et unum diem. Et per residuum temporis dum sol vadit a primo gradu Arietis usque ad primum gradum Librae est sol

Polar regions.

semper super horizonta eorum; et hoc est per dimidium annum: quia horizon eorum est aequinoctialis circulus. Et ideo sex signa septentrionalia sunt semper super horizonta, ut apparet in sphaera ad sensum. Et ideo dum sol est in illis signis habent diem manifestum. Et nihilominus duo crepuscula in quibus apparet claritas solis super horizonta simul sumpta continent tres menses et quindenam et duos dies, et ideo respectu crepusculorum et diei habitatores illius loci sub polo parum habent de nocte per annum.

Haec omnia scripsi sequendo principaliter Ptolemaeum et Alfraganum[1], et tabulam de longitudinibus et latitudinibus civitatum. Nam in latitudinibus climatum et spatiorum ante climata et post secutus sum sententiam Ptolemaei in Almagesti. Expansionem autem climatum per sua milliaria et spatiorum ante climata et post, atque quae civitates et regiones continentur in illis climatibus et spatiis, descripsi secundum Alfraganum principaliter, nisi quod ipse non tangit praecise computationes quas pono per examinationem majorem, atque sequendo alios auctores aliquod immuto aliquando et addo, secundum quod oportet propter certitudinem majorem: ut tetigi de civitate Syene.

Different senses in which East and West are taken.

Quod si objiciatur, quod in canonibus astronomiae et in tabulis aliis invenitur aliter in longitudinibus et latitudinibus civitatum, ut evidenter patet de Toleto, ad cujus meridiem factae sunt tabulae[2] Toletanae; dicendum est quod aliter et

[1] Ahmad Ibn Muhammad Ibn Kathir, commonly known as Alfreganus or Alfraganus, lived early in the ninth century. His works on astronomy, geography, and chronology were translated in the twelfth century by Joannes Hispalensis. This version was very inaccurate, and a more correct one was made afterwards by the Jew Jacobus Anatolius. The mathematician, Regiomontanus, gave lectures on Alfraganus at Padua in the fifteenth century. His works were printed at Nurimberg in 1537, together with a very interesting lecture on the history of Mathematics by Regiomontanus. Another edition was published, apparently from the corrected version, at Frankfurt, in 1590. The seven zones described by Alfraganus, extend from the equator to lat. 45°. The parallels are arranged according to the length of the day at the Summer solstice: the length in each exceeding that of the parallel to the south of it by half an hour. See the two editions referred to: also Cantor, vol. ii. p. 238, and Jourdain, p. 115

[2] Alfonso X of Castile appointed a committee of astronomers to construct new astronomical tables. In this work the Rabbi Isaac Ebn Sid Hazan played an important part. 'Their principal defect was the introduction of an inequality in the motion of the fixed stars in longitude by which this motion appeared

aliter accipiunt oriens et occidens. Nam oriens et occidens de quibus hic loquimur sunt ultima terrae habitabilis. Sed uno modo accipiuntur oriens et occidens sub aequinoctiali circulo in medio terrae; ita quod principium Indiae sub aequinoctiali est oriens habitabilis; et finis Hispaniae ulterioris, si extenderetur usque ad aequinoctialem circulum, esset occidens; sed non tantum extenditur, immo est magnum spatium terrae a meridiana parte Hispaniae ulterioris usque ad aequinoctialem. Finis igitur istius spatii a parte occidentis est occidens habitabilis. Sed cum terra a parte occidentali habeat magnam latitudinem, scilicet ab aequinoctiali usque ad montem Atlantem et Gades Herculis, et citra per totum circuitum Hispaniae totius et Hiberniae, possunt diversi diversimodo accipere occidens; unde aliqui accipiunt a Gadibus, aliqui a monte Atlantis, aliqui a fine habitabilis sub aequinoctiali. Sed cum accipitur sub aequinoctiali, certius accipitur, quia illud uno modo est et melius, quia hoc est in medio mundi inter duos polos, et ideo est verum occidens; et eodem modo est de oriente. Tabula vero latitudinum et longitudinum non accipit longitudines ab occidente sub aequinoctiali, ut certum est. Nam sic longitudo Toleti est viginti novem graduum ab occidente, et secundum tabulam non est nisi undecim. Et auctor illius tabulae sumpsit occidens ei notum et certum et respectu situs suae regionis. Considerandum vero hic est quod oriens et occidens in divisione terrae non possunt sumi nec debent nisi penes ultima terrae, ubi mare quod vocatur Oceanus deficit, qui extensus est inter Indiam et Hispaniam ulteriorem et regiones alias occidentales et orientales, quae sequuntur Indiam et Hispaniam. Unde non debent accipi penes oriens et occidens secundum fines alicujus orientis, secundum quod nos dicimus aliquando oriens et occidens penes ortum et occasum solis. Nam

The Alphonsine Tables.

to be sometimes accelerated and sometimes retarded. It was known as the trepidation of the fixed stars, or the motion of the eighth sphere.' This imaginary inequality continued to disfigure the best tables till the time of Copernicus. See *Library of Useful Knowledge*, History of Astronomy, p. 38.

Previous tables had been constructed at Toledo in the eleventh century by Arzachel. It does not appear quite certain whether it is to those that reference is made, or to those of Alfonso.

infiniti sunt horizontes tam obliqui quam recti. Et ideo oriens et occidens in divisione terrae non sumuntur penes horizonta; tunc enim quod esset oriens unius horizontis esset occidens alterius horizontis, et medium ejus.

Determination of longitude and latitude.

Et notandum quod verum occidens et oriens sub aequinoctiali circulo est, ut dictum est: et hoc est ad ultimum habitabile ulterioris Hispaniae pro occidente et ad ultimum Indiae ulterioris pro oriente. Si igitur volumus sumere distantiam civitatis ab occidente nunc dicto ducemus unam lineam ab hoc occidente aequidistantem civitati, et linea intercepta inter civitatem et dictam aequidistantem denotabit longitudinem ejus ab occidente. Et similiter ducatur una linea ab Arym[1] civitate in medio mundi usque ad polum arcticum, et ab ea ducatur linea recta usque civitatem, quae notabit distantiam civitatis a medio mundi. Sed aliquando accipitur distantia ab occidente respectu finis habitabilis in directo illius civitatis: et quia hoc variatur infinitis modis apud diversos habitantes, variatur sumptio longitudinum. Sed melior acceptio est ab occidente sub aequinoctiali, quia hoc est uno modo. Et quia aliarum civitatum a Toleto longitudines et latitudines non inveniuntur congregatae nisi in hac tabula, ideo secutus sum eam in hac parte. Quamquam major certitudo requiratur, quia nondum apud Latinos certificatae sunt longitudines et latitudines civitatum et regionum; nec unquam certificabuntur nisi per apostolicam auctoritatem vel imperialem, aut per auxilium alicujus regis magni praebentis philosophantibus adjutorium. Secundum igitur praedicta praesentem affero descriptionem in albiori parte pellis[2], ubi civitates notantur per circulos rubros; nam in alia parte pellis alia descriptio poterit assignari propter evidentiam majorem locorum mundi. Et hanc secundam descriptionem addo propter summam utilitatem locorum.

Bacon's map.

More accurate determination of

Quoniam igitur locorum mundi cognitionis maxima utilitas est, ideo aliam descriptionem oportet afferri. Nam res mundi sciri non possunt nisi per notitiam locorum in quibus

[1] O. has ab arcu civitatis, which is clearly wrong. See note on p. 310.

[2] Another reference to the lost map included in the writings sent by Bacon to the Pope.

continentur. Locus enim est principium generationis rerum, ut dicit Porphyrius; quia secundum diversitatem locorum est diversitas rerum; et non solum naturalium, sed moralium et scientialium, ut videmus in hominibus quod secundum diversitatem regionum habent mores diversos et occupant se in artibus et scientiis diversis. Quia igitur philosophia intromittit se rebus mundi, multum ei deest adhuc apud Latinos, postquam non habet certificationem locorum mundi. Sed haec certificatio stat in cognitione longitudinis et latitudinis cujuslibet loci; tunc enim sciremus sub quibus stellis est quilibet locus, et quantum a via solis et planetarum, et quorum planetarum et signorum loca recipiant dominium, quae omnia faciunt diversas complexiones locorum: quae si scirentur, possit homo scire complexiones omnium rerum mundi et naturas et proprietates quas a virtute loci contrahunt.

latitudes and longitudes necessary (1) for knowledge of man and of nature.

Et hoc requirit non solum philosophia, sed sapientia Dei cujus series tota decurrit per loca mundi. Unde sensus literalis stat in cognitione locorum mundi. ut per convenientes adaptationes et similitudines sumptas ex rebus eliciantur sensus spirituales. Nam haec est propria expositio scripturae, ut superius manifestavi in exemplo. Et haec cognitio locorum mundi valde necessaria est reipublicae fidelium et conversioni infidelium et ad obviandum infidelibus et Antichristo, et aliis. Nam propter diversas utilitates reipublicae et propter praedicationem fidei mittuntur homines ad loca mundi diversa, in quibus occupationibus valde necessarium est proficiscentibus ut scirent complexiones locorum extraneorum, quatenus scirent eligere loca temperata per quae transirent. Nam valentissimi homines aliquando ignorantes naturam locorum mundi seipsos Christianorumque negotia peremerunt, eo quod loca nimis calida in temporibus calidis aut nimis frigida in frigidis transierunt. Receperunt etiam pericula infinita, eo quod nesciverunt quando intraverunt regiones fidelium, quando schismaticorum, quando Saracenorum, quando Tartarorum, quando tyrannorum, quando hominum pacificorum, quando barbarorum, quando hominum rationabilium. Deinde qui loca mundi ignorat, nescit non solum quo vadit, sed quo

(2) For spiritual government of the world.

tendat; et ideo sive pro conversione infidelium proficiscatur, aut pro aliis ecclesiae negotiis, necesse est ut sciat ritus et conditiones omnium nationum, quatenus proposito certo locum proprium petat; ne, si velit Paganos adire, cadat in idololatras, vel si illos intendat, scismaticos invadat, vel pro scismaticis obedientes Romanae ecclesiae eligat, aut indifferentes utrique parti, cujusmodi sunt populi qui vocantur Aas; quatenus etiam Nestorianos desiderans Nicholaitas declinet; et sic multis gentibus sectarum diversarum ne unam pro alia eligat oberrando. Quamplurimi enim a negotiis Christianorum maximis sunt frustrati eo quod regionum distinctiones nesciverunt.

(3) For knowledge of the ten tribes and of Antichrist.

Deinde non modica necessitas sciendi loca mundi oritur ex hoc, quod oportet ecclesiam optime scire situm et conditiones decem tribuum Judaeorum, qui exibunt in diebus futuris. Nam Orosius in libro de Ormesta Mundi ad Augustinum libro tertio dicit; Ochus, qui est Artaxerxes, plurimos Judaeorum in transmigrationem egit, atque in Hyrcania ad Caspium mare habitare praecepit; quos usque in hodiernum diem amplissimis generis sui incrementis consistere atque exinde quandoque erupturos opinio est. Et magister in historiis addit, quod Alexander magnus conclusos ibi invenit, et ob malitiam eorum arctius eos constrinxit, quos tamen egressuros circa finem mundi testatur, et magnam stragem hominum esse facturos. Et in Cosmographia sua Ethicus astronomus [1] dicit gentes varias debere exire circa dies Antichristi, et eum vocabunt Deum Deorum, prius mundi regiones vastaturi. Et Hieronymus hoc confirmat in libro quem transtulit de sapientiis hujus philosophi.

[1] Under this title two geographical writers of the early middle ages are often confounded. One is Julius Honorius, sometimes spoken of as Julius Aethicus. The work of the other is known as *Cosmographia Aethici Istrici*, written in barbarous Latin, and fabulously said to have been translated from the Greek by St. Jerome. To the first is perhaps due (or on the other hand his work may be a modification of) the summary of geography prefixed to the history of Orosius. The second, purporting to be written by a mysterious traveller who visited all parts of the known, and many of the unknown, world, is a treatise wholly without value, except on the ground of its popularity as a text-book of geography in the middle ages. (Cf. Bunbury, vol. ii. pp. 692-3, and pp. 703-5.)

Et Alexander magnus cum eis pugnavit, sed superare non potuit, sicut iste Ethicus testatur et refert Hieronymus, ideo ingemuit et ait, 'Gentes rationabiles et sapientes vastavimus, attrivimus populum inclytum et sublimem ac sinceram gentem. Quid utilitatis aut necessitatis fuit? cum omnes inferorum daemones et adversariorum phalangas hic relinquimus in humana specie latitantes. Item ne quando audiant vel percipiant terram mellifluam et uberrimam mundi gloriam; ne forte irruant in universam superficiem terrae, et quasi panem cuncta decerpant atque deglutiant. O terra, mater draconum, nutrix scorpionum, fovea serpentum, lacus daemonum, facilius fuerat in te infernum esse quam tales gentes parturire. Vae terrae fructiferae et mellifluae, quando ingruent tot serpentes et bestiae in eam. Vae habitatoribus orbis, cum istae coeperint triumphare: contra quos nisi ad tempus posuisset Alexander remedia, nulla gens aut populus oppressionem illorum ferre potuisset,' ut scribit Hieronymus. Quando igitur hae nationes inclusae in locis certis mundi exibunt in desolationem regionum et obviabunt Antichristo, multum deberent Christiani et maxime ecclesia Romana considerare situm locorum, ut posset percipere hujusmodi gentium feritatem et per eos percipere tempus Antichristi, et originem: nam debent obedire ei: ergo si illi ex una parte mundi veniant, ipse ex contraria procedet, quando autem claustra Alexandri rupta sunt. Nam aliqua fuerunt ante Isidorum quassata, quoniam scribit de eis. Atque frater Willielmus[1], quem dominus rex Franciae misit ad Tartaros anno Domini 1253, quando fuit ultra mare, scripsit regi nunc dicto quod transibat cum Tartaris per medium portarum quas Alexander construxit. Quando enim non potuit vincere has gentes, tunc ut scribit Ethicus et confirmat Hieronymus, Alexander immolavit hostias Deo et deprecatus est tota nocte et die Dei misericordiam et consilium, et divina potentia affuit terrae motus magnus, et convenerunt montes adversus montes, et approxi-

Attempts of Alexander to control incursions of dangerous tribes.

[1] More will be said on this subject in a subsequent note. For an account of Alexander's gates, and the nations enclosed by them, see Aethiopian version of the pseudo-Callisthenes, edited by Mr. Wallis Budge, pp. 230–238 (1896). See also Marco Polo, i. 4 (vol. i. pp. 49–50, Yule's ed.).

maverunt per stadium unum usque ad spatium unius quadrigae, et ipse tunc erexit portas mirae magnitudinis et linivit eas bitumine incognito, quod nec igne nec ferro nec aqua nec aliqua re dissolvi potest, nisi solo terrae motu violento.

Special geographical description, partly from Pliny and other writers,

Quoniam igitur infinita est utilitas cognitionis locorum hujus mundi pro philosophia et theologia et ecclesia Dei, volo adhuc alium sermonem de hujusmodi locis componere et divisiones regionum evidentiores assignare; et sequar Plinium abundantius, quem omnes sancti et sapientes sequuti sunt. Ubi autem aliquod certum per alios auctores invenero tam per sanctos, ut Hieronymum, Orosium, Isidorum, quam per alios auctores, non negligam quae necessaria sunt assignare. Sed regionum nobis notarum divisiones particulares non oportet annotare, similiter nec loca singula in aliis regionibus, sed magis notabilia et famosiora in scriptura et philosophia; de quibus gentes tyrannicae venient et venerunt, quae mundum referuntur de praeterito vastasse aut aliquando vastaturae. Et assignabo ritus et sectas gentium, ut qui sunt Pagani, qui idololatrae, qui Tartari, et sic de aliis, ut certior apprehensio locorum pateat perlegenti. Haec autem via, qua procedam, non est per certificationem astronomiae, scilicet per veras longitudines et latitudines[1] locorum respectu coeli; quia nondum habent eam Latini, sed est sumpta ex auctoribus qui mundi regiones describunt secundum quod quilibet potest loca natalis soli describere, et per alios de locis extraneis edoceri.

partly from reports of travellers.

Non nunquam tamen plura reperiuntur scripta, quae ex

[1] In Ptolemy's Atlas, lines of longitude and latitude are drawn, so as to give an appearance of precision to his maps, which in reality they do not possess. His conception of what a map ought to be was derived from Hipparchus, and is thoroughly scientific. Having made these maps, he drew up a list of places with the precise latitude and longitude (to the twelfth part of a degree) assigned to each. But these positions are in the immense majority of cases purely hypothetical, and are determined by the best estimate he could get from travellers' reports.

Considering the immense reputation of Ptolemy during the middle ages, it is extremely remarkable that Bacon should have seen so clearly the fallacious character of this side of Ptolemy's work, and should show himself so clearly aware of the necessity for accurate determination of the latitude and longitude of each place by astronomical observation. Cf. p. 300.

rumore magis quam per experientiam habuerunt auctores. Nam et Plinius minus bene dixit mare Caspium oriri ex mari Oceano, et Ptolemaeus in Almagesti de plano erravit de situ Britanniae majoris et minoris, sicut manifestum est cuilibet, et sic isti de aliis multis, et caeteri auctores similiter. Propter quod recurram ad eos qui loca hujus mundi pro magna parte peragrati sunt. Et maxime in regionibus aquilonaribus sequar fratrem praedictum, quem Dominus Rex Franciae Lodovicus misit ad Tartaros anno Domini 1253, qui perlustravit regiones orientis et aquilonis et loca in medio mundi his annexa, et scripsit haec praedicta illustri regi; quem librum diligenter vidi, et cum ejus auctore contuli, et similiter cum multis aliis, qui loca orientis et meridiana rimati sunt. Sicut tantam priorem descriptionem feci, magis propter exemplar et excitationem Gloriae Vestrae, ut compleatur suo tempore per sapientes hujus mundi, quam propter certificationem, sic et hanc divisionem noto ut Vestra Sapientia recognoscat quod major labor hic requiritur, quam praesens persuasio debeat continere. Scriptura enim perfecta quam requiritis habet utramque descriptionem perficere.

Inhabited regions south of the Equator.

Expediendo quidem naturalium et experimentatorum insuper sanctorum sententias circa partes habitabiles non oportet nos coarctari in his quae mathematici apud Latinos certificarunt; nam pauca sunt haec; sed latius progredientes auctoritate et experientia multiplici roborati dicimus quod non solum septem climata habitantur, sed quarta terrae, et plus longe quam quarta, hominum continet nationes. Nam invenimus apud Plinium et alios, quod quaedam loca sunt in nostro habitabili, quae ascia dicuntur, id est sine umbra, ab *a* quod est *sine*, et *scia* quod est umbra, et haec multis modis variantur. Nam in aliquibus locis res non habent umbram aquilonarem nec meridianam in solstitio aestivali. Nam quando sol fertur supra capita eorum in meridie, non est aliqua umbra neque ad septentrionem vel ad meridiem, neque ad orientem vel occidentem, et hoc est in insula Nili quae dicitur Syene, quae est in suprema parte Aegypti in confinio illius et Aethiopiae, sicut patet ex Plinio in libro secundo et quinto; et Lucanus dicit 'umbram nullo flectente Syene'; id est, in meridie solstitii aestivalis, et hoc est versus finem secundi climatis, de qua in Ezechiele pluries sermo fit. Alia loca dicuntur ascia, quia bis in anno umbram non habent. Nam quaedam in aestate nostra projiciunt umbram in austrum, solem ad aquilonem habentes. Et in hyeme nostra, faciunt umbram ad aquilonem, quoniam sol est in austro, secundum Plinium secundo libro: et haec alternatio umbrae fit per senos menses, sicut Plinius dicit sexto libro, capitulo decimo nono. Quod impossibile est fieri, nisi sub aequinoctiali, quoniam licet ei qui habitant inter tropicum Cancri et aequinoctialem habeant varietatem multiplicem de umbris projiciendis nunc ad aquilonem nunc ad meridiem, tamen non possunt habere hanc varietatem per menses sex, sed plus projicient umbram in aquilonem quam in meridiem; quia solem plus habent ad meridiem quam aquilonem; sed illi qui sub aequinoctiali circulo sunt, habent aequaliter solem ad septentrionem et meridiem, scilicet per senos menses utrumque. Et haec gens in India Orestes dicuntur et Monedes et Simari, apud quos est mons Malcus nomine, in quo umbrae sic per senos menses variantur, ut dicit Plinius secundo et sexto libro. Sed, quod plus est, invenimus per eum habitationem fieri sub tropico Capricorni ultra. Nam regio Pathalis in India dicitur habens portum, ut dicit, celeberrimum, ubi umbrae solum in meridie cadunt; ergo habitatores ejus habent semper solem ad Aquilonem. Et idem refert sexto libro de insula Taprobane in India, de qua homines, cum Romam venerunt Claudii principatu, mirati sunt quod umbrae eorum cadebant in Aquilonem et quod sol a meridie oriebatur; et ideo apud eos umbrae cadunt semper in meridiem, et sol semper oritur eis aquilonaris [1].

The habitable region may extend beyond the tropic of Capricorn.

Et ideo verum est quod Ptolemaeus dicit libro de dispositione sphaerae, quod natura exigit ut sint duo genera Aethiopum sub duobus Tropicis. Quod si sol habeat eccentricum, tunc licet quantum ad naturalem coeli dispositionem erit locus inhabitabilis in superficie terrae propter calorem, quando sol

[1] Pliny, *Hist. Natur.* lib. vi. cap. 22 'Sed maxime mirum iis erat umbras suas in nostrum caelum cadere, non in suum: solemque a laeva oriri et in dexteram occidere potius quam e diverse.' As Ceylon is in north latitude these men must have come from far more distant and southerly regions.

venit ad Sagittarium et Capricornum propter appropinquationem ejus ad terram, et propter confusionem aequidistantium, et propter casum radiorum ad angulos rectos; et inhabitabilis propter frigus, quando est in Geminis et in Cancro, quia sol tunc nimis recedit ab eis, et cadunt radii ejus ad angulos obliquos; tamen propter alias dispositiones locorum accidentales, tum propter altitudinem montium prohibentium calorem solis, tum propter alias causas, possunt aliqua loca ibi esse habitabilia et maxime in locis subterraneis, quando sol est circa oppositum Augis. Et alibi potest esse talis planities inter eos et solem, et a posteriori talis montium laevitas et figuratio ad modum speculorum concavorum et comburentium, quod loca possunt esse habitabilia quando sol est in Auge non obstanti frigore, sicut prius dictum est de locis circa polos mundi[1]. Quod si ponamus solis concentricum cum epicyclo, sicut possibile est, ut dicit Ptolemaeus in Almagesti, tunc de facili potest salvari habitatio illius loci, quia tunc sol non tantum accedit ad terram ut omnino comburat eam sub tropico hyemali, nec tantum recedit ab ea ut eam frigore destituat. Quod si nec epicyclum nec eccentricum ponamus ut naturales aestimant, tunc nullum inconveniens sequitur de habitatione. Et ideo quamcunque opinionem in hac parte sequamur potest Plinius salvari. Et necesse est, quia experientia certa docuit eum saltem per eos qui de Taprobane insula venerunt in Romanam civitatem, et per eos qui ad insulam perducti sunt. Sed quamvis locus ultra tropicum Capricorni sit optimae habitationis, quia est superior pars in mundo et nobilior per Aristotelem[2] et Averroem primo Coeli et Mundi, tamen non invenimus apud aliquem auctorem terram illam describi, nec homines illorum locorum vocari, nec quod ad nos venerunt, nec nostri ad eos. Et ideo opinio aliquorum est quod ibi sit paradisus, cum sit locus nobilissimus in hoc mundo secundum Aristotelem et Averroem secundo Coeli et Mundi. Sed non solum philosophi, sed sancti, ut Ambrosius

[1] See p. 134.

[2] *De Coelo*, ii. 2, § 8 *τῶν δὲ πόλων ὁ μὲν ὑπὲρ ἡμᾶς φαινόμενος τὸ κάτω μέρος ἐστίν, ὁ δ' ἡμῖν ἄδηλος τὸ ἄνω.* We see here the origin of the view as to the position of the terrestrial Paradise described in the *Purgatory* of Dante.

in Hexaemeron et Basilius in hac umbrarum diversitate concordant. Nam quarto libro dicit Ambrosius, 'sunt qui per duos totius anni dies sine umbra sunt in partibus meridianis, eo quod solem habentes super verticem suum undique per circuitum illuminantur; unde ascii dicuntur, id est, sine umbra; et amphiscii, id est, circum umbrati; et hi sunt sub aequinoctiali et circiter ab utroque latere, qui quando sol non est super eorum capita projiciunt aliquando umbram in aquilonem aliquando in meridiem, secundum quod sol nunc est ad aquilonem, nunc ad meridiem eorum'; et addit 'quod sunt in hoc quem nos incoluimus orbe terrarum circa meridiem positi, qui in australem plagam videntur umbram transmittere.' Quod signanter dicit de umbris meridianis, potest intelligi de his qui solas habent umbras meridianas, scilicet sub tropico Capricorni et ultra; quia semper sol est eis ab aquilone, nisi quando semel est super capita eorum in tropico Capricorni.

Pliny and others have shown how far it reaches northward.

Quantum vero habitetur versus aquilonem Plinius ostendit quarto libro per experientiam et auctores varios. Nam usque ad locum illum habitatur ubi extremi cardines mundi sunt; et ubi est dies per sex menses et nox per tantum. Et Martianus in sua mundi descriptione concordat; unde volunt quod sit ibi gens beatissima, quae non moritur nisi satietate vitae, ad quam cum venerit praecipitat se alto saxo in mare; et vocantur hyperborei de Europa et Arumphei in Asia. Haec igitur dicta sunt secundum latitudinem regionum quae est ab aequinoctiali citra vel ultra, ut videamus quod habitatio excedit quartam secundum latitudinem regionum.

East and west, it is far larger than is commonly thought.

Quod similiter potest patere secundum longitudinem, quae consideratur ab oriente in occidens. Nam, ut Plinius sexto libro naturalium scribit, sola India est pars tertia [1] habitabilis. Nam gentes habet centum et decem et octo. Et Hieronymus dicit ad Rusticum monachum, 'Navigantes rubrum mare multis difficultatibus ac periculis ad urbem maximam perveniunt.

[1] Pliny, *Hist. Natur.* lib. vi. cap. 17, 18. He states this on the authority of writers who accompanied Alexander. 'Alexandri Magni comites ... scripserunt ... Indiam tertiam partem esse terrarum omnium, multitudinem populorum innumeram, probabili sane ratione.' Pliny had said (lib. iii. cap. 1), somewhat inconsistently, that Europe was perhaps half the habitable world.

Felix cursus, si post sex menses supradictae urbis portum teneant, a quo se incipit aperire oceanus, per quem vix anno perpetuo ad Indiam pervenitur.' Ergo a portu rubri maris versus nos exigitur navigatio usque Indiam per annum et dimidium. Et Hieronymus dicit libro locorum, quod classis Solomonis per triennium ab India deportabat commercia, ut anno et dimidio navigarent usque ibi, et tanto tempore redirent. Sed immensa est distantia rubri maris usque ad finem ulterioris Hispaniae circa montem Atlantem. Manifestum est igitur quod a fine occidentis usque ad finem Indiae supra terram erit longe plus quam medietas terrae; ut cadamus necessario in opinionem Esdrae et Aristotelis et Averrois, quae superius tacta est de magnitudine habitationis inter oriens et occidens, quae quartam terrae secundum longitudinem excedit. Cum vero dicat Plinius, Europam esse majorem Asia, non includit ibi Indiam sub ea propter Indiae magnitudinem; cum sit tertia pars habitabilis secundum eum, ut dictum est.

Special descriptions. South India and Ceylon.

His consideratis in summa circa magnitudinem habitationis, nunc partes aliquas magis famosas in scriptura et in philosophia notare dignum est; et quas utile est sciri a Christianis propter infidelium conversionem, et propter negotia diversa cum diversitate gentium tractanda, ac propter utilitates ecclesiae contra furorem Antichristi et eorum qui tempora ejus praevenire creduntur, ut mundum primo vastent donec tribulatio maxima veniat per Antichristum. Et hic non solum necessaria est depictio locorum et figuratio, sed narratio eorum quae depingi debent; neutrum enim sufficit. Sequar vero auctores et experimentatores ea diligentia qua valeo quantum modo sufficit, usquequo perfecta locorum doctrina requiratur. Et incipiam a meridianis et orientalibus partibus, praecipue propter scripturam, quae loca illa magis frequentat. Dico igitur secundum praedicta, quod frons Indiae meridianus pellitur ad tropicum Capricorni propter regionem Pathalis[1] et terrarum

[1] The port of Patale is noted by Pliny (ii. 73) as a place where the sun was in the north at noon, and the shadows were thrown southwards. It may be noted here that Bacon's general method in this description is to pass from East to West in the equatorial region, then to return from West to East in a higher

vicinarum, quas alluit brachium maris magnum descendens a mare oceano, quod est inter Indiam et Hispaniam ulteriorem seu Africam, de quo superius dictum est secundum Aristotelem. Quod enim mare tangat meridiana Indiae hoc expresse dicit Plinius, et per Hieronymum patet, et Alfraganus similiter hoc testatur. Et illud mare decurrit per regiones Indiae meridianas, et annua navigatione protrahitur donec concurrat mare rubrum, sicut planum est per Hieronymum et Plinium et caeteros; et in illo mari ad Eurum respectu Indiae est insula Taprobane distans per mare Nadosii septem dierum navigatione in qua non videntur septentriones et vergiliae. Auro et argento et gemmis pretiosis abundant, majoresque opes eorum sunt quam Romanae. Sed apud Romanos major usus opulentiae, ut recitat Plinius. Et eligitur apud eos rex senex et clemens, liberos non habens; quod si postea habeat filios, non fit regnum haereditarium. Regi dantur rectores triginta, quorum utatur consilio in regimine populi; qui si delinquat criminaliter, morti judicatur, ita tamen quod nullus manum ei imponat: sed alimenta et omnia ei denegantur, etiam nullus cum eo loquitur, unde per se deficit. Vita hominis apud eos centum annorum modica est[1].

Arym, or Syene.

Meridianum vero latus Indiae descendit a tropico Capricorni, et secat aequinoctialem circulum apud montem Malcum et regiones ei conterminas et transit per Syenem, quae nunc Arym[2] vocatur. Nam in libro cursuum planetarum dicitur quod duplex est Syene; una sub solstitio, de qua superius, alia sub aequinoctiali circulo, de qua nunc est sermo, distans per nonaginta gradus ab occidente, sed magis ab oriente elongatur propter hoc, quod longitudo habitabilis major est quam medietas coeli vel terrae, et hoc versus orientem. Et ideo Arym non distat ab oriente per nonaginta gradus tantum. Sed mathematici ponunt eam in medio habitationis sub aequi-

latitude; then from East to West in a still more northerly parallel. He does not, however, always abide strictly by this plan.

[1] See Pliny, vi. 22.

[2] Arym was supposed to be situated on the equator, midway between West and East. Apparently some of the tables of longitude used by Arabian geographers took it as a point of departure. It was of course quite distinct from Syene on the Nile, which in Ptolemy's map was not more than 60° E.

noctiali distans aequaliter ab occidente et oriente, septentrione et meridie. Nec est contradictio, quia mathematici loquuntur de habitatione eis nota secundum veras comprehensiones longitudinum et latitudinum regionum : et hoc non est tantum quantum notum est per experientiam itineris et navigationis apud Plinium et alios naturales. Secundum vero quod praetendunt auctores et maxime Plinius, mare Indicum decurrens per latus Indiae a tropico Capricorni secans aequinoctialem transit per latus meridianum Indiae; deinde immensum spatium terrae absorbens flectit se versus occidens per meridiem, donec recipiat fauces maris rubri et ostia ejus, decurrens versus aequinoctialem ultra meridiana Aethiopiae in mare occidentis [1].

Et inter fauces maris rubri et mare Aethiopum incipit Aethiopia. Et ubi latitudo regionis est circiter sedecim gradus, et longitudo diei prolixior fere tredecim horis secundum Ptolemaeum in Almagesti, et Plinius satis concordat sexto libro et secundo, est Saba, insula Nilo inclusa, civitas regalis Aethiopiae, de qua in Isaia 'labor Aegypti, negotiatio Aethiopiae et Sabaim,' super quod decimo tertio libro dicit Hieronymus, quod est gens Sabaea quae est trans Aethiopiam. Et haec est Meroe, quae est in ultimo Aethiopiae et in fine habitabilis notae, secundum quod prius dictum est, et in Ezechiele xxvii, quam Josephus primo antiquitatum a Cambyse rege Meroen cognominatam ex sororis vocabulo refert, et hoc Hieronymus testatur libro locorum. Quae civitas distat a mari Aethiopico circiter septingenta milliaria secundum Plinium sexto libro. Et est haec civitas in primo climate, a qua nominatur clima Diameroes. Regnavit ibi Candax foemina, unde nomen Candacis multis jam annis ad reginam transit, ut dicit Plinius, et infert, cum potirentur rerum Aethiopes, insula ea magnae claritatis fuit. Tradunt armatorum 250 millia dare solitum, artificum 400 millia alere [2]. Actus autem apostolorum memoriam faciunt de eunucho Candacis reginae Aethiopum,

Aethiopia.

Meroe. First parallel.

[1] It will be noticed that Bacon does not adopt the Ptolemaic view that the Indian ocean was an enclosed sea.

[2] These numbers are wrongly given in the MSS. They are given here as in Pliny, vi. 29.

quem Philippus baptizavit: unde Candax est nomen dignitatis, sicut Caesar, Ptolemaeus, Pharao, Antiochus, Abimelech. Nam Abimelech in Philistiim, Antiochi in Syria, Ptolemaei in Aegypto post mortem Alexandri, Pharaones ibidem tempore antiquo, sicut Caesares et Augusti in Romano imperio, ut dicit Hieronymus nono libro super Ezechielem. Sed ubi latitudo regionis est quasi consimilis super ripam rubri maris ad orientem, est Ptolemais civitas a Ptolemaeo Philadelpho condita ad primos venatus elephantorum, ubi quasi per quadraginta quinque dies ante solstitium et post non sunt umbrae in meridie, sicut dicit Plinius, et in illis nonaginta diebus, aut circiter, cadit umbra in meridiem, quia sol est ad aquilonem, et post hos dies cadit umbra per residuum anni in aquilonem. Et hic habitant sub circuitione medietatis Tauri et medietatis Leonis. Unde sol transit bis in anno per capita eorum in illis medietatibus signorum.

Central Ethiopia. Berenice.

Deinde ulterius in eadem latitudine sed versus occidens inter Ptolemaidem et Meroen per 4,820 stadia distans a Ptolemaide, ut Plinius dicit et Beda libro temporum contestatur, est Berenice urbs Troglodytarum Aethiopum, in qua sol similiter bis in anno pertransit, et umbrae consimiles sunt sicut in Ptolemaide. Oportet enim quod Troglodytarum regio declinet in parte versus occidens, ut exponetur inferius. Et ideo non erit in orientali Aethiopia, sed magis in medio. De istis vero Troglodytis meminit Scriptura secundo Paralipomenon duodecimo capitulo, qui venerunt cum Selac rege Aegypti in adjutorium. Hi, ut Plinius narrat libro quinto specus excavant, haec illis domus, victus serpentum carnes, stridorque non vox, sermonis commercio carent. Et in sexto libro dicit, 'gens Troglodytarum mirae velocitatis a venatu dicta velociores equis.' Unde Isidorus libro nono, 'Troglodytae gens Aethiopum ideo nuncupati, quod tanta celeritate pollent ut feras cursu pedum assequantur.' Post hos ad orientem sunt Aethiopes de Nubia et ultimo illi qui vocantur Indi propter approximationem ad Indiam, a quibus incipit Plinius describere gentem Aethiopum. Nam secundum Isidorum nono libro, Aethiopum tres sunt populi principales; Hesperi, Garamantes, et Indi. Hesperi sunt occidentales, Garamantes in medio, Indi in oriente. Cum Garamantibus Troglodytas involvit, qui conjuncti sunt. Meroe vero quae est domina gentium inter Nubienses et Indos et Garamantes secundum Alfraganum collocatur. Garamantes a Garama oppido, quod est caput regni eorum nominati, matrimoniorum exsortes passim cum foeminis degunt. Hesperi vero circa Hispaniam commorantur. Nam Hispania dicitur Hespera: unde illi qui supra ulteriorem Hispaniam inhabitant Hesperi dicuntur. Multi tamen alii sunt Aethiopes his tribus populis diversis locis copulati, etiam a natura hominis debita multum depravati, quorum nomina et regiones et mores assignare non est praesentis persuasionis. Quae omnia ex libris Plinii et aliorum satis patent, et in scriptura principali debent annotari.

Aethiopia vero terminatur inferius ad mare rubrum a parte orientis, ad Africam ex parte occidentis, ad Aegyptum in medio inter haec extrema; in quo medio est civitas Syene, de qua Ezechiel xxix et xxx capitulis expresse loquitur, dicens, 'quod a terra Syene usque ad terminos Aethiopiae non transibit pes hominis.' Syene vero est terminus Aethiopiae inferior et pars suprema Aegypti, sicut dicit Hieronymus super Ezechielem nono libro. Meroe vero est terminus superior notae habitationis secundum Plinium secundo libro. Nam a meridie ponit Meroen principium habitationis notae. Et dicit Plinius sexto libro quod a Syene ex utroque latere orientis et occidentis, scilicet Arabiae et Africae, non remansit oppidum nec castrum nec villa usque Meroen. Nam omnia deleta sunt bellis continuis, ut Scriptura sacra verificetur. A Syene vero usque ad Meroen, secundum Plinium secundo libro, sunt quinque millia stadiorum. In sexto vero libro ponit 972 millia passuum. Latitudo autem hujus Syenes dicta est. Nam sub tropico Cancri collocatur, et ab ea clima secundum denominatur clima Diasyenes.

Syene. Second parallel.

Non possunt vero sequentia expediri, nisi hic aperitur descriptio Aegypti et Africae et Nili decursus. Caput vero Aegypti est Syene, ut dictum est. Sed Aegyptus est duplex, scilicet superior et inferior; quae vero inferior vocatur, concluditur infra Nilum ad modum insulae triangularis, sicut

Egypt.

litera Graeca quae delta dicitur: et ideo Aegyptus vocata est Delta antiquitus. Et haec habet ab oriente terram Philistinorum, a septentrione mare Mediterraneum, ab occidente Africam, a meridie superiorem Aegyptum. Et a parte Palaestinae est ostium Nili quod vocatur Pelusium, ubi cadit unum latus trianguli, scilicet unum brachium Nili in mare. De hoc Pelusio habetur Ezechielis xxx, ubi dicitur, 'effundam indignationem meam super Pelusium robur Aegypti'; et Hieronymus libro nono ait, 'quod robur Aegypti dicitur eo quod habeat portum tutissimum, et negotiationes maris ibi maxime exerceantur.' Aliud quidem ostium vocatur Canopium, ubi aliud brachium trianguli cadit in mare versus Africam, inter quae ostia Nili est basis trianguli super littus maris continens cx milliaria, ut dicit Plinius quinto libro. Et a concursu brachiorum Nili in vertice trianguli usque ad Canopium ostium sunt cxlvi, et ad Pelusium ostium cclvi milliaria. Superior vero Aegyptus est contermina Aethiopiae, ut dicit Plinius, et Thebais vocatur, et incipit a Syene, quae est civitas Thebaidis, ut dicit Hieronymus libro Locorum. Et habet a meridie Aethiopiam, et ab oriente partem Arabiae, ut inferius magis clarescet; ab occidente superiorem partem Africae. Et haec est regio Thebaidis, in qua est Thebae civitas. Thebas autem Aegyptias, ut dicit Isidorus quinto decimo libro, Cadmus aedificavit: quae inter Aegyptias urbes numero portarum nobiliores habentur, ad quas commercia undique Arabes subvehunt. Deinde Cadmus in Graeciam profectus Thebas Graecorum condidit in Achaia, quae nunc dicitur terra principis Amoreae.

Alexandria. Third parallel.

Aegyptus vero inferior habet in parte Africae super mare Alexandriam, nobilem urbem ab Alexandro magno conditam, quae ab illo tempore caput Aegypti constituta est. Et Alexandria est in tertio climate, quod ab ea nominatur clima Dialexandrios, et distat a Syene secundum Plinium secundo Naturalium quinque millia stadiorum. Et ab hac civitate versus orientem super littus maris per circiter centum leucas, ut experientia itineris docet, est Memphis civitas, quondam arx Aegypti et caput, quae nunc Damiata vocatur. A qua per unam dietam est Tampnis, ubi Pharao habitavit et Moises fecit miracula, secundum quod dicit Hieronymus nono libro super Isaiam. Et in extremis finibus Aegypti, ut Hieronymus dicit in epistola de mansionibus, est Ramesse urbs versus orientem, quam aedificaverunt filii Israelis. Olim autem, ut dicit Hieronymus in libro Locorum, universa provincia sic vocabatur, in qua habitavit Jacob cum liberis suis; et haec est terra Gessen, ut Genesis liber testatur, et Hieronymus dicit libro praedicto quod in ea habitavit Jacob cum filiis prope Memphim. Et non multum a Tampne est Heliopolis civitas solis, Arabiae conterminum, ut dicit Plinius, et magnae claritatis oppidum, in qua, ut dicit Hieronymus libro praefato, fuit Phutifar sacerdos, cujus filiam accepit Joseph, sicut in Genesis legimus xli capitulo. Tana vero est civitas Aegypti, ut dicit Hieronymus in libro Locorum, in qua habitaverunt Judaei qui ob metum Babyloniorum cum Jeremia fugerunt, et non solum in ea, sed in Memphi et in terra Phatures et Magdalo, sicut habetur in Jeremia xliv capitulo. Phatures, ut dicit Hieronymus in libro Locorum, est regio Aegypti in qua profugi habitavere Judaei. Sed Socoth et Ethan et Phiaroth et Magdalon, de quibus xiii et xiv Exodi legitur, non sunt in Aegypto sed prope extra, ut patet ex epistola Hieronymi de mansionibus, et hoc versus orientem ad mare rubrum; ad quae loca profecti sunt filii Israelis de Aegypto, antequam transirent mare rubrum, sicut Exodus declarat.

Descriptione Aegypti facta quantum hic est necesse, nunc ad Africam accedendum est. Et licet Plinius et multi multa hic scribunt, sententia tamen Sallustii in Jugurthino certior est et planior, quam principaliter exponam; quoniam Hieronymus dicit libro Locorum et Hegesippus in Historia Hierosolymitana quod Sallustius est auctor certissimus. Et insistam attentius circa hanc provinciam, quia licet sit prope nos, tamen minus nota est quam Europa et Asia; atque sacra scriptura et dicta sanctorum et historiae multa requirunt de hac regione. Africa quidem dicitur ab uno posteriorum Abrahae, ut dicit Hieronymus super Genesim, qui vocatur Affer: qui dicitur duxisse exercitum adversus Libyam, et ibi victis hostibus consedisse, ejusque posteros Afros et terram Africam nuncupasse, quae

Africa.

tamen prius Libya et ante omnia Phuticensis regio a filiis Cham nuncupata fuit, ut inferius exponetur.

Getulians and Libyans.

Africam autem, ut dicit Sallustius, in initio habuere Getuli et Libyes. Getuli vero, ut Isidorus aliquando et Hugucio, a Getis seu Gotis ab aquilone venerunt per mare. Sed Hieronymus super Genesim auctor est quod ab Evila, filio Chus, filio Cham, filio Noe venerunt: nec est verisimile quod advenae primo inhabitarunt terram debitam uni nationi, unde Africa debebatur filiis Cham, sicut Aegyptus et Aethiopia. Libyes vero fuerunt filii Labaim, filii Mesraym, filii Chus, filii Cham, ut Hieronymus dicit super Genesim. A quo Labaim Libya appellata est, quamvis secundum Hieronymum super Genesim et super ultimum capitulum Isaiae, Libya prius dicebatur Phuth vel Phutensis regio, a filio Cham qui Phuth dicebatur. Est enim adhuc fluvius in Libya, qui Phuth dicitur, omnisque regio Phutensis dicta est. Getuli vero magis versus Aegyptum et Libyes versus occidentem habitabant, utrique oberrantes latius propter regionum amplitudinem. Et aliquando tota Africa dicebatur Libya, ab una sua parte dominante nominata, et Libyes vocati sunt populi terrae, de quibus secundo Paralipomenon xii et xvi, et Naum tertio, et alibi pluries invenitur; sed, ut ait Sallustius, postquam in Hispania Hercules interiit, exercitus ejus compositus ex variis gentibus dilapsus est. Ex eo numero Medi et Persae et Armeni navibus in Africam transvecti proximos mari nostro locos occupavere; sed Persae[1] intra oceanum magis et plus approximabant Aegypto et Italiae quam caeteri prope mare habitantes sub Getulis. Nam Getuli ad solem magis approximabant, et Aethiopiae vicini. Hi paulatim copulaverunt se Getulis per connubia. Et infert, quia saepe temptantes agros alia deinde loca petiverant, semetipsos Numidas appellavere; id est, sine oppido vagos et errantes, sicut dicit Isidorus libro nono. Medi vero et Armeni supra littus maris nostri a Numidis usque ad Gades inhabitabant sub Libyis, qui supra eos ad meridiem versus Aethiopes coarctati sunt. Sed Medorum paulatim Libyes nomen corrupere, barbara lingua Mauros pro Medis appellantes.

[1] J. has per se.

Et hi omnes ab oceano et Gadibus usque ad provinciam Carthaginensium habitabant. Nam secundum Sallustium postea Phoenices imperii cupidine moti a Tyro et Sidone venientes loca Africae invaserunt, et Numidas ac Getulos et caeteros Africanos coarctaverunt, et Carthaginem sive provinciam firmaverunt, in qua sunt Punicae, id est, Phoeniceae civitates nobiles, scilicet Hippo, civitas beati Augustini, Utica Catonis magni, Carthago quae tanquam Roma altera [fuit]. Cujus imperium extendebatur versus Aegyptum usque ad aras Philenorum, quam posuerunt lxxii interpretes in Ezechiele, ubi Hebraeum habet Tharsus, sicut dicit Hieronymus libro Locorum, et non solum ibi, sed xxiii Isaiae et pluries alibi. Et xxvii Ezechielis habetur de Carthaginensibus. Carthage.

Deinde sequitur regio Tripolitana eorum qui Byzantium incolunt, quam Tyrii et Sidonenses occupabant, unde vocatur Africa vel Libya Phoenices, quia Phoenices ibi habitabant scilicet Tyrii et Sidonenses. Terra haec est fertilitatis eximiae, nam centesimam frugem reddit, ut dicit Plinius. Ibique est illa Leptis magna inter duas syrtes, minorem a parte Carthaginensium et majorem a parte Aegypti: quae, Sallustio referente, sunt loca vadosa et arenosa, quae cum excitantur ventis et fluctibus maris eructant pulverem infinitum et arenas copiosas, unde syrtes dicuntur a trahendo seu a tractu arenarum et pulverum. Syrma enim Graece est tractus Latine, et syro idem est quod traho: unde habitantes loca terrae proxima perturbant et confundunt. Tripolis.

Deinde sequitur provincia Pentapolitana regio, quae Cyrenensis dicitur in scriptura, ubi sunt quinque civitates magnae, quarum caput Cyrene dicitur, de qua in evangelio Lucae, et in Marco, ac Matthaeo habetur. Cum Dominus ducebatur ad passionem apprehenderunt Simonem quendam Cyrenensem, &c. Et in Actibus Apostolorum, Surrexerunt de synagoga quae appellabatur Libertinorum et Cyrenensium, &c., et quarto regum 26 dicitur, quod rex Assyriorum transtulit Damascenos in Cyrenen; et Amos quinto et nono habetur de ea. Quum vero arae Philenorum in multis scripturis sanctorum et historiarum reperiuntur et multotiens corrupte leguntur, ut dicantur arae Philistinorum, propter hoc ad vitandum errorem non est Cyrene.

inutile considerare quod Sallustius dicit in hac parte. Cum enim Cyrenenses et Carthaginenses multa bella commiserant, nec fuerunt fines certi inter eorum imperia, ob bonum pacis decreverunt ut legati eodem die et eadem hora ab utraque civitate mitterentur, et ubi sibi obviarent invicem, ibi fines regnorum deberent constitui. Sed legati Cyrenenses casu impediti non potuerunt in tantum procedere sicut vellent; finxerunt igitur quod Carthaginenses citius recesserunt a loco suo quam debebant, et dixerunt eis quod vivi obruerentur in loco quem attigerant, si eum pro termino regni habere vellent; vel quod eis permitterent transire quo vellent; aut ipsi mortem ibi eligerent. Legati vero Carthaginensium consenserunt, et principales fuerunt duo fratres, qui Philenes vel Phileni dicebantur, qui pro republica sua voluntarie vivi obruti sunt: in quorum memoriam Carthaginenses aras erexerunt quae vocantur arae Philenorum usque in hunc diem. Sub Cyrenensium vero provincia totum usque ad Aegyptum secundum multos auctores computatur: sed Plinius parvulam provinciam per se constituit, quae vocatur ab eo Libya Mareotis. Et sic terminatur extensio totius Africae a Gadibus in Aegyptum cum distinctione suarum provinciarum.

Aethiopian tribes south of Egypt and Africa.

Supra vero Aegyptum et Africam ad meridiem extenditur Aethiopia ab oriente in occidens usque ad mare Aethiopicum, et principales regiones eorum sunt, ut dixi, Indi, Sabaei, qui sunt Meroenses, Nubienses, Troglodytae, Garamantes, Hesperides. Pars autem Troglodytarum flectit se versus occidentem super syrtes majores et partes vicinas, a quibus distare videtur secundum Plinium libro quinto per xviii dietas. Et ideo licet major pars gentis Troglodytarum vergat ad mare rubrum, tamen aliqua se inclinat in partem occidentalem super Africae regiones. Et ultra eos versus occidens est regio Garamantum in directo syrtium minorum et Carthaginensium: orientalis tamen pars Garamantum vergit in directum Cyrenaicae regionis secundum Plinium libro sexto, ut tandem Hesperides occidentales Atlanticas partes attingant.

The Nile; its probable source.

Quoniam vero Nilus Aegyptum et Aethiopiam alluit et earum provincias multipliciter distinguit, et Scriptura multiplicem facit mentionem de eo, et in philosophia et in historiis vulgatus est, satis conveniens est aliquid notabile recitari de eo. Oritur autem de paradiso, ut scriptura refert, sed ubi erumpat nostrum habitabile, aestimatur diversimode a diversis. Verisimile tamen est quod in littore incipientis maris rubri in Aethiopia oritur, secundum quod affirmat Orosius in libro de Ormesta Mundi ad beatum Augustinum, et Seneca tertio Naturalium[1] satis concordat. Nam refert Neronem imperatorem misisse duos centuriones ad explorandum Nili ortum, et cum venirent ad primum regem Aethiopum fuerunt per eum instructi et adjuti, ut reges caeteri Aethiopiae darent eis conductum. Et tandem venerunt ad paludes vadosas et herbosas, quarum magnitudo ignorabatur ab incolis: de qua etiam certificanda desperabant. Nam neque navigio propter aquae parvitatem potuerunt homines explorare; nec humus limosa potuit pondus hominis sustinere. Credebant igitur incolae ibi esse caput Nili: Et ideo quod Plinius dicit Nilum oriri in finibus occidentis juxta montem Atlantem non procul a mari, non est credendum. Duplex enim testimonium fortius est hic quam singulari, et experientia Neronis principis multum operatur.

Deinde fluvius Africanus tendit in regionem, quae Libya Aegyptia dicitur, ad immensum lacum[2], quo finitur, sicut dicit Orosius. Et ad hoc est quod paradisus est in oriente. Et ideo verisimilius est quod Nilus erumpat in oriente quam in occidente; nec est unus et idem fluvius Africae et Nilus, licet

A river in Africa flows into a vast lake with which the Nile may be connected.

[1] This reference is to *Nat. Quaest.* vi. 8. The marshes are in all probability those found by modern explorers on the White Nile above its junction with the Sobat in 9° N. Ptolemy describes this river as formed 2° south of the equator, by the junction of two rivers flowing from two lakes still further south. He describes also the Astapus, answering to the Blue Nile, which joined the main stream 4° 30′ south of Meroe and originated in Lake Coloe, now known as the lake Tzana in Abyssinia. [Bunbury, vol. ii. pp. 612-13.]

The views of Aristotle as to the inundations of the Nile, which had been carefully studied by Bacon, though with his usual independence, will be found in Frag. 212 and 213. In *De Animalibus*, lib. viii. cap. 12, he speaks of the marshes of the Upper Nile; and in *Meteorol.* i. cap. 13 he alludes to its origin in the Silver Mountain. The passage in *Post. Analyt.* ii. cap. 13, as to the lessening of the flow at the end of the month, does not seem to represent any real opinion held by him.

[2] There was great confusion in antiquity, and throughout the middle ages, as to the connexion of the Nile with other African rivers; nor, indeed, was it

consimiles pisces et monstra conformia et crocodilos enutriant, sicut arguit Plinius, quia flumina in diversis regionibus videmus consimilia animalia secundum speciem enutrire: atque secundum ipsum Plinium et alios, flumina in India nutriunt crocodilos, sicut Nilus. Sed quod allegat Nilum Aegypti augeri per incrementa imbrium et nivium fluminis Africani, concedendum est hoc, quia lacus dictus quo absorbetur hoc flumen Africae parum distat a Nilo secundum Orosium et occulto meatu in alveum Nili potest transfundi, sicut nos videmus multotiens in diversis regionibus consimile evenire.

Its subsequent course.

Decursus autem Nili a sua origine per Aethiopiam et Aegyptum traditur a Plinio et aliis, licet in ortu ejus discordet ab illis. Fluit igitur ab origine sua, ut ait Orosius, diu ad occasum solis, et transit per medium Aethiopiae spargens multas insulas, quarum omnium nobilissima est Meroe, quae et Saba dicitur. Deinde flexus ad septentrionem inter Meroen et Syenem, ut exprimit Plinius, et montibus inclusus cataractas inter occursantes scopulos aperit, ut non fluere videatur sed ruere, ubi nimio fragore auditum in accolis tollit; propter quod ad loca convenientiora se transtulerunt, ut dicit Seneca libro octavo Naturalium. Istud dico propter hoc, quod Macrobius Pythagoricus volens docere qualiter infinitum sonum ex

dissipated till our own time. Pliny (v. 9) speaks of the belief, held by many in his time, that the ultimate source of the Nile was near Mount Atlas in North Western Africa, that, proceeding eastward, it sank underground and reappeared at a distance of several days' journey and finally reached Eastern Aethiopia and took a northerly direction. Pliny distinguished the Niger as distinct from the Nile, but thought that both arose from the same source. Orosius also maintained (lib. i. cap. 2) the existence of an African river which, though distinct from the Nile, yet probably had some subterranean connexion with it (occulto meatu in alveum ejus qui ab oriente descendit eructat).

The great Arabian geographer of the twelfth century, Edrisi, describes at great length a Western Nile which flowed through the Soudan, and at last found its way into the Atlantic. He speaks of it as the Nile of the black people; and notes the names of many of the cities on or near it, especially Gana, a great commercial centre, probably identical with the modern Kano, described in Mr. Robinson's recent book on the Hausas. This Western Nile Edrisi believed to originate in the same range of mountains as the Egyptian Nile, 16° S. latitude. His description obtained from the reports of Mohammedan travellers consulted by Roger of Sicily, Edrisi's patron, corresponds on the whole with the Niger; but some parts of it seem to relate to the northern tributaries of the Congo.

motu coelorum sustinemus illaesi auribus, ponit vanum exemplum de gente quae sustinet propter consuetudinem rugitum Nili pacifice. Sed exemplum suum falsum est, sicut Plinius et Seneca docent, et exemplatum esse non potest, sicut docet Aristoteles secundo Coeli et Mundi. Hic locus est prope Syenem secundum Hieronymum nono libro super Ezechielem ; et dicit quod usque ad Syenem est Nilus navigabilis a mari Italico. Et ulterius septentrionalis effectus Nilum includit, et tandem fauces ejus in mare quod est inter Aegyptum et Italiam projicit ; ostia ejus duo, scilicet Pelusiacum et Canopium. Sicut autem ait Hieronymus quarto libro, super xix capitulum Isaiae, Nilus ante Caesarem Augustum unus alveus fuit, sed tunc divisus est in septem, unde ad Pelusium una pars descendit, et ad Memphim quae est Damiata decurrit: alteraque extenditur in altum versus meridiem usque ad Kayr et Babyloniam, a qua nunc dicitur Soldanus Babyloniae, per circiter tres dietas a Damiata ; et ab illa Damiata extenditur a latere una pars fluminis per unam dietam quasi inter meridiem et orientem ad villam, quae vocatur Lancassor[1], ubi exercitus Christianorum devictus fuit, quando Dominus Lodovicus filius Lodovici filii Philippi illustris rex Franciae primo crucem arripuit ad partes transmarinas. Alii autem rivi fluminis Nili descendunt prope Tampnem et Alexandriam et caetera loca Aegypti.

Nili vero proprietas secundum Plinium et caeteros est, quod inundat certis temporibus et plana Aegypti rigat: secundum cujus egressiones stat Aegyptiaca fertilitas vel negatur. Nam si fines suos naturales egrediatur solum per xii cubitos, tunc Aegyptus famem sustinet, in tredecim non esurit, xiv cubiti hilaritatem afferunt, xv securitatem, xvi delicias. Si plus, temperate tamen affluat, ad abusum deliciarum excitat indigenas. Sed si creverit super debitum nocet, ut dicit Seneca. Incipit autem crescere, ut dicunt, luna existente nova quacunque post solstitium sensim, id est paulatim modiceque, Cancrum sole transeunte, abundantissime vero in Leone, et residet in Virgine. Et iisdem quibus crevit

Inundation of the Nile.

[1] O. has Lamassor.

modis revocatur intra ripas dum sol est in Libra et centesimo die a principio fluxus. Causas vero hujus inundationis et crementi difficile est assignare, quia valde mirabilis est, eo quod fit in fervore aestatis, quum aquae plus consumuntur quam in aliis temporibus. Caeterum nullus fluvius sic inundat secundum Aristotelem in tractatu de Nilo, secundum Plinium nisi Euphrates; possumus tamen tertium addere, scilicet Ethiliam, qui est major Euphrate faciens mare Caspium, de quo superius tactum est. Istud docent illi qui apud Tartaros fuerunt, ut frater Willielmus et alii. Aristoteles vero et Plinius locuti sunt secundum suas experientias.

Difficulty of explaining it.

Hujusmodi igitur natura singularis, quae in aliis mundi fluminibus paucissimis reperitur, est satis mirabilis. Deinde sapientum discordia in causis istius incrementi quasi infinita inducit in nobis perplexitatem, ut in quam partem vertamur non sit perspicuum; praecipue cum multi tam probabiles sententias reprobent quam sunt illae quas affirmant. Seneca etiam omnibus auctoribus, excepto si volumus Aristotele, ubicunque figit intentionem certior, tamen in proposito solum certas dat reprobationes in libro suo de Nilo, qui est octavus Naturalium, nullam praesumens firmare sententiam, victus difficultate hac, licet alias victor gloriosus existens. Aristoteles etiam quamvis opinionem spargat, multis tamen contradictionibus semper turbari potest. Quae vero magis aestimo digna relatu propter hanc persuasionem usque ad tractatum principalem inferam tolerabili brevitate.

Wind theory of Thales.

Sapientes vero Latini negligentes experientiam in hac parte adhaerent opinioni Thaletis, qui fuit primus de septem sapientibus famosis, quae consistit in hoc, quod venti annuales determinati flant contra ostia Nili revolventes fluctus et arenas maris, quibus obstruuntur ostia, et aquae Nili penitus redeunt, et sic exit suas ripas. Sed auctoritate et experimento refellitur. Nam secundum Aristotelem et Senecam et secundum quod docet experientia eorum qui in Aegypto fuerunt, aquae Nili incipiunt fluere a superiori parte Aegypti ab Aethiopia. Nam et Aegyptii prae gaudio exultantes ascendunt naves et obviam Nilo defluenti occurrunt cum cantu et instrumentorum musicalium genere multiplici; unde ab ostiis ejus prius dictis non incipit fluere, sed a superiori parte in ostia.

Anaxagoras vero philosophus inducit opinionem magis probabilem omnibus, et dicit nivem liquefieri in aestate in montibus Aethiopiae, et sic Nilum augmentari, sicut Rhodanus, et Padus, et Danubius, et hujusmodi flumina prope Alpes crescunt ex nivium liquore. Sed hanc opinionem Aristoteles et Seneca negant et reprobant. Aristoteles vero redarguit per hoc, quod ex multa nive parum de aqua generatur, sed augmentum Nili est maximum, quia regiones infundit latissimas et aliquando usque ad triginta cubitos altitudinis. Iterum aliter arguit; aquae quae fluunt a longe sunt fortiores in fine, sicut venti qui a longinquo veniunt; et qui de propinquo fluunt sunt impetuosiores in principio. Cujus causa est in aquis, quia in magna distantia concurrunt multae aquae in unum et imbres multi fiunt et vapores multi resultant ex terra, et ideo fortificantur omnia flumina versus finem et sunt ampliora quam in principio; et sic de ventis propter multiplicationem vaporum ex diversis locis confluentium in unum corpus propter distantiae quantitatem. Sed inundatio Nili augmentatur in principio et magis ac magis fortificatur ad tempus, donec tepescat in fine, ut Aristoteles dicit et Plinius, et experientia docet. Ergo non veniunt aquae istae ex longinquo, sed et montes Aethiopiae, ubi probabilius posset dici quod nives abundent, distant secundum orbem a Nilo per iter quinque mensium. Ergo non causabitur inundatio Nili a nivibus. Major propositio istius argumenti est digna valde; nam multam habet scientiam, quicquid sit de minore propositione. Item dicit quod in plenilunio omnia congelata tabescunt et resolvuntur magis, sed Nilus crescit in fine mensis, ergo non est ex nivibus. Item Nilus magis fluit quando flat Boreas, quam quando Auster; sed Auster magis dissolvit nives quia calidus est. Caeterum Aristoteles dicit quod in Aethiopia propter calorem maximum, qui comburit omnia, non possunt esse nives; quod facile credendum est. Et his Seneca concordans addit, quod nives dissolvuntur in tempore verno et liquescunt tactae temperato calore et faciunt crementum fluviorum; sed calor nunquam temperatus est in Aethiopia antequam erit in

Snow theory of Anaxagoras refuted by Aristotle.

hyeme. Nilus autem post solstitium crescit aestivale. Quid vero Pythagoras, quid Diogenes, quid Democritus, quid alii multi et magni philosophi senserint, non est necesse in hac persuasione praeambula recitare.

Aristotle's rain theory.

Sed inferatur sententia Aristotelis pro omnibus dicentis, quod in Aethiopia in aestate nostra sunt multae pluviae et in hyeme nullae, et Nilus in illis regionibus oritur, unde augmentantur ejus paludes et lacus; et addit, quod venti annuales determinati in aestate et orientales flant, et fugant nebulas ad regiones unde Nilus fluit, quae dissolvuntur in lacus ejus. Causa autem quare augmentatur in fine mensis datur ab Aristotele secundo Posteriorum, quia finis mensis lunaris est frigidior, et frigus multiplicat humorem et augmentatur per Boream, quia ille ventus fugat nubes naturaliter propter impetum suum, quia venit de prope propter hoc quod habitatio est in septentrionali quarta terrae, et impetuosus est in hac quarta, ut Aristoteles docet, et propter hoc fugat ante se nubes, ut congregari possint in paludibus Nili, quae sunt quasi incomprehensibilis magnitudinis, ut superius tactum est, et ideo multum capiunt de loco nubium, et possunt sic per resolutionem nubium in pluvias aquae multiplicari.

Objections to it.

Sed sicut objectum est contra alias positiones, sic contra istam fieri potest. Nam cum terra sit inhabitabilis propter calorem, id est, pessimae habitationis cum sit combusta, quomodo potest ibi esse abundantia pluviae et maxime in aestate, sicut nec nives, ut ipse Aristoteles objicit contra secundam positionem? Et contra primam dicit quod idem accideret in aliis fluminibus, et quod venti annuales non semper veniunt suo tempore. Cum ergo pluviae sunt in regionibus multis ubi sunt flumina magna et venti annuales, et tamen non videmus hujusmodi augmentum ibi contingere; et Boreas similiter fugat nubes cum majori impetu in regionibus prope, quia minus a sua origine distat; igitur magis augmentarentur flumina in nostris regionibus et similiter in fine mensis; sed non videmus hujusmodi augmentum. Dubia igitur est Aristotelis positio, sicut aliorum. Quapropter difficillimum est reddere causas hujus augmenti singularis, quod non est nisi in Nilo secundum Aristotelem, vel saltem secundum Plinium non est nisi in Nilo et Euphrate. Est tamen adhuc in tertio flumine superius tacto, ut fuit in Jordane ante subversionem Sodomae et vicinarum civitatum, testante scriptura Geneseos. Propter igitur hujus rei difficultatem sufficit in hac persuasione, ut sententias philosophorum videamus quatenus per eas excitati in principaliori [1] tractatu certius veritas inquiratur.

Arabia; its boundaries.

Revertendum igitur est ad descriptionem regionum. Et inveniemus apud Plinium sexto libro, et Alfraganus concordat, et Lucanus quod antiqui vocabant partem Arabiae totum quod habitatur a mari Aethiopico et meridie descendendo per Meroen et Syenem, ita etiam quod Heliopolis Aegypti, de qua dictum est, in Arabia computetur; et ideo totum quod habitatur a Meroe et Syene et Heliopoli versus orientem inter mare Rubrum et mare Aethiopicum sub Arabia continetur; unde Alfraganus in primo climate ponit insulam Arabum, et in secundo; quae insula est in mari Aethiopico vel circa ortum maris Rubri. Et Lucanus dicit,

'Ignotum vobis Arabes venistis in orbem,
Umbras mirati nemorum non ire sinistras.'

Hoc dicit de Arabibus, qui venerunt Romam in adjutorium Pompeii, qui mirati sunt umbras sinistras et septentrionales non ire, id est, non mutari in dextras seu meridianas. Nam in terra sua, quae est inter tropicum Cancri et aequinoctialem, habent in aliqua parte anni meridianas umbras quando sol transit ultra eos versus tropicum Cancri, quia tunc sol est in aquilone respectu eorum; et quando transit ultra eos versus aequinoctialem, tunc oportet quod habeant aquilonares, quia sol est in meridie eorum. Tota igitur haec pars Aethiopiae citra Meroen et Syenem et Heliopolim versus orientem sub Arabia continetur. Et non solum hoc, sed quicquid est circa linguam, id est, extremitatem maris Rubri et super littus ejus versus orientem a cuspide linguae usque ad sinum ejus Persicum. Et extendit se a mare Rubro usque ad Pelusium Aegypti ad occidentem, et dilatat se ad septentrionem per totum

[1] It will be noticed that this discussion of the Nile problem contains three references to the systematic work in which this and many other questions were to be more thoroughly investigated; the *Opus Majus* being merely a *persuasio praeambula.*

desertum, in quo vagati sunt filii Israelis usque ad terram Philistinorum super mare nostrum conterminam Aegypto, et extensam ad orientem donec occurrat Amalechitarum regio, quae est ad orientem terrae Philistiim, et usque ad terram Edom, seu Idumaeam, quae est ad orientem Amalech et usque ad terram Moab. Deinde flectit se magis versus septentrionem per terram Seon regis Esebon, et Og regis Basan usque ad montem Galaad et Libanum, et adhuc magis flectit se ad septentrionem orientalem usque ad Ciliciam et Syriam Comagenam, et usque ad Euphratem.

Desert of Sur.

Unde Arabia large sumpta est terra magna valde, et continet in se primo Desertum Sur seu Ethan, nam Ethan dicitur solitudo ex utraque parte maris Rubri, et in extremitate ejus juncta Aegypto et Palaestinae. Quoniam in Exodo habetur quod filii Israelis metati sunt castra in Ethan; et deinde transierunt mare Rubrum et venerunt iterum in Ethan. Nam Scriptura dicit quod post transitum maris Rubri venerunt in desertum Sur, et ibi fixerunt tabernacula in Mara, et ambulaverunt per tres dies antequam fixerunt tabernacula, et primo fixerunt tabernacula in Mara, deinde in Helim. Sed Hieronymus dicit in epistola de mansionibus, quod desertum Sur et Ethan sunt idem. Et in hac Arabia prope Sur versus orientem ultra transitum filiorum Israelis super littus maris Rubri est Elamitarum regio secundum Plinium et Hieronymum in libro interpretationum; ubi est Elam civitas urbs ultima Palaestinorum. Nam in hac parte juxta desertum Sur flectit se angulus Palaestinae ad Rubrum mare secundum Hieronymum, ut dicit Plinius, ibi prope est Stagnos [1], insula maris Rubri, quam canes non intrant expositique cuncta littora errando moriuntur.

Desert of Sin.

Post desertum vero Sur versus orientem sequitur Sin desertum, ubi fuerunt quinque mansiones filiorum Israel secundum Hieronymum in epistola de mansionibus, quarum prima non habetur in Exodo, sed xxxiii Numerorum, unde dicitur, 'profectique sunt de Helim ad mare Rubrum quod vocatur Jamsuph.' Et Hieronymus quaerit quomodo reversi sunt ad mare Rubrum, et solvit dupliciter uno modo quod potuit esse

[1] This appears to be a mistake for Sygaros, of which Pliny says (vi. 28) 'Sygaros insula quam canes non intrant expositique circa errando litora moriuntur.'

quidam sinus maris Rubri qui extenderetur in deserto Sin a principali corpore maris, nam Jam est Mare et Suph dicitur Rubrum. Sed convenientius solvit dicens quod Suph potest signare rubrum vel scirpum, et hic non habet sumi pro rubro sed pro scirpo ; unde dicit quod possumus suspicari quod venerunt ad paludem quandam et lacum, qui carecto et juncis plenus fuit. Quod autem Scriptura sacra omnem congregationem aquarum mare vocet, nulla dubitatio est; hic igitur secundum Hebraicam veritatem significat paludem juncorum. Sed quia etiam nimis vulgatum fuit in antiqua translatione quod haec mansio vocetur mare Rubrum, ideo Hieronymus dimisit in nostra stare ut fuerat in antiqua, sicut multa alia per universum corpus scripturae. Ultima vero istarum quinque, quae est undecima ab egressu filiorum Israel de Aegypto, est Raphidim in directo ad septentrionem, cujus est Amalechitarum regio, quae pugnavit cum filiis Israel in deserto Raphidim, et victa est ab eis. Deinde magis versus orientem est desertum Sinai, ubi est mons Sina, qui est mons Dei Oreb, ut dicit Hieronymus in libro Locorum. Sed non est petra Oreb in Raphidim, de qua Moyses aquam eduxit. Nam Horeb pro monte Sina scribitur per Heth literam, sed Oreb non. Et post sunt sepulcra concupiscentiae et Asseroth, quae sunt duae mansiones post montem Sina in hoc deserto Sinai.

Deinde magis ad orientem est desertum Pharan, ubi incipit regio Ismaelitarum tendens versus mare Rubrum ad orientem. Ad septentrionem vero Pharan est Ebron civitas David, et ubi Adam magnus et Abraham, et Isaac, et Jacob, sepulti sunt. Per viam deserti inter Pharan et Ebron misit Moyses Josuen, et Caleph, et alios exploratores. Et in hoc deserto Pharan, ut dicit Hieronymus in libro mansionum, fecerunt filii Israel octodecim mansiones, a quinque usque ad triginta duo, computando extremas cum mediis, ut ultima sit Asyongaber; unde Pharan desertum est latissimum. Et hic percussi sunt ab Amalechitis et Chananaeis, et hic dat Dominus judicia, et orta est seditio Chorae, et fronduit virga Aaron, et multa alia facta sunt, sicut a tertio decimo capitulo Numerorum in vicesimum patens est.

Desert of Pharan.

Postquam exiverunt desertum Pharan, transiverunt magis ad

Another desert known as Sin.

orientem usque ad desertum Sin, quod est Cades Barne, ubi murmuravit populus ad aquas contradictionis. Sed istud non est illud desertum Sin. Nam istud desertum multo plus elongatum est a Rubro mari, cui annexa est terra Edom ad orientem septentrionalem. Nam ab hoc loco miserunt filii Israel nuntios ad Edom dicentes, Ecce in urbe Cades, quae est in extremis finibus tuis, positi obsecramus ut nobis transire liceat, sicut dicitur Numerorum xviii. Quod si quis contendat quod desertum Sin superius dictum possit se extendere aliquando ad hunc locum, patet quod non ; per Hieronymum in epistola de mansionibus, tam per interpretationem quam per Scripturam ; quoniam Sin primum scribitur per Samech et interpretatur rubus vel odium ; hoc autem per Sade et vertitur in mandatum. Et ideo filii Israel declinaverunt per viam quae ducit ad mare Rubrum circumeuntes terram Edom, et venerunt ad montem Or in ultimis finibus terrae illius, in quo monte mortuus est Aaron.

Moab.

Et in finibus terrae Edom fecerunt adhuc alias tres mansiones secundum Hieronymum usquequo venerunt ad terminos Moab, unde Moab est in oriente Edom. Nam egressi de finibus Edom fixerunt tentoria in deserto quod respicit Moab contra orientalem plagam, sicut dicitur Numerorum xxi. Deinde, ut habetur Deuteronomiae xi, transierunt urbem Moab nomine usque ad confinium terminorum Ammon. Et in his locis incipit ab oriente terra Seon regis Amorraeorum, et terra Moabitarum et Ammonitarum, et ideo diligenter consideranda sunt haec loca. Nam hic incipit terra filiorum Israel, et de his multum loquitur Scriptura et sancti. Est igitur in confinio horum locorum rupes in sublime porrecta, quae vocatur Arnon, et haec rupes distinguit extrema filiorum Ammon et Moab et Seon regis Amorraeorum ; et ideo ibi incipit terra filiorum Israel. Sed sub hac rupe versus occidentem est vallis quae vocatur Arnon, juxta quam a latere meridiano est civitas Ar, quae est caput regni Moabitarum, quae postea dicta est Acropolis ex Hebraeo vocabulo et Graeco composita, id est, civitas adversarii, sicut dicit Hieronymus quinto libro super Isaiam prophetam. A rupe autem hac descendit torrens versus occidentem, quae vocatur torrens

Arnon, super cujus ripam situatur oppidum, quod vocatur Aroer prope Arnon. Haec patent ex libro Locorum Hieronymi, et ex textu in libro Numerorum xxi et Deuteronomiae ii et iii et Josue xiii et Judicum xi et multis aliis locis Scripturae.

Terra igitur Moab ascendit ab Arnon in occidentem usque ad Edom et usque ad mare mortuum, et ubi fuerunt civitates submersae, et usque ad Jordanem contra Jericho, sicut patet ex locis praedictis Scripturae et Hieronymi. Et infra terram Moab prope Arnonem et Ariopolim est Madian civitas Jethro soceri Moysi, sicut dicit Hieronymus libro Locorum. Et necesse est ut ibi sint Madianitae. Nam ut patet Exodi xxii et Numerorum xxiii et xxiv et xxv, Balac rex Moabitarum vocavit Balaam hariolum ut malediceret Israeli, qui dedit eis consilium ut offerrent eis filias Madianitarum, et peccaverunt cum eis et interfecti sunt multi de Israele. Insuper Madianitae destructi sunt et deleti postea per filios Israel. Ex altera parte torrentis Arnon incipiebat terra filiorum Ammon versus septentrionem et orientem tendens versus Euphratem, et ad occidentem accedit ejus angulus versus Jordanem usque ad vadum seu torrentem Jaboc, quem transivit Jacob quando venit de Mesopotamia Syriae, post cujus transitum luctabatur angelus cum eo sicut xxxii Genesis recitatur. Et quod ad hunc torrentem Jaboc sit terminus filiorum Ammon patet per Deuteronomiam iii capitulo. Et hic est terminus Ammon et Seon regis Amorraeorum et Og regis Basan, sicut patet ex Judicum xi. Nam a Jaboc incipiebat terra Seon, ut ibi habetur; et ubi terminatur ejus terra, incipit terra Og regis Basan, et descendit etiam usque prope torrentem Arnon usque ad confinia Esebon urbem regis Seon Amorraeorum. Et ideo terra quae fuit propria Seon est inclusa a meridie Moabitis et ab oriente Ammonitis, et ab occidente habet Jordanem fluvium, et a septentrione habet terram Og regis Basan. Sed iste Seon potentior factus irrupit in regiones Moab et Ammon, et abstulit terras eorum. Nam quod abstulit terram filiorum Ammon habetur Judicum xi. Et quod medietatem terrae suae amiserunt filii Ammon habetur Josue xiii. Moab etiam multum amisit sicut patet ex xxi Numerorum.

Midian, Ammon, Seon.

Other regions of Arabia.

Postquam vero inventae sunt hae regiones a lingua maris Rubri per mansiones filiorum Israel, adhuc considerandum est quod in desertis quae sunt inter mare Rubrum et terras jam nominatas sunt aliae regiones magnae, quae extenduntur ab Euphrate in circuitu terrarum jam dictarum, scilicet terra filiorum Ammon et Moab, et desertum Pharan, usque ad terram Elamitarum, quam superius esse positam dixi super littus maris Rubri a transitu filiorum Israel usque versus orientem. In hac igitur regione pergrandi similiter et in Pharan habitaverunt filii Keturae et Agar, quos generavit Abraham, de quibus fit mentio xxv capitulo Genesis. Et primo ab Euphrate incipit regio Nabathena a filio primo Ismaelis, qui vocatur Nabaioth, sicut dicit Hieronymus, super Genesim xxv capitulo, et nunc dicto concordat Plinius primo libro, nisi quod unam partem Nabathenorum vocat Nomades, qui vagantur circa Euphratem prope Chaldaeos; post hos versus desertum Pharan est Cedar regio, quae ab altero filiorum Ismael nominatur qui Cedar vocatus est. Et quamvis aliae regiones filiorum Ismael nominentur usque Sur, nam habitavit ab Evila usque Sur, sicut dicit Scriptura, tamen omnes vocantur Cedar, sicut vult Hieronymus quinto libro super illud Isaiae xxi Onus in Arabia, dicens, Hic loquitur pro Cedar, quae est regio Ismaelitarum, qui dicuntur Agareni et Saraceni nomine perverso; et septimo libro super capitulum Isaiae lx dicit de his regionibus Cedar et Nabathena, quod Cedar est regio Saracenorum, qui in Scriptura vocantur Ismaelitae, et Nabaioth est unus filiorum Ismael, quorum nominibus solitudo appellatur, quae frugum inops, pecorum plena est. Evila vero est pars regionis Ismaelis distans a Pharan oppido in deserto Pharan per tres dietas, ut dicitur in libro Locorum. Est vero et alia regio Evila in India juxta flumen Gangem de qua primo Genesis dicitur.

Saba, in which lies Arabia Felix.

Et inter Cedar et Elamitarum regionem supradictam extenditur Saba regio super littus maris Rubri secundum Plinium quinto libro. Et haec regio est thurifera et aromatibus plena, et habet tres partes. Una vocatur Arabia Eudaemon, quae includitur inter sinum Persicum maris Rubri et sinum Arabicum, secundum Orosium libro de Ormesta Mundi, et secundum

Isidorum libro xiv. Alia est Madian, nominata ab uno filio Abrahae ex Kethura. Tertia est Epha, regio nominata ab uno filio Madian, ut patet ex xxv Genesis. Et quod hae duae regiones sint de regno Saba vult Hieronymus expresse super xvii Isaiae, dicens, Madian et Epha sunt regiones fertiles camelorum, omnisque provincia appellatur Saba; unde fuit regina Sabae sicut dicit. Et quod Arabia Eudaemon sit pars Sabae patet, quia illa juncta est Chaldaeis immediate, sicut dicit Orosius. Et Chaldaei et Sabaei eorum vicini simul invaserunt armenta Job sanctissimi, ut patet ex capitulo primo libri sui; et illud patet expresse per Isidorum decimo quarto libro, qui dicit sic, Arabia appellatur sacra, eo quod regio sit thurifera odores creans. Hanc etiam Graeci Eudaemon, nostri Beatam nominaverunt, in cujus saltibus et myrrha et cinnamonum provenit; ibi nascitur et phoenix; ipsa est et Saba appellata a filio Chus, filii Cham, filii Noe, qui filius Chus Saba appellatus est; et hoc dicit Hieronymus in Hebraicis quaestionibus. Et ideo cum Madian et Epha numerantur lx Isaiae, adjungitur etiam Saba, cum dicit, Omnes de Saba venient. Nam Arabia Eudaemon nominatur specialiter Saba et principaliter, licet tota regio illa et Madian et Epha dicantur Saba. Et ideo tota regio a Chaldaeis supra mare Rubrum usque ad Elam dicitur Sabaea.

Different acceptations of Arabia.

Et considerandum est hic propter aliquas contrarietates, quod Arabia large sumpta habet omnes regiones dictas ex utraque parte maris Rubri secundum Plinium et Alfraganum et philosophos antiquos. Specialius tamen sumpta accipitur solum pro regione quae est a lingua maris Rubri usque ad Euphratem et sinum Persicum versus orientem, et usque ad Palaestinam Idumaeam versus septentrionem ex una sui parte, cujus alia pars septentrionalis et magis orientalis extenditur ad montem Libani, et comprehendit totam regionem Moab et filiorum Ammon et regna Seon et Og regis Basan, et aliqua eis conjuncta. Et sic utitur Scriptura ut in Isaia cum dicit, Onus in Arabia, ubi Cedar comprehenditur sub ea; et mons Sina sic est in Arabia, secundum quod apostolus dicit ad Galatas iv. Tertio modo magis stricte accipitur, ut excludatur Pharan et Cedar et Madian et Epha et Saba Eudaemon.

Nam sic in tempore Hieronymi et deinceps accepta est Arabia, quoniam ipse dicit in libro Locorum Pharan trans Arabiam esse; quarto et decimo septimo libro super Isaiam dicit Madian et Epha et Cedar et Nabatheam regiones esse trans Arabiam. Quod vero Saba thurea Eudaemon et tota regio Saba sit distincta similiter contra Arabiam isto tertio modo dictam patet per Hieronymum in Hebraicis quaestionibus. Nam dicit quod cum in Psalmis dicitur, reges Arabum et Sabae dona adducent, illa est Saba thurea et thurifera praedicta, de qua adducit auctoritatem Virgilii, Solis est, inquit, thureis virga Sabaeis. Nam, ut Hieronymus dicit et patet cuilibet scienti Hebraeum, in Hebraeo dicitur, reges Saba et Saba dona adducent; sed primum Saba interpretatur Arabia et scribitur per sin litteram, secundum scribitur per samec, et est illa Saba thurifera, et de hac Saba venerunt magi qui adoraverunt Christum, non de Saba quae est in Ethiopia. Nam illa est in meridie. Ab oriente enim secundum Evangelium venerunt magi. Et isti sunt reges Sabae, sive reges Arabum et Sabae.

Syria. Post haec sequitur regio praegrandis quae vocatur Syria, quae, secundum Scripturam et Plinium et antiquos auctores continet omnes provincias a Tigri flumine ab oriente usque ad Arabiam a meridie et usque ad mare nostrum seu magnum, quod dividit Italiam et Syriam et Aegyptum ab occidente et a Cilicia et Tauro monte altissimo a septentrione. Prima ejus et principalis regio est Mesopotamia seu Assyria, nam idem sunt secundum Plinium; et Hieronymus dicit tertio libro super Isaiam, quod tota regio inter Tigrim et Euphratem est regio Assyriorum. Similiter Mesopotamia inter Tigrim et Euphratem continetur. Unde dicitur a meson, quod est medium, et potamus, quod est fluvius, quasi inter duos fluvios contenta, Tigrim scilicet et Euphratem. Et ideo idem fuit antiquitus Mesopotamia et Assyria. Et haec Mesopotamia seu Assyria ab oriente habet Tigrim, ab occasu Euphratem, a meridie mare Persicum, quod est sinus Persicus maris Rubri, a septentrione Taurum montem, cujus longitudo est circiter octingenta milliaria, latitudo trecenta, secundum Plinium. In hac Mesopotamia sunt Ninive et Babylon, et tota terra Chaldaeorum, et Babel turris constructa in terra Senaar. In hac

Mesopotamia. (margin note)

autem Mesopotamia sunt civitates quas Nimroth construxit, scilicet Arad, id est, Edissam, et Archad, quae nunc Nisibis, seu vulgariter Nisibin dicitur, et Calampne, quae postea a Seleuco rege dicta est Seleucia, sicut Hieronymus exponit super decimum capitulum Genesis. Et praeterea in Mesopotamia est Aram, sicut in Genesi habetur, quae adhuc nomen suum retinet. Et Aram distat per duas dietas ab Euphrate, et Hieronymus dicit quod Aram est trans Edissam; ergo Edissa est inter Aram et Euphratem, et Ninive est circiter decem dietas ab Aram, id est, ad orientem super Tigrim fluvium, secundum quod vult Scriptura quod Tigris currit contra Assyrios. Ninivitae enim principaliter vocati sunt Assyrii. Ab Aram usque ad Baldac versus meridiem sunt dietae circiter xxvi. Et est Baldac civitas regia, in qua Caliph Dominus Saracenitae sectae sedem suae dignitatis constituit. In illis vero partibus est turris Babel, et ruinae Babylonis magnae, quae fuit caput regni Babyloniorum et Chaldaeorum, qui cum fuerant Mesopotamii et Assyrii a principio, eo quod tota terra inter Tigrim et Euphratem dicebatur Mesopotamia et Assyria, tum quia Babylon Chaldaicarum gentium caput summam claritatem obtinuit inter urbes toto orbe. Reliqua pars Mesopotamiae Assyriaeque Babylonia appellata est, ut dicit Plinius, ita quod tandem Babylonii praevaluerunt. Nam, ut patet ex libris Regum et Paralipomenon primo nominati sunt reges Assyriorum, ut Salmanasar et Sennacherib et caeteri. Deinde Nabugodonosor, rex Babylonis, et ejus successores vastaverunt Assyrios, et dominati sunt per totum inter Tigrim et Euphratem. Noe quidem et filii ejus primo post diluvium habitaverunt in Babylonia, sicut dicit Albumazar quinto libro in majori introductorio astronomiae. Nam cum ipsi fuerunt sapientes astronomi et docuerunt primo Chaldaeos astronomiam, ut ibidem dicit, sciverunt quod quartum clima est temperatissimum, in quo Babylonia est, et ideo ad eam declinaverunt.

The Tigris and the Euphrates.

Quoniam vero Tigris et Euphrates sunt duo de quatuor principalibus fluminibus mundi et Nilo connumerati, ideo de illis aliqua sunt dicenda. Varius autem est ortus eorum. Nam secundum veritatem primo oriuntur de Paradiso, ut vult

Scriptura. Deinde secundum Plinium Tigris erumpit in Armenia majore et postea cadit in lacum omnia illata pondera sustinentem et nebulis exhalantem, cui unum genus piscium est, qui aquae transcurrentis non miscentur alveo, sicut neque a Tigri pisces in lacum transnatant. Deinde occurrente Tauro monte in specum mergitur, et ab altero montis latere erumpit in lacum, et postea ad formam fluminis revertitur et Euphrati jungitur, et transit per Ninivem, et post longa spatia currit in maris Rubri sinum, qui Persicus vocatur. Euphrates vero secundum Plinium quinto libro in Armenia majore oriens separat Cappadociam ab illa, deinde occurrit ei mons Taurus postea fluenti in occasum. Iterum se flectit in meridiem, et funditur in duo brachia. Unum cadit in Tigrim includens a septentrione Mesopotamiam, aliud alluit eam ab occidente et currit per mediam Babyloniam, ut dicit Orosius ad Augustinum, deinde fluit in paludes, et tandem in mare Persicum. Nam Chaldaei sunt inter Babyloniam ad meridiem versus Persicum mare, et Euphrates eos, sicut caeteros Mesopotamios et Assyrios, alluit ab occidente, separans eos a caeteris Syriae regionibus et ab Arabia. Euphrates vero, ut dicit Plinius, crescit Nili modo parum ab eo differens. Nam Mesopotamiam inundat sole obtinente xx partem Cancri, et incipit minui in Virgine, Leone transgresso. In totum vero remeat xxxix parte Virginis. Quod autem Boetius quinto de consolatione et Sallustius dicunt, quod Tigris et Euphrates uno se fonte revolvunt, potest intelligi de fonte Paradisi ; nam hoc verum est secundum Scripturam, quam Boetius saltem bene scivit, et Sallustius ex revolutione historiae Scripturae credere potuit ; aut hoc verum est de ortu eorum in Armenia, quoniam uterque ibi oritur secundum Plinium ; aut intelligi poterit de ortu eorum citra Taurum montem, nam occursu ejus absorbentur in terram, et ex altera sui parte erumpunt.

Other parts of Syria.

Ab Euphrate qui currit in oriente est Arabia, de qua dictum est, versus meridiem et mare Rubrum. Et versus septentrionem sunt reliquae regiones Syriae, scilicet Syria Comagena, Syria Coele seu Coele-Syria, et Syria Phoenicis, et Syria Palaestinae, quae includunt provincias a Judaeis possessas, scilicet Judaeam, Samariam, et Galilaeam citra

Jordanem, et per eam trans Jordanem, quam possederunt tribus Ruben et Gad, ac dimidia tribus Manasse, ubi est regio Decapolis et Ituraea seu Traconitis regio.

Et in his provinciis reperiuntur omnia loca sacra, quae calcaverunt primo sancti patriarchae et prophetae, deinde Dominus ipse et Mater ejus et apostoli sacrati, et in quibus crevit ecclesia primitiva et quae resonant evangelia, in quibus majora mysteria continentur quam auris mortalis possit audire aut mens humana intelligere, ut vult Origenes xviii Josue. Quapropter de his diligentius est dicendum. Sacred geography.

Primo igitur super mare nostrum, quod dividit Italiam et Aegyptum et Syriam, collocandae sunt civitates famosae vel prope illud, ut sic superius versus orientem facilius deprehendantur loca quae volumus. Primo igitur non super mare sed prope, quasi per tres leucas in confinio Aegypti et Palaestinae et Judaeae est Gaza civitas insignis Palaestinae, ut dicit Hieronymus libro Locorum. Deinde sunt novem leucae usque ad Ascalonem, quae est metropolis Palaestinae super mare sita. Deinde xii leucae usque ad Joppen. Deinde usque ad Acon xxiv, vel duae dietae. Nam duae leucae vel tres usque Assur, quod antiquitus Adotus dicebatur, et postea decem vel novem usque Caesaream Palaestinae, quae antiquitus turris Stratonis vocabatur, in qua Petrus baptizavit Cornelium, ut dicit Hieronymus multis locis, deinde usque castrum peregrini quinque leucae; et postea usque Caiphas tres; et quatuor usque Acon; et postea usque Tyrum in corde maris sunt novem leucae; et postea quatuor vel quinque usque ad Sareptam Sidoniorum, ubi vidua pavit Heliam prophetam; deinde tres vel quatuor usque ad Sidonem; postea novem vel octo usque Berithum, quod Barut vocatur. Deinde novem usque Biblium, quod nunc Gibeleth dicitur, unde xxvii Ezechielis dicitur, Senes Biblii et prudentes ejus habuerunt nautas. Postea novem leucae usque Tripolim; postea est una dieta usque Tortosam, quae ante Radum dicitur. Deinde usque Laodiceam sunt dietae circiter tres. Nam usque Valaniam sunt leucae circiter decem. Alii tamen dicunt, quod a Tortosa usque ad Margat est una dieta; et a Margat usque Laodiceam una dieta; et a Laodicea usque Antiochiam duae dietae. Sed Cities on the Syrian coast.

Antiochia est per quinque leucas a mari infra terram. Ab Antiochia usque ad Tarsum Paulo apostolo gloriosam, metropolim Ciliciae, sunt dietae circiter tres. Sed tamen usque principium Ciliciae sunt dietae circiter duae, vel una et dimidia.

Palestine; its dimensions and boundaries.

Redeundum est igitur ad loca infra terram. Abraham et Isaac frequentabant loca Gerarae. Gerara enim, a qua Gerarchica regio, ut dicit Hieronymus libro Locorum, fuit terminus olim Palaestinorum inter Cades et Sur, et ibi est Bersabe, qui vocatur puteus juramenti, ubi Abraham et Isaac foedus cum Abimelech iniverunt. Et ab isto loco incipit terra Hebraeorum ab eis possessa. Nec ultra hunc locum ad austrum amplius possederunt, sicut dicit Hieronymus in epistola ad Dardanum de terra repromissionis. Quamvis terra repromissa eis a Deo incipiebat a torrente Aegypti. sicut dicit Hieronymus octavo libro super Isaiam et primo libro similiter. Ille enim torrens, ut dicit Hieronymus, ibi est fluvius turbidus in finibus Aegypti versus Palaestinam et Judaeam, nec habens perpetuas aquas, non procul a Nilo, sed juxta castrum quod Rinocorura dicitur, quod LXX interpretes posuerunt loco torrentis sicut in eorum translatione habetur in Isaia. A Bersabe vero viginti milliaria versus septentrionem est secundum Hieronymum libro Locorum Ebron metropolis olim Philistinorum, sed decorata nobilium sepultura quatuor patriarcharum scilicet Adae maximi, Abrahae, Isaac, et Jacob. Et facile est tunc advertere loca vicina, scilicet convallem Mambre et ilicem Abrahae et campum Damascenum a meridie Ebron, qui sic nominatur a Damasco servo Abrahae. Unde non est Agar ille prope civitatem illam magnam Damascum, quae caput Syriae est. Nam ab illo loco distat circiter per quinque dietas, sed est juxta Ebron, in quo Adam plasmatus est, et in quo Cain interfecit fratrem suum; sic dicit magister in historiis et super Genesim. Carmelus, ubi quondam Nabal Carmelus fuit, et nunc villa est Carmela nomine in sexto lapide oppidi Ebron ad orientalem plagam, sicut dicit Hieronymus libro Locorum. Et prope Carmelum ad orientem in octavo milliario ab Ebron Ziph vicus ostenditur, ubi absconditus est David, juxta quem mons squalidus idem nomen habet, scilicet Ziph, in quo sedit David prope Carmelam ut dicit Hieronymus.

Quartodecimo vero milliario versus septentrionem est Bethleem, civitas in qua Dominus natus est. Et secundum Hieronymum sexto milliario a Bethleem versus septentrionem fuit Jerusalem longe clarissima urbium orientis, ut dicit Plinius[1]. Haec autem civitas distat a Joppe per duodecim leucas, et ab Acon per dietas circiter tres, et ad orientalem plagam distat Jericho a Jerusalem per novem leucas; et inter Jericho et Jordanem sunt duae leucae. Et Thecua villa Amos prophetae distat versus Eurum inter orientem et austrum per duodecim milliaria, ut dicit Hieronymus secundo libro super Jeremiam.

Et secundum Hieronymum in epistola de epitaphio sanctae Paulae prope illam fuit Pentapolis regio continens quinque civitates maledictas, scilicet Sodomam, Gomorram et caeteras. Nam ab illo loco describit Paulam reversam Jerusalem, et primo per Thecuam proximam. Orosius autem primo de Ormesta Mundi dicit quod Pentapolis regio in confinio Arabiae et Palestinae sita est; mediamque vallem quam Jordanus irrigaverat nunc mare superinfusum tegit. Et hoc est mare Mortuum, et mare salis, et mare salinarum et lacus bituminis, et vallis salsa, et vallis salinarum, et mare Araba, id est, deserti. Unde in libris Regnorum scribitur, Ab introitu Emath usque ad mare Araba et mare Asphalti, id est, bituminis, secundum Hieronymum in libro Locorum, et super Genesim. Nam in valle salinarum fuerunt putei bituminis ante subversionem civitatum, sed post pluviam sulphuris in mare Mortuum versum est, quod stagnum bituminis appellatur. Quatuor autem civitates submersae sunt, et quinta quae Bale post vocabatur remansit ad preces Loth, ut in ea post ruinam aliarum posset morari, quae postea Segor dicebatur, et nunc Zoara Syriace nuncupatur, sicut dicit Hieronymus super Genesim et in multis locis. Et haec civitas, licet non fuerit cum suis sociabus igne sulphuris consumpta, tamen post lapsum temporis tertio terrae motu subversa est, ut ait Hieronymus, quae reparata Zoara dicitur ab incolis, qui Zoari nuncupantur. Et haec civitas est in termino maris Mortui ad occidentalem plagam, a qua non longe super mare Mortuum ab occidente

Pentapolis.

[1] Pliny, v. 14 'Hierosolyma longe clarissima urbium Orientis, non Judaeae modo.'

est oppidum Engaddi urbs palmarum fertilis, unde balsamum venit et opobalsamum. Nam arbor est distillans balsamum in vineis Engaddi, de quibus Solomon meminit in Canticis. Et haec civitas vocatur in Genesi Asasontamar, quod in lingua nostra dicitur urbs palmarum; tamar quippe palma dicitur, sicut Hieronymus dicit. Quamvis vero multi multa scribunt de conditionibus istius maris et locorum subversorum, tamen hic principaliter introducam Hegesippum[1] in quarto libro historiae de subversione Jerusalem, quia plura scribit quam alii, et multi aliorum receperunt ab eo quae recitant, ac si sua essent quae narrant.

The Dead Sea.

De mari vero Mortuo dicit, quod omnia viventia resiliunt et statim excutiuntur nec mergi possunt in eo, quod aqua ipsa amara et sterilis est, nihil recipiens generum viventium, denique neque pisces, neque assuetas aquis et laetas mergendi usu patitur aves. Lucernam accensam ferunt aquis supernatare, sine ulla conversione extincto demergi lumine, et quamvis demersum difficile haerere in profundo. Denique Vespasianum imperatorem praecepisse ferunt nandi ignaros revinctis manibus in profundum dejici, eosque omnes illico supernatasse quasi spiritu quodam venti levatos, et ad superiora vi magna repulsos resiluisse. Vagari super aquas bituminis glebas certum est atro liquore, quas scaphis appropinquantes colligunt, quibus id muneris est; haerere sibi fertur bitumen; ut ferro haudquaquam vel alia peracuta metalli specie recidatur; sanguini sane cedit mulierum quo menstrua solventes lavari feruntur, cujus attactu, ut allegant quibus experiendi usus fuit, interrumpi proditur. Utilis ad compagem navium fertur, et corporibus hominum salubris admixta medicamentis. Longitudo lacus hujus ad Zoaros Arabiae dirigitur stadiis dlxx. Latitudo in stadiis cl usque ad viciniam Sodomorum, qui quondam uberrimam regionem inhabitabant.

[1] The Hegesippus here spoken of is not the Christian writer of the second century, but the unknown author of a work probably written in the fourth century, *De bello Judaico*, or *De excidio urbis Hierosolymitanae*. It was supposed to have been translated from Greek into Latin by St. Ambrose. The book is principally composed of extracts from Josephus. Some have thought that Hegesippus was a misreading of an early copyist who had before him the words, Ex Josippo. (See Smith's *Dict. of Christ. Biog.*)

Quatuor itaque urbes incensae sunt, quarum umbra quaedam et species in favillis videtur. Arsere terrae, ardent aquae, in quibus coelestis ignis reliquiae adhuc manent. Aspicias illic ad speciem poma viridantia, formatos uvarum racemos, ut edendi gignant spectantibus cupiditatem; si carpas, fatiscunt et resolvuntur in cinerem, fumumque excitant quasi adhuc ardeant. Quicquid vero Isidorus et Solinus de mirabilibus mundi et Hieronymus decimo quarto libro super Jeremiam et Plinius et alii multi dicunt, sub sententia Hegesippi comprehenduntur. Nam si ipse aliqua summatim perstringat, dicens nihil vivum ibi posse nutriri nec mergi, alii specificant. Nam Plinius et Solinus dicunt quod nullum corpus animalium recipit; tauri camelique fluitant[1]. Et Hieronymus dicit, quod propter amaritudinem maris nihil spirans ibi invenitur, unde nec pisces nec serpentes, sed et pisces a Jordanis inundatione delati in hoc mare statim moriuntur. Isidorus vero in Etymologiarum libro verba Hegesippi exponens dicit, quod nec ventis movetur, resistente turbinibus bitumine, quo aqua omnis stagnatur, nec ullam materiam sustinet, nisi quae bitumine illustratur. Et extenditur inter Jericho et Zoaram, quae est Segor.

The Jordan.

Quoniam vero Jordanes influit in mare Mortuum, et ibi nomen suum et virtutem amittit, atque regiones Hebraeorum et multae aliae notificantur per Jordanem, ideo de illo nunc dicendum est. Quamvis vero Hieronymus eleganter et veraciter loquatur de ortu manifesto et discursu Jordanis, atque Plinius et Isidorus et alii, tamen omnibus merito praeponendus est Hegesippus, qui in tertio libro explicat certius et abundantius istius fluminis originem. Omnes enim auctores praeter Hegesippum, qui per experientiam procedit, aestimant Jordanem fluvium oriri ex duobus fontibus ad radices montis Libani juxta Daneadem, quae nunc Caesarea Philippi nominatur; quorum fontium unus Jor, et alius Dan, qui sua profluvia divisim protendentes tandem concordant in unum, unde Jordanes efficitur; et per aliquod spatium terrae

[1] Plin. v. 16 'Nullum corpus animalium recipit: tauri camelique fluitant; inde fama nihil in eo mergi.'

fluminis jura defendens se postea transmittit in lacum qui Genesar dicitur cui contiguatur lacus Tiberiadis et deinde Jordanes erumpens decurrit ad orientem Jericho et fluit in Mortuum mare supra dictum. Sed totum illud praeter ortum ejus certum est. Ortus etiam manifestus et vulgatus se habet, ut dictum est, a duobus fontibus. Sed Hegesippus probat quod ibi non est primus ejus ortus, immo a Phiala fonte, qui est ex altera parte Jordanis in Traconitide regione distans ab urbe Caesarea cxx interjacentibus stadiis. Ab hoc igitur fonte labitur aqua subterraneis meatibus, et ubi Caesarea est rursum ebullit. Nam Philippus tetrarches Traconitidis regionis misit paleas in Phialam, quas ad Caesaream fluvius subterraneus ebullivit; unde liquet non exordium primum esse Jordanis in Caesarea, sed fluentum, ut paleis resurgentibus manifestatum est. Addit etiam ad decursum ejus, quod a Paneade seu Caesarea, non latente jam neque abdito per cava terrae meatu, sed visibili per terras atque aperto flumine incipiens se superfundere Semeconitim lacum, paludesque ejus intersecat. Inde quoque cursus suos dirigens centum viginti stadiis progreditur usque ad urbem, cui Julias nomen est. Postea lacum istum qui Genesar dicitur in medio transit fluento, quibus ex locis plurima evagatus deserta Alfacio suscipitur lacu atque in eum conditur. Itaque duos lacus victor egressus tertio haeret.

Cities and districts of Palestine.

Notificato Jordane urbes et regiones distinguendae sunt. Nam vicinia Jericho se extendit ad urbem Scythopolim a parte septentrionali Jericho secundum Hegesippum, quae civitas in biblia vocatur Bethsan, ut dicit Hieronymus in libro Locorum, et est oppidum in tribu Manasse, a quo accolas pristinos non potuerunt filii Manasse expellere. Descendendo vero ad occidentem usque in septentrionem Jerusalem est civitas sacerdotalis et insignis ortu Jeremiae prophetae, quae vocatur Anathoth, tribus milliaribus ab Hierosolymis separata, sicut dicit Hieronymus quinto libro super Jeremiam. Deinde magis ad septentrionem per xii leucas a Jerusalem, et per xii a Caesarea Palaestinae in directo Caesareae est Samaria civitas famosa, metropolis decem tribuum, quae nunc Sebaste dicitur. Super autem regionem Caesareae Palaestinae, et mari magno usque ad fines Ptolemensis regionis incumbit mons Carmelus per dietas circiter duas in longitudine distensus, in quo oravit Helias propheta, oleis consitus, et arbustis ac vineis, sicut vult Hieronymus in quinto et primo libro super Jeremiam. Post viciniam Samariae ad orientem septentrionalem sequitur planities Sabae nunc vocata, sed antiquitus dicebatur campus magnus Estrelon de quo[1] . . . . Et campus Magido in quo Josias rex optimus fuit occisus . . . . Et per fines ejus ad septentrionem decurrit torrens Fison usque in mare magnum inter Caiphas et Acon. Deinde ad septentrionem illius campi, et ab Acon versus orientem ab ea septem leucis distans est Nazareth beata civitas Domini Salvatoris. Deinde ulterius ad orientem per duas leucas est mons Tabor gloriosus, in quo Dominus suam gloriam discipulis tribus et Moysi et Heliae ostendit.

Lakes of Tiberias and Gennesaret.

Et deinde ad orientem est Tiberias civitas, quae antiquitus Cenereth dicebatur, ut dicit Hieronymus decimo quarto libro super Ezechielem, et prope hoc mare Tiberiados et Cenereth dicitur lacus, super quem sita est civitas, qui secundum Isidorum decimo tertio libro, 'omnibus aquis in Judaea est salubrior, et circuit stadia centum sexaginta, cui conjungitur lacus Genesar amplissimus in Judaea; longitudine clx stadiis extenditur, latitudine xl diffunditur, aquis crispantibus aura non ventis, sed de se ipso sibi excrispans, unde et Genesar dicitur Graeco vocabulo quasi generans sibi auram. Deinde per diffusiora spatia lacus frequentibus auris spirantibus agitatur. Unde et purior haustus ejus et ad potandum dulcis et habilis.' Haec Isidorus, qui distinguit hos lacus quantitate et naturali proprietate, quamvis glossa sexto Matthaei dicit quod idem dicitur stagnum Genesareth, et mare Tiberiadis, et lacus salinarum. Sed lacus salinarum secundum omnes, ut prius dictum est, mare Mortuum vocatur, et ideo haec glossa magistralis magis ex rumore locorum habita est quam ex auctoritate sanctorum vel experientia. Quod tamen dicit unum esse mare Tiberiadis et Genesar, hoc potest referri ad vicinitatem eorum. Nam contigua sunt et cohaerent, et ideo pro uno computantur. Quoniam diversos esse lacus evangelium Johannis ostendit, quia sexto capitulo dicit, quod abiit Jesus

[1] A hiatus occurs here in all the MSS.

trans mare Galilaeae quod est Tiberiadis, et deinde venit in desertum Bethsaidae secundum Lucam; et postea venerunt discipuli Bethsaidam, secundum Marcum; deinde ascenderunt mare ut irent Capharnaum, secundum Johannem, quod non erit proprie mare Tiberiadis, quia ultra illud prius transierant. Ergo hoc est mare Genesareth, et ideo diversa sunt, sed cohaerentia. Deinde versus septentrionem, deserto interjacente, in quo Dominus pavit quinque millia hominum ex quinque panibus hordeaceis et duobus piscibus, est Bethsaida civitas principis apostolorum et Andreae ac Philippi; deinde Capharnaum. Et quod hic sit ordo istorum locorum patet ex evangeliis. Nam ante miraculum de panibus dicit Johannes, abiit Jesus trans mare Galilaeae quod est Tiberiadis; et tunc occurrebat multitudo quam pavit; et post discipuli ascenderunt in navem, ut transfretarent in Capharnaum, sicut dicit Johannes; sed antequam illuc devenerunt, et antequam pavit multitudinem, venit in desertum locum, qui est Bethsaida, cum discipulis suis, ut Lucas refert, et Marcus dicit quod venerunt ad Bethsaidam. Quapropter primo est Tiberias, deinde ultra mare Tiberiadis versus septentrionem est desertum Bethsaidae, et juxta illud est Bethsaida, et postea lacus Genesareth et tandem Capharnaum super oram ejus, et haec omnia docentur in glossa magna sexto Marci. Deinde post Capharnaum est Julias oppidum de quo supra; deinde Caesarea Philippi ad radices montis Libani.

Cities of North Palestine.

Amplius ab Acon in orientem versus septentrionem magis quam Nazareth est Cana Galilaeae, in qua Dominus mutavit aquam in vinum. Et distat Cana ab Acon per quinque leucas. Et inter Canam et Nazareth sunt duae leucae. Item ab Acon in oriens septentrionale per novem leucas est Sapheth civitas Tobiae ultra Cana quasi per quinque leucas. Deinde per leucam et dimidiam est Corazaim civitas, et inter Corazaim et Tiberiadem sunt leucae circiter duae. Et in his locis scilicet Tiberiade, Bethsaida, Corazaim, Capharnaum, Cana, Nazareth, Dominus maxime conversatus est praedicando et faciendo miracula, sicut evangelia referunt. Mons vero Libani extenditur a Paneade seu Caesarea a regione Tyri et Sidonis, et Baruch, et Biblii, et Tripolis, per mille quingenti stadia, ut dicit Plinius, et contra Tyrum descendit aqua de Libano sub terra, et currit prope Tyrum ad unam leucam, et ibi erumpit in puteum latissimum, et altitudinis ad modum turris, qui per canalem descendit in terram et irrigat loca vicina. Et hic puteus est puteus aquarum viventium, qui fluit cum impetu de Libano, de quo in Canticis habetur. Contra vero Tripolim est fons hortorum juxta colles Libani et fluit usque ad montem peregrinum, a quo flumen emanat, et cadit in mare inter Tripolim et Tortosam, prope tamen Tripolim. Sed et ab eodem fonte peregrini currit aquaeductus in Tripolim. Fons hortorum et Libanus distant a Tripoli per tres leucas. Sidon distat a collibus Libani per leucas circiter tres, a monte magno per leucas circiter quinque.

Cities beyond Jordan.

Et cum nunc tactae sunt civitates et montes citra Jordanem et versus mare nostrum, possunt aliqua addi versus Euphratem. Et primo super mare Mortuum est Machaerus, quondam arx secunda post Hierosolymam, ut dicit Plinius, a qua incipit tribus Ruben a meridie; et in septentrione super Jordanem juxta montem Libanum est Pella[1] civitas, quae est ultimus terminus terrae Hebraeorum ultra Jordanem in septentrione prope Caesaream Philippi. Est autem ad orientem finis istius terrae secundum Hieronymum libro Locorum civitas Philadelphia, quae dicitur in Biblia Rabath filiorum Amon. Et in termino regni Seon, et Og regis Basan, est Jaboc fluvius ut dictum est, ubi est Ramoth-Galaad, de quo satis habetur in bellis regum Israel. Circa Pellam et Caesaream Philippi est mons Hermon Caesareae imminens; et e regione Libani ad orientem terrae filiorum Israel, ut dicit Hieronymus libro Locorum, de quo aestivae nives Tyrum ob delicias deferuntur. Sed, ut dicit Plinius, valle interjacente post Libanum, huic par mons exoritur qui Antilibanus nominatur, et est in orientalem partem montis Libani, ut dicit Hieronymus libro Locorum. Et secundum Hieronymum libro Locorum, Galaad mons est ad terminum Phoenicis et Arabiae collibus Libani copulatus ad meridiem, extenditurque per desertum usque ad locum illum, ubi trans Jordanem habitavit olim Seon rex

[1] Pella is twenty miles south of the sea of Tiberias, and sixty miles away from Caesarea Philippi.

Amorraeorum. Et cecidit hic mons in sortem tribuum Ruben, Gad, et dimidiae tribus Manasse: sed et Jeremias loquitur, 'Galaad, tu mihi caput seu principium Libani.' Antilibanus vero, secundum Hieronymum libro Locorum, extendit se circa regiones urbis Damasci, quae cecidit in sortem tribus Manasse, et haec est Damascus, quae in libris Regnorum caput Syriae nominatur, quae distat a Jerusalem per dietas circiter quatuor, ab Acon circiter tres, a Tripoli circiter duas, a Baruch per unam. Deinde a Damasco per dietas circiter septem vel octo in septentrionem est civitas famosa Alap, quae antiquitus mansio fuit Abrahae, quae ab Euphrate distat circiter per duas dietas et ab Antiochia per dietam et dimidiam. Deinde in fine terrae repromissionis ad septentrionem in oriente est et Amath civitas, ut habetur Numerorum xxxiv, de qua etiam scribitur secundo Regum viii, et primo Paralipomenon xviii, et pluribus aliis locis. Et Hieronymus dicit libro Locorum quod diligenter inquisivit de hac civitate et invenit quod vocatur Epiphania. Deinde est Comaga civitas prope Ciliciam, ubi malta bitumen ardens reperitur, quod projectum super militem armatum comburit eum, nec est remedium per aquam nec per aliud liquidum, sed per terram apponendum. Diu vero turbatus est exercitus Romanus et confusus per hoc bitumen projectum in milites, usquequo remedium per terrae pulveres sparsos super locum a bitumine tactum didicerunt. Haec Plinius narrat libro secundo [1].

Division of Syria into provinces.

Postquam civitates, et montes, et aquae, et caetera loca particularia sunt assignata, nunc juxta haec facilius provinciae et regiones capi possunt. Haec autem tota Syria citra Euphratem situm habet porrectum in immensam longitudinem, sed in lato angustior est, ut dicit Isidorus, et Plinius ait quod ejus longitudo a Cilicia usque ad Arabiam tenet ccccl xx milliaria; habet autem multas provincias, quae omnes nomine Syriae continentur. Nam in ea nominatur Syria Comagena, Syria-Coele, Syria Phoenicis, Syria Palaestinae, Galilaea, Samaria, Judaea. Nam hae regiones ad Syriam pertinent Palaestinam secundum auctores. Syria vero Comagena, ut dicit Isidorus decimo quarto libro, nominatur a vocabulo

[1] Pliny, ii. 104.

Comagae urbis quae quondam ibi metropolis habebatur. Haec ab oriente habet Euphratem, a septentrione Ciliciam et Cappadociam, ab occasu mare nostrum, a meridie Syriam-Coele, quae per diphthongum scribitur, et Coele-Syria nuncupatur. In hac est caput et principalis civitas Antiochia in occidente, cui super mare adnectuntur Laodicea, et Ateradum, et vicinae civitates usque ad provinciam Phoenicis, et in oriente est Emath. Nam Hieronymus dicit libro Locorum, quod diligenter investigans invenit quod haec civitas Emath fuerit in Coele-Syria, et hoc dicit Plinius. Et similiter Alap, quae est prope Antiochiam et multum distans a Damasco, quae est in Syria Phoenicis. Habet igitur haec ab occidente mare magnum, a septentrione Syriam Comagenam, ab oriente Euphratem, a meridie Syriam Phoenicis, quae incipit in septentrionali termino montis Libani. Nam Plinius dicit quod porrigitur hic mons usque in Coele-Syriam et hoc est circa Tripolim; in qua sunt Tripolis, et Tyrus, et Sidon, et Acon, usque ad Caesaream Palaestinae.

Nam Plinius dicit, quod in ora Phoenicis est Ptolemais, Phoenicia. quae Acon dicitur, et Caesarea Philippi est de provincia Phoenicis, ut dicit Hieronymus, et totum citra Jordanem usque ad Palaestinam; atque continet montem Libani et Antilibanum et Damascum cum sua regione, et totum ultra Jordanem continet usque Pellam et montem Hermon et montem Galaad, et illas terras filiorum Israel super Jordanem. Quod autem contineatur Damascus in Syria Phoenicis patet. Nam Hieronymus dicit libro Locorum, Damascus est nobilis urbs Phoenicis. Et super Genesim comprehendit Damascum sub Syria Phoenicis, cum dicat, quod Hus filius Aran possedit Damascum, et usque ad Coele-Syriam. Principales tamen civitates Phoenicum sunt Tyrus et Sidon. Nam, ut dicit Isidorus decimo quarto libro, Phoenix Cadmi frater de Thebis Aegyptiorum in Syriam profectus apud Sidonem regnavit, eamque provinciam suo nomine Phoenicen appellavit. Et isti similiter condiderunt Tyrum; a quibus tota terra in circuitu vocatur Phoenicea. Haec tamen in duas partes principales distribuitur, scilicet in regionem Tyriorum, Syriorum, et Sidoniorum, et Aconensium, et totam terram inter Libanum et Tripolim; et alia ejus pars principalis est Syria Damasci, quae civitas Damascus quantum ad regnum Syriae inter Euphratem et montem Libanum usque ad terram Hebraeorum vocabatur caput Syriae. Nomen enim Syriae in tempore regum Israel attribuebatur Damasco et regioni ejus. Haec igitur provincia Syriae Phoenicis habet terram Hebraeorum a meridie et terram Philistinorum; sed terra Philistinorum incipit a finibus Aconensis territorii usque ad turbidum fluvium Aegypti, et antiquitus continebat fere totam terram Judaeorum citra Jordanem.

Judaea. Quia tamen Judaei multa occupabant de regione Philistinorum et coarctabant eos in civitatibus maritimis, scilicet Caesarea, Joppe, Ascalone, et Gaza, et caeteris; ideo distinguendum est hic quod citra Jordanem sunt tres regiones principales Judaeorum, scilicet Galilaea, Samaria, Judaea specialiter dicta, secundum quod xix Matthaei dicitur in glossa. Omnis Judaeorum provincia ad distinctionem aliarum gentium Judaea dicitur, sed specialiter meridiana plaga, in qua Hierosolyma, ad differentiam Samariae, Galilaeae, Decapolis et caeterarum regionum ejusdem provinciae. Hanc autem terram Judaeorum totam trans Jordanem et citra Josephus in antiquitatum libro distinguit in partes et ordinat; quem secutus est Hegesippus tertio libro, et exponit ea quae obscurius apud Josephum reperiuntur. Tota autem regio trans Jordanem vocatur ab eis Pera. Longitudo ejus a Macheronte super mare Mortuum usque ad Pellam prope Caesaream Philippi et montem Hermon; latitudo ejus a Philadelphia usque ad Jordanem. Duas vero partes principales invenimus in ea; una est Decapolis regio continens decem civitates, quarum una est Pella, ut dicit Plinius, et caeterae ei annexae ad meridiem Libani et Antilibani secundum Plinium versus Philadelphiam, quam cingunt, ut dicit, duae tetrarchiae, scilicet Paneas seu Caesarea Philippi ab occidente et Traconitis regio a meridie super Jordanem. Et ideo Decapolis haec est prope Libanum et Caesaream Philippi, et jungitur finibus Tyri et Sidonis, secundum quod Marci viii habetur; 'Exiens Jesus de finibus Tyri venit per Sidonem ad mare Galilaeae inter medios fines Decapoleos.' Et post hanc ad meridiem super Jordanem est Ithuraea seu Traconitis tetrachia, in qua est Phiala fons Jordanis, ut prius dictum est, parum distans a Caesarea, et Gerasenorum regio, cujus caput est Gerasa. Et dicit glossa Marci quinto, Gerasa est urbs Arabiae trans Jordanem juncta monti Galaad in tribu Manasse non longe a stagno Tiberiadis, in quo porci praecipitati sunt. Hic igitur habitabant Geraseni seu Gergaseni, ut ex hoc loco accipitur. Deinde versus meridiem est alia pars Ithuraeae seu Traconitidis regionis, de qua dicit Hieronymus libro Locorum. Traconitis regio sive Ithuraea, cujus tetrarches Philippus juxta evangelium Lucae, est trans Bosram Arabiae civitatem in deserto contra australem plagam, et in septentrionem respicit versus Damascum. Quae Bosra est in deserto trans Jordanem, quae cecidit in tribu Ruben ad orientalem plagam Jericho et extenditur versus Macherontem et terminos Moabitarum. Sed ne fiat quaestio de Bosra, dicendum est quod alia est civitas in Idumaea, de qua dicit Isaias: 'Quis est iste qui venit de Edom, tinctis vestibus de Bosra?' sicut distinguit Hieronymus libro Locorum. Et quod hic Ithuraea extendatur usque prope Decapolim et Caesaream patet per hoc, quod Plinius dicit quod Ithuraea cingit Decapolim.

Regiones vero citra Jordanem sic dividuntur. Nam Galilee. secundum Josephum et Hegesippum primo est tota Galilaea, quae habet a septentrione fines Tyri et Sidonis, ab occasu Aconense territorium cum monte Carmelo, ab ortu Decapolim, a meridie Samariam et Scythopolim superius assignatam. Duplex tamen est Galilaea, una gentium et superior, quae finibus Tyri et Sidonis adhaeret. Quae ideo dicitur Galilaea gentium, quia Solomon dedit in ea Hiram regi Tyri viginti civitates; et ideo quia gentes in hac Galilaea mixtae sunt cum Judaeis, et quia est prope nationes gentium a septentrione, et ortu solis, et occidente, vocatur Galilaea gentium; et haec terminatur circa Tiberiadem, et infra campum magnum Esdrelon. Et in illis locis incipit Galilaea inferior, quae dicitur Galilaea Judaeorum, quae est in tribu Zabulon. Sed hic cavendum est, ne putetur esse trans Jordanem, ut multi aestimaverunt propter verbum Isaiae et evangelium Matthaei, cum dicit, 'terra Zabulon et terra Nepthalim via maris trans

Jordanem, Galilaea gentium.' Sed glossa super Matthaeum dicit Galilaea gentium quae est in tribu Nepthalim vicina Tyriis; et ideo cum tribus Nepthalim sit citra Jordanem, tunc haec Galilaea similiter, et hoc dicit Hieronymus in libro Locorum; et Josephus, et Hegesippus, et omnes volunt hoc. Sed considerandus est modus loquendi in hac parte. Nam pluries invenimus hunc modum loquendi in evangeliis. Quoniam sexto Marci dicitur, 'et ascendentes in navim abierunt in locum desertum, et viderunt eos abeuntes et cognoverunt multi, et pedestres de omnibus civitatibus concurrerunt illuc et praevenerunt eos.' Ubi dicit glossa, non ad aliam maris ripam sive Jordanis pervenerunt, sed transito aliquo freto vel stagno a Domino et discipulis ejus, proximos ejusdem regionis locos indigenae adierunt quo pedestres pervenire potuerunt. Ex quo accepimus quod trans Jordanem hic significat fieri trans ejus aliquam partem, non totum. Similiter cum dicatur sexto Johannis, 'Post haec abiit Jesus trans mare Galilaeae quod est Tiberiadis,' non transivit usque ad aliud littus, ubi est regio Gerasenorum, sed angulum maris transivit ex eadem parte, scilicet citra Jordanem, et ideo hic sicut prius ponitur pars pro toto. Et cum eodem capitulo dicat, 'Venerunt trans mare in Capharnaum,' adhuc in eadem regione et citra Jordanem semper fuerunt. Et ideo non totum mare accipitur, nec fit transfretatio ad aliud littus, sed pars pro tota accipitur ex eadem parte Jordanis: et ita hic, cum dicitur trans Jordanem Galilaea gentium, sumitur pars pro toto. A loco enim Isaiae, in quo haec dicebat, fuit magna pars Jordanis usque ad Galilaeam gentium, quam in eundo de uno loco ad alium oportet transmeare, et ideo dicit trans Jordanem, id est, trans magnam partem Jordanis, quae extendebatur a loco Isaiae usque ad Galilaeam superiorem.

Samaria. Deinde ad meridiem Galilaeae Judaeorum est Samaria, quae non est solum civitatis nomen, sed regionis, quae incipit in campo magno et extenditur usque ad Judaeam, cujus Judaeae latitudo est a Jordane usque ad Joppen secundum Josephum et Hegesippum, cujus longitudo usque ad Bersaben extenditur.

Extent of the — Et nunc in fine solvenda est quaestio notanda, ut sciamus quantum est terra repromissionis et quantum est possessum a Judaeis. Sed Hieronymus certitudinaliter hoc in epistola de terra repromissionis determinat, dicens quod nec David, nec Solomon, nec alii possederunt unquam nisi a Dan usque Bersaben, licet post victoriam multos inimicos et tributarios receperunt. Et hujus terrae longitudo inter Dan et Bersaben vix continet centum et sexaginta milliaria, ut ipse dicit et infert. Pudet dicere latitudinem, nam a Joppe usque ad viculum nostrum Bethlehem sunt quadraginta sex milliaria, et a Bethlehem usque ad Jordanem est circiter una dieta; quapropter parum fuit possessum a Judaeis. Sciendum tamen est, quod hoc quod nunc tactum est, solum fuit possessum citra Jordanem. promised land.

Ultra tamen habuerunt suas possessiones duae tribus et dimidia, ut patet, sicut Hieronymus etiam in hac epistola exponit. Sed repromissum fuit eis ab Euphrate ex parte orientis, usque ad mare nostrum ex parte occidentis, et a Cilicia et Tauro monte a parte aquilonis usque ad fluvium turbidum Aegypti, et usque ad terram Edom, et Moab, et Ammon, ex parte meridiei. Nam octavo libro super Isaiam dicit Hieronymus quod ab Euphrate usque ad rivulum Aegypti fuit Judaeis repromissum, immo usque ad Nilum. Nam ille rivus est prope Nilum. Et Euphrates est ab oriente istius terrae. Rivus autem Aegypti cum mari nostro, in quod influit, est ad occidentem, et in primo libro dicit illud idem. Et addit quod ex parte aquilonis fuit eis repromissum a Cilicia et Tauro monte, et decimo quarto libro super Ezechielem dicit, quod plaga septentrionalis incipit a mari nostro usque ad Zephirum Ciliciae oppidum, et usque ad Taurum montem altissimum, et usque Emath, quae est civitas Coele-Syria, quae nunc Epiphania dicitur, et in occidente durat a torrente Rinocorurae urbis influente in mare magnum super mare istud usque ad eum locum maris qui est contra Emath urbem Syriae, de qua supra. Meridiana vero pars incipit a torrente Aegypti, ubi cadit in mare magnum, ascendendo per desertum Sin et Cades, et per terram Edom, Moab et Ammon, usque ad Euphratem. Nam si Euphrates est in oriente, et mare in quod cadit rivus Aegypti in occidente, tunc meridiana pars extenditur inter illud mare et Euphratem. Hoc sequitur Judaea beyond Jordan.

necessario. Sed non habetur hoc una auctoritate, sed ex multis colligitur et sequitur ex praedictis. Quod enim xxxiv Numerorum et xiv super Ezechielem multis locis habetur, quod mare Cenereth et Jordanis et hujusmodi sunt in oriente, hoc est verum respectu terrae possessae a Judaeis citra Jordanem. Sed ultra Jordanem multa possederunt, ut patet per duas tribus et dimidiam, et plura fuerunt eis repromissa, quoniam usque ad Euphratem.

Jerusalem. [1] In medio Judaeae est Hierusalem variarum opum dives, unde secundum gratiam elementorum putaverunt Judaei eam promissam fluentem lac et mel, cum hinc eis Deus resurrectionis praerogativam pollicetur. Scissura decem tribuum nomen dedit Judaeis, nam ante Hebraei sive Israel nominabantur. Ex quo autem in duo regna Dei populus est divisus, duae tribus quae de stirpe Judae reges habebant Judaei dicti sunt. Reliquae decem tribus quae in Samaria regem sibi constituerunt Israel dictae sunt.

The Tauric range, including Caucasus. Ad caeteras regiones accedendo oportet describere Taurum montem [2], quoniam ipse disterminat regiones infinitas. Ab oriente et Indico mari incipit, et transit in occasum per fines Indiae et Parthorum regna, et Mesopotamiam et Syriam, quas provincias relinquit versus meridiem, et ab aquilone totas Scytharum regiones et partem Armeniae majoris et Cappadociam relinquit et transit in Ciliciam. Sed secundum varietatem regionum diversa sortitur nomina. Aliquando enim Caucasus, ubi altior est propter abundantiam nivium, nam illorum lingua quibus attollitur Caucasus significat candidum; alibi Caspius vocatur, alibi Taurus, et aliquando Hyrcanus; et multis aliis nominibus, pluribus quam viginti, secundum

[1] This paragraph, which is missing in the other MSS., is found in a MS. of Corpus Christi Coll., Cambridge, containing the geographical section of the *Opus Majus*, which I was kindly permitted to collate.

[2] This conception of the continuity of the great Asian chain seems to be due to Eratosthenes. 'He considered it as a vast range of mountains occupying in parts a width of not less than 2,000 stadia, and stretching in a direct line nearly parallel with the equator from the interior of Lycia and Cilicia . . . north of the plains of Mesopotamia and Assyria as well as of the table-land of Persia, and the plains of India, till they ultimately ended in the unknown ocean that formed the eastern boundary of Asia.' Bunbury, *Hist. of Ancient Geography*, vol. i. p. 641. Cf. Pliny, v. 27.

Plinium. Nam novem nomina habet ab oceano Indico in oriente, antequam vocetur proprio nomine Taurus, deinde Caucasus, et postea tria sortitur nomina extranea, et iterum Taurus vocatur. Sed ubi se aperit et portae fiunt Caspiae, ibi vocatur Caspius, et Hyrcanus, et multis aliis nominibus, de quibus non est curandum ad praesens. Haec secundum Plinium, quamvis Orosius velit aliter, et multi alii aliter quam ille. Unde totus ille mons a multis vocatur ab India Caucasus et postea Taurus. Alii e converso, sed non est curandum, quia diversae imaginationes et diversa vocabula hic sunt circa rem eandem. Quod tamen magis est in usu sapientum describentium mundi regiones, est quod vocetur in orientali parte Caucasus, deinde Caspius et Hyrcanus, postea Taurus, et iterum Caucasus quia ubi se maxime attollit vocatur Caucasus, et tum postea adhuc Taurus nominatur. Totus etiam Caucasus dicitur, et totus Taurus secundum diversas considerationes.

Media and Parthia.

Revertendum est igitur ad regiones orientales supra Mesopotamiam, Assyriam et Babylonem, et dicendum est secundum Plinium et omnes, quod ibi sunt regna Medorum et Persarum et Parthorum. Et haec regna habent ab occidente Tigrim fluvium, ab oriente Indum fluvium, a meridie mare Persicum, sive sinum Rubri maris Persicum, a septentrione habent Armeniam et Taurum, et Caucasum montes, et portas Caspias, seu Hyrcanias, et terram Hyrcanorum et mare Hyrcanicum, quod est et Caspium. Nam idem est mare Caspium et Hyrcanium, sicut dicit Plinius. Et Persarum regna nunc Parthorum intelligimus, ut dicit Plinius. Veruntamen illa pars quae est super mare Persicum est prope Persida; nam ab illa mare Persicum nominatur. Sunt enim Parthorum regna octodecim, ut dicit. Undecim dicuntur superiora et septentrionalia, quae incipiunt a confinio Armeniae et Caspiae littoribus seu Hyrcanicis, et hi proprie dicuntur Parthi, et pertinent ad Scythas, cum ex aequo vivunt qui adhaerent montibus et mari Caspiis et Hyrcanicis. Alia septem regna Parthorum sunt meridionalia, et adhaerent mari Persico et proprie vocantur regna Persarum, et illi sunt Elamitae, id est principes Persidis, sicut Hieronymus dicit super Genesim et in libro Locorum. Nam Elam est civitas principalis

Persarum, in qua fuit Susis vel Susa castrum, de quibus habetur octavo Danielis, ubi fuit caput regni Persarum. Et Plinius dicit, quod prope Tigrim et a mari Persico per ccl milliaria est Susa Persarum regia, a Dario Hydaspis filio condita in septentrionali Tigris alveo. Et ibi prope est oppidum ubi mortalium soli aurum in odium contrahunt, idque fodiunt ne cui in usu sit. Medi vero sunt vicini tam Parthis quam Persis. Nam una pars Medorum scilicet septentrionalis est sub Parthis et Caspiis, et incipit recte a portis Caspiis, et in confinio Armeniae. Et ideo habent hi Parthos ab oriente, et a septentrione Armeniam et portas Caspias, ab occidente vero Tigrim, quia Parthi sunt supra eos versus Indum fluvium. Alia pars Medorum meridiana flectit se inter regna Parthica superiora et inferiora, ita quod inferiora, scilicet regna Persidis habeat non recte ab ortu, sed magis a meridie vergente in occasum secundum intentionem Plinii. Nam Media includit utraque regna Persarum et involvit secundum Plinium.

India. Supra vero Indum fluvium versus orientem est tota India usque ad mare [1] Scythicum, quod est ab aquilone, et montes Himanus, Hemodus, et multi qui sunt partes Caucasi; et extenditur usque ad mare Eoum, quod est orientale, et usque ad mare Indicum meridianum, in quod cadit Indus fluvius, ut dicit Plinius, quia jam evanuit mare Rubrum. Unde India habet Indum fluvium ab occidente, et regna Persarum et Medorum; et habet mare Scythicum, montes Caucasum et Taurum a septentrione et Scytharum regna, et Indicum mare a meridie, et Eoum ab oriente; cujus dispositio in principio tacta est in multis, quia ibi est principium habitabilis. Et ideo ab ea fuit incipiendum, ut ulterius curreret stylus per longitudinem habitabilis usque ad occidens per regiones Aethiopum, et iterum ab occidente revertendo secundum longitudinem.

Indus and Ganges. Tactae sunt regiones succedentes prioribus usque in Indiam, de qua adhuc aliqua dicenda sunt. Habet enim flumina maxima, inter quae praecipue sunt Indus et Ganges, de quo loquitur Scriptura. De magnitudine vero Indi, dicit Plinius, Alexander magnus nullo die minus sexcenta stadia navigavit in Indo, nec potuit ante quinque menses enavigare totum,

[1] Reg. has Scythicum: O. Sericum.

adjectis paucis adhuc diebus. Et tamen Ganges major est ut dicit, et hic fluvius, sicut Scriptura dicit, circuit omnem terram Evilat, ubi aurum nascitur optimum. Nam a montibus Caucasi oriens in septentrione dividit Indiam, decurrens ad orientem ubi sunt ejus ostia magna, quibus in mare Eoum, id est, orientale delitescit.

Bragmani vero, de quibus habetur in epistola Hieronymi Bibliae praeposita, sunt in India. Et quia sancti et philosophi et historiae narrant mirabilia de eis plus quam de aliis gentibus, ideo aliqua hic inseram de illis, et praecipue adducam ea quae volo de scripturis beati Ambrosii propter majorem certitudinem. Dicit igitur in epistola ad Palladium de vita Bragmanorum quod habitant juxta flumen Gangem, ubi in mare oceanum quod est Eoum ingreditur, sed viri ultra flumen versus oceanum, foeminae vero citra inhabitant propter insignia castitatis. Nam solum propter spem prolis conveniunt mares cum foeminis certis temporibus, scilicet Julio et Augusto, ut refert sanctus praedictus. At ubi quadraginta cum foeminis suis compleverunt dies, mox ad propria revertuntur. Et cum fuerit alicujus uxor enixa edideritque unum et alterum partum, non transit ulterius ad ipsam ejus maritus, singulis enim filiis loco patrum substitutis per totam in reliquum ab hujusmodi coitu abstinent vitam. Si autem acciderit, ut quisque sterilem sortiatur uxorem, usque ad annos quinque maritus ejus transit et cum uxore propria dormit, quae si gravida per illud tempus omnino non fuerit, mox abstinet penitus ab ea. Et sicut ex hac epistola patet et ex libro principali, quem de vita Bragmanorum scribit beatus Ambrosius, illi habent aerem temperatissimum, ita ut vestimentis non utuntur, sed foliis arborum se tegunt. Nec colunt terras, nec arbores, nec panem habent, nec vinum; sed herbis et foliis et fructibus sponte nascentibus vescuntur, et aquis optimis sitim extinguunt. Sani sunt sine infirmitate, et vitam protendunt longissimam.

Brahmins.

Ad aquilonarem vero partem Indiae, ut dictum est, sunt mare Scythicum et montes illi magni, qui Caucasus et Taurus et multis aliis nominibus vocantur secundum diversitatem locorum et gentium. Et ad occidentem est Persida seu Parthia, et Media. Deinde sub eis ad occidentem est Mesopotamia

The Caspian.

et tota Syria, ut dictum est. Sed in confinio Mediae et Parthiae est porta ferrea Alexandri, quae est civitas denominata a portis, et illae portae dicuntur Caspiae[1], non Caucasiae, ut dicit Plinius. Nam aliae sunt portae Caucasi, ut postea dicetur, quia in littore maris Caspii sunt hae portae. Est enim mare quoddam, quod fit ex concursu maximorum fluminum venientum ab aquilone, et Caspium vocatur, atque Hyrcanicum secundum Plinium. Nam Caspii et Hyrcani super littora illius maris inhabitant; non igitur est hoc mare veniens ab oceano[2], ut Isidorus et Plinius, et omnes auctores occidentales scribunt. In hoc enim casu non habuerunt experientiam certam per se nec per alios, sed ex rumore scripserunt. In libris autem de moribus Tartarorum, ut per fide dignos qui in illis regionibus fuerunt patet, quod hoc mare fit ex concursu fluminum, et est mare satis magnum.

[1] This pass is 'still traversed by the most frequented route from Teheran to Meshed and Herat. The identity of this pass with the one now known as the Sirdar pass, between Veramin and Kishlak in Khowar, has been fully established by modern travellers.' Bunbury, *Anc. Geog.* vol. i. p. 477.

The pass called by Bacon Caucasian, and carefully distinguished by him from the Sirdar pass (cf. p. 364), is described by Marco Polo (i. ch. 4) when speaking of Georgia. 'This is the country beyond which Alexander could not pass when he wished to penetrate to the region of the Ponent, because that the defile was so narrow and perilous, the sea lying on the one hand and on the other lofty mountains impassable to horsemen. Alexander caused a very strong tower to be built there to prevent the people beyond from passing to attack him, and this got the name of the Iron Gate. This is the place that the book of Alexander speaks of when it tells us how he shut up the Tartars between two mountains; not that they were really Tartars, however, for there were no Tartars in those days; but they consisted of a race of people called Comanians and many besides.' 'This,' says Yule, in his note on this passage, 'refers to the Pass of Derbend. still called in Turkish Demir Kápi or the Iron Gate, and to the ancient wall that runs from the Castle of Derbend along the ridges of Caucasus, called in the East Sadd-i-Iskendr, the Rampart of Alexander. Bayer thinks the wall was built originally by one of the Antiochi, and renewed by the Sassanian Kobad or his son Naoshirvan.'

The mediaeval legend, bearing the name of Callisthenes, as to the imprisonment by Alexander of twenty-two wild tribes behind those gates, Gog and Magog among them, is repeated by Aethicus, and by many other mediaeval writers.

[2] This is one of the most important corrections made in the geography of Asia by Bacon, in consequence of the attention paid by him to the reports of the two Franciscan missionaries, Rubruquis and Carpini. Previously to their travels the Caspian had usually, though not always, been held to be a gulf communicating with the Scythic, or Arctic, ocean.

Nam circuitus ejus est quatuor mensium. Et Hyrcania prope est super littus meridionale illius maris in termino Parthiae; et ubi Parthia conjungitur Mediae in portis illius extenditur a portis Caspiis versus orientem, ut dicit Plinius; deinde contra residuum Mediae ad ejus septentrionem, et ad occidentem Hyrcaniae est major Armenia, et eam dividit Euphrates a Cappadocia, ut dicit Plinius, propter quod Cappadocia est ad occidentem Armeniae majoris.

Deinde versus Syriam et mare nostrum est Cilicia, quae dicitur Armenia minor. Unde jacet partim in meridie, partim in occidente Cappadociae, et ejus principium non distat ab Antiochia per duas dietas. Sub Cilicia ad septentrionem super mare comprehenditur Pamphylia, ut dicit Plinius, neglecta gente Isaurica, vel non computata per se propter parvitatem, sed sub eis comprehensa. In Cilicia est Tarsus metropolis ejus, in quo natus fuit beatus apostolus Paulus. Et durat Cilicia a meridie in septentrionem per Tarsum in latitudine circiter quatuor dietas versus Turkiam; nam ad septentrionem Ciliciae est Lycaonia, ubi est Iconium celeberrima civitas, a qua Lycaonia dicitur quasi Iconia; unde princeps eorum vocatur Soldanus Iconii et Turkiae, nam Lycaonia vocatur nunc Turkia. Et a finibus Armeniae usque in Iconium sunt octo dietae. Nomina provinciarum in his regionibus sunt multum mutata propter guerras. Nam Turkia multas terras occupat, quae apud auctores habent nomina antiqua, ut est pars Asiae minoris, et Phrygia, et Lydia. Asia vero major continet plus medietate mundi, et totum praeter Europam et Africam, unde continet hanc Asiam minorem. Et haec nunc vocatur apud Graecos Anatolia, id est orientalis Graecia, in qua est Galatia, unde Galatae quibus scribit apostolus, Ilium, quod et Troja dicitur, illa famosissima civitas. Et sunt aliae multae, ut Ephesus, et septem ecclesiae Apocalypsis, et Nicaea, unde synodus Nicaena, et multae aliae. Et ab Iconio usque Nicaeam sunt viginti dietae in aestate; et ab ea usque ad brachium sancti Georgii, quod apud antiquos vocatur Hellespontus, sunt circiter septem, quod brachium exit a mari inter Italiam et Antiochiam. Et ibi terminatur Asia minor, et hoc ex parte occidentis. A parte vero meridiei habet mare illud

Asia Minor.

quod est inter Italiam et Graeciam et Antiochiam et Aegyptum. Ab oriente habet Phrygiam. Nam, ut dicit Plinius curialiter, Phrygia Troiadi superjecta septentrionali sua parte Galatiae continua est, meridiana Lycaoniae, ab oriente Cappadociam habet, et dicit quod Lydia Phrygiae ex ortu solis vicina est, unde fuit Croesus ditissimus rex Lydorum. Brachium vero sancti Georgii est strictum multum et habet Constantinopolim ab occidente in Europa, et extenditur a mari magno quod est inter Asiam et Aegyptum, Syriam et Italiam, circiter centum leucas versus aquilonem usque ad aliud mare quod vocatur Ponticum, et mare majus. Et istud mare habet formam arcus Scythici, multas regiones disterminans.

The Tartars.

Unde hic incipiunt regiones aquilonares, de quibus philosophi meridiani parum sciverunt, secundum quod Ethicus astronomus refert in suo libro; sed hic perambulavit omnes has regiones, et mare oceanum septentrionale cum insulis suis navigavit. Volo igitur ipsum sequi, et nihilominus libros de moribus Tartarorum, et praecipue fratrem Willielmum [1], quem Dominus rex Franciae, Lodovicus, in Syria existens misit ad terram Tartarorum anno Domini 1253, qui frater scripsit Domino regi situm regionum et marium.

The Black Sea.

Extenditur autem hoc mare majus ab occidente, scilicet a

[1] William de Rubruquis, to whom several allusions have already been made in the course of this work, was a Flemish Franciscan sent by Louis IX to the Emperor of Tartary in 1253, with letters inviting him to cease from his wars of extermination. In Joinville's memoirs we read how, when Louis IX was at Cyprus on his way to Egypt, 'envoia le grant roy des Tartarins ses messages à li, et li manda moult débonnairement paroles. Entre les autres li manda que il estoit prest de li aidier à conquerre la terre sainte, et de délivrer Jhérusalem de la main aus Sarrazins.' The king sent back many presents, including a tent embroidered with Christian emblems, 'et touz les autres poins de la foy. Et ces choses leur envoia-il par deux frères préescheurs qui savoient le Sarrazinois, pour eulz moustrer et enseigner comment ils devoient croire.' Part of the narrative of Rubruquis was published by Hakluyt in 1598, and is to be found in Purchas's, Bergeron's, and other geographical collections. But two-thirds remained unpublished till 1839, when the whole was carefully edited by the Paris Geographical Society (*Recueil de Mémoires*, vol. iv.).

Of Rubruquis, Yule remarks (Marco Polo, vol. i. p. cxxx), 'His narrative in its rich detail, its vivid pictures, its acuteness of observation and strong good sense, seems to me to form a book of travels of much higher claims than *any one series* of Polo's chapters; a book indeed which has never had justice done to it, for it has few superiors in the whole history of travels.'

Constantinopoli in oriens per mcccc milliaria in longum, et in medio ejus coarctatur ex utraque parte in angulos, et in angulo meridiano est castrum et portus soldani Turkiae, quod vocatur Sinopolis. A parte vero aquilonis habet aliud castrum in angulo, quod vocatur Soldaia, et est in provincia, quae nunc Cassaria dicitur vel Cessaria, et sunt trecenta milliaria inter Sinopolim et Soldaiam, et haec est latitudo maris inter illos angulos. Et ista castra sunt duo portus famosi, a quibus transeunt homines a meridianis regionibus ad aquilonares et e converso. Et ab istis castris versus occidentem seu Constantinopolim extenditur mare per septingenta milliaria in longum et latum, similiter ad orientem per septingenta. Et ista provincia Cassaria circumdatur mari a tribus lateribus. Nam in occidente habet partem maris Pontici, ubi est civitas Kersona, in qua fuit sanctus Clemens martyrizatus. Et prope eam est insula, in qua est templum, quod dicitur angelicis manibus praeparatum, in quo corpus sancti sepultum fuit. Et a Kersona usque ad Soldaiam sunt quadringenta castra, quorum quodlibet fere habet proprium idioma. Et sunt ibi multi Gothi, qui omnes loquuntur Teutonicum.

Et a parte meridiei Cassariae extenditur mare Ponticum, et in oriente ejus cadit flumen Tanais in mare, ubi habet latitudinem duodecim milliariorum, ubi est civitas Matrica. Et flumen illud versus aquilonem facit mare quoddam habens septingenta milliaria in longitudinem et latitudinem, nusquam habens profunditatem ultra sex pedes. Et hoc mare est palus Maeotis famosissima, de qua philosophi et historiae et poetae loquuntur. Et flumen Tanais extenditur ultra illam paludem versus aquilonem usque ad montes Riphaeos, qui sunt in ultimo aquilonis, a quibus montibus oritur hoc flumen et descendit per longum terrae tractum in paludem supradictam, faciens eam. Et ultra eam digreditur et fluit in mare Ponticum, ut praedixi. Et hoc flumen famosum dividit Europam ab Asia in locis illis, et palus dicta et plures paludes sunt contiguae, sed quasi pro una computantur; et vocantur paludes Maeotis, vel Maeotidae in adjectivo. Paludes igitur illae, quae vocantur mare illud vadosum, sunt in oriente Cassariae, et pars fluminis Tanais quae est inter paludes et mare Ponticum.

The Don.

Cumanians destroyed by Tartars.

Et haec provincia Cassaria habet ab aquilone vastam solitudinem, quae extenditur a flumine Tanai in oriente usque ad Danubium in occidente, itinere duorum mensium velociter equitando, sicut equitant Tartari ; et hoc est una die quantum est ab Aurelianis Parisius. Unde durat haec terra circiter quatuor menses secundum quod alii homines communiter equitant [1]. Et haec terra fuit tota Cumanorum, qui dicebantur Captac ; sed Tartari destruxerunt eam totaliter et interfecerunt Cumanos praeter partem, quae fugit ad regnum Hungariae, qui sunt ei tributarii, et a Teutonicis dicitur Valana, a Plinio et Isidoro et caeteris Alania occidentalis. Et haec provincia habet Danubium et Poloniam et Hungariam ab occidente.

Russia and Scandinavia.

Et ab aquilone istius provinciae est Russia magna, quae similiter a Polonia in una parte sua extenditur ad Tanaim ; sed in magna sui parte habet Leucoviam in occidente, quae est terra ita magna sicut Alemannia. Ad cujus partem occidentalem sunt multae terrae in circuitu maris cujusdam, quod mare fit per multa brachia maris oceani, quae veniunt per medium Daciae ; et ultra eam versus orientem dilatat se in magnum mare, quod ab occidente habet Daciam [2] et Sweciam. Sed Swecia est ad aquilonem Daciae, declinans aliquantulum in orientem ultra Daciam ; ultra quas ad aquilonem est Norguegia. Deinde, mari magno interjecto, sunt Scotia et Anglia, et mari parvo interposito, Hibernia. Notae sunt hae regiones, sed tango eas propter notitiam aliarum. Si ergo a finibus occidentalibus ex parte aquilonis ascendamus versus orientem, primo est Hibernia, secundo Britannia major, quae continet Angliam et Scotiam, deinde Norguegia, Swecia, Dacia. Et postea orientem versus est magnum mare praedictum, quod vocatur mare orientale, quia

The Baltic.

[1] The distance from Orleans to Paris is seventy miles ; eight weeks of such riding would make the distance from the Don to the Danube 4,000 miles ; about four times the real distance between Buda Pesth and the easterly bend of the Don. Obviously the unit of measurement is extremely lax. But cf. p. 366, in which the Cumani are described as extending eastward far into Central Asia, prior to the Tartar conquests.

[2] *Dacia* is often found instead of *Dania* in mediaeval maps, e.g. in the Catalan map of 1378 given in Lelewel's Atlas, Plate xxix (*Géographie du Moyen Age*, ed. 1850).

oceanus non se extendit ultra illud mare. Sed super latus aquilonare istius maris immediate post angulum Sweciae est Estonia; deinde Livonia versus orientem illius maris; deinde Curonia seu Curlandia declinando ad meridianum latus; postea Prussia magna terra in meridiano latere; deinde Pomerania; postea Lubec, portus magnus et famosus in confinio Daciae et Saxoniae. In medio vero istius maris est insula quaedam quae vocatur Gothlandia. Et super Livoniam ad orientem est Semi-Gallia. Et istas terras, scilicet Estoniam, Livoniam, Semi-Galliam, Curoniam, circumdat Leucovia praedicta, et eam circumdat Russia magna ex utraque parte dicti maris, et terminatur in parte meridiana ad Prussiam et Poloniam. Polonia vero jacet ad meridiem Prussiae; et ad meridiem ejus est Bohemia, deinde Austria. Et ad occidentem terrarum istarum est Alemannia, et postea Francia et Hispania; sed hae sunt notae, facio tamen mentionem propter alias. Ad orientem Austriae et Bohemiae est Hungaria, ad quam occidentalis pars Albaniae descendit. Nam cadit super Danubium qui fluit per medium Hungariae, et ultra eam in mare Ponticum per duodecim ostia magna. In termino vero Hungariae orientalis a parte aquilonis occurrit haec Albania, contra quam ad meridiem Danubii sunt Balchi et Bulgari, et Constantinopolis, quae terrae antiquitus vocabantur Thracia. Albania igitur occidentalis extenditur a Danubio post finem Hungariae versus orientem, usque ad flumen Tanaim, habens Cassariam a meridie et Balchiam, et Bulgariam, et Constantinopolim; ab occasu vero Hungariam et Poloniam, et extremitatem Russiae; ab aquilone habet totam longitudinem Russiae.

Poland and Prussia.

Hungary.

Thrace.

Et ultra Russiam ad aquilonem est gens Hyperborea, quae sic nominatur a montibus magnis, qui vocantur Hyperborei. Et haec gens propter aeris salubritatem vivit in sylvis, gens longaeva usquequo fastidiant mortem, optimarum consuetudinum, gens quieta et pacifica, nulli nocens, nec ab alia gente molestatur. Sed alii confugiunt ad eam sicut ad asylum. Qualiter autem potest regio ibi esse temperatissima, tetigi prius in complexionibus locorum mundi. Et sic habemus regiones aquilonares in Europa notabiles.

The Hyperboreans.

Religions of Northern nations.

Ritus vero istarum gentium sunt diversi. Nam Prusceni, Curlandi, Livonii, Estonii, Semigalli, Leucovii sunt pagani. Alani vero non sunt, quia Tartari terram illam invaserunt, et fugaverunt Cumanos usque ad Hungariam; et Cumani sunt pagani, et Alani fuerunt similiter, sed deleti sunt. Rusceni sunt Christiani et sunt schismatici, habentes ritum Graecorum, sed non utuntur lingua Graeca, immo lingua Sclavonica, quae est una de linguis quae plures occupat regiones. Nam Rusciam, Poloniam, et Bohemiam, et multas alias nationes tenet. Tartari vero a Danubio inhabitant terram Alanorum seu Cumanorum, et ulterius usque fere ad partes ultimas orientis; et alias nationes eis vicinas ad aquilonem et meridiem subjugaverunt pro maxima parte. Nam aliquae gentes sunt in montanis et locis tutissimis, quas non possunt debellare, licet sint eis vicini, quia sunt inexpugnabiles.

Sclavonic languages.

The Don.

Tanais quidem fluvius descendit a montibus Riphaeis altissimis, qui sunt in vero aquilone, nec est ultra illos ad aquilonem habitatio. Et in termino orientali Rusciae et Alaniae, ubi mercatores et alii qui venerunt de Hungaria et Cassaria, et Polonia, et Russia, est quoddam casale, ubi navigio pertransitur flumen Tanais. Et est Tanais ibi ad latitudinem Secanae Parisius. Et ultra flumen illud est Albania superior usque ad aliud flumen magnum, quod vocatur Ethilia, quae major est in quadruplo quam Secana, et est de majoribus fluminibus mundi, et crescit in aestate sicut Nilus[1]. Et a parte aquilonis distat hoc flumen a Tanai per decem dietas, sed versus meridiem multum separantur. Nam Tanais cadit in mare Ponticum, et Ethilia in mare Caspium, et facit illud mare cum aliis fluminibus multis, quae veniunt de Perside, et aliis locis. Nam a Pontico mari secundum Plinium, sunt ccclxxx milliaria usque ad mare Caspium.

The Volga.

Tartar principalities.

Et in hac terra habitabant Cumani, sed Tartari deleverunt omnes, sicut ex altera parte Tanais fecerunt usque ad Danubium, ut dictum est. Et Tartari habent pecora infinita

[1] The Tartar name for Volga is Ethel, which means bountiful. In May and June this river is swollen by melting snow, and inundations result. By Marco Polo the Volga is not spoken of by this name, but under that of Tigeri (cf. i. 9, ed. Yule). See p. 322.

et habitant in tentoriis, non habentes villas nec castra nisi rarissime. Et unus princeps cum exercitu, et cum suis armentis gregum vagatur inter duo flumina, ut unus inter Danubium et Tanaim, alius inter Tanaim et Ethiliam, et sic ultra versus orientem, quod semper divisi sunt per pascua et aquas. Et a Januario incipiunt adire partes aquilonares infra flumina usque ad Augustum, et tunc redeunt versus meridianas propter frigus aquilonis in hyeme. Et versus aquilonem distat Ethilia a provincia Cassariae per unum mensem et tres dietas, sicut equitant Tartari.

Haec vero terra Tartarorum inter Tanaim et Ethiliam habet ab aquilone quasdam gentes. Et primo est gens Arumphea prope montes Riphaeos, quae est similis Hyperboreis in omnibus. Et hae duae gentes sunt prope polum in aquilone; sed remotius ab aquilone ultra Tanaim est primo gens quae vocatur Moxel subjecta Tartaris; et sunt adhuc sine lege pure pagani, civitatem non habent, sed casulas in sylvis. Dominus illorum et magna pars eorum fuerunt interfecti in Polonia per Polonos et Alemannos et Bohemos. Nam Tartari duxerunt eos ad bellum cum Polonis. Et multum approbant Polonos et Alemannos de strenuitate, sperantes adhuc liberari a servitute Tartarorum per eos. Si mercator veniat inter eos, oportet quod ille in cujus domo primo hospitatur, det ei expensas quantum vult ibi morari. Haec enim est consuetudo illius regionis. Post hos ad orientem est gens quaedam, quae vocatur Merduim subjecta Tartaris. Sed sunt Saraceni habentes legem Mahometi. Post eos est Ethilia flumen praedictum, quod descendit a majori Bulgaria, de qua postea dicetur.

Northern tribes subject to Tartars.

A meridie vero istius regionis Tartarorum super mare Ponticum sunt Hiberi et Georgiani. Et in Georgia est metropolis civitas, quae vocatur Thephelis, in qua fratres praedicatores habent domum. Et ulterius versus orientem est terra Corasiminorum, sed deleti sunt a Tartaris. Et in his locis solebant antiquitus esse Amazones secundum Plinium et Ethicum astronomum. Amazones enim, ut refert Ethicus, fuerunt mulieres ducentes exercitum magnum ex mulieribus sine viris collectum, quae advocantes viros certis temporibus

Georgia.

The Amazons of antiquity.

anni conceperunt; sed masculos genitos interfecerunt foeminas reservando; quarum in juventute mamillas dextras per artificium chirurgiae abstulerunt, ne sagittando reciperent impedimentum per mamillas. Et minotauros ac centauros monstra ferocissima nutriverunt a juventute mamillis suis; unde antecedebant eas sicut matres suas et omnem exercitum premebant magis per hujusmodi monstra quam per arma. Et similiter elephantes a juventute nutriebant et assuescebant ad proelia; et sic per centum annos vastabant meridianas partes Asiae et Graeciae, donec ab Hercule fuerunt seductae et destructae. Haec vero loca Georgianorum et Corasiminorum habent terram Soldani Turkiae et Cappadociam a meridie. Nam in latere meridiano maris Pontici est terra Soldani usque ad Sinopolim, de qua prius dictum est. Et post eam in eodem latere maris versus occidentem est terra Vastachii, scilicet Graecia orientalis. Nam occidentalis vocatur ubi Constantinopolis est, et regiones ei annexae citra brachium sancti Georgii in Europa.

Armenia. Sed Armenia major est super Cappadociam ad orientem; et ideo illa Armenia etsi meridiana sit respectu Georgiae, tamen in orientem tendit et extenditur usque Mediam et Mesopotamiam; et haec terra tota aestimatur a multis esse terra Ararath, propter hoc, quod in Isaia dicitur quod filii Senacherib, interfecto patre, fugerunt in Ararath; et in libro regum dicitur quod fugerunt in Armeniam. Sed Hieronymus secundo libro super Isaiam solvit hoc, dicens Ararath regio est Armenia campestris per quam Araxes fluit, incredibilis ubertatis ad radices monti Tauri, qui usque illuc extenditur. Quapropter Ararath non est tota Armenia, sed regio determinata, magna tamen est. Nam Araxes fluvius, a quo Ararath regio nominatur, extenditur a capite suo per iter trium mensium et amplius. Caput autem ejus est fons in monte Armeniae, ubi prope oritur Euphrates ex parte aquilonis et Tigris ad aliud latus montis versus meridiem. In montibus autem Armeniae, testante Scriptura, requievit arca Noe; sed non in quibuscunque montibus, quoniam non est in istis ubi oriuntur ista tria flumina magna, sed in altissimo Tauri montis cacumine, ubi Ararath regio est, secundum quod dicit Hieronymus secundo super Isaiam, arca, in qua liberatus est Noe cessante diluvio; non ad montes generaliter Armeniae, sed ad montes Tauri altissimos, qui Ararath imminent campis.

Destruction of Armenian churches.

Et prope illos montes est civitas, quae maxima fuit antequam eam destruxerunt Tartari. Nam fuerunt in ea octingentae ecclesiae Armenorum; et in tempore fratris Willielmi, quando transivit per eam, non fuerunt nisi duae parvulae. Et ibi prope fuerunt martyrizati beati Bartholomaeus, et beatus Judas, et Thaddaeus. Et ibi sunt duae prophetiae. Una est beati Methodii martyris, qui fuit de gente illa, et plane prophetavit de Ismaelitis, quae prophetia impleta est in Saracenis. Alius propheta eorum vocatur Akaton, qui prophetavit de Tartaris et de destructione eorum. Dicit enim 'quod ab aquilone veniet gens sagittaria, quae omnes nationes orientis subjugabit, et venient in regno occidentis, scilicet ad Constantinopolim, et ibi per principes occidentis destruentur. Et tunc convertentur omnes gentes ad fidem Christi, et erit tanta pax ubique, quod vivi dicent mortuis, Vae vobis, quod non vixistis usque nunc. Et imperator Christianus ponet solium suum in Taurino in Perside.' Et Armeni habent istam prophetiam sicut evangelium. Et haec civitas memorata vocatur nunc Naxuam[1], quae solebat esse caput regni; et est versus partes aquilonares Armeniae. Nam a festo sancti Clementis ascendit frater Willielmus juxta Araxem, ubi terminatur versus aquilonem. Et in festo natalis venit ad illam civitatem, et in octavis epiphaniae recessit ab ea, et per Araxem ivit usque ad caput ejus, ad quod pervenit in secunda dominica quadragesimae; et ideo longe plus est a civitati illa ad ultimum Armeniae a parte meridiei quam ad aquilonem.

Regions between Armenia and Caspian.

Deinde superius ad orientem sunt montes Alanorum et Aas, qui sunt Christiani, et recipiunt indifferenter omnes

[1] This is Naxuana or Nachitschuan on the Araxes. It is now in Russian dominion, about 100 miles to the East of Lake Van, between 60 and 70 miles south east of Mount Ararat. On his return journey from Tartary, Rubruquis after passing through the Caucasian gates at Derbend on the Caspian travelled up the valley of the Araxes. He mentions the town Vaxua (probably = Naxua) of which he says, 'solebant esse in ea octingentae ecclesiae Hermenorum; modo non sunt nisi duo parvulae. Sarraceni enim destruerunt eas.' (See his journal published by the French Geographical Society, vol. iv. 1839, p. 385.)

Christianos tam Latinos quam Graecos, unde non sunt schismatici, et pugnant cum Tartaris; et similiter Alani. Post eos ad orientem sunt Saraceni, qui vocantur Lelgi[1], qui propter terrae fortitudinem pugnant cum Tartaris. Post illos ad orientem sunt portae Caspiae super mare Caspium, quas Alexander magnus construxit in concursu montium. Nam cum voluit expugnare gentem aquilonarem, non potuit propter illius gentis ferocitatem et multitudinem. Et, ut dicit Ethicus, stetit per annum unum et menses tres, ut defenderet se ab eis, et ingemuit quod talis gens pessima fuisset in partibus aquilonis, et exclamavit ad Deum ut apponeret remedium, ne mundus destrueretur per eos. Sed licet non fuit dignus exaudiri, tamen Deus sua bonitate et propter salutem generis humani jussit fieri terrae motum maximum, et montes per stadium distantes conjuncti sunt usque ad latitudinem unius portae.

Alexander's Caspian gates.

Et Alexander tunc fecit fundi columnas aereas mirae magnitudinis, et erexit portas et linivit eas bitumine quod nec igne, nec aqua, nec ferro dissolvi posset, quod adquisivit ex insulis maris, nec potuerunt dirui aliquo modo nisi per terrae motum: et jam dirutae sunt. Nam frater Willielmus transivit per medium earum cum Tartaris. Et est ibi civitas quae vocatur porta ferrea Alexandri: a qua versus orientem incipit Hyrcania super Hyrcanium mare, quod est Caspium, ut superius dictum est. Nam Hyrcania jacet super latus meridionale illius maris et extenditur usque ad terminos Indiae, a cujus latere meridionali sunt Media et Parthia, sicut prius fuit annotatum. Hae vero portae non sunt Caucasiae sed Caspiae, ut dicit Plinius[2], nec Caucasiae sunt Caspiae. Nam Caucasiae distant a Caspiis per ducenta milliaria versus mare Ponticum; et a mari Pontico distant per clxxx milliaria circa partes Hiberiae et Georgiae. Et ista loca cum montibus interjacentibus vocantur claustra Alexandri, per quae cohibuit gentes aquilonares ne irruerunt terras meridianas devastando eas. Nam Alexander multa bella gessit cum eis, ut refert

[1] J. has Belgi: O. Lelgi.

[2] Pliny in the eleventh chapter of his sixth book describes the Caucasian gates: in the fourteenth chapter, the Caspian gates. See notes on pp. 303 and 354.

Ethicus, et aliquando infra tres dies ceciderunt ex utraque parte decies centena millia hominum. Alexander tamen magis arte et ingenio vicit quam armorum potestate. Et cum fuerunt excitati sicut ursi de cavernis suis, non potuit eos reprimere per violentiam, sed Deus juvit per terrae motum et clausuras montium. Nunc autem ruptae sunt, et diu est quod fractae sunt sive terrae motu sive vetustate.

These regions should be carefully examined.

Et considerandum est diligenter de locis istis. Nam Gog et Magog, de quibus Ezechiel prophetavit et Apocalypsis, in his locis sunt inclusi, secundum quod dicit Hieronymus secundo libro super Ezechielem. Gog Scythica gens trans Caucasum et Maeotim et Caspium mare ad Indiam usque tenduntur; et a principe Gog omnes qui subditi sunt Magog appellantur et Judaei similiter, quos Orosius et alii sancti referunt exituros. Atque sicut Ethicus scribit, Alexander inclusit viginti duo regna de stirpe Gog et Magog, exitura in diebus Antichristi, qui mundum primo vastabunt et deinde obviabunt Antichristo, et vocabunt eum Deum Deorum, sicut et beatus Hieronymus confirmat. O quam necessarium est ecclesiae Dei, ut proelati et viri catholici haec loca considerent, non solum propter conversionem gentium in illis locis, et consolationem Christianorum captivorum ibidem, sed propter persecutionem Antichristi, ut sciatur unde venturus sit, et quando, per hanc considerationem et alias multas.

The Caspian is an inland sea.

A portis vero Caspiis incipit mare Caspium[1] extendi in longum ad orientem; et in latum ad aquilonem, et est non minus quam Ponticum mare, ut dicit Plinius, et habet spatium quatuor mensium in circuitu. Frater vero Willielmus in

[1] The changes of opinion in antiquity as to this sea are curious. Herodotus and Aristotle held the true view that it was surrounded by land. See Arist. *Meteor.* ii. 1, § 10. But Patrocles, an officer of Antiochus I, maintained that it was a gulf of the northern ocean, and that it would be possible to sail from it round to the Indian sea (Strabo, xi. 11, § 6). Pliny went so far as to say that Patrocles had actually accomplished this voyage (*Hist. Nat.* vi. 17). And notwithstanding that Ptolemy reverted to the belief that the Caspian was an inland sea (*Cosmographia*, vii. 5, § 4), Pliny's view continued to be reasserted through the Middle Ages, till Bacon, on the authority of Rubruquis, confirmed the fact that it was surrounded by land. Bacon however did not free himself from Ptolemy's error of supposing its greatest length to be from east to west, rather than from north to south.

redeundo ab imperatore Tartarorum circuivit latus occidentale; et in eundo ad eum perambulavit latus aquilonare, ut ipse retulit Domino regi Franciae qui nunc est, anno Domini 1253. Et a parte aquilonis habet vastam solitudinem, in qua sunt Tartari. Et ultra eos sunt multae regiones aquilonares antequam perveniatur ad Oceanum; et ideo non potest illud mare esse sinus maris Oceani, quod tamen fere omnes auctores scribunt; sed experientia hujus temporis facta per fratrem Willielmum et homines alios fideles docet, quod non venit a mari, sed fit per flumina magna et multa, quorum congregatione fit hoc mare Caspium et Hyrcanicum. Tota vero haec terra Tartarorum a Tanai usque ad Ethiliam fuit Cumanorum, qui vocabantur Canglae, qui omnes sunt deleti per Tartaros. Et tota ista vocabatur Albania antiquitus. Et hic sunt canes maximi, ita ut leones perimant, tauros premant, et homines ponunt eos in bigis et aratris.

The third and principal of the Tartar kingdoms; Caracathaia.

Deinde ultra Ethiliam est tertius principatus Tartarorum; et destructae sunt gentes indigenae ab eis, et fuerunt Cumani Canglae, sicut prius. Et durat ille principatus a flumine in orientem per iter quatuor mensium ex parte meridionali usque veniatur ad terram principalem imperatoris. Sed ex parte aquilonari durat per iter duorum mensium, et decem dierum. Ex quo patet quod Cumania fuit terrarum maxima. Nam a Danubio usque ad terram hanc, in qua imperator residet, habitabant Cumani, qui omnes sunt destructi a Tartaris praeter eos qui fugerant ad regnum Hungariae. Et hic principatus habet ab aquilone primo majorem Bulgariam[1], a qua venerunt Bulgari qui sunt inter Constantinopolim, et Hungariam, et Sclavoniam. Nam haec quae in Europa est minor Bulgaria, habens linguam illorum Bulgarorum qui sunt in majori Bulgaria, quae est in Asia. Et isti Bulgari de majori Bulgaria sunt pessimi Saraceni. Et hoc est mirabile; quoniam illa terra distat a porta ferrea seu a portis Caspiis triginta diebus et plus per transversum solitudinis; et est in fine

Great Bulgaria.

[1] Marco Polo (i. 9, ed. Yule) mentions the city of Bolgara. Yule remarks on this, 'It was the capital of the region sometimes called Great Bulgaria, on the Volga. . . . latitude about 54° 54′, 90 miles below Kazan. It was captured by the Mongols in 1225.'

aquilonis; unde mirum est valde quod ad eos tam distantes a Saracenis pervenit secta Mahometi. Et de hac Bulgaria venit Ethilia, de qua dictum est. Post eam ad orientem est terra Pascatyr quae est magna Hungaria, a qua exiverunt Huni, qui postea Hungri, modo dicuntur Hungari; qui colligentes secum Bulgaros et alias nationes aquilonares ruperunt, sicut dicit Isidorus, claustra Alexandri. Et solvebatur eis tributum usque in Aegyptum, et destruxerunt terras omnes usque in Franciam; unde majoris potentiae fuerunt quam adhuc sunt Tartari; et magna pars eorum resedit in terra quae nunc vocatur Hungaria, ultra Bohemiam et Austriam, quae est modo apud Latinos regnum Hungariae. Et juxta terram Pascatyr sunt Balchi[1] de Balchia majore, a quo venerunt Balchi in terra Assani inter Constantinopolim, et Bulgariam et Hungariam minorem. Nam populus ille dicitur nunc a Tartaris Ilac, quod idem est quod Blac. Sed Tartari nesciunt sonare *b* literam. A parte vero meridiei istius solitudinis Tartarorum est mare Caspium. Et deinde montes Caucasi usque in orientem.

Et extenditur hic principatus ab Ethilia usque in Cathaiam nigram, unde vocatur Caracathaia. Et Cara idem est quod nigra; et dicitur Cathaia nigra ad differentiam alterius Cathaiae, quae est longe ultra hanc ad orientem multis regionibus distantem ab ista Cathaia nigra; de qua postea dicetur. Et haec terra cum sibi adjacentibus est terra imperatoris Tartarorum praecipua, in quibus vagatur semper cum curia sua, ascendendo in aestate ad frigidas regiones, et in hyeme ad calidas. Et haec Cathaia nigra fuit terra Presbyteri Johannis, sive regis Johannis, de quo tanta fama solebat esse, et multa falsa dicta sunt et scripta.

Hic vero oportet exprimi originem Tartarorum, non solum propter evidentiam distinctionis regionum majorem, sed propter ipsam gentem, quae nunc famosissima est, et subdit mundum pedibus suis. Sciendum igitur est, quod quando bellum fuit Antiochiae tunc regnavit in terra ista Coir Cham[2]. Origin of the Tartar power.

[1] J. has Blaci and Blacia; O. has Balchi and Balchia.

[2] Yule, in his notes on Marco Polo, i. 46, remarks that during the last two-thirds of the twelfth century there was a great empire in Northern Asia called Kara

Nam in historia Antiochena legitur, quod Turci miserunt pro succursu contra Francos ad regnum Coir Cham, qui tenebat monarchiam in regionibus aquilonis tempore illo quo capta fuit Antiochia, qui fuit de Caracathaia. Coir vero est nomen proprium, et Cham est nomen dignitatis, et sonat idem quod divinator. Nam principes ibi regunt populum per divinationes et scientias quae instruunt homines in futuris, sive sint partes philosophiae, ut astronomia et scientia experimentalis, sive artes magicae, quibus totum oriens est deditum et imbutum.

Prester John.

Omnes igitur imperatores Tartarorum vocantur Cham, sicut apud nos vocantur imperatores et reges. Mortuo vero isto Coir, fuit quidam pastor Nestorinus in terra illa potens et dominus super populum, qui populus vocatur Naiman, qui erant Christiani Nestorini, qui sunt mali Christiani, et tamen dicunt se esse subjectos Romanae ecclesiae. Et isti Nestorini non solum sunt ibi in terra Naiman, sed per omnes regiones usque in orientem sunt dispersi. Iste vero pastor erexit se in regem, et vocatus est Presbyter et Rex Johannes. Huic Johanni erat frater quidam pastor potens, nomine Unc [1], habens sua pascua ultra fratrem suum per iter trium septimanarum, et erat dominus cujusdam villae, quae dicitur Caracarum [2], quae

Khitai: founded by a Prince of the Khitan dynasty of Liao, who had escaped from North China on the overthrow of that dynasty by the Kin or Niuché about 1125. His empire extended over Eastern and Western Turkestan. He took the title of Gur-khan, said to mean universal khan; and fixed his centre of government at Bala Sagun north of Thian San. This Gur-khan is the Coir Cham of Rubruquis.

[1] Prester or Presbyter John is a shadowy and almost mythical personage whose *habitat* shifts between Abyssinia and the wall of China. Marco Polo (i. 46) identifies him with Unc Cham. Rubruquis, a somewhat earlier authority, regards him as his brother. Unc, or Ung, Cham, says Yule, is called Tuli by the Chinese and Togrul by the Persian historians. The Kin sovereigns of North China had conferred on him the title of Wang (=king) of which probably Aung or Ung is a corruption. He was the king of the Keraites who had professed a nominal form of Christianity since the eleventh century. The wide diffusion of Nestorian Christianity through Central Asia is very emphatically marked in Rubruquis's narrative. It was a somewhat colourless creed.

[2] Near the upper course of the river Orkhon, about 250 miles south of lake Baikal. After the overthrow of Prester John, Okkodai the successor of Chinghis established his capital here. 'It continued to be the Mongol head-quarters till 1256, when Mangu Khan decided to transfer the seat of government to Kaiping, far north of Peking.' [Yule's ed. of Marco Polo, vol. i. note to p. 204.]

est nunc civitas imperialis, et major in terra imperatoris, et tamen non est ita bona, sicut sanctus Dionysius juxta Parisius in Francia, sicut scripsit frater Willielmus domino regi. Et ultra pascua illius circiter duodecim dietas erant pascua Moal, qui erant pauperes homines et stulti ac simplices sine lege. Juxta quos fuerunt alii pauperes, qui vocabantur Tartari, similes eis. Mortuo igitur Johanne rege elevavit se in regem Unc frater ejus, et vocavit se Cham, unde Unc Cham dicebatur qui misit armenta sua versus terminos Moal.

Chinghis Khan.

Inter quos Moal erat faber quidam nomine Cingis[1] furans animalia et rapiens ipsius Unc Cham, quo congregante exercitum Cingis fugit ad Tartaros qui dixit ad eos et ad Moal, Quia sumus sine duce ideo vicini nostri opprimunt nos. Et factus est dux eorum; et congregato exercitu irruit super Unc Cham et vicit eum, et factus est princeps in terra, et vocavit se Cingis Cham, et accepit filiam Unc et dedit eam filio suo in uxorem, ex qua natus est Mangu Cham qui regnum divisit istis principibus Tartarorum, qui nunc regnant et discordant ab invicem. Ad hunc enim Mangu Cham fuit frater Willielmus missus. Et ipse Cingis Cham ubique praemittebat Tartaros in pugna, unde exibat fama Tartarorum, qui fere deleti sunt per crebra bella. Et licet propter hanc causam nos istam gentem vocemus Tartaros apud quos est imperium et dominium, tamen semper sunt imperatores et principes de gente Moal. Nec volunt vocari Tartari, sed Moal; quia primus eorum imperator, scilicet Cingis Cham, fuit natione Moal. Unde non habuerunt adhuc ante istos

[1] Cingis, or Chinghis, was born in 1155. His original name was Temújin. The resemblance of the name to Temurji, the Turkish word for blacksmith, probably gave rise to the legend of his origin (Marco Polo, ed. Yule, i. 234). In the list of Mongol sovereigns Bacon, following Rubruquis, omits Chinghis's son, Okkodai, who succeeded him. Okkodai was succeeded by his son Kuyuk, the Keu of Rubruquis; Kuyuk by his cousin Mangu, a son of Tuli, and grandson of Chinghis. Mangu was followed by his brother Kublai, in whose time Marco Polo visited the Tartar court. Another grandson of Chinghis, Batu son of Juji, held dominion between the Volga and the Don at the time of the Franciscan mission; he had led the great Mongol invasion of Europe 1240–2. It need hardly be said that Moal is Bacon's reading, after Rubruquis, for Mongol.

The war between Chinghis and Ung Cham is described at somewhat greater length by Marco Polo.

qui nunc regnant, nisi tres reges, scilicet Cingis Cham, et Keu Cham, et Mangu Cham: qui Keu fuit filius Cingis, et Mangu filius ejus.

Present extent of Tartar dominion.

Haec igitur gens Moal est stultissima ab origine prima et pauperrima, quae tamen permissione divina paulatim omnes nationes vicinas subjugavit, et totam mundi latitudinem in parvo tempore prostravit. Quae si esset concors, primo egressu Aegyptum et Africam vastaret, ut sic ex omni parte Latinos circumdaret. Nam nunc a parte aquilonis regnant usque in Poloniam, quia tota Ruscia est eis subjecta; et tota terra ab oriente usque ad Danubium et ultra Danubium, scilicet Bulgaria et Blachia sunt eis tributariae. Ita quod usque ad terram Constantinopolitanam tenet eorum imperium. Et Soldanus Turkiae, et rex Armeniae, et princeps Antiochiae, et omnes principes in oriente usque in Indiam sunt eis subjecti, praeter paucos, qui aut nimis distant aut habent loca in montanis tutissima quae expugnari non possunt. Primo igitur in terra ubi imperator moratur est Cathaia nigra, ubi fuit Presbyter Johannes. Post eam terra fratris sui ultra per iter trium septimanarum, deinde terra Moal et Tartarorum ultra eos per iter quasi duodecim dietarum. Sed tota haec terra est in qua moratur imperator vagans per diversa loca. Terra tamen, in qua primo fuerunt Moal, vocatur Oznam Kerule; et ibi est adhuc curia Cingis Cham; sed quia Caracarum ejus cum regione fuit prima adquisitio eorum, ideo civitatem illam habent pro imperiali, et prope illam eligunt suum Cham, id est imperatorem. Deinde post Moal et Tartaros ad orientem sunt fortes homines, qui vocantur Tangut [1], qui ceperunt primo Cingis Cham in bello; sed, pace facta, iterum subjugavit eos. Isti homines habent boves fortissimos, habentes caudas plenas pilis, sicut equi, quorum vaccae non permittunt

[1] 'Tangut is a name very conspicuous in the Mongol era. The name Tanggod is properly a Mongol plural designating certain tribes of Tibetan blood called by the Chinese Tanghiang, who established an independent kingdom on the North-West frontier of China with their capital at Ninghia on the Yellow River. . . . This kingdom, called by the Chinese Hia and by the Mongols Tangut, was several times invaded by Chinghis, and it was on the campaign of its final conquest that he died in 1227. In a general way Tangut corresponded to the modern province of Kansuh.' [Yule's ed. of Marco Polo, vol. i. note to p. 186.]

se mulgeri, ni cantetur eis; et si isti boves vident hominem indutum rubris pannis, insiliunt in eum volentes interficere.

Post istos ad orientem sunt homines qui vocantur Thebeth, qui solebant comedere parentes suos causa pietatis, ut non facerent eis alia sepulchra nisi viscera sua, de quibus scribunt philosophi, ut Plinius et Solinus, et alii, et frater Willielmus testatur in libro suo, ac frater Johannes de Plano Carpini[1] similiter in libro quem composuit de Tartaris, inter quos fuit anno Domini 1246, missus a Domino papa in legationem ad imperatorem Tartarorum. Sed quia fuerunt ex hoc facto abominabiles omni nationi, ideo jam mutaverunt hunc ritum, et tamen adhuc de ossibus capitum faciunt scyphos, in quibus bibunt propter memoriam parentum suorum. Post istos ad orientem sunt parvi homines et fusci sicut Hispani, et dicuntur Solangi, quorum nuntii quando veniunt ad curiam alicujus habent in manu tabulam de ebore, in qua inspiciunt quando narrant ea quae volunt ac si omnia essent ibi scripta. Et ultra istos est quidam populus, quorum animalia non appropriantur alicui nec habent custodem, sed si aliquis indigena vult habere aliquod animal, stat super collem et vociferat sicut vult, et veniunt ad vocem suam, et accipit quod vult. Quod si extraneus veniret, odore suo fugaret omnia et faceret ea sylvestria. Et ideo quando homines veniunt extranei, claudunt eos in domo, et dant eis necessaria vitae, donec habeant

Nations to east of Tartars.

[1] Carpini was a native of Perugia, and was amongst the earliest disciples of St. Francis, who sent him in 1221 to Germany to assist Caesar of Spires as provincial of that country. At the Council of Lyons, held by Pope Innocent IV in 1245, it was decided to send a mission to the Tartar emperor (at that time Kuyuk, the grandson of Chinghis) inviting him to cease from slaughter and adopt Christianity. Carpini was charged with the mission. Starting in the spring of 1246 from Poland, he reached the Khan's headquarters near Kara Korum on July 22. The story of his journey is full of interest, though not equal to that of Rubruquis. It is to be found complete in the fourth volume of the Memoirs of the Paris Geographical Society, 1839. Carpini devotes much attention to the military problem of arresting Tartar invasions. His description of the marvellous manœuvring of their huge hordes of cavalry is most vivid. He was much struck by their discipline and loyalty, and their abstinence from internal quarrels. 'Sunt magis obedientes dominis suis quam aliqui homines qui sunt in mundo sive religiosi sive saeculares, et magis reverentur eosdem; neque de facili mentiuntur eis. Verbis ad invicem raro aut nunquam contendunt; factis vero nunquam.'

responsum de eis pro quibus veniunt; nec permittunt eos vagari per regionem.

Great Cathay, or China.

Sed ultra hos est magna Cathaia, quae Seres dicitur apud philosophos, et est in extremitate orientis a parte aquilonari respectu Indiae, divisa ab ea per sinum maris et montes. Et hic fiunt panni serici optimi, et in magna copia, unde ab hac terra deferuntur ad alias regiones. Et hic populus aspirat multum per nares, et sunt optimi artifices in omni arte, et sunt boni medici apud eos in omnibus praeterquam de urina, cujus judicio non utuntur, sed per pulsum et alia signa optime dijudicant, et bene cognoscunt vires herbarum, et totius medicinae potestatem. Multi ex eis sunt apud Tartaros. Et istorum Cathaiorum moneta vulgaris est carta de Bambasio[1], in qua imprimunt quasdam lineas. Nec mirum, cum Rusceni qui prope nos sunt habent pro moneta faciem Hesperiolorum. Et ista Cathaia non distat per octo et viginti dietas[2] a terra in qua moratur imperator. Et in illa terra sunt rupes exaltae, in quibus habitant quaedam creaturae habentes per omnia formam humanam, non tamen genua flectunt, sed ambulant saltando; sed non sunt longitudinis majoris quam cubiti, et vestitur totum corpus crinibus, et non loquuntur. Et venatores portant cerevisiam et faciunt foveas in rupibus ad modum scyphorum, et illa animalia veniunt et bibunt cerevisiam et inebriantur, et dormiunt, et sic capiuntur. Et venatores ligant eis manus, et pedes, et aperiunt venam in collo, et extrahunt tres vel quatuor guttas sanguinis, et dissolvunt ea ac dimittunt ea abire, et ille sanguis est pretiosissimus pro purpura.

Religions of Tartary.

Sciendum vero est quod a principio Cathaiae nigrae usque in finem orientis sunt principaliter idololatrae[3], sed mixti sunt

[1] Not Gombasio, as in J. For an account of the Chinese paper-money see Marco Polo, Book ii. ch. 24. It was made of the inner bark of the mulberry tree. Its use began in the ninth century A.D. and rapidly extended.

[2] 28 days journey at 20 miles would come to 560 miles, which is nearly the distance between Caracorum and the nearest point of the Great Wall. The reading of some of the MSS., cxxviii, is quite inadmissible.

[3] Marco Polo, like Rubruquis and Bacon, usually denotes Buddhists by this term. We do not know, says Yule (Marco Polo, vol. i. p. 187), whether the Buddhism found here at this time was a recent introduction from Tibet, or

inter eos Saraceni et Tartari, et Nestoriani, qui sunt Christiani imperfecti habentes patriarcham suum in oriente, qui visitat regiones et ordinat infantes in cunabulis ad sacros ordines, quia ipse solus ordinat, et non potest venire ad unum locum nisi quasi in quinquaginta annis; et ille dicit se habere auctoritatem a Romana ecclesia ab antiquo, et paratus est obedire, si via esset aperta. Et isti docent filios nobilium Tartarorum evangelium et fidem et alios quando possunt; sed quia parum sciunt, et sunt malorum morum, ideo Tartari despiciunt eos. Et consecrant in missa unum panem latum ad modum palmae, et dividunt primo in duodecim partes secundum numerum apostolorum; et postea illas partes dividunt secundum numerum populi, et sacerdos dat unicuique corpus Christi in manu sua, et tunc quilibet assumit de palma sua cum reverentia. Sed idololatrae praevalent in multitudine in omnibus his regionibus, et omnes conveniunt in hoc, quod habent templa sicut nos et campanas magnas. Et ideo ecclesiae Graecorum, Armeniae, et totius orientis nolunt habere campanas propter idololatras. Rusceni tamen habent et Graeci in Cassaria. Omnes sacerdotes eorum radunt totum caput et barbam, et servant castitatem ex quo radunt caput, et vivunt centum vel ducenti in una congregatione. Diebus quibus intrant templum ponunt duo scamna, et sedent e regione chorus contra chorum in terra, habentes libros in manibus, quos aliquando deponunt super illa scamna; et habent capita discooperta quamdiu sunt in templo, legentes in silentio, et nullo modo loquerentur in templo, nisi verba officii sui. Habent etiam in manibus quocunque vadunt quandam chordam centum vel ducentorum nucleorum, sicut nos pater noster, et dicunt semper haec verba, On man baccan, id est Deus tu nosti[1]. Haec sunt communia omnibus idololatris. Sed tamen Ingeres, qui habitant in terra ubi imperator moratur, differunt ab aliis. Nam alii non ponunt unum Deum, sed plures, et creaturam adorant. Isti vero

a relic of the old Buddhism of Khotan and other Central Asian kingdoms, but most probably it was the former.

[1] This is evidently a perversion of the Buddhist formula, often imprinted in flags and revolved in prayer-mills, Om Mani Padme Hum, Ah, the jewel is in the lotus (the self-creative force is in the Cosmos).

propter viciniam Christianorum et Saracenorum ponunt unum Deum. Et sunt optimi scriptores, unde Tartari acceperunt literas eorum, et illi sunt magni scriptores Tartarorum. Et isti scribunt a sursum in deorsum, et a sinistra in dextram multiplicant lineas et legunt. Thebeth scribunt sicut nos, et habent figuras similes nostris. Tangut scribunt a dextra in sinistram, sicut Arabes, sed multiplicant lineas ascendendo. Cathai orientales scribunt cum punctorio quo pingunt pictores, et faciunt in una figura plures literas comprehendentes unam dictionem, et ex hoc veniunt characteres qui habent multas literas simul; unde veri characteres et physici sunt compositi ex literis, et habent sensum dictionum.

Various modes of writing.

Et tota terra a Danubio usque in orientem vocatur apud antiquos Scythia, a quo Scythae et omnes regiones Tartarorum sunt de Scythia, et etiam Ruscia, et totum usque ad Alemanniam.

Brief notice of Southern Europe.

Descripsi igitur regiones omnes Asiae et Africae, et aquilonares Europae. Nunc orientales et meridianas Europae et occidentales breviter annotabo. Nam fere omnes notae sunt omnibus. Dictum quidem est quod Albania occidentalis terminatur ad Danubium sub mari Pontico, et ad Hungariam minorem extenditur. Sed ex altera parte Danubii sub eodem mari antiquitus prima regio occurrens dicebatur Thracia, in qua est Constantinopolis. Cui ad occidentem super Danubium jungebatur Moesia. Sed nunc in illis locis sunt Balchia et Bulgaria minor. Deinde Hungaria ad occidentem. Et postea Moravia, quae est sub regno Bohemiae. Cui a meridie est Histria, quibus in occidente succedunt Bohemia Moraviae, et Austria Histriae. Deinde tota Alemannia ad occidentem, et postea Francia, quae notae sunt.

Greece.

Post Thraciam versus meridiem est Macedonia regibus inclyta maximis, Antigono, Philippo, Alexandro magno; quae alio nomine dicebatur Emathia. Post eam ad meridiem est Magnesia; deinde Thessalonia, quibus apostolus scripsit. Deinde adhuc ad meridiem est Boeotia, ubi Thebae civitas famosa, a qua in orientem xviii milliaribus est civitas nota, quae Niger Pons vocatur. Sub istis terris ad occidentem primo contermina Macedoniae Thessaloniae et Bulgariae est

Sclavonia. Sub Boeotia est Attica dicta ab Athenis civitate, quae est nutrix philosophorum, ubi Socrates et Plato et Aristoteles et caeteri famosi docuerunt. Et haec est Attica pars Arcadiae secundum Ethicum philosophum, quae est illustrissima provincia non solum propter Athenas, quam vocat Ethicus umbilicum Graeciae, sed propter potentiam militarem. Nam haec a Scycione rege dicebatur Scycionia, unde regnum Scycioniorum dicitur, unum de quatuor principalibus regnis mundi quae fuerunt a principio, scilicet regnum Scytharum sub Reu atavo Abrahae exortum; et regnum Aegyptiorum sub Seruch proavo Abrahae: et regnum Assyriorum, et Scycioniorum sub Nachor avo Abrahae; sicut scribit Beda libro Temporum; et hoc est regnum Graecorum. Nam, ut ait Ethicus, tota Graecia conspiravit in nomen Scycioniae quoniam regnum Graecorum dicebatur regnum Scycioniorum, propter hoc quod principale robur militiae fuit ad Scycionios, qui sunt Arcadi. Haec gratis expressi abundantius, quia historiae omnes faciunt magnam mentionem de regno Scycioniorum; et apud solum Ethicum philosophum manifeste potest inveniri ratio vocabuli, et quid sit quod per nomen designatur. Secundum vero Plinium postea est Achaia, postea Peloponnensis provincia, in qua est Corinthus nobilis. Deinde Locris et postea Epirus, et sic est terminus Graeciae.

Sub qua ad occidentem est Dalmatia, in qua Duracium. Et sub ea Illyria, a qua dicitur mare Illyricum, quod extenditur a flumine Arsiae usque ad flumen Dirinum, quod totum alio nomine Libnia dicitur. Sed proprie circa flumen Dirinum sunt Illyrii, et ejus longitudo est a flumine Arsia usque ad flumen Dirinum; et jacet super mare Adriaticum, quod est mare Venetiarum. In cujus extremitate occidentali sunt Venetiae. Et habet haec Illyria multas parvas insulas, et habet haec longitudo dxxx milliaria, latitudo vero cccxv milliaria. Haec de Illyrico recitavi verbo ad verbum secundum Plinium, propter hoc, quod nos moderni nescimus intelligere verbum apostoli, cum dicit se replevisse evangelium usque ad Illyricum. Et in multis historiis invenimus Illyricum et Illyricos, et non intelligimus quae dicuntur; unde Illyrii jacent inter Dalmatiam et Histriam. Et ideo ubi nunc

The Adriatic, Italy, Spain.

Sclavonia est et forum Julii et circa partes Venetiarum solebant esse Illyrii. Et hae omnes nationes sunt inclusae inter brachium sancti Georgii, et mare magnum ab oriente; et inter Danubium qui vocatur Hister in magna sui parte a septentrione; et inter mare Adriaticum a meridie. Distantia vero penes milliaria et dietas potest in aliquibus notari. Nam a Venetiis per littus maris sunt plus quam cccc milliaria usque ad Duracium. Deinde superius ascendendo usque ad Patras civitatem famosam sunt xl, a qua usque Corinthum lx, a qua usque ad Athenas xl, a qua usque Thebas xl, deinde usque Nigrum pontem xviii, a quo per mare usque Constantinopolim quingenti milliaria; et a Nigro ponte usque Cretam insulam ccc milliaria. Deinde ex altera parte maris Adriatici, inter ipsum et mare quod ab Adriatico decurrit in Hispaniam, jacet tota Italia, et deinde Provincia, et tandem Hispania. Sed quia notae sunt, non oportet nunc plus de istis regionibus dici. Haec igitur est historia, quam volui hic texere secundum experientiam naturalium et itinerantium de locis et gentibus totius habitabilis, usquequo Vestra Reverentia requirat principalem scripturam [1].

Outlines of an astrological treatise.

Post locorum descriptionem deberent sequi alia quatuor, scilicet certificatio naturarum stellarum fixarum et planetarum, ut certior haberetur per haec certitudo complexionum omnium locorum et rerum locatarum, quatenus posset fieri judicium de praesentibus, praeteritis, et futuris, ut tandem fierent, quinto [2], opera quae promoverent omnia rei publicae utilia et nociva excluderent. Sed cum non potui propter impedimenta

[1] 'Usquequo . . . scripturam' omitted in C. C. C. (Cambridge) MSS. In O. are the words: 'Finitur quarta pars majoris operis.' Nevertheless the MS. proceeds to the astrological section that follows, and indeed divides this section into chapters as though it were a distinct treatise. The reference to *principalem scripturam*, so often spoken of before, will be noted.

[2] The four previous points are (a) the nature of the zodiacal signs; (b) of the planets; (c) the influence of each over places and things; (d) judgement as to past and future events founded on these facts.

perficere descriptionem locorum in figura secundum vias naturalium, et haec quatuor sequuntur, oportuit ab horum tractatu cessare; volo tamen hic in summa tangere intentionem tractatus horum quatuor, ac si factus esset, sicut de aliis feci de quibus tractavi; quatenus videat Vestra Sapientia quid requirendum sit pro utilitate rei publicae procuranda, et qualiter scripturas et opera sapientiae petatis a quocunque.

Qualities of the Zodiacal signs.

Superius quidem dictum est quod sunt 1,022 stellae fixae, quarum quantitas potest deprehendi per instrumenta astronomiae, quae habent virtutes varias in calore frigore humore et siccitate et omnibus aliis passionibus et alterationibus naturalibus. Inter quas sunt principales stellae duodecim signorum, per quas omnia alia specialiter alterantur. Signa vero sunt, Aries, Taurus, Gemini, Cancer, Leo, Virgo, Libra, Scorpio, Sagittarius, Capricornus, Aquarius, Pisces; quae sic nominantur, quia stellae in coelo habent dispositionem rerum sic nominatarum. Et haec signa Aries, Leo, et Sagittarius sunt effective ignea; Taurus, Virgo, et Capricornus sunt terrea; Gemini, Libra, et Aquarius, sunt aerea; Cancer, Scorpio et Pisces sunt aquatica. Et Aries, Cancer. Libra et Capricornus sunt mobilia, quia renovantur in eis quatuor complexiones principales omnium rerum, scilicet calida et humida in Ariete; calida et sicca in Cancro; frigida et sicca in Libra; frigida et humida in Capricorno. Taurus et sui sequaces sunt signa fixa, quia dictae complexiones figuntur in eis et perficiuntur. Gemini et sui sequaces dicuntur signa communia, quia jam declinat complexio in eis ad novam, quae renovatur in signo sequente. Habent autem signa haec quamplures alias varietates et proprietates, quas caeteri tractatus habent determinare.

Qualities of the planets.

Planetarum prima diversitas est in virtutibus propriis. Nam Saturnus est frigidus et siccus, et omnis pigritiae et mortificationis et destructionis rerum causativus per egressum siccitatis et frigoris. Mars vero est corruptivus propter egressum caliditatis et siccitatis. Et isti duo planetae nunquam faciunt bonum nisi per accidens; sicut aliquando venenum est bonum per accidens, ut Scammonea, quae purgat materiam morbi, sed tamen per se laedit naturam. Et vocantur isti planetae inaequales, et infortunia, et malevoli. Jupiter vero et Venus habent caliditatem et humiditatem; sed Jupiter magis et melius: et isti duo planetae dicuntur aequalis fortunae, et benevoli. Mercurius est mediocris inter bonum et malum, et convertibilis naturae. Nam cum bonis est bonus, et malus cum malis. Luna vero frigida est et humida. Sol habet generativum calorem et vitalem, quia est causa vitae et generationis in omnibus, unde licet sit calidus et siccus, suus tamen calor non est corruptivus, sed generativus, et sua siccitas non est mortificativa, et ideo aliter est quam in Marte.

Their power varies with their position.

Habent autem planetae virtutes alias a signis; nam quando sunt in signo calido, habent virtutem calefaciendi, et sic de aliis. Sed speciales praeter has habent virtutes, de quibus superius dictum est, quae sunt domus, exaltatio, triplicitas, terminus et facies, et secundum has dignitates habent effectus mirabiles. Et iterum penes aspectus, qui sunt conjunctio, oppositio, &c. Conjuncti dicuntur planetae, quando sunt in eodem signo; oppositi, quando unus est in septimo ab alio. Trinus vero aspectus est, quando per quatuor distant signa. Quartus, quando per tria. Et sextilis, quando per duo. Et oppositio et quartus aspectus sunt mali ex natura sua. Trinus et sextilis sunt boni, et conjunctio similiter. Aspectus autem isti considerantur in quinque planetis, praecipue respectu Lunae et Solis. Quando ergo malus planeta aspicit in malo aspectu, detestabile est, quia tunc duplicat malitiam; quando vero bonus planeta aspicit in malo aspectu, tolerabile est; et si malus in bono aspectu aspicit, mitigatur ejus malitia. Quando vero malus opponitur aut conjungitur malo, tunc magnum malum est.

It varies in different parts of their orbit.

Habent etiam planetae magnam varietatem actionum penes eccentricos et epicyclos. Nam quando sunt in partibus superioribus istorum, faciunt operationes fortissimas, quando vero in inferioribus debiles; quia quando sunt in superioribus partibus suorum circulorum, qui vocantur auges, tunc moventur circa mundum motu diurno in circulis magnis, et tunc velociter feruntur, et velocitas motus inducit fortitudinem actionis in rebus quae natae sunt moveri, cujusmodi sunt stellae. Habent etiam effectus varios penes partes revolutionum suarum, et penes totas revolutiones. Nam secundum quod transeunt quartam coeli et medietatem et tres quartas, et totum circulum, habent sensibiles varietates in effectibus. Et non solum sic, sed penes quamlibet revolutionum determinatarum; sicut Saturnus, completis decem revolutionibus suis, inducit magnam mundi alterationem, sicut prius tactum est in sectis. Domus etiam duodecim, in quas dividitur totum coelum, quae superius tactae sunt, maxime considerantur respectu virtutum planetarum. Nam varias potestates contrahunt planetae ex eis, et varia per virtutes harum domorum operantur planetae in hoc mundo: et ideo has domos considerare maxime docent mathematici. Haec igitur sunt radices principales in actionibus stellarum, quae habent ramos et flores, et fructus infinitos. Haec igitur est intentio tractatus quem in hoc opere facere proposui de virtutibus et actionibus stellarum.

The effects of celestial forces on terrestrial things.

Tertio considerandum est quomodo loca mundi et res alterantur in complexionibus per jam dictas radices. Quidam autem universales sermones facti sunt de hac materia in prioribus, ut oportuit, sed in hac parte operis deberent plura certificari in particulari. Et certum est per Aristotelem, quod coelum non solum est causa universalis, sed particularis, omnium rerum inferiorum. Nam Aristoteles dicit secundo[1] de Generatione quod elementa debilius agunt respectu coeli quam organa respectu artificis, ut securis et dolabrum respectu carpentatoris; et ideo licet per qualitates elementares fiant omnia hic inferius, non erunt tamen respectu coeli nisi sicut instrumenta respectu artificis. Et ideo si artifex est agens principale universale et particulare respectu operis, multo magis coelum respectu rei generandae. Et Averroes dicit in septimo Metaphysicae quod idem facit virtus coeli in materia putrefica, quod virtus patris in seminibus; et ideo quae generantur per putrefactionem, licet sunt animata, fiunt tamen immediate a coelo, et multo fortius alia inanimata. Et non solum in his sed in generatis secundum propagationem in

[1] *De Generatione*, ii. 9. In this chapter Aristotle shows that the elements left to themselves are like tools without a workman; in the following chapter the revolution of the heavens is indicated as the efficient cause of generation.

plantis; quia dicit Aristoteles in libro Vegetabilium, quod Sol est pater plantarum et Luna mater. Et in hominibus et animalibus dicit, quod homo generat hominem et sol[1]. Sed Averroes dicit quod sol plus facit quam homo ad productionem rei. Nam virtus solis continuatur in semine a principio generationis usque ad finem, virtus autem patris non, sed fit semel, scilicet solum in seminis decisione, et ideo nihil faceret nisi esset virtus coeli continue multiplicata et infusa, regulans totam generationem.

Four operations to be distinguished.

Per coelum ergo complexiones omnium rerum habentur, et non solum regiones diversificantur per coelum, sed res ejusdem regionis et partes ejusdem rei, et non solum in generatione recta sed monstruosa, et peccatis ac erroribus naturae. Nam per diversitates horizontum, secundum quod singula puncta terrae sunt centra in horizontibus novis, patet omnia variari, ut superius notatum est de herbis diversarum specierum, quae nascuntur quasi in eodem puncto terrae, et de diversitate gemellorum in eadem matrice; propter hoc quod coni diversarum pyramidum continentium virtutes stellarum et partium coeli super capita habitantium veniunt ad singula puncta terrae, ut cadat diversitas plena in rebus. Sed alia diversitas est per elongationem a polis et a medio mundi. Et hujus causa duplex est. Una est causa universalis, scilicet solis distantia vel propinquitas secundum quod superius in locis mundi est memoratum. Alia est causa particularis, scilicet diversitas stellarum fixarum super capita habitantium. Nam specialiter per has variantur res naturales in diversis regionibus, et homines, non solum in naturalibus, sed in moribus, in scientiis, in artibus, in linguis, et in omnibus. Tertia est ex praedominanti virtute duodecim signorum. Nam diversa signa dominantur in diversis regionibus, aut quia in mundi principio fuerunt in directo illarum creaturarum recipientium virtutes primas, et quod nova testa capit inveterata sapit; aut quia sunt similes in natura cum stellis quae super capita habitan-

[1] *Nat. Auscult.* ii. 2 ἄνθρωπος γὰρ ἄνθρωπον γεννᾷ καὶ ἥλιος. The reference to the apocryphal work *De Plantis* seems to be incorrectly given. See Lib. i, cap. 6 Ἀναξαγόρας . . . ἔφη πρὸς Λεχίνεον ὅτι ἡ γῆ μήτηρ μέν ἐστι τῶν φυτῶν, ὁ δ' ἥλιος πατήρ.

tium revolvuntur. Et quarta causa est per planetas. Nam planetae assignantur diversis regionibus per dominium sicut signa, et hoc ex duplici causa dicta. Nunc de signis, quomodo aut in quibus regionibus assignentur signa et planetae, difficile est certificare, nam auctores diversificantur. Tractatus tamen de his habet hic expedire.

These things best understood by Hebrew astrologers.

Sed non est plena certitudo de his, sicut non de aliis difficultatibus multis, nisi ex libris Hebraeorum, quos primi composuerunt astronomi qui revelationem a Deo habuerunt in omnibus specialem. Nec solum quidem accidit diversitas complexionis in diversis regionibus per stellas, sed in rebus ejusdem regionis. Quaedam enim sunt de complexione solis, et quaedam de complexione lunae, et sic de aliis; quae secundum quod sui planetae habent quinque dignitates suas antedictas, vigorantur, confortantur, augmentantur, et roborantur. Quando vero sunt in oppositis locis suarum dictarum dignitatum, deteriorantur et destituuntur a naturali vigore. Et similiter accidit de signis, nam res diversae sequuntur complexiones diversorum signorum. Nam secundum quod quaedam sunt ignea, et secundum quod planetae sunt ignei, id est, calidi et sicci, penes hoc sunt res aliquae de complexione ignis, et hujusmodi res vocantur martiales a planeta, et de complexione Arietis, Leonis, et Sagittarii, quae sunt signa calida et sicca; et sic de aliis rerum complexionibus et signorum et planetarum. Nominare tamen et signare res singulas respectu suorum planetarum et signorum, est satis difficilis certificationis et impossibilis, nisi per libros Hebraeorum, ut ad fontem certitudinis recurratur. Ista vero quae nunc dicta sunt de assignatione signorum et planetarum rebus diversarum regionum et ejusdem regionis, sunt de majoribus difficultatibus philosophiae, et ubi major diversitas est auctorum Latinorum.

Action of these forces on special parts of the body.

Et non solum sic variantur res ejusdem regionis, sed partes ejusdem rei et maxime in homine, quia omnia sunt propter hominem. Nam caput est de complexione Arietis; collum de complexione Tauri; et humeri et brachia de complexione Geminorum, et sic ulterius. De qua divisione satis concordant auctores et ratio sufficiens dictat, et experientia, quod plus est. Nam si Luna sit in Geminis, quod est signum respondens brachiis et humeris, periculum est tangere hujusmodi membra ferro, ut minutione, scarificatione, seu ventosa, et maxime cultello vel ense. Et non solum sic, sed de omnibus medicaminibus chirurgiae; difficultatem enim magnam recipiunt et languorem, et aliquando mortem; quae non fierent si Luna esset in alio signo. Nam Ptolemaeus dicit in Centilogio, Luna existente in signo respondente membro, periculum est tangere membrum ferro. Et Haly medicus, dans causam hujus, dicit, quod tunc propter existentiam Lunae in tali signo confluunt humores ad tale membrum, et generant nocumentum; et hoc experientia tota die probat. Nam in anno praecedente medicus famosior in Francia tibiae fratris sui fecit chirurgum dare medicamen; sed astronomus peritus prohibuit ei ne faceret, quia Luna fuit in Aquario, quod est signum respondens tibiis. Et accidit quod continue invaluit infirmitas et languor usque ad mortem. Hujusmodi casus accidunt saepe, sed propter ignorantiam astronomiae non percipiuntur.

Planetary action on special hours and days.

Sed res ejusdem regionis non solum sic variantur in complexionibus per diversitatem signorum et planetarum in eodem tempore, sed in diversis. Nam omni horae dominatur unus planeta specialiter, et medietas signi oritur super horizonta in qualibet hora, ex quibus variantur horae semper et quartae dierum similiter et ipsi dies, qui sicut in omni lingua dies nominantur a planetis, sic habent diversitatem ab eis, ut omnes concordant sapientes. Et hoc est a divina ordinatione a principio mundi. Unde in prima hora diei sabbati dominatur Saturnus, et toti diei principaliter, quia per ejus qualitatem alteratur prima hora, quae est principalis, et principium unde derivatur virtus Saturni in totum diem. Et tunc Jupiter habet aspectum ad horam secundam, Mars ad tertiam, Sol ad quartam, Venus ad quintam, Mercurius ad sextam, Luna ad septimam, et iterum Saturnus ad octavam, et sic ulterius, ut dominetur in xv, et tertio in xxii, ut Jupiter dominetur in xxiii, et Mars in xxiv, et sic completur dies naturalis ipsius sabbati, ut incipiat Sol dominari in prima hora diei dominici, qui ideo vocatur dies solis apud omnes nationes. Et per hanc artem currit variatio omnium dierum septimanae et horarum.

Ex his patent complexiones quartarum diei et noctis, et quae sunt horae sanguinis, quae cholerae, quae phlegmatis, et quae melancholiae. Sed medici, qui nesciunt astronomiam, non judicant de qualitate horarum nisi per intentionem et remissionem caloris solaris, vel per absentiam. Nam solum considerant quod sex horae diei naturalis sunt calidae et humidae, et reliquae sex sunt calidae et siccae ; aliae sex sunt frigidae et siccae, et ultimae sex frigidae et humidae. Sed cujus complexionis sit signum quod oritur in duabus horis, et quis planeta dominetur in hora secundum regulam praedictam, insuper quis et qui planetae oriantur super horizonta, non considerant, cum tamen haec et quaedam alia necessario debeant considerari. Haec tamen diversitas dierum et horarum bene debet ab omni homine sciri, et maxime a medicis, et a quolibet qui vult suam sanitatem servare. Nam secundum diversitatem complexionis cujuslibet et secundum diversitatem aetatis et occupationis, oporteret eligere diversas horas pro minutionibus, pro medicinis, et omnibus aliis actibus vitae, et pro negotiis et occupationibus et aggressionibus operum, seu sapientia aliorum, quia virtutes coelorum in his horis et diebus diversimode movent complexiones sanorum et infirmorum diversas. Et ulterius secundum quod complexiones variantur, excitantur animi et ducuntur ut gratis velint sequi complexionis motum, etiam in omnibus operibus voluntariis, salva in omnibus arbitrii libertate. Unde in die sabbati et maxime in hora prima, non esset minuendum, nec aliquod dignum incipiendum quod ad complexionem pertinet, sive primo in naturalibus, sive secundario ut in voluntariis. et hoc propter malitiam Saturni, qui in omnibus generat infortunium quantum est ex parte sui, nisi aliud impediat. Et tamen ex errore omnes homines faciunt se minui in die sabbati, quod faciunt propter quietem diei dominici ab occupationibus et laboribus.

Application of this knowledge to Medicine.

Et non solum in horis et diebus accidit variatio magna consideranda per coelestia, immo in septimana, et hoc praecipue per aetatem Lunae. Nam dum crescit Luna, omnia crescunt ; dum decrescit, omnia decrescunt, sive decrementum patiuntur. Hoc non solum astronomi, sed medici considerant; ut Galenus

Influence of the moon.

in libro Dynamidiarum[1], et alii considerant in superiori experientia, ut in maribus, cerebrisque hominum, medullis, et conchyliis et omnibus. Nam augmentantur et plena sunt cum Lunae plenitudine, et deficiente Luna deficiunt. Nam et lapis Lunae, qui vocatur Selenitis, crescit et decrescit secundum Lunam, quo varietates lunationis paschalis festi sancti antiquitus probaverunt, ut suo loco in opere recto. Nam quaedam malva in prima Luna ascendit super ipsam, et crescit usque ad plenam Lunam, et deinde cum Luna decrescit. Et est herba mirificae virtutis; cum prima Luna duo folia nascuntur ex utroque latere stipitis, et in secunda Luna duo alia, et sic usque ad plenilunium, et deinde ordine consimili cadunt secundum quod Luna decrescit. Caeterum ut Ptolemaeus ait, et Haly exponit lvi propositione Centilogii, humores in corporibus prima septimana Lunae, et tertia, recedunt ab interioribus ad exteriora, sicut flumina fluunt a suis canalibus. In secunda vero septimana et in quarta retrocedunt ab exterioribus ad interius, et in hoc est magna rerum transmutatio et mirabilis. Ex quo sequitur satis veridice, quod in prima septimana et in tertia utendum est evacuationibus exterioribus, ut est phlebotomia, et in secunda et in quarta septimana medicina laxativa; quia Hippocrates in prima particula aphorismorum dicit, undecumque reperit natura, inde ducere per convenientem regionem, id est per partem corporis ad quam humores naturales decurrunt. Sed hoc medici astronomiam ignorantes non attendunt, saepius in praejudicium patientum.

The lunar mansions.

Et etiam est alia alteratio magna rerum per mansiones Lunae, quae sunt xxviii. Mansio quidem est spatium zodiaci quod pertransit Luna in die. Istae autem mansiones distinguuntur, quia quaedam sunt temperatae, quaedam distemperatae in sicco, frigido, calido vel humido. Et secundum hoc quolibet die si homo aspiciat mane in qua mansione sit Luna, poterit de pluvia et alterationibus aeris judicare, sicut docet Albumazar in majori introductorio, et patet per experientiam. Et praecipue considerandum est de dispositione

[1] The two books, *De Dynamidiis*, are among the spurious works attributed to Galen.

Lunae in novilunio; nam talis debet esse dispositionis, saltem prima septimana vel quindena, et aliquotiens totus mensis; igitur in principio secundae septimanae, tertiae, et quartae, currit eadem observatio. Nam in istis quadraturis fortissima operatio Lunae est, ut Galenus in libro de Crisi et criticis diebus, secundum quod transit quartas sui circuli. Nec solum septimanae, sed menses integri variantur praecipue per motum solis in singulis signis, secundum quod recedit a nobis vel accedit ad nos. Sed quartae anni mutantur manifeste in complexionibus suis, quum ex his communiter oriuntur quatuor complexiones principales, ut calida et humida in vere, calida et sicca in aestate, frigida et sicca in autumno, frigida et humida in hyeme. Nam fortissima est mutatio stellarum in quadraturis suorum circulorum, ut auctores docent et experientia certificat.

Influence of Mars, Jupiter, and Saturn.

Annorum vero mutatio est major per hos caeteros planetas, qui secundum suas revolutiones in quadraturis suorum circulorum et in totius circuli revolutione mutant annos; et maxime tres superiores, scilicet Saturnus, Jupiter et Mars; ut in rebus maximis, quae tarde veniunt, quia hi planetae sunt tardi motus, et ideo effectus eorum non sunt quotidiani vel mensivi sicut Lunae vel Solis. Hujusmodi autem effectus sunt diluvia, terrae motus, et pestilentia, fames valida, cometarum et caeterorum ignium in aere apparitio, sicut docet Albumazar in Conjunctionibus, et omnes sapientes concordant; unde cum anno Domini 1264 mense Julii fuit apparitio cometae horribilis, per virtutem Martis fuisse generatus approbatur. Nam Mars cum fuerit tunc in Tauro et cometa ortus sit in Cancro, non cessavit decurrere ad suam causam, scilicet Martem; sicut ferrum currit ad magnetem. Sicut enim magnes trahit ferrum, sic planeta, cujus virtute sublimatur vapor ignitus et assimilatur coelesti naturae, attrahit cometam, et ideo cum ad Martem motus est, et ibi delituit, necesse fuit eum a Marte causari. Deinde cum natura Martis sit ignea, cujus natura est multiplicare choleram, et per consequens excitare homines ad iram discordiam et bella, ideo accidit quod cometa ita significabat super iras et discordias hominum et bella, sicut sapientes astronomi docent. Sed verius experientia totius ecclesiae comprobavit per bella Angliae, Hispaniae et Italiae, et aliarum regionum, quae circiter tempus illud et a tempore illo acciderunt. O quanta utilitas ecclesiae Dei potuisset procurari, si coeli qualitas istorum temporum fuisset praevisa a sapientibus, et praelatis et principibus cognita, et pacis studio mancipata! Nam non fuisset tanta Christianorum strages nec tot animae positae in infernum.

Planetary conjunctions.

Et non tantum dico hoc propter revolutiones planetarum generantes cometas et hujusmodi, sed magis propter conjunctiones superiorum planetarum a tempore memorato. Conjunctiones autem planetarum multiplicatae sunt nimis his temporibus; sed in capitulo superius de sectis et Christi professione confirmanda adducta est praeclara astronomorum scientia de istis conjunctionibus, quomodo consuetae naturae jura mutantur, et animi hominum excitantur ad seditiones et immutationes consuetudinum et legum, ad discordias et bella, et mutantur dominatus et regna, et vices principum renovantur. Albumazar et caeteri sapientes docent hoc evidenter. Et si haec exempla sint nimis sublimia, converte stylum ad exempla vulgata et utilia corporibus humanis. Nam Ptolemaeus dicit in nona decima propositione Centilogii, si quis purgationem receperit Luna conjuncta cum Jove, abbreviabitur opus ejus et effectus minuetur. Et dicit Haly, quod hujus causa est, quia ex beneficio Jovis in tantum confortatur natura, quod impedit effectum medicinae. Et si Luna conjungatur Saturno, potionatus patietur torsiones et in periculo erit propter malitiam Saturni, nec se poterit expedire propter Saturnalem siccitatem; qua retinebitur medicina in corpore, quia siccum est bene retentivum.

Medical authorities as to their effects on health.

Sed qualiter cognoscuntur infirmitates penes hujusmodi conjunctiones et aspectus caeteros et dignitates planetarum, egregie designant auctores medicinae; et praecipue Hippocrates in libro suo de alterationibus corporum humanorum, quae contingunt ex motu Lunae suscipientis conjunctiones et aspectus caeteros planetarum. Et Galenus dicit tertio de Crisi et criticis diebus[1], 'Et ego redii ad rem, quam consideravi

[1] The next three sentences are quoted from Galen. The passage will be found in the treatise *De Diebus Decretoriis* (lib. iii. cap. 4 and 5).

cum studio et perscrutatione vehementi, et inveni eandem veram et certam quae non perit, et est res quam invenerunt astrologi Aegyptii. Et est quod Luna signat super dies qualiter erit dispositio hominis in aegritudine et sanitate sua; quod est, si applicentur Lunae quaedam stellae aequales, et sint stellae quae nominantur fortunae, erunt illi dies illi homini convenientes. Et si applicentur illi quaedam stellae egredientes ab aequalitate, erunt illi dies laesionis et tristitiae.' Et sumit hic applicationes pro conjunctionibus et caeteris aspectibus, et multa dicit hic de hac materia. Atque Haly medicus in libro de Regimine Regali docet praeclare quod medici indigent consideratione hujusmodi, unde ait, medicina eget astronomia, quae medicaminibus electis utitur temporibus electis, quibus est Luna bonis et felicibus contemperata planetis et competentibus figuris zodiaci. Et omnes auctores concordant in istis. Et propter has conjunctiones et revolutiones planetarum mutuas Avicenna secundo libro de Animalibus dicit, totam generationem regulari et augmentari, et initium vitae et mortis. Unde ait, opus est ut termini vitae et mortis essent positi in specie, donec sit vita concordans cum revolutione alicujus stellae aut multarum stellarum. Et primus terminus est dies cum sua nocte: deinde est alius, quem Luna constituit in quarta sui circuli; deinde est mensis, qui est circulus, qui completur in conjunctione Solis et Lunae; deinde annus planetarum; deinde annus mediarum conjunctionum et reversionum ad eas figuras in quibus fuerunt istae conjunctiones, quae faciunt ad augmentum naturalium et diminutionem innaturalium. Et non solum haec accidunt in recta generatione, sed in monstruosa et peccatis naturae, secundum quod ipse Avicenna docet eodem libro, dicens, Si embryo potest recipere humanitatem, potest recipere animalitatem, sicut in rebus monstruosis, ut quando filius hominis habuit caput arietis, et agnus habuit caput tauri, quoniam virtus in eo induxit formas et figuras coelestes, quae accidunt unicuique.

Planets in the *apsides* of their orbits.

Nec solum his modis accidunt alterationes complexionum rerum inferiorum, sed propter ascensiones planetarum ad auges suorum circulorum, et postea descensiones ad oppositionem augium propter fortitudinem motuum in augibus et debilitatem in oppositis augium, sicut nos videmus per experientiam. Omnia enim terrae nascentia circa solstitium aestivale quando Sol est in auge sui eccentrici, plus crescunt et vigorantur per unum diem quam alias per septimanam. Et quando Luna est in augibus suorum circulorum, ut in novilunio et plenilunio, tunc sunt fortiores operationes ejus, ut patet in fluxibus maris et in piscibus; tunc enim pisces meliores sunt, tunc saniora sunt conchylia, sicut auctores docent, et experientia; et sic est de omnibus planetis.

House of the planet.

Sed ultimum quod hic volo tangere est de domibus duodecim, quae per divisionem coeli in duodecim partes oriuntur, scilicet quarum consideratio maxime docetur a philosophis. Nam Albumazar in libro de Floribus dicit, quod certissime discutiendae et examinandae sunt. Nam per has primo consideratur, quis planeta dominatur complexioni totius anni; quae consideratio fit in principio veris, quando Sol ingreditur primum minutum Arietis in aequinoctio vernali; et per has significationes scitur, quomodo planetae inducunt complexiones singulorum dierum et singularum horarum, et omnium rerum, secundum quod ascendunt super horizontem. Nam ostendunt hoc manifeste astronomi et experientia manifestat. Hae igitur sunt radices de alteratione complexionum rerum per virtutes stellarum, in cujus tractatu debent explicari rami et fructus et flores cum tabulis et figurationibus nobilissimis, ex quo manifestior est potestas figurarum et numerationum, nec non praegrandis utilitas et sapientiae pulchritudo.

Interpretation of these things with reference to human events.

Post haec sequitur quartum, quod est consideratio de judicio et cognitione praeteritorum praesentium et futurorum. Nam si causa complexionum rerum est coelestis constellatio, poterit hujusmodi effectus sciri per hanc causam. Et quomodo vel quando, certificatur hoc per sanctos, et praecipue per patriarchas a principio mundi, a quibus habuerunt philosophi possibilitatem judicandi. Et considerandum est quod omnes philosophi in hoc concordaverunt, et Aristoteles certificavit in libro suo de hac materia, et experientia docet. Nec est contradictio, nisi per ignorantiam infinitam, quae currit in vulgo et in eis qui praesunt eis, qui omnia quae ignorant reprehendunt et negligunt. Modus autem universalis judicandi per planetas consistit in hoc, quod homo sciat per tabulas et instrumenta aequare motus coelorum et invenire loca planetarum; et considerare quis eorum habeat plures et majores fortitudines ex loco suo secundum illas quinque famosas, quae sunt domus, exaltatio, triplicitas, terminus et facies; et similiter cum hoc videre fortitudines quae contingunt ex aspectibus illorum et ex motu in augibus suorum circulorum, et ex domibus duodecim. Et cum bene examinaverit et viderit haec et eis annexa, tunc si Sol habuerit plura testimonia et fortitudines hujusmodi, debet judicare secundum complexionem solarem in rebus, eo quod illa dominatur in eis; et sic de aliis. Particularis vero consideratio et specialis in singulis habet suas leges determinatas secundum rerum conditionem. Et potest homo revolvere historiam ad tempora praeterita, et considerare effectus coelorum a principio mundi, ut sunt diluvia, terrae motus, pestilentiae, fames, cometae, monstra, et alia infinita, quae contigerunt tam in rebus humanis quam in naturalibus. Quibus comparatis, revolvat tabulas et canones astronomiae, et inveniet constellationes proprias singulis effectibus respondere. Deinde consideret per tabulas consimiles constellationes in futuro tempore propinquo vel remoto sicut vult; et poterit tunc pronuntiare in effectibus, qui consimiles erunt sicut fuerunt in praeterito, quia posita causa ponitur effectus.

Authorities on the subject.

Et istorum judiciorum introductoria habentur sufficienter ex libro Albumazar proprio ad hoc, et ex libro Alkabiz, et Haly Abenragel. Sed radices judicandi ponuntur in centum verbis Ptolemaei, qui est potentior in philosophia speculativa. Rami vero extrahuntur in quadripartito Ptolemaei, ut idem dicunt. Flores vero et fructus in multis aliis libris colliguntur: sed praecipue in libro Conjunctionum Albumazar, secundum quod pervenit hujus scientiae notitia ad manus Latinorum; hoc dico, quia nondum habent scientiam completam de his, quia Latini nihil quod valet habent nisi ab aliis linguis; et pauci fuerunt interpretes et mali ut ostenditur. Libri enim Hebraeorum certificant de his, quam certificationem Aristoteles consideravit et redegit in scriptura Graeca. Nam Averroes dicit super quartum Coeli et Mundi quod Aristoteles certificavit hoc in libro suo de impressionibus coelestibus, qui liber est melior tota philosophia Latinorum, et potest per vestram jussionem transferri.

Our conduct should be modified by this knowledge, as advised by Aristotle.

Quintum vero annunciatum superius est principale inter haec, nam nominata omnia sunt propter hoc ; quia non sufficit rei publicae ut cognoscantur omnia, sed oportet quod utilia cognita promoveantur, et nociva evacuentur. Sed hic licet geometria, et arithmetica, et musica dent maxima et quamplurima juvamenta, tamen astronomia regulat omnia, propter hoc quod omne opus magnificum debet fieri in temporibus electis. Et ideo operata aliarum scientiarum requirunt tempora idonea, quae consideratione astronomi eligi cognoscuntur, et ideo praeest omnibus scientiis in hac parte, quamvis etiam specialia remedia habeat infinita, in quibus aliae scientiae non habent potestatem. Et quia tam universalis est potentia istius scientiae, ideo Aristoteles sapientissimus philosophus docet Alexandrum in libro Secretorum, quod nec comedat, nec bibat, nec aliquid faciat sine consilio astronomi, quia tempora electa sunt ad omnia ; omnia enim tempus habent, sicut dicit Salomon sapientior Aristotele. Et Aristoteles sibi objicit dicens, Deus praevidit omnia ab aeterno, ergo non potest astronomus ea impedire vel mutare. Et ipse respondet, quod ea quae Deus praevidit sunt immutabilia. Sed tamen effectus eorum quae Deus posuit in potestate hominis ex sua provisione aeterna, potest homo mutare ut vult, quia in contingentibus ad utramlibet. Et in rebus humanis Deus non posuit necessitatem, licet sciverit ab aeterno in quam partem debet terminari contingens, et humani libertas arbitrii. Et ideo potest homo omnes utilitates suas procurare, et impedimenta removere, si fuerit peritus in hac scientia. Et ponit exemplum. Si enim, ut dicit, necessario erit superfluitas frigoris in tempore futuro, astronomus praevidens potest apponere remedia, ut sine gravamine patiatur frigus illud, quod alii incaute non potuerunt sustinere : potest enim astronomus praeparare loca calida, nutrimenta calida, vestimenta calida et medicamina multa, ita ut frigus ei in nullo nocebit, quamvis alii qui haec non praeviderint prae frigore moriantur. Et ideo in potestate hominis periti in hac scientia est ut evadat nociva et consequatur utilia.

Et hoc egregie docet Isaac in libro Febrium, capitulo de pestilentia; nec solum de infirmitatibus, sed etiam universaliter in omnibus, dicens quod anima rationalis dignior est stellis, nam actio earum minime ei nocet nisi cum ignorantia teneatur. Dum enim ipse sciverit earum actionem omnino cavere poterit; et infert, Testificatur autem id quod videmus ab optimis factum mathematicis, qui cum certissime noverint dies mortis in eis qui mare ingressuri sunt, satis possunt eos tueri; similiter et eos qui in bello morituri sunt; et addit, Homo differt ab animalibus, nam rationis et discretionis est super inquirere et desiderare bona et respuere mala, et eis repugnare, quia non est subditus actioni naturali, nisi cum ignorantia tentus sit. Et ideo dicit Ptolemaeus in quinto verbo Centilogii, astrologus optimus multo magis prohibere poterit quod secundum stellas venturum est; et in octavo dicit, anima sapiens adjuvabit opus stellarum, quemadmodum seminator fortitudines naturales, unde et nociva poterit repellere, et utilia promovere. Et Isaac ponit exemplum de pestilentia futura, et facit sibi quaestionem juxta ignorantiam medicorum, sicut Aristoteles fecit. Quaerit ergo utrum hujusmodi putrefactio veniet necessario propter coelum, quia tunc medicus non potest prohibere, et ideo inutile videtur hoc praecognosci. Sed solvit, dicens quod corruptio aeris non nocet nisi corporibus praeparatis ad eam; et ideo cum medicus videret significationem aeris corrupti in unoquoque tempore anni venturo, poterit mundificare naturam, et malam corporis dispositionem auferre, ut sciat venienti corruptioni repugnare. Nam si viderit per motum stellarum fieri in aere calorem et siccitatem nimiam et ustionem, festinet purgare choleram eorum in quibus inest complexio cholerica, det frigida et humida calori venturo repugnantia. Et Haly dat consimilem sententiam in expositione quintae propositionis Centilogii. Ex his patet quod astronomus potest dare remedia contra nociva et promovere utilia, non solum in infirmitatibus, sed in aliis universaliter, cum ratione electionis temporis apti ad omnia opera, cum ratione operum utilium, cum remotione nocivorum; et hoc specialiter in curis corporum humanorum, quo nihil melius est quantum ad bonum civium particulare.

By Isaac the physician,

confirmed in public matters by example of Moses and of Solomon,

Sed tamen respectu boni communis civitatis et regnorum, majora possunt fieri opera per virtutem astronomiae quam aliquis potest dicere, et quam aliquis vult per scripturam explanare; nam haec sunt multa de secretissimis operibus sapientiae. Sed propter rerum magnitudinem et ignorantiam multitudinis studentium, qui de operibus sapientiae non curant, et propter fraudes multorum et malitias qui abusi sunt his, occultantur semper a vulgo et a capitibus ejus; de quibus tamen Moyses et Aaron, Solomon, Aristoteles, Ptolemaeus et caeteri sapientes operati sunt nimis admirande. Unde cum hoc sit unum de maximis, et quasi potens quodammodo in omnia, non decet ut Vestrae Gloriae occultetur. Et quia praecepistis ut scriberem de sapientia philosophiae, recitabo Vestrae Clementiae sententias sapientum, praecipue cum ecclesiae Dei sit omnino necessarium contra furiam Antichristi. Cum vero Moyses in juventute sua fuisset dux exercitus Aegyptiorum contra Aethiopes, et ob bonum pacis acceperat Aethiopissam, pro qua contra eum duodecimo Numerorum objurgati sunt Aaron et Maria, quae fuit regina Aethiopum, nolens dimittere eam ut rediret in Aegyptum propter amoris superfluitatem, cum esset peritus astronomus, fecit duas imagines in annulis, unam oblivionis quam dedit mulieri, et alteram memoriae quam sibi detinuit, et sic ab ea libere cum exercitu et sine bello recessit. Haec Josephus primo Antiquitatum libro, et magister in historiis, et multi confirmant. Mira res fuit haec, quae animum mulieris immutavit. Salomon vero ordinavit de hujusmodi multis, quae praeter solitum cursum naturae facta sunt, ut dicit Josephus octavo libro, qui cum fuit sapientior omnibus praecedentibus et sequentibus, ut scriptura sacra et Josephus docent, non potuit haec opera negligere, et ideo scripta multa in aenigmate reliquit, quae postea per magicos in malos usus conversa sunt atque male interpretata, in quibus multa enormia sunt per fraudulentos admixta. Sed sapientes sciunt eligere grana de paleis, et theriacam de serpenti separare.

and of Aristotle.

Aristoteles vero summus philosophorum docet haec opera fieri in libro Secretorum, ut quaeque magnifica utiliter promoveantur et omnia nociva destruantur. His enim operibus Alexandrum magnum fecit Aristoteles cum minus quadraginta millibus armatorum mundum superare. Is cum in lecto mortali jacebat, et jam esset in januis mortis, civitatem suam et regionem ab inimicis eripuit. De primo facto notum est in gestis Alexandri, de secundo in Valerio Maximo, libro quinto. Sed constat eum non posse hoc fecisse potentia corporali, sed magna sapientiae virtute, quam in libris Secretorum reliquit solum sapientissimis indagandam. Et cum propter pondus senectutis, devicto Dario, mundum tradidit Alexandro, recessit in terram suam, dixitque Alexandro quod ei scriberet quae vellet, et ipse consilium daret opportunum. Et cum Alexander invenit gentes habentes mores pessimos, scribens Aristoteli quid faceret de eis, ipse princeps philosophiae respondit, Si potes alterare aerem ipsorum, permitte eos vivere; si non, interfice omnes. O quam occultissima responsio est, sed plena sapientiae potestate! Nam intellexit quod secundum mutationem aeris, qui continet coelestes virtutes, mutantur mores hominum; eo quod alios mores habent Gallici, alios Romani, alios Hispani, et sic de singulis regionibus. Voluit ergo quod Alexander in bonum mutaret qualitatem aeris illarum gentium, ut secundum mutationem illam mutarentur mores, et excitarentur ad honestatem morum sine contradictione liberi arbitrii; sicut quaelibet natio excitatur ad proprios mores per aerem proprium habentem virtutes stellarum quae sunt super capita hominum, et secundum quod signa vel planetae dominantur singulis regionibus. Ostendo enim in tractatu quem mitto [1], quod induci potest aliquis ad bonum et malum, tam publicum quam privatum, per virtutes coelorum sine coactione, sicut exemplariter videmus quod homines mutant suas voluntates per dominos, amicos, et socios, et rerum praesentiam novarum, et infinitis modis sine coactione. Quod satis explano in capitulo de potestate judicandi de rebus humanis per coelum, in quo capitulo est ratio

[1] This does not refer to the treatise *De Multiplicatione Specierum*, several times mentioned in the *Opus Tertium* as being sent together with the *Opus Majus*. The subject of astrology is not dealt with there. The reference may be to p. 251 *et seq*. If so, the fact that the *Opus Majus* should be in this place spoken of as a distinct work shows that the whole of this second treatment of astrology would have been elaborated into a special treatise had opportunity permitted.

omnium quae tanguntur de hac materia [1]. Philosophus igitur voluit quod faceret opera sapientiae per debitas constellationes ad modum Moysis, qui excitavit animam mulieris per coelestes virtutes receptas in materia. Nam qua ratione potuit illa mulier mutari ad castitatem et oblivionem viri per imagines, potuit et ad alios mores moveri non solum ipsa, sed quicunque. Ptolemaeus in nona propositione Centilogii docet, quod vultus hujus seculi sunt subjecti vultibus coelestibus. Et dicit Haly, quod in hoc capitulo vult Ptolemaeus imaginum secreta patefacere. Et patet intentio eorum in universali, quia si contingit temporibus electis eas sculpere ad vultus coelorum, omnia nociva repelli possunt et utilia promoveri. Thebit [2] inter omnes Christianos summus philosophus, qui in multis addidit ad opera Ptolemaei et aliorum astronomorum tam in speculativis quam in practicis, specialiter hanc scientiam ampliavit, et alii sapientissimi super his insistebant.

Theologians avoid this subject; fearing the imputation of magic.

Sed quia haec opera videntur vulgo studentium esse supra humanum intellectum, quia vulgus cum suis doctoribus non vacat operibus sapientiae, ideo vix est aliquis ausus loqui de his operibus in publico. Statim enim vocantur magici, cum tamen sint sapientissimi qui haec sciunt. Utique theologi et decretistae non instructi in talibus, simulque videntes quod mala possunt fieri sicut et bona, negligunt haec et abhorrent, et computant inter magica. Vident enim quod magici et abusores documentorum sapientiae his utuntur, et ideo aestimant indigna Christianis. Sed non debet veritas damnari pro ignorantia, nec utilitas pro malitia, quamvis eadem fieri valeant; nam tunc deberent homines carere cultellis in mensa, quoniam comedentes possunt interficere socios eodem ferro quo scindunt cibaria, et tunc arma debent deleri ab ecclesia et brachium seculare, quia multa mala possunt per haec fieri, et

[1] See pp. 246–53.

[2] Thebit, or Thabit ben Corra ben Zahrun el Harrani was born A.D. 836, at Harran in Mesopotamia, and died A.D. 901. Most of his life was spent at Bagdad. He was a friend of the great algebraist Mahommed ben Musa; and he translated into Arabic the works of Archimedes, Apollonius, and Euclid; also the Physics and Analytics of Aristotle, and many of the works of Hippocrates and Galen. He was not a Mahommedan; but his Christianity is more than doubtful.

quotidie peraguntur. Et nihilominus jura debent esse in usu bonorum hominum, quamvis per cavillationes et fraudes multi juristae juribus abutuntur. Et nunc videmus quod plus est in usu studentium quicquid de cavillationibus sit scriptum quam verus usus demonstrationum ; non tamen propter hoc studium sapientiae damnandum est. Nam semper mali in malis convertuntur, ex quibus boni utilitates varias producunt.

Operative force of words and written characters.

Et cum opus animae rationalis praecipuum sit verbis uti efficaciter et ex intentione formatis, potest astronomus formare verba temporibus electis quae ineffabilem habebunt potestatem. Nam ubi intentio desiderium et virtus animae rationalis, quae dignior est stellis, concurrunt cum potestate coeli, necesse est sive verbum sive aliud opus produci mirae virtutis in alteratione rerum mundi hujus, ut non solum naturalia, sed animi inclinentur ad ea quae velit sapiens operator, salva arbitrii libertate, quia potest animus sequi coelestes virtutes gratis sine coactione, ut tractatum est suo loco et declaratum.

Ex hac radice incepit apud sapientes usus characterum et carminum. Nam characteres sunt sicut imagines, et carmina sunt verba ex intentione animae rationalis prolata, virtutem coeli in ipsa pronunciatione recipientia; unde de mira potestate literarum ego facio mentionem in tertia parte[1]. Per hanc enim potestatem sanantur corpora, fugantur animalia venenosa, advocantur ad manum bruta quaecunque, et serpentes de cavernis et pisces de profundis aquarum. Alteratur enim mundi materia ad multas passiones mirabiles, si recte fierent ista, et ideo contra malevolos homines et inimicos reipublicae possunt mirabiliter valere, sicut et caetera opera stellificata, ut oportet. Sed magici maledicti induxerunt summam infamiam in hac parte, quum non solum in malis abusi sunt characteribus et carminibus scriptis a sapientibus contra nociva, et pro utilibus maximis, sed adjunxerunt mendosa carmina et characteres vanos et fraudulentos quibus homines seducuntur. Insuper daemones temptaverunt multos et tam mulieres quam daemones docuerunt multa superstitiosa, quibus omnis natio plena est. Nam ipsae vetulae ubicunque faciunt characteres et carmina et conjurationes, ac

[1] See pp. 93–4.

ipsi magici utuntur invocationibus daemonum et conjurationibus eorum, et sacrificia eis faciunt. Sed haec omnia sunt maledicta et extra vias philosophorum, immo contra sententias eorum ; et per haec defamatur philosophiae potestas. Itaque theologi nunc temporis et Gratianus et sancti plures multas utiles et magnificas scientias reprobaverunt cum his magicis, non attendentes differentiam inter magicam et philosophiae veritatem propter causas quinque de quibus in prima parte dictum est ; quod capitulum necessarium est omni homini ad magnalia sapientiae aspiranti et volenti verum a falso separare. Nam Gratianus, sicut multa scripsit jura quae nunc abrogata sunt, sententia saniore praevalente, sic, cum de scientiis locutus est, multa dixit quae debent in partem alteram commutari, ut inferius abundantius explicabo.

In this action upon nature the mind co-operates with stellar forces.

Redeo igitur ad verba et opera sapientum, stellarum virtute et animae rationalis potestate formata, reddens rationem in summa de his secundum quod docuerunt sapientes. Sicut enim puer natus, et aeri novo tanquam mundo recenti expositus, recipit impressionem virtutum coelestium, a quibus habet complexionem radicalem quam nunquam potest amittere, quia quod nova testa capit, inveterata sapit ; sic est de omni re noviter facta, quae recipit virtutem coeli in prima ejus existentia, et illam quam in principio recipit nunquam amittit donec a suo esse naturali destituatur et corrumpatur. Et ideo in istis imaginibus, carminibus, et characteribus, constellatione debita compositis, recipiuntur virtutes stellarum et retinentur, ut possint per eas agere in res hujus mundi, et recedente constellatione, in qua fuerunt res hujusmodi compositae, recedunt. Et quoniam anima rationalis dignior est stellis, ideo sicut stellae et omnia faciunt virtutes suas et species in rebus extra, de quibus speciebus et virtutibus scribo satis in quarta parte, ubi de geometricis agitur, potest ergo anima rationalis, quae est substantia maxime activa inter omnia post Deum et angelos, facere et facit continue speciem suam et virtutem in corpus, cujus est actus, et in res extra ; et maxime quum ex forti desiderio et intentione certa et confidentia magna operatur. De quibus operibus Avicenna in sexto Naturalium potenter eloquitur. Et ideo hujusmodi opera et verba de quibus loquor non solum recipiunt virtutem a coelo, sed ab anima rationali, quae est nobilior, et propter hoc possunt habere magnam virtutem alterandi res mundi hujus.

Et si dicatur, quod sicut hujusmodi opera recipiunt virtutem coeli, sic alia omnia quae sunt in eadem regione, et in tempore compositionis hujusmodi verborum et operum; et ita omnia[1] debent habere has virtutes, et homines, et boves, et equi, et arbores, nam radii coelestis constellationis attingunt infinita in eodem tempore; dicendum, quod non valet objectum, quia non sunt res in eodem horizonte. Nam singula puncta terrae sunt centra diversorum horizontum, ad quae coni diversarum pyramidum virtutum coelestium veniunt, ut possint producere herbas diversarum specierum in eadem particula terrae minima, et gemellos in eadem matrice diversificare in complexione et moribus, et in usu scientiarum, et linguarum, et negotiorum, et caeteris omnibus; tum quia res aliae jam factae prius ante compositionem imaginis, licet sint[2] cum illa, tamen ab origine sua receperunt influentiam propriam radicalem secundum quam operantur, et ideo virtus coeli ad hanc horam, de qua loquimur, non habet effectum naturalem in rebus hujusmodi prius factis, sicut habet in istis operibus et verbis nunc de novo factis; tum quia ab anima rationali non sunt facta, aut non tunc; aut si tunc, multa fiunt[3] ab homine alia, non tamen ea intentione, nec desiderio, nec confidentia, nec ordinantur ad hujusmodi actiones, et ideo hujusmodi aut non habebunt virtutem alterandi, aut non ita notabiles operationes. Si dicatur, quod saltem omnis res per virtutem quam recipit in sua origine a coelesti constellatione, cum in illa maneat complexio radicalis juxta fortitudinem constellationis, alterabit res extra se et immutabit sensibiliter, et maxime in novitate sua; et quod per processum temporis, per continuas virtutes coelorum alias, remittatur et debilitetur virtus prima paulatim donec deficiat; dicendum est, quod hoc est verum, et secundum quod oriuntur tales res, fiunt aliquando

Objection. Reply. Second Objection. Reply.

[1] J. has, *omnia animalia*: but *animalia* is omitted in O. and the sense is improved by the omission.

[2] J. fuit.

[3] J. has *fuerint*: but *fiunt*, the reading of O., is evidently better.

magnae mutationes, licet non consideremus unde tales alterationes contingunt, sicut accidit de cometis et aliquibus aliis.

Fascination.

Et iterum potest elici quid intelligendum est de fascinatione, quae licet habeat verbum suspectum, et potest modo intelligi ut vetulae sortilegae et magici considerant; tamen certum est quod multi homines sunt malae complexionis et corruptae compositionis et infirmae, ita quod contagiosi sint, sicut homines infirmi multis morbis contagiosis. Et praecipue accidit complexioni hujusmodi virtus infectiva et contagiosa per malam constellationem in qua conceptus est puer vel natus. Nam a sanissimis parentibus et optimae complexionis nascuntur infiniti qui sunt pessimae complexionis; et isti, sicut omnes res suam speciem et virtutem producentes, contaminant res praesentes, maxime teneras aetate et complexione, et praecipue per oculos; quia porosi sunt et rari, et exeunt vapores et spiritus corrupti et inficiunt res. Unde Virgilius, 'Nescio quis teneros oculus mihi fascinat agnos.' Sicut mulier menstruosa, si aspiciat speculum novum et politum, maculat ipsum nube sanguinea, ut Aristoteles dicit secundo de Somno et Vigilia [1], et experientia docet, et ita inficit alias res, licet non ita appareat: et sic leprosi inficiunt. Et haec omnia naturalia sunt. Quod si ulterius aliqua anima maligna cogitet fortiter de infectione alterius, atque ardenter desideret et certitudinaliter intendat, atque vehementer consideret se posse nocere, non est dubium quin natura obediet cogitationibus animae, ut docet Avicenna octavo de Animalibus et quarto de Anima, quatenus fortior sit multiplicatio speciei et infectio violentior; sicut si leprosus intendat desideret et confidat multum nocere alicui quem odio habet, longe plus laedit eum quam alium cui malum non cogitat. Omnis enim operatio hominis est fortior et impetuosior, quando ad eam est multum sollicitus et voluntarius, et fixo proposito firmat intentionem, et sperat firmiter se posse consequi quod intendit. Nam dicunt sapientes, quod sicut scammonea acuit medicinam et dat ei vigorem, sic intentio et desiderium et confidentia se habent in operibus humanis; et adhuc non est magicum.

Nam Plinius docet septimo Naturalium, quod multi homines

[1] See p. 142.

sunt talis complexionis, qui alios inficiunt et laedunt ex sua praesentia, et per visum et per verbum [1]. Nam quia verbum ab interioribus membris naturalibus generatur et formatur ex cogitatione et sollicitudine, et delectatur homo in eo, et propriissimum est instrumentum animae rationalis, ideo maximam efficaciam habet inter omnia quae fiunt ab homine, praecipue cum ex intentione certa, desiderio magno, et vehementi confidentia profertur. Cujus signum est, quod omnia fere miracula quae facta sunt per sanctos a principio fiebant per virtutem verborum, unde in verbis summa est potestas, sicut explicavi. Et si hujusmodi multiplicatio speciei et verbi prolatio fiant in constellatione debita, necesse est quod operatio valida consequatur; et in his omnibus nihil est magicum vel insanum. Et si hoc vocetur fascinatio, nomen, si volumus, possumus mutare. Sed res in nullo habet calumniam apud sapientes. Feci vero mentionem de fascinatione [2]; sed hic uberius explicatur. Sed quia magici et vetulae maledictae non considerant fascinationem per has vias, immo asserunt alterationes subitas indifferenter fieri per homines quoscunque, cum videtur res in mane vel sero, vel loquuntur de illa, ideo vana est eorum sententia et a sapientibus reprobata.

The eye or the spoken word have power for good or evil.

Non solum pro consideratione sapientiali haec scribo, sed propter pericula quae contingunt et contingent Christianis et ecclesiae Dei per infideles, et maxime per Antichristum, quia ipse utetur potestate sapientiae, et omnia convertet in malum. Et per hujusmodi verba et opera stellificanda, et magno desiderio malignandi componenda cum intentione certissima et confidentia vehementi, ipse infortunabit et infascinabit non solum personas singulares, sed civitates et regiones. Et per hanc viam magnificam faciet sine bello quid volet, et obedient homines ei sicut bestiae, et faciet regna et civitates pugnare ad invicem pro se, ut amici destruant amicos suos, et sic de mundo faciet quod desiderabit.

Et sic poterunt Tartari et Saraceni facere. Nam constat, quod Tartari plus vacant astronomiae quam alii, eo quod licet in multis nationibus sint astronomi sapientes, tamen

The enemies of Christendom may

[1] Plin. vii, 2.

[2] Cf. p. 143.

avail themselves of this knowledge, to our hurt.

principes reipublicae[1] non reguntur nisi per eos. Sed illi qui sunt apud Tartaros sicut praelati apud nos, sunt astronomi. Mangu Cham, imperator Tartarorum existens anno Domini 1253, quando dominus rex Franciae Lodovicus misit fratrem Willielmum de ordine Minorum ad Tartaros, dixit Christianis congregatis coram eo, praesente dicto fratre; 'Nos habemus legem a Deo per divinos nostros, et facimus omnia quae ipsi dicunt. Et vos Christiani habetis legem a Deo per prophetas, sed non facitis eam.' Hoc dixit, quia in oriente sunt mali Christiani, ut Nestoriani et alii multi, qui non vivunt bene secundum legem Christi. Et ille frater scripsit domino regi, quod si parum scivisset de astris, ipse fuisset bene receptus apud eos; sed quia nescivit terminos astronomiae, ideo despexerunt eum. Et ideo Tartari procedunt in omnibus per viam astronomiae, et in praevisione futurorum et in operibus sapientiae. Cujus signum est evidens, quod cum sit gens habens parvos et debiles homines, et quasi non comedentes nec bibentes quod naturam confortet, et nihil pedibus valentes, et inermes si proprie loquamur, nisi quod habent sagittas ad terrorem quibus persequantur fugientes, nec unquam de prope confligunt potestate bellandi, ut acies contra acies ordinentur, cum jam totam latitudinem mundi prostraverunt ab aquilone ad orientem, et ab oriente usque ad meridiem, solum eis deficiunt duo anguli terrae Christianorum, scilicet Aegyptus et Africa. Nisi enim Dominus reprimeret eos et permitteret seminari discordias inter eos frequentes, jam totum mundum occupassent. Et ideo oportet quod per opera sapientiae procedant magnifica quibus mundum conculcant. Nam narrat dictus frater in libro suo de moribus Tartarorum quem Domino

[1] Principes et respublicae, J. The report of William Rubruquis, already referred to, contains (p. 362) a full account of the Mongol diviners. 'Ipsi sunt multi et semper habent unum capitaneum tanquam pontificem qui semper collocat domum suam ante majorem domum ipsius Mangu Cham . . . Istorum aliqui sciunt de astronomia, maxime iste princeps, et praedicant eis eclipsim solis et lunae . . . Praedicant dies festos vel infestos ad omnia negotia agenda ; unde nunquam faciunt exercitum nec ineunt bellum sine dicto eorum.' Mangu, in his interviews with Rubruquis, frequently dwelt on the inconsistency of Christians with their principles. 'Vobis,' he said 'dedit Deus scripturas, et non custoditis eas; nobis autem dedit divinatores, et nos facimus quod ipsi dicunt nobis, et vivimus in pace' (pp. 359-60).

regi transmisit, quod quatuordecim millia Tartarorum devicerunt Soldanum Turkiae, qui habuit secum ducentos millia equitum sine peditibus. Sed hoc facere non poterant per potentiam bellandi, ut planum est, et ideo per opera sapientiae, et maxime per astronomiam, qua profitentur se regi et duci in omnibus. Similiter Saraceni multum utuntur astronomia, et sciunt sapientes inter eos facere haec opera. Et timendum est valde quod Tartari et Saraceni in suis regionibus commorantes ad Christianos mittant homines, quibus opera astronomiae conficiant ad infortunandum et ad ponendum discordias inter principes, quia ad hoc maxime nituntur inimici Christianorum, ut ponant inter eos bella et discordias.

Proofs of this danger are not wanting.

Nam multa facta sunt hujusmodi, licet multitudo stulta non consideret unde accidant. Forsan vidistis aut audistis pro certo quod pueri de regno Franciae semel occurrebant in infinita multitudine post quendam malignum hominem, ita quod nec a patribus nec a matribus nec amicis poterant detineri, et positi sunt in navibus et Saracenis venditi; et non sunt adhuc quatuor et sexaginta anni[1]. Similiter in temporibus nostris magister Pastor totam Alemanniam et Franciam commovit, et cucurrit post eum multitudo hominum, et gratiam habuit coram toto vulgo laicorum in contemptu cleri et ecclesiae confusionem. Et dixit Dominae Blanchiae, quod iret ad filium suum ultra mare, talibus verbis sapientissimam mulierem decipiens. Non dubitent sapientes, quin ipsi fuerunt nuntii Tartarorum aut Saracenorum, et quin habuerunt aliqua opera unde fascinabant plebem. Et vidi cum oculis meis portare patenter in manu sua quiddam tanquam esset res

[1] Matthew Paris relates this under the year 1213, 54 years before Bacon wrote. The Shepherd, or leader of shepherds, of whom Bacon goes on to speak was, according to Matthew Paris, a renegade Hungarian, who, in 1251 gathered round him troops of shepherds and other poor people offering to lead them to the Holy Land, professing to have a mandate for that purpose from the Virgin. 'Addidit fidem dictis suis eloquentia et manus suae indissolubilis clausura, in qua se mentitus est beatae Virginis habuisse chartulam et mandatum.' He seems to have imposed upon Blanche the queen-regent, until a disorderly attack made by the leader and his followers upon Orleans undeceived those who had previously trusted him. This impostor had two successors. Matthew Paris thought, like Bacon, that they were emissaries from the Mahommedan camp.

sacra, ac si homo deferret reliquias, et ivit nudis pedibus, et erat circa eum multitudo armatorum, ita tamen dispersa in campis, quod ab omnibus occurrentibus potuit videri cum illo quod portabat in manu cum magna ostentatione.

Papal authority may react against these dangers.

Quicquid sit de Tartaris et Saracenis, certum est de Antichristo et suis, quod haec operabuntur. Et nisi ecclesia occurrat per sancta consilia ad impediendum et destruendum opera hujusmodi, aggravabitur intolerabiliter flagellis Christianorum. Et creditur ab omnibus sapientibus quod non sumus multum remoti a temporibus Antichristi, sicut in capitulo de sectis per astronomiam in uno revolutis patet. Si igitur Christiani scirent haec opera auctoritate papali facienda ad impedienda mala Christianorum, satis esset laudabile, et non solum propter mala repellenda, sed ad promotionem quorumcunque utilium. Et quia personae, et civitates, et regiones secundum praedicta possunt alterari in melius, et ut vita quantum sufficit prolongetur, et omnes res utiliter procurari, atque multo majora fieri, quam praesenti scripturae debeant commendari, non solum in naturalibus, sed in moralibus scientiis et artibus, sicut patuit per Moysen et Aristotelem. Et haec praecipue fieri possunt, quando cum virtute et specie coeli currit a parte operantis species et virtus animae rationalis, quae est dignior coelo, ita ut adsit fortis cogitatio, et ardens desiderium, certa intentio et plena confidentia, et maxime sanctitas vitae; et quia natura obedit cogitationibus et affectibus animae, et maxime sanctitati. Nam octavo de Animalibus Avicenna ponit exemplum de gallina strenua, quae vicit gallum unum miserum, et ex gloria victoriae statim crevit ei cornu in crure, et dicit ibi Avicenna. in hoc cognoscimus quod natura obedit cogitationibus animae sensitivae. Et ponit exemplum in quarto de Anima, de eo qui ligno posito super aquam ambulans cadit, quia cogitat de casu et desperans est, et natura obedit cogitationibus et affectibus animae; unde ait, quod provenit calor non ex calore, et frigus non ex frigore, sed ex sola animae cogitatione et sic fit infirmitas, et omnis alteratio proprii corporis. Et proculdubio multa fieri possunt in corpore alieno; quia quoddam genus luporum est, quod reddit hominem raucum, si primo aspiciat eum, ut vulgatum est apud auctores. Et multo fortius anima rationalis potest propter nobilitatem suae speciei multa immutare cum quinque conditionibus supradictis, et maxime cum sanctitate, quia nobilior est sanctitas quam cogitatio vel desiderium, et ideo animae sanctae plus obediet materia mundi quam aliis quatuor conditionibus. Et hoc videmus adimpleri in sanctis qui fecerunt miracula, quibus elementa mundi obediebant. Et Avicenna quarto de Anima docet, quod anima sancta et munda a peccatis potest universale et elementa alterare, ut ejus virtute fiant pluvia, tempestates, et omnes alterationes corporum mundi. Verum est autem quod gratia Dei multum facit, sed anima sancta cooperatur gratiae gratum facienti, ut non salvetur homo per solam gratiam, sed oportet animam cooperari gratiae tali; multo ergo fortius poterit cooperari gratiae gratis datae quae est gratia miraculorum. Sed in aliis mirabilibus melius est cogitare quam scribere, donec apostolica auctoritate requiratur major certitudo, et haec possibilia sunt. Et si vos et successores Vestrae Beatitudinis velint, poterunt omnia adimpleri. Hae sunt radices in summa tractatus, quem proposui de hujusmodi conscribere; sed propter impedimenta non potui plura scribere [1].

[1] With regard to the influence of the Moon on human action, on which so much has been said in the foregoing pages, the following extract from Bacon's Commentary on the Secretum Secretorum (Tanner MSS. 116, fol. 86), may be of interest. It represents Lunar action as modified by each of the planets, under various aspects.

DE ASPECTU LUNAE AD PLANETAS.

	SEXTILIS ASPECTUS	TRINUS ASPECTUS	QUADRATUS ASPECTUS	OPPOSITIO	CONJUNCTIO
VENUS	Dies laudabilis jungi juvenibus et militibus et quaerere dilectionem et ornamentum.	Dies laudabilis. Bonum est indui novis vestibus, ornari et nubere et omne ornamentum quaerere causa dilectionis.	Dies laudabilis. Bonum est emere servos et indui novis vestibus, adducere operarios, ornari ornamentis, uti balneis, et delectari.	Dies laudabilis, bonum emere servos, indui novis vestibus, uti medicaminibus digestivis, si in scorpione opponitur.	Dies laudabilis omnibus operibus et peregrinationi, uti militibus, emere subtilia, et petere uxorem.
MERCURIUS	Dies laudabilis. Bonum est in ea computare scribere et jungi consulibus et scriptoribus et disponere opera.	Dies laudabilis. Bonum jungi scriptoribus, firmare pactum, jungi consulibus, et disponere opera.	Dies laudabilis. Bonum mittere cartulas, tractare de scripturis, sedere cum sapientibus, et his similibus.	Dies mediocris in conventione scriptorum, regum, et horum similium.	Dies laudabilis. Bonum jungi scriptoribus et computatoribus, et agere omnia opera.

	Sextilis aspectus	Trinus aspectus	Quadratus aspectus	Oppositio	Conjunctio
SATURNUS	Dies fortunata; bonum est virginibus consulibus et senibus, et agereres cum eis.	Dies laudabilis; bonum est in ea aedificare plantare et res antiquas agere.	Dies timenda; cavendum est in ea jungi regibus et potentibus seu divitibus omnino.	Dies cavenda in omnibus operibus et nullum in ea est bonum opus faciendum.	Dies ista cavenda est in omnibus et nullum bonum est in ea.
JUPITER	Dies laudabilis. Bonum est jungi justis judicibus, et quaerere quod jus est et laudabile in omni opere.	Dies laudabilis ad augenda justa judicia et omnem justam inquisitionem et omne opus bonum et amicitiam secundum Deum.	Dies laudabilis ad justa agenda, et inquisitionem sapientiae, et rem quam vult occultari, et maxime si fuerit ipse occultatus sub radiis solis.	Dies laudabilis in omnibus operibus; bonum est jungi potentibus et petere veritatem et quod justum est.	Dies laudabilis et propria in minutione sanguinis et medicaminibus et caeteris similibus.
MARS	Dies laudabilis. Bonum jungi ducibus et regibus, disponere exercitus, uti armis, et tractare de bello.	Dies laudabilis fortunata. Bonum jungi regibus et emere bestias, et tractare de bello.	Dies cavenda; non bonum jungi ducibus et regibus. Cavendum in ea omne opus in quo speratur bonum.	Dies cavenda in omnibus operibus; et in ea nullum bonum est.	Dies cavenda in omnibus operibus, quia nullum bonum est in ea.
SOL	Dies fortunata. Bonum jungi consulibus et senibus, ac regibus potestatibus principibus et petere principatum.	Dies laudabilis. Bonum est in ea jungi senibus et regibus et colere terras et disponere omnia opera.	Dies laudabilis omni operi si fuerit luna recepta. Si non, nullum bonum, sed cavenda est in omni opere.	Dies cavenda est in omnibus operibus quia nullum bonum est in ea.	In hac die erit luna sub radiis. Nullum bonum nisi in his quae necesse sunt occultari et contegi.

Printed in the United States
By Bookmasters

Printed in the United States
By Bookmasters